THE ANCHOR DICTIONARY OF ASTRONOMY

VALERIE ILLINGWORTH has a degree in physics from Bristol University and a master's degree in radiation physics from London University. After several years' research at Monash University, Melbourne she joined Laurence Urdang Associates, Ltd., in England and later became senior editor. Now a freelance science writer, she has edited or written several books, including *Physics, Junior Science Encyclopedia, New Dictionary of Physics, Dictionary of Physics,* and *Dictionary of Electronics.*

The Anchor Dictionary of Astronomy

Edited by

Valerie Illingworth

Anchor Books

Anchor Press/Doubleday
Garden City, New York
1980

This book was originally published in hardcover by Facts on File Inc.

Anchor Books Edition: 1980

ISBN: 0-385-15936-6
Library of Congress Catalog Card Number 79-6538

COPYRIGHT © 1979 BY LAURENCE URDANG ASSOCIATES LTD.
ALL RIGHTS RESERVED
PRINTED IN THE UNITED STATES OF AMERICA

PREFACE

There has been an explosive growth of astronomy in the last few years with many exciting developments and discoveries. The *Dictionary of Astronomy,* written principally in the past eight months, has tried to reflect this. It is concerned both with established theories and with current research and ideas. It contains descriptions of the different types of celestial bodies, of the instruments used to investigate them, and the concepts underlying these studies. The astronomical achievements of space missions are also considered. The Dictionary should also be of value to anyone with a keen interest in the universe and its contents.

Astronomy has various branches, often closely associated with other scientific disciplines. The most important words and terms used in these fields of study appear as entries in the Dictionary. The set of entries for each field has been written by one scientist or by a group of scientists working together.

The Dictionary contains several long entries in which a word of primary importance is defined and discussed and in which closely associated words, printed in italics, are also defined. In these and in the shorter entries, an asterisk before a word or group of words indicates an entry that the reader should consult for further information. Alternative names and abbreviations of entries are placed in brackets immediately following the alphabetical listing of the entry. Tables at the back of the Dictionary contain general information while tables and illustrations in the text relate to specific entries.

I would like to thank all those who have contributed to the Dictionary and also those, like Dr. Garry Hunt of University College, London, who have read and commented on parts of the Dictionary.

Valerie Illingworth
April 1979

A

AAT. *Abbrev. for* Anglo-Australian telescope.

aberration. 1. (aberration of starlight). The apparent displacement in the position of a star due to the earth's orbital motion around the sun. It was discovered in 1729 by the English astronomer James Bradley. Light appears to approach the observer from a point that is displaced slightly in the direction of the earth's motion. The angular displacement, a, is given by the ratio v/c, where v is the earth's orbital velocity and c is the speed of light. Over the course of a year, the star appears to move in a small ellipse around its mean position; the ellipse becomes a circle for a star at the pole of the *ecliptic and a straight line for one on the ecliptic. The maximum displacement, i.e. the semimajor axis of the ellipse, is 20.5 arc seconds.

The aberration due to the earth's orbital motion is sometimes termed *annual aberration* to distinguish it from the very much smaller *diurnal aberration* that results from the earth's rotation on its axis. *Compare* annual parallax.

2. A defect in the image formed by a lens or curved mirror, seen as a blurring and possible false coloration in the image. Aberrations occur for all light rays lying off the optical axis and also for those falling at oblique angles on the lens or mirror surface. The four principal aberrations are *chromatic aberration (lenses only), *spherical aberration, *coma, and *astigmatism. *Curvature of field and *distortion are other aberrations. Chromatic aberration occurs when more than one wavelength is present in the incident light beam. For light of a single wavelength, only the latter five aberrations occur. These image defects may be reduced—but not completely eliminated—in an optical system by a suitable choice of optical materials, surface shape, and relative positions of optical elements and stops. *See also* achromatic lens; correcting plate.

3. A defect in the image produced by an electronic system using magnetic or electronic lenses.

absolute magnitude. *See* magnitude.

absolute temperature scale. *See* thermodynamic temperature scale.

absolute zero. The zero value of the *thermodynamic temperature scale, i.e. 0 K (-273.15°C). Absolute zero is the lowest temperature the-

absorption, atmospheric

oretically possible. At absolute zero molecular motion almost ceases.

absorption, atmospheric. *See* atmospheric absorption.

absorption lines, bands. *See* absorption spectrum.

absorption nebula. *See* dark nebula.

absorption spectrum. A *spectrum that is produced when *electromagnetic radiation has been absorbed by matter. Typically, absorption spectra are produced when radiation from an incandescent source, i.e. radiation with a *continuous spectrum, passes through cooler matter. Radiation is absorbed (i.e. its intensity is diminished) at selective wavelengths so that a pattern of very narrow dips or of wider troughs—i.e. *absorption lines* or *bands*—are superimposed on the continuous spectrum.

The wavelengths at which absorption occurs correspond to the energies required to cause transitions of the absorbing atoms or molecules from lower *energy levels to higher levels. In the hydrogen atom, for example, absorption of a photon with the required energy results in a 'jump' of the electron from its normal orbit to one of higher energy (*see* hydrogen spectrum).

The absorption lines (or bands) of a star are produced when elements (or compounds) in the outermost layers of the star absorb radiation from a continuous distribution of wavelengths generated at a lower level in the star. Basically the same elements occur in stars. Since each element has a characteristic pattern of absorption lines for any particular temperature (and pressure) range, there are several different types of stellar spectra depending on the surface temperature of the star. *See* spectral types. *See also* emission spectrum.

abundance. *See* cosmic abundance.

Academy of Sciences Astrophysical Observatory. The Soviet observatory near Zelenchukskaya, Northern Caucasus, at an altitude of 2100 metres. Its instruments include the world's largest *reflecting telescope, six metres in diameter, and one of the largest *radio telescopes, *RATAN 600*. The six metre telescope is the first large optical telescope to have an *altazimuth mounting. Seeing conditions are at best one arc second but this should be improved to 0″.5 with a new mirror due to be installed in 1983. At present blue objects of magnitude 24 can be observed with exposures of 30 to 40 minutes; for red objects the limiting magnitude is 23-23.5 for a 90 minute exposure. Final tests on the telescope were completed in early 1977.

The radio telescope RATAN 600 has a diameter of 600 metres (about 2 km in circumference) with a total collecting area of 10 000 m². It is made up of aluminium mirrors that can move laterally and rotate on a horizontal and vertical axis, the movement being computer controlled. The reflecting area can be split into four sections, each of which can act as a telescope: the north and south sections were in use by 1976. It is designed to operate at short wavelengths (8 mm-30 cm).

Acamar. See Eridanus.

acceleration. 1. (linear acceleration). Symbol: a. The rate of increase of velocity with time, measured in metres per second per second (m s^{-2}), etc.
 2. **(angular acceleration).** Symbol: α. The rate of increase of *angular velocity with time, usually measured in radians per second per second.

acceleration of gravity. Symbol: g. The acceleration to the centre of a planet (or other massive body such as a natural satellite) of an object falling freely without air resistance, i.e. acceleration due to downward motion on a gravitational field. It is equal to GM/R^2, where G is the gravitational constant and M and R are the mass and radius of the planet. The acceleration is thus independent of the mass of the accelerated object, i.e. it is the same for all bodies (neglecting air resistance) falling at the same point on the surface of the planet, satellite, etc.

On earth the standard value of the acceleration of gravity is 9.806 65 metres per second per second. The value varies from place to place on the earth's surface because of different distances to the earth's centre and greater acceleration towards the equator. In addition, it is affected by local deposits of light or heavy materials.

accretion (aggregation). The increase in mass of a body by the addition of smaller bodies that collide and stick to it. Once a large enough body forms, its gravitational attraction accelerates the accretion process. Accretion is now assumed to have had an important role in the formation of the planets from swarms of dust grains. See solar system, origin. See also mass transfer.

accretion disc. See black hole; mass transfer.

Achernar (α Eri). A conspicuous bluish-white star that is the brightest in the constellation Eridanus. m_V: 0.5; M_V: -2.2; spectral type: B5 V; distance: 35 pc.

Achilles. The first member of the *Trojan group of *minor planets to be discovered, in 1906. It lies east of Jupiter and precedes the latter in its orbit of the sun. See Table 3, backmatter.

achondrite. Any of a class of *stony meteorites that lack the characteristic *chondrules of the *chondrites. They are usually more coarsely crystallized than the chondritic stones and are more similar chemically and mineralogically to some terrestrial rocks. They contain very little iron and nickel.

achromatic lens (achromat). A two-element lens—a doublet—that greatly reduces *chromatic aberration in an optical system. The components, one positive, one negative, are of different types of glass (i.e. they have different *refractive indices); the combination focuses two selected colours, say red and blue, at a common image plane with a small spread in focal length for other colours (see illustration). The difference in optical power (reciprocal of focal length) for the two colours in one element must cancel that in the other element. By a suitable choice of glasses and surface curvatures, the doublet can be *aplanatic as well as achromatic, so that three major aberrations are minimized.

Residual colour effects in an achromat can be further reduced by using a compound lens of three or more elements—an *apochromatic lens;* each element has an appropriate shape and dispersive power so that three or more colours can be focused in the same image plane.

Achromatic doublets

acronical rising (or setting). The rising (or setting) of a star at or just after sunset. When the rising is acronical the setting is *cosmic* and vice versa.

Acrux. *See* Alpha Crucis.

active galaxy. Any galaxy that is emitting unusually large amounts of energy from a very powerful source. This violent activity is often seen in the nucleus of the galaxy and takes various forms, all of which are relatively short-lived. It is thought that the energy output of some active galaxies could be accounted for by a central supermassive *black hole, tearing stars apart by tidal forces. Alternative energy sources could be supermassive stars or star clusters. It has been suggested that these latter sources, together with accreting black holes, might represent successive stages in the evolution of the active galaxy. *Seyfert galaxies, certain *radio galaxies, *N galaxies, and *quasars are active galaxies.

active prominence. See prominences.

active-region filament. See prominences.

active regions. Regions of intense localized magnetic field on the sun that extend from the *photosphere, through the *chromosphere, to the *corona. They may encompass a variety of phenomena, such as *sunspots, *faculae/*plages, *filaments (or *prominences), and *flares, and are characterized by enhanced emission of radiation at x-ray, ultraviolet, and radio wavelengths.

active sun. The term applied to the sun around the maximum of the *sunspot cycle, when the profusion of *active regions ensures a high level of *solar activity. *Compare* quiet sun.

Adhara (ε CMa). A very luminous remote blue-white giant that is the second brightest star in the constellation Canis Major. It is a visual double star with an 8th-magnitude companion at a fixed separation of 8″. m_V: 1.5; M_V: -5.0; spectral type: B2 II; distance: 200 pc.

adiabatic process. A process that takes place in a system with no heat transfer to or from the system. In general, a temperature change usually occurs in an adiabatic process.

Adonis. A member of the *Apollo group of *minor planets. Discovered in 1936 when it passed 0.015 astronomical units (AU) from the earth, it was not observed again until it was recovered in 1977. Its *perihelion distance of 0.44 AU is the fourth smallest among the minor planets. *See* Table 3, backmatter.

advance of the perihelion. The gradual movement of the *perihelion of a planet's elliptical orbit in the same direction as that of the planet's orbital motion. This advance results from the slow rotation of the major axis of the planet's orbit due to gravitational disturbances by other planets and to the curvature of *space-time around the sun. The small contribution from the curvature of space-time was predicted by Einstein's general theory of *relativity.

The value of this relativistic contribution towards the advance of the perihelion of Mercury is about 43 arc seconds per century. This agrees almost exactly with the discrepancy between the experimentally determined value for the advance and that predicted by classical Newtonian mechanics. It was therefore an important confirmation of general relativity. Recent measurements of the advance of the perihelia of Venus and Earth have also confirmed Einstein's predicted values for those planets.

aeon. period of one thousand million years: 10^9 years.

aerial. *See* antenna.

aerolite. *See* stony meteorite.

aeronomy. The physics and chemistry of the upper atmosphere of the earth, i.e. its temperature, density, motions, composition, chemical processes, reactions to solar and cosmic radiation, etc. The term has been extended to include the physics and chemistry of the atmospheres of the other planets.

aerospace. The earth's atmosphere and the space beyond.

Ae stars. Hot stars of *spectral type A that in addition to the normal absorption spectrum have bright emission lines of hydrogen. These lines are thought to arise in an expanding atmospheric shell of matter lost from the star. Like *Be stars they are probably similar to *T Tauri stars except that they have larger masses.

afocal system. An optical system in which both the object and the secondary image are infinitely distant. It is the usual adjustment in a simple refracting telescope and is produced when the objective and eyepiece lenses are separated by the sum of their focal lengths: the lenses are then *confocal*.

AG catalogue. See AGK.

age of the earth. The oldest rocks found in the earth's crust, in West Greenland, have been assigned ages of 3.59 thousand million years from radioisotope studies, but other properties of the rocks support the belief that the planet shares a common origin with the rest of the solar system, about 4.6 thousand million years ago. The cycling of the crustal material by plate tectonics (*see* earth) and the eroding effects of ice, water, and wind mean that none of the earliest rocks survive intact and that evidence of an early meteoritic bombardment, so apparent on the moon, Mercury and Mars, is absent. *See* earth; solar system, origin.

age of the universe. The observed expansion and evolution of the universe suggest that it has a finite age, considered as the time since the *big bang, i.e. the time of infinite density. The inverse *Hubble constant, $1/H_0$, gives a measure of the age if the rate of expansion has always been constant. Since gravitation tends to diminish the expansion rate, H_0 can only give an upper limit. Using the value of H_0 of 55 km s^{-1} Mpc^{-1} gives an upper limit of 18 thousand million years.

In the standard *cosmology (solutions of Einstein's field equations without a cosmological constant) with *deceleration parameter q_0, the age is given as one of the three alternatives:

$$H_0^{-1}q_0(2q_0-1)^{-3/2}[\cos^{-1}(q_0^{-1}-1) - q_0^{-1}(2q_0-1)^{1/2}]$$
$$2/3H_0^{-1}$$
$$H_0^{-1}q_0(1-2q_0)^{-3/2}[q_0^{-1}(1-2q_0)^{1/2} - \cosh^{-1}(q_0^{-1}-1)]$$

The choice depends on whether q_0 exceeds, equals, or is less than ½ (and > 0), i.e. on whether the universe is closed, infinite, or open, respectively.

Ages of 10 and 18 thousand million years thus correspond to values of q_0 of 1 and 0, for H_0 equal to 55 km s^{-1} Mpc^{-1}.

Lower limits to the age of the universe, other than through measurements of H_0 and q_0, can be found from *radioactive dating of the earth and Galaxy from studies of globular clusters. The relative abundances of radioactive elements, such as uranium, and their decay products yield an estimate of the time since formation of that body of material. The age of a globular cluster may be estimated by comparison of the observed main-sequence *turnoff point in the Hertzsprung-Russell diagram for the cluster with theoretical models. Both of these ages should be less than the age of the universe and in particular they must be less than H_0^{-1}. Various errors in the estimate of H_0 have in the past meant that the above inequality has sometimes not been obeyed. Present estimates are consistent and may lead to constraints on q_0.

aggregation. *See* accretion.

AGK (AG catalogue). *Abbrev. for* Katalog der Astronomischen Gesellschaft. A general catalogue of star positions originally proposed by F.W.A. Argelander in 1867. The first version, known as *AGK 1,* covered the northern sky and listed about 190 000 stars: by 1912 zones between *declinations 90° and -18° had been published. A revision, known as *AGK 2,* was based on photographic measurements begun in the 1920s and was published 1951-58. A further set of photographic plates was produced in the 1950s to determine *proper motions, which appeared in the *AGK 3* version; this was distributed on magnetic tape from 1969.

airglow. The faint everpresent glow arising in the earth's atmosphere that is light emitted during the recombination of ionized atoms and molecules following collisions with high-energy particles and radiation, mainly from the sun. This dim light, often termed *nightglow* when seen at night, interferes with observations of faint celestial bodies.

Air Pump. Common name for *Antlia.

air shower (extensive air shower). *See* cosmic rays.

Airy disc. The bright disc-like image of a point source of light, such as a star, as seen in an optical system with a circular *aperture. The disc is formed by *diffraction effects in the instrument and is surrounded by faint diffraction rings that are only seen under perfect conditions (see illustration). About 87% of the total light intensity lies in the disc. The disc diameter, first calculated by George Airy in 1834, is the factor limiting the *resolving power of the telescope.

Light distribution in Airy disc image of single point source (left) and two just resolvable point sources

AI Velorum stars. *See* dwarf Cepheids.

Alba Patera. Probably the largest central-vent volcanic structure on Mars. Located north of the *Tharsis Ridge, it has a low dome containing a central caldera and enclosed by a ring of fractures 600 km in diameter. Associated lava flows cover an area more than 1500 km across. *See* Mars, volcanoes.

albedo. The reflecting power of a nonluminous object, such as a planet or planetary surface feature. It is the ratio of the total amount of light reflected in all directions to the amount of incident light: an albedo of 1.0 indicates a perfectly reflecting surface while a value of 0.0 indicates a totally black surface that absorbs all incident light.

A comparison of albedo observations with measures of terrestrial substances may be made to infer the surface characristics of planets, satellites, or minor planets. Volcanic rocks, whether on the earth or the moon, typically have very low albedos: the albedos of the moon and Mercury are 0.07 and 0.06 respectively. Clouds, such as those of Venus and the *giant planets, have albedos of 0.7 or more. *See also* Table 1, backmatter.

Albireo (β Cyg). An orange giant in the constellation Cygnus. It is a beautiful double star with a deep blue companion 35" away. m_V: 3.2 (A), 5.5 (B); spectral type: K5 II (A), B8 V (B).

Alcaid or **Alkaid (Benatnasch;** η **UMa).** A blue-white star that is one of the seven brightest in the *Plough. m_V: 1.86; M_V: -2.3; spectral type: B3 V; distance: 70 pc.

Alcor (80 UMa). A white star that lies in the *Plough and forms an optical double with

*Mizar. m_V: 4.1; spectral type: A5 V; distance: 27 pc.

Alcyone (η Tau). A bluish-white double, possibly multiple, star in the constellation Taurus that is the brightest star in the *Pleiades. m_V: 2.86; M_V: -2.7; spectral type B; distance: 125 pc.

Aldebaran (α Tau). A conspicuous red giant that is the brightest star in the constellation Taurus and lies in the line of sight of but much nearer than the Hyades. It is a slow irregular variable. It is also both a visual binary with an 11th-magnitude red dwarf companion at 31″ distance and an optical double with an independent binary system 122″ away. m_V: 0.9; M_V: -0.8; spectral type: K5 III; radius (by interferometer): 45 times solar radius; distance: 21 pc.

Alfvén's theory. A theory proposed by the Swedish physicist Hannes Alfvén in 1942 for the formation of the planets out of material captured by the sun from an interstellar cloud of gas and dust. As atoms fall towards the sun, they become ionized and subject to the control of the sun's magnetic field. Ions are concentrated in the plane of the solar equator where the planets coalesce. The theory, in its original form, had difficulty in accounting for the inner planets, but was important in suggesting the role of *magnetohydrodynamics in the genesis of the solar system. *See* solar system, origin.

Algeiba (γ Leo). An orange giant that is the second-brightest star in the constellation Leo. It is a multiple star, forming a visual binary (separation 4″.4, period 619 years) with a 4th-magnitude yellow companion. m_V: 2.02 (A), 1.84 (AB); M_V: -0.5; spectral type: K0p III (A), G7 III (B); distance: 32 pc.

Algenib (γ Peg). A remote bluish-white subgiant that is one of the brighter stars of the constellation Pegasus and lies on the Great Square of Pegasus. m_V: 2.8; spectral type: B2 IV; distance: 143 pc.

Algol (Demon Star; Winking Demon; β Per). A white star that is the second-brightest one in the constellation Perseus. It was the first *eclipsing binary to be discovered, being the prototype of the *Algol variables, although its variations in brightness were known to early astronomers. The theory of a darker companion periodically cutting off the light of the brighter star (Goodricke, 1782, Pickering, 1880) was confirmed spectroscopically in 1889. The brighter star (Algol) is about three times the sun's diameter; the fainter orange 3rd-magnitude component is about 20% larger than Algol. It revolves around Algol in a period of 2.87 days, the magnitude varying from about 2.2 to 3.47 (see overleaf). A third star revolves around the other two in a period of 1.87 years. There is possibly a massive unseen fourth component, period 190 years. Algol is also an erratic radio source: m_V: 2.2; M_V: -0.5; spectral type: B8 V; distance: 35 pc.

Algonquin Radio Observatory. The observatory at Lake Traverse, Ontario, Canada run by the National Research Council. Its principal instrument is a 46 métre (150 foot) fully steerable radio dish, in operation since 1967.

Light curve of Algol

Algol variables (Beta Persei stars). A large class of *eclipsing binaries, named after *Algol, with periods greater than two days and components that are usually sufficiently well separated to retain a spherical or near spherical shape with no distortion or *mass transfer. The components are generally much larger than the sun and have very different luminosities: this results in a characteristic *light curve in which the secondary minimum is of negligible depth. *Compare* Beta Lyrae stars; W Ursae Majoris stars.

Alhena. *See* Gemini.

Alioth (ε UMa). A white *spectroscopic binary, period 4.15 years, that is the brightest star in the constellation Ursa Major. It is a *spectrum variable star. m_V: 1.78; M_V: -0.02; spectral type: A0p; distance: 25 pc. *See also* Plough.

Alkaid. *See* Alcaid.

Allende. *See* meteorite.

Almach (γ And). An orange giant that is the third-brightest star in the constellation Andromeda. It is a triple star, forming a very fine visual binary with its apparently greenish companion (B), 10″ distant, with which it has a common proper motion; B is a spectroscopic binary, period 56 years. m_V: 2.28 (A), 5.1 (B), 2.16 (AB); M_V: -2.3; spectral type: K3 II (A), A0p (B); distance: 80 pc.

Almagest. (Arabic: the Greatest). An astronomical work compiled by Ptolemy of Alexandria in about AD 140. It was translated into Arabic in the ninth century and became known in Europe when it was translated into Latin in the late 12th century. Its 13 volumes cover the whole of astronomy as conceived in ancient times, with a detailed description of the *Ptolemaic system of the solar system. It also included a star catalogue giving positions and *magnitudes (from 1 to 6) of 1022 stars.

almucantar. 1. A *small circle on the *celestial sphere parallel to the horizon.
2. An instrument for measuring altitude and azimuth.

Alnilam. *See* Orion.

Alnitak. *See* Orion.

Alpha (α). The first letter of the Greek alphabet, used in *stellar nomenclature usually to designate the brightest star in a constellation or sometimes to indicate a star's position in a group.

Alpha Capricornids. *See* Capricornids.

Alpha Centauri (Rigil Kent; α Cen). A triple star that is the brightest star in the constellation Centaurus, the third-brightest star in the sky, and the second-nearest star to the sun. The two brighter components, A and B, form a yellow-orange *visual binary (separation 17″.7, period 80.1 years) and are similar in mass and size to the sun. *Proxima Centauri is the third component. m_V: 0.0 (A), 1.38 (B), -0.27 (AB); M_V: 4.4 (A), 5.8 (B), 4.2 (AB); spectral type: G2 V (A), K5 V (B); mass: 1.09 (A), 0.89 (B) times solar mass; radius: 1.23 (A), 0.87 (B) times solar radius; distance: 1.31 pc.

Alpha Crucis (Acrux; α Cru). A conspicuous white subgiant that is the brightest star in the constellation Crux. It is a visual binary (separation 6″), both components being spectroscopic binaries. Alpha and Gamma Crucis point towards the south celestial pole. m_V: 1.6 (A), 2.1 (B), 0.83 (AB); M_V: -3.7; spectral type: B1 IV (A), B3 (B); distance: 80 pc.

alpha particle (α particle). The nucleus of a helium atom, i.e. a positively charged particle consisting of two protons and two neutrons. It is thus a fully ionized helium atom. Alpha particles are very stable. They are often ejected in nuclear reactions, including *alpha decay* in which a parent nucleus disintegrates—or breaks up—into an alpha particle and a lighter daughter nucleus.

Alphard (α Hya). An orange giant that is the brightest star in the constellation Hydra and lies in a fairly isolated position. m_v: 2.05; M_v: -0.7; spectral type: K4 III; distance: 35 pc.

Alphecca or **Alphekka** (**Gemma**; α CrB). A blue-white star that is the brightest one in the constellation Corona Borealis. It is an *eclipsing binary (separation 42″, period 17.36 days). m_v: 2.23; M_v: 0.5; spectral type: A0 V (A), G6 (B); distance: 22 pc.

Alpheratz (α And). A bluish-white giant that is the brightest star in the constellation Andromeda. It is both a spectroscopic binary (period 96.7 days) and a *spectrum variable. It originally lay in the constellation Pegasus, as Delta Pegasus, and is still considered part of the Great Square of Pegasus. m_v: 2.07; M_v: -0.5; spectral type: B9p III; distance: 31 pc.

ALSEP. *Abbrev. for* Apollo Lunar Surface Experiments Package. Any of the experimental packages carried to the moon by *Apollos 11-17, set up by the astronauts, and left there to transmit information back to earth. The content of each package differed, becoming more sophisticated as the programme progressed. In the Apollo 17 ALSEP, a central station and thermal generator provided the main power; the experiments included an analysis of any residual atmosphere, detection and measurement of any lunar ejecta and meteorite impacts, measurement of any gravitational anomalies, lunar surface studies, and seismic measurements.

Altair (α Aql). A nearby conspicuous white star that is the brightest star in the constellation Aquila. It is an optical double. m_v: 0.77; M_v: 2.3; spectral type: A7 IV-V; distance: 5.1 pc.

Altar. Common name for *Ara.

altazimuth mounting (azimuthal mounting). A *mounting in which the telescope swings in azimuth about a vertical axis and in *altitude about a horizontal axis (see illustration). It is easy to make and use and needs no counterpoise weights to balance the telescope. It provides a very firm support and is well adapted for terrestrial observations and for following rapidly moving objects such as artificial satellites. Its great disadvantage is the need to adjust both altitude and azimuth to follow the diurnal motion of a heavenly body. The application of very precise computer-controlled drive mechanisms to altazimuth mountings should lead to their wider use in large telescopes. The *Multiple Mirror Telescope and the Soviet six metre telescope at

Altazimuth mounting

Zelenchukskaya have altazimuth mountings, as does the radio telescope at *Jodrell Bank. *Compare* equatorial mounting.

altitude. The angular distance of a point or celestial object above or below the *horizon, or of an object, such as an artificial satellite, above mean sea level. Altitude and azimuth are coordinates in the *horizontal coordinate system.

aluminizing. A process whereby a very thin but perfectly uniform coating of aluminium is deposited by evaporation on a suitable base. It is used in astronomy to produce the reflective layer of a mirror. The aluminizing is done in an evacuated chamber. The aluminium layer is usually protected by a silicon coating that is applied in the same way and that oxidizes to silica when exposed to air. The surface is harder and more stable than silver and is also able to reflect shorter wavelengths than silver.

Amalthea. The innermost satellite of Jupiter, discovered in 1892. It has been greatly distorted by the gravitational pull of Jupiter, becoming markedly elongated in shape; its long axis (290 km) points towards Jupiter. Photographs from *Voyager I show it to have crater-like features and a low surface reflectivity (i.e. albedo). *See* Jupiter's satellites; Table 2, backmatter.

American Ephemeris and Nautical Almanac. *See* Astronomical Ephemeris.

Amor group. A group of more than 16 known *minor planets that come within the *perihelion distance of Mars but not within the orbit of the earth. The group is named after the minor planet Amor, discovered in 1932. *Compare* Apollo group. *See* Table 3, backmatter.

amplitude. 1. Symbol: Δm. The difference between the maximum and minimum *magnitudes of a *variable star, i.e. the range in magnitude of the star. The amplitude of pulsating variables is related to the logarithm of the period.
2. The maximum instantaneous deviation of an oscillating quantity from its average value.

Am stars. *Peculiar main-sequence stars of spectral types from A0 to F0 in which there is an over-abundance of heavier elements and rare earths and (less so) of iron, and an apparent under-abundance of calcium. They rotate slower than normal A stars and are usually short-period spectroscopic binaries.

Ananke. A small satellite of Jupiter, discovered in 1951. *See* Jupiter's satellites; Table 2, backmatter.

Andromeda. A constellation in the northern hemisphere close to *Pegasus, the brightest stars being *Alpheratz (α), *Mirach (β), and the fine visual binary *Almach (γ). It contains the *Mira stars R and T Andromedae, the spiral *Andromeda galaxy, and the bright planetary nebula NGC 7662. Abbrev.: And; genitive form: Andromedae; approx. position: RA 1h, dec +40°; area: 722 sq deg.

Andromeda galaxy (M31; NGC 224). The largest of the nearby galaxies, visible to the unaided eye as a faint oval patch of light in the constellation Andromeda. It is an intermediate (Sb) spiral (*see* Hubble classification), orientated at an angle of about 15° from the edge-on position, and has a bright elliptical-shaped nucleus. Its distance is currently estimated as 670 kiloparsecs (2.2 million light years). With a total mass of about 3×10^{11} solar masses (roughly double that of our own galaxy) and an overall diameter of approximately 38 000 parsecs, M31 is the largest of the established members of the *Local Group. It has two dwarf elliptical satellites: NGC 205 and NGC 221 (M32).

Andromedids. *Another name for* Bielids.

Anglo-Australian telescope (AAT). The 3.9 metre (154

inch) reflecting telescope sited at Siding Spring, New South Wales, Australia at an altitude of 1165 metres. It was funded jointly by the governments of Australia and the UK, each country having a half share in observing time. It began regular operation in 1975. It works at a *focal ratio of f/3.5 at the prime focus, f/8 and f/15 at the Cassegrain foci, and f/35 at the coudé focus. It has a limited magnitude of 23-25 (depending on wavelength and technique used), the maximum field of view covering about one square degree of sky. The telescope is computer controlled so that the pointing and tracking are exceptionally accurate.

angstrom. Symbol: Å. A unit of length equal to 10^{-10} metres. It is mainly used to specify the wavelength of light and ultraviolet radiation.

angular acceleration. *See* angular velocity.

angular diameter. *See* apparent diameter.

angular distance. *See* apparent distance.

angular measure. Units of angle or length given in terms of degrees (°), arc minutes ('), and *arc seconds ("):

$$60'' = 1'$$
$$60' = 1°$$

$360° = 2\pi$ radians (a full circle)
In the case of *right ascension, angular measure is in terms of hours (h), minutes (m), and seconds (s).

angular momentum (moment of momentum). Symbol: L. A property of any rotating or revolving system whose value depends on the distribution of mass and velocity about the axis of rotation or revolution. It is a vector directed along the axis and for a body orbiting about a point it is given by the vector product of the body's linear *momentum and *position vector r, i.e. it is the product $m(v \times r)$. It is also expressed as the product of the *moment of intertia (I) and *angular velocity (ω) of the body.

In a closed system, such as the solar system or an isolated star, there is always *conservation of momentum.

angular resolution. *See* resolving power.

angular separation. *See* apparent distance.

angular-size redshift test. *See* cosmology.

angular velocity. Symbol: ω. The rate at which a body or particle moves about a fixed axis, i.e. the rate of change of angular displacement. It is expressed in radians per second and is a vector quantity. For a point of *position vector r, an angular velocity ω about the origin imparts to the point a velocity $\omega \times r$ at right angles to the position vector. *Angular acceleration* is the rate of change of angular velocity.

annihilation. A reaction between an *elementary particle and its *antiparticle. An electron and a positron, for example, interact and produce two gamma-ray photons. Hadrons, such as the proton and the antiproton, also undergo annihilation to produce mesons.

annual equation (annual inequality). The periodic *inequality in the moon's motion that arises from variations in solar attraction due to the *eccentricity of the earth's orbit. Its period is the *anomalistic year and the maximum displacement in longitude is 11'8".9. *See also* evection; variation; parallactic inequality.

annual inequality. *See* annual equation.

annual parallax (heliocentric parallax). The *parallax of a celestial body that results from the change in the position of an observational point during the earth's annual revolution around the sun: nearby stars are seen to be displaced in position relative to the more remote background stars. The instantaneous parallax of a given star is the angle formed by the radius of the earth's orbit at the star; it varies through the year as the orbital radius varies. If the positions of a star are determined during one year they are found to describe an ellipse—a *parallactic ellipse*—on the celestial sphere. The annual parallax, π, of the

Annual parallax

star is equal to and can be measured from the semimajor axis of this ellipse.

Annual parallax is the maximum parallactic displacement of the star and occurs when the angle earth-sun-star is 90° (see illustration). It is thus expressed by the relation sin π = a/d, where a is the mean orbital radius of the earth and d is the star's distance from earth. Since π is extremely small (<1"), this reduces to π = a/d. If a is assumed to be unity (i.e. one *astronomical unit) then a measurement of π in arc seconds will be equal to the reciprocal ($1/d$) of the distance in *parsecs.

The determination of distance from measurements of annual parallax is termed *trigonometric parallax*. Due to its tiny value annual parallax can only be measured (photographically) with reasonable accuracy for stars within about 30 parsecs (π>0″.03). Friedrich Bessel published the first account of a successful distance measurement from parallax—that of 61 Cygni—in 1838. The distances of over 6000 stars have since been measured thus, the nearest star—Proxima Centauri—having the greatest value (0″.76) of annual parallax.

annular eclipse. *See* eclipse.

anomalistic month. The time interval of 27.554 55 days between two successive passages of the moon through the *perigee of its orbit.

anomalistic year. The time interval of 365.259 64 days between two successive passages of the earth through the *perihelion of its orbit. The anomalistic year is longer than the *sidereal and *tropical years because of the *advance of the perihelion caused (mainly) by planetary perturbations.

anomaly. Any of three related angles by means of which the position, at a particular time, of a body moving in an elliptical orbit can be calculated. For a body S moving around the focus, F, of an orbit (see illustration), the *true anomaly* is the angle v made by the body, the focus, and the point, P, of nearest approach—the periapsis. For a body orbiting the

Anomaly

sun, P is the perihelion. The angle is measured in the direction of motion of S. If an auxiliary circle is drawn centred on the midpoint, C, of the major axis of the elliptical orbit, then the *eccentric anomaly* is the angle E between CS′ and CP, where S′ lies on the circle and is vertically above S. The *mean anomaly* is the angle M between P, F, and a hypothetical body with the same period as S but moving at a constant velocity. It can thus be given as

$$M = (2\pi/T)(t-t_0)$$

where T is the period, t the time of computation, and t_0 the time at which the body S passes through the periapsis.

The eccentric and mean anomalies are related by *Kepler's equation:*

$$E - e \sin E = M$$

where e is the *eccentricity of the orbit. The coordinates (x,y) of the body S can be found from the equations

$$x = a(\cos E - e)$$
$$y = a \sin E \sqrt{(1 - e^2)}$$

where a is the semimajor axis of the orbit. *See also* orbital elements.

ANS. *Abbrev. for* Astronomical Netherlands Satellite.

ansae. The parts of *Saturn's rings that are visible on either side of the planet as viewed from the earth.

Antalgol stars. *Obsolete name for* RR Lyrae stars.

antapex (solar antapex). *See* apex.

Antares (α Sco). A huge remote but conspicuous red supergiant that is the brightest star in the constellation Scorpius. It is in an advanced stage of evolution. It is a visual binary (separation 3″), its 7th-magnitude B-type companion with which it has a common proper motion appearing green by contrast. The companion is a peculiar hot radio source. m_V: 0.94; M_V: -4.7; spectral type: M1 Ib; diameter: about 285 times solar diameter; distance: 130 pc.

antenna (aerial). A device used for the transmission or reception of radio waves. When connected to a transmitter, the oscillating electric currents induced in the antenna launch radio waves into space. When connected to a *receiver, incoming radio waves from a distant source generate oscillating electric currents in the antenna, which are detected by the receiver. A practical antenna neither radiates nor receives equally in all directions, which may be demonstrated by its *antenna pattern*—a plot of relative *gain as a function of direction. The plot may be made in terms of the antenna's voltage response or power response, giving a *voltage pattern* or a *power pattern*. By the principle of reciprocity the antenna pattern for transmission is identical with that for reception.

A number of distinct *lobes* may often be identified in the antenna pattern. The lobe corresponding to the direction of best transmission or reception is the *main lobe* or *main beam;* all the others are called *side lobes* and are usually unwanted. The ratio of the power received in the main beam to the total power received by all the lobes is called the *beam efficiency*. The angle between the two directions in the main beam at which the power response has fallen to half its maximum value is called the *beamwidth;* it is a measure of the antenna's *directivity. Some highly directional antennas have *pencil beams,* which are narrow main beams of circular cross section. Others have *fan beams* where the cross section of the main beam is greatly extended in one direction. *See also* array; dish; radiation resistance.

antenna pattern (field pattern; polar diagram). *See* antenna.

antenna temperature. The apparent temperature of the *radiation resistance of an antenna: for a lossless antenna (no energy dissipation) this is not the same as its physical temperature but is determined by the temperature of the distant emitting regions in the antenna's beam. It is thus a measure of the strength of a signal from a *radio source. A resistance at temperature T generates a noise power that, in bandwidth B, is given by $P = kTB$, where k is the Boltzmann constant. The antenna temperature, T_A, is equal to the (weighted) average *brightness temperature of the distant emitting regions so that the radiation resistance generates a noise power $P = kT_AB$; this is passed to the *receiver. A radio source whose angular size is smaller than the antenna beam and whose *flux density is S generates an antenna temperature equal to $SA_e/2k$, where A_e is the effective area (see array) of the antenna.

*Synchrotron emission from *cosmic rays in the Galaxy produces a diffuse radio emission centred broadly on the galactic plane. Any practical observation of a radio source has to be made against this background emission, which at low frequencies limits the *sensitivity of the *radio telescope. The antenna temperature of the diffuse emission is called the *background temperature;* the electrical *noise it produces in radio receivers is often called *cosmic noise.*

antineutrino. *See* neutrino.

antiparticles. A pair of *elementary particles, such as the electron plus positron or the proton plus antiproton, that have an identical rest mass but a charge, strangeness, and other fundamental properties of equal magnitude but opposite sign (positive rather than negative and vice versa). Thus, the electron has a negative charge and the positron an equal positive charge. A reaction of a particle with its antiparticle is called *annihilation.

Matter composed entirely of antiparticles is called *antimatter*. Although individual antiparticles are produced in cosmic-ray showers and in high-energy particle accelerators, the search for antimatter in the universe has so far proved unsuccessful. *See also* pair production.

Antlia (Air Pump, Pump). A small inconspicuous constellation in the southern hemisphere near Centaurus, the brightest stars being of 4th magnitude. Abbrev.: Ant; genitive form: Antliae; approx. position: RA 10h, dec -35°; area: 239 sq deg.

Antoniadi scale. *See* seeing.

apastron. The point in any orbit around a star that is farthest from the star. The nearest point is the *periastron*. The terms are mainly used in connection with a component's orbit about the primary of a binary star.

aperture. The clear diameter of the objective lens in a refracting telescope or of the primary mirror in a reflector. As the aperture is increased the telescope gathers more light, and so will discern fainter objects: the light-gathering power depends on area, i.e. on the square of the aperture. A larger aperture also produces a smaller *Airy disc and so has greater *resolving power.

aperture ratio (relative aperture). The ratio d/f of the effective diameter (i.e. *aperture), d, of a lens or mirror to its *focal length, f. The ratio f/d is the *focal ratio.

aperture synthesis. A technique used in radio astronomy to obtain the high *resolution of a very large aperture with a number of small *antennas that are moved to cover the large aperture. The antennas are connected as interferometer pairs (see radio telescope) and the *amplitudes and *phases of the signals from a *radio source are recorded at all the spacings necessary to cover the required area. The information is then processed in a computer where an inverse *Fourier transform is carried out to yield a radio map of the source. In practice, the rotation of the earth is often used to move the antennas in space; the technique is then called *earth-rotation synthesis*.

In the *Five-Kilometre telescope* at Cambridge, England, for example, four moveable and four fixed *dishes are mounted on an east-west line to give 16 simultaneous interferometer pairs, each moveable dish with each fixed dish. The dishes are set to track a radio source for 12 hours during which time the earth, and hence the *baseline of the instrument, moves round by half a revolution. The data collected from each interferometer pair corresponds to a strip of an elliptical aperture and by moving the four moveable dishes each day half the aperture can be filled. The information for the other half of the aperture is the same as that already collected (except for a reversal of phase) so that a 12-hour run each day suffices. The major axis of the synthesized aperture equals the maximum baseline of the instrument, five km, while the minor axis depends on the sine of the *declination of the source; this limits the resolution that may be achieved for sources near the equator.

The *Very Large Array (VLA)* near Socorro, New Mexico, uses 27 dishes mounted in a **Y** formation to offset this difficulty. The dishes, each 26 metres across, are arranged around a central control room. The VLA can synthesize a maximum aperture of 27 km diameter and when fully operational (about 1981) will produce radio maps, relatively quickly, with a very high resolution. The VLA is part of the *National Radio Astronomy Observatory.

In the computer processing, the data from each interferometer pair are not necessarily given equal weight but are scaled according to a *grading function:* this reduces the levels of the side-lobe responses in the synthesized antenna beam. Some loss of resolution results from this process and in practice a compromise is achieved between high resolution and low side-lobe levels. It is often not necessary to fill in the entire aperture, a few equally spaced strips being sufficient. The final radio map will, however, then have *grating rings* on it—concentric elliptical responses at equal intervals from the middle of the observed area of sky. Their presence does not matter provided that all of the structure in a particular radio source is within the first grating ring.

apex (solar apex). The point on the celestial sphere towards which the sun and the solar system are moving relative to the stars in our vicinity. The apex lies in the constellation Hercules at a position RA 18h, dec +30°, close to the star Vega. The point diametrically opposite to the direction of this relative *solar motion* is the *antapex* (or *solar antapex*); it lies in the constellation Columba. The position of the apex can be determined from an analysis of the proper motions, radial velocities, and parallaxes of stars, based on the assumption that solar motion is equal and opposite to the group motion of stars. The velocity of this motion has been calculated as about 19.5 km s^{-1} with respect to the *local standard of rest, when all available information on proper motion and radial velocity is used. The value is lower when only the stars in the sun's immediate neighbourhood are considered. Solar motion produces the *secular parallax of stars.

aphelion. Symbol: Q. The point in the orbit of a planet, comet, or artificial satellite in solar orbit that is farthest from the sun. The earth is at aphelion on about July 3.

aplanatic system. An optical system that is able to produce an image free from *spherical aberration and *coma.

apoapsis. *See* apsides.

apochromatic lens. *See* achromatic lens.

apocynthion. The point in the lunar orbit of a satellite launched from earth that is farthest from the moon. The equivalent point in the lunar orbit of a satellite launched from the moon is the *apolune*. *Compare* pericynthion.

apogalacticon. The point in a star's orbit around the Galaxy that is farthest from the galactic centre. The nearest point is the *perigalacticon*.

apogee. 1. The point in the orbit of the moon or an artificial earth satellite that is farthest from the earth and at which the body's velocity is at a minimum. Strictly the distance to the apogee is taken from the earth's centre. At apogee the moon's distance from earth is 406 700 km, about 42 000 km farther than that at *perigee. When a spacecraft or satellite reaches apogee, its *apogee rocket* can be fired to boost it into a more circular orbit or to allow it to escape from earth orbit.

2. The highest point above the earth's surface attained by a rocket, missile, etc.

apojove. The point in the orbit of a satellite of Jupiter that is farthest from Jupiter. The nearest point is *perijove*. Corresponding terms for satellites of Mars are *apomartian* and *perimartian* and for satellites of Saturn *aposaturnian* and *perisaturnian*.

Apollo. The American space programme for landing men on the *moon and returning them safely to earth. A manned lunar landing, to be achieved before 1970, was proposed by Congress in 1961 by J.F. Kennedy following the first manned flight by the USSR. The first six Apollos were used in flights designed to test the Saturn 1B and Saturn V launch vehicles, the *command and service module (CSM)*, and the *lunar module (LM)*. After three astronauts were killed in January 1967 during ground tests, the programme was delayed for 18 months while the *command module (CM)* was redesigned.

The method selected for the flight programme was *lunar orbit rendezvous*. The astronauts were to travel to and from the moon in the command module, which contained the controls and instruments. Rocket engines and fuel supplies were housed in a separate *service module (SM)*. On entering lunar orbit the command module pilot would remain in the CM while the commander and lunar-module pilot made the landing in the LM. On completion of the mission on the moon's surface, the descent stage was to remain on the moon: the ascent stage of the LM was to carry the astronauts into lunar orbit and rendezvous with the command and service modules. The craft would then embark on its return journey, the LM being jettisoned. The SM was to be jettisoned just prior to re-entering the earth's atmosphere.

Details of the Apollo flights are given in the table. Apollo 7 was the first manned test flight of the CSM. Apollo 8 was the first to be launched by Saturn V and to enter lunar orbit. The LM was tested in earth orbit by Apollo 9, in lunar orbit by Apollo 10, and landed on the moon for the first time by Apollo 11 on 20 July 1969. Apollo 12 landed with higher accuracy close to *Surveyor 3, parts of which were returned to earth. Both missions returned basalts from the *maria. Apollo 13 was aborted safely after an explosion in the service module and its target was taken over by Apollo 14. This was to collect rocks from Cone crater (25 million years old) in the *Fra Mauro formation.

Mobility was increased by the use of a *Modular Equipment Transporter but the terrain was unexpectedly rugged. Complex nonmare basalt breccias were returned.

The last three Apollos visited sites with multiple objectives, collected deep (two metre) drill cores (in addition to wide cores, soils, pebbles, and larger rocks), and used a *Lunar Roving Vehicle. Apollo 15 returned anorthosite and green glass from the Apennine *mountains and mare basalts from *Hadley Rille. Apollo 16 landed in the vicinity of South Ray and North Ray craters (2 and 50 million years old respectively) in the *highlands to sample the *Cayley and *Descartes formations, but returned only anorthositic breccias rather than volcanic rocks. Apollo 17 included a scientist-astronaut for the first time. It landed in a dark and *light mantled mare-filled valley between high massifs on the borders of Mare *Serenitatis. Several boulders were sampled and unusual orange glass was found near a 30 million year old *dark halo crater, Shorty.

Experiments performed on the moon by Apollo included *solar-wind, lunar-atmosphere, *cosmic-ray and neutron detection, heat-flow and magnetic-field measurements, active and passive seismometry, and laser ranging. Apollos 15, 16, and 17 were equipped with experiments in the service module, including metric and panoramic photography, laser altimetry, radar sounding, magnetometry, and gamma-ray, ultraviolet, and alpha-particle spectrometry. Subsatellites were ejected from the SM for particle and field measurements. Spent boosters were crashed as seismic sources. After Apollo 17, the remaining hardware of the programme was used in *Skylab and the *Apollo-Soyuz test project.

See also Mercury project; Gemini project; Ranger; Lunar Orbiter probes; Luna probes; Zond probes; moonrocks.

Apollo-Amor objects. *See* Apollo group.

Apollo group. A group of 28 or more known *minor planets, including *Hermes and *Icarus, that have *perihelia inside the orbit of the earth. Like members of the *Amor group, most are very small bodies and can be observed only when close to the earth. The group is named after the one km diameter minor planet *Apollo*, which was discovered when it approached within 0.07 astronomical units of the earth in 1932 but was lost because of uncertainties in its orbit; it was recovered in 1973. Many of their spectra closely resemble that of *chondrite meteorites. Members of the Apollo and Amor groups are often classed collectively as *Apollo-Amor objects*, the perihelia of such bodies usually being taken as smaller than 1.3 AU. See Table 3, backmatter.

Apollo Missions

Mission	Astronauts** (Commander) (LM pilot) (CM pilot)	Date (of landing)	Landing site coordinates (accomplishment)	EVA time (hr)	Distance traversed (km)	Sample weight (kg)
Apollo 7	Schirra Eisele Cunningham	October 1968	(first test of CSM in earth orbit)	—	—	—
Apollo 8	Borman Lovell Anders	December 1968	(first men in lunar orbit)	—	—	—
Apollo 9	McDivitt Scott Schweickart	March 1969	(test of LM in earth orbit)	—	—	—
Apollo 10	Stafford Young Cernan	May 1969	(full dress rehearsal for lunar landing)	—	—	—
Apollo 11	Armstrong* Aldrin* Collins	(20) July 1969	M. Tranquillitatis 0°67'N 23°49'E (first landing)	2.2	0.5	21.7
Apollo 12	Conrad* Bean* Gordon	(19) November 1969	Oc. Procellarum 3°12'S 23°23'W (Surveyor 3)	7.7	1.3	34.4

Apollo 13	Lovell Swigert Haise	April 1970	(Mission aborted: explosion in service module)	—	—	—
Apollo 14	Shepard* Mitchell* Roosa	(31) January 1971	Fra Mauro formation 3°40'S17°28'E (Cone crater)	9.2	3.4	42.9
Apollo 15	Scott* Irwin* Worden	(30) July 1971	Hadley Rille 26°06'N3°39'E (first test of LRV)	18.3	27.9	76.8
Apollo 16	Young* Duke* Mattingley	(21) April 1972	Cayley-Descartes 8°59'S15°31'E (lunar highlands)	20.1	27.0	94.7
Apollo 17	Cernan* Schmitt* Evans	(11) December 1972	Taurus-Littrow 20°10'N30°46'E (first scientist)	22.0	30.0	110.5

** Astronauts marked with an asterisk landed on the moon.

Apollo-Soyuz test project (ASTP). The first international manned space flight, finally agreed to in 1972 and achieved by the US and the USSR in 1975. An American *Apollo spacecraft—Apollo 18—and a Soviet *Soyuz craft—Soyuz 19—were launched into earth orbit on July 15, rendezvoused on July 17 at an altitude of 225 km, and successfully docked. The crews of two cosmonauts (Leonov and Kubasov) and three astronauts (Stafford, Slayton, and Brand) visited each other's craft and conducted joint experiments and surveys. The mission involved major design modifications in both spacecraft.

Apollo Telescope Mount (ATM). The complex telescope mount on NASA's orbiting space laboratory *Skylab. It was sited at the centre of the cross-shaped solar panels, being about 4.5 metres in overall length and about 3.5 metres in diameter; the experimental rack was almost three metres long. The instruments that it carried were larger and more complex than those previously shot into space. Free from the distortion and absorption effects of the earth's atmosphere they were able to make observations of very high accuracy and resolution.

The six principal instruments operated in the x-ray and/or ultraviolet spectral regions, not accessible to ground-based instruments. Two x-ray telescopes (0.3-6, 0.6–3.3 nanometres [nm] and a white-light coronagraph (350-700 nm) were used in studying the solar *corona. An XUV spectroheliograph (15-62 nm), a UV spectroheliometer (30-140 nm), and a UV spectrograph (97-394 nm) provided data on the *chromosphere. There were also two hydrogen-alpha telescopes, operating at 656.3 nm, which observed the low chromosphere. The observations, made between May 1973 and Feb. 1974, led to great advancements in solar astronomy.

apolune. *See* apocynthion.

apomartian. *See* apojove.

aposaturnian. *See* apojove.

apparent. Denoting a property of a star or other celestial body, such as altitude or brightness, as seen or measured by an observer. Corrections must be made to obtain the true value.

apparent diameter (angular diameter). The observed diameter of a celestial body, expressed in degrees, minutes, and/or seconds of arc. *See also* stellar diameter.

apparent distance (angular distance; angular separation). The observed distance between two celestial bodies, expressed in degrees, minutes, and/or seconds of arc.

apparent equinoxes. *See* true equinoxes.

apparent magnitude. See magnitude.

apparent noon. The time at which the sun's centre is seen to cross the meridian. See apparent solar time.

apparent position. See true position.

apparent sidereal time. See sidereal time.

apparent solar time. Time measured by reference to the observed (apparent) motion of the sun, i.e. time as measured by a sundial. The *apparent solar day* is the interval between two successive *apparent noons. The time on any day is given by the *hour angle of the sun plus 12 hours. Because the sun's motion is nonuniform (see mean sun) apparent solar time does not have a constant rate of change and cannot be used for accurate timekeeping. The difference between apparent time and *mean solar time, i.e. clock time, is the *equation of time.

apparition. The period of time during which a particular planet, star, etc., can be observed.

appulse. 1. The apparent close approach of one celestial body to another. Often it refers to the close approach of a planet or minor planet to a star, without the occurrence of an *occultation.
2. A penumbral *eclipse of the moon.

April Lyrids. See Lyrids.

apsidal motion. Rotation of the line of *apsides (i.e. the major axis) of the elliptical orbit of a single or a binary star so that the orbit is not closed and there is a gradual shift in the position of periapsis. It can occur if the gravitational field of a third body can produce a significant effect on a two-body situation, as with the earth, moon, and sun, or if one or both stars of a binary system is appreciably distorted from a spherical shape.

apsides (*sing.* apsis; apse). The two points in an orbit that lie closest to—*periapsis*—and farthest from—*apoapsis*—the centre of gravitational attraction. The prefixes *peri-* and *apo-* are used in describing the apsides of specific orbits. The velocity of an orbiting body at the apsides is either maximal (at periapsis) or minimal (at apoapsis). The *line of apsides* is the straight line connecting the two apsides and is the major axis of an elliptical orbit.

Ap stars. *Peculiar main-sequence stars of spectral types from B5 to F5 in which the spectral lines of certain elements (mainly Mn, Si, Eu, Cr, and Sr) are selectively enhanced and are sometimes of varying intensity (*see* spectrum variables). The enhancement and its variation is apparently associated with strong and often variable magnetic fields. These spectral peculiarities could also relate to stellar temperature, the *manganese stars* being hotter than the *silicon stars,* which in turn are hotter than the *europium-chromium-strontium stars*. See also magnetic stars.

Apus (Bird of Paradise). A small inconspicuous constellation in the southern hemisphere near Crux, the brightest stars being of 4th magnitude. Abbrev.: Aps; genitive form: Apodis; approx. position: RA 16h, dec -75°; area: 206 sq deg.

Aquarids. Either of two active *meteor showers: the *Eta Aquarids,* radiant: RA 336°, dec 0°, maximize on May 4, having a peak duration of 10 days and a *zenithal hourly rate of 20; the *Delta Aquarids,* double radiant: RA 339°, dec -17° and 0°, maximize on July 19, have a peak duration of about 20 days, and a ZHR of about 30. The Eta Aquarids have an orbit that is closely aligned with the post-perihelion path of *Halley's comet.

Aquarius (Water Bearer). An extensive zodiac constellation in the southern hemisphere near Pegasus, the brightest stars being of 3rd magnitude. It contains the planetary nebulae NGC 7293 (the *Helix nebula*) and NGC 7009 (the *Saturn nebula*), and the globular clusters M2 and M72. Abbrev.: Aqr; genitive form: Aquarii; approx. position: RA 22h, dec -10°; area: 980 sq deg.

Aquila (Eagle). An equatorial constellation near Cygnus, lying in the Milky Way, the brightest stars being the 1st-magnitude *Altair and some of 3rd magnitude. Eta Aquilae is a bright *Cepheid variable. Abbrev.: Aql; genitive form: Aquilae; approx. position: RA 19.5h, dec +3°; area: 652 sq deg.

Ara (Altar). A small constellation in the southern hemisphere near Centaurus, lying in the Milky Way, the brightest stars being of 3rd magnitude. Abbrev.: Ara; genitive form: Arae; approx. position: RA 17h, dec -55°; area: 237 sq deg.

Archer. Common name for *Sagittarius.

arch filament. See prominences.

arc second (arc sec; second of arc). Symbol: ″. A unit of angular measure equal to 1/3600 of a degree. It is used, for example, to give the apparent diameter of objects, the angular separation of binary stars, and the proper motion and annual parallax of stars. An *arc minute,* symbol: ′, is 60 arc seconds, i.e. 1/60 of a degree.

Arcturus (α Boö). A conspicuous red giant that is the brightest star in the constellation Boötes and also in the northern sky. The three stars Alioth, Mizar, and Alcaid in the handle of the Plough point in its direction. It is a high-velocity star and shows spectral peculiarities. m_v: -0.05; M_V: -0.3; spectral type: K2 IIIp; radius (by interferometer): 28 times solar radius; distance: 11 pc.

areal velocity. The velocity at which the *radius vector sweeps out unit area. It is constant for motion in an elliptical orbit. *See* Kepler's laws.

Arecibo radio telescope. The radio telescope, 305 metres (1000 ft) in diameter and spherical in shape, sited in a natural hollow at Arecibo, Puerto Rico. It is the world's largest single-dish telescope. It is operated by Cornell University and has been in use since 1963. The original metallic mesh lining was replaced in the early 1970s by tens of thousands of reflecting panels in order to improve reception of lower wavelengths, including the 21 cm line of hydrogen. Although not steerable some directionality can be obtained by moving the feed antenna, which is suspended above the dish.

Arend-Roland comet (1975 III). A *comet discovered by S. Arend and G. Roland in the course of routine work on *minor planets. It was unusual because it had a sunward spike: many of the solid particles ejected from the *nucleus near *perihelion remained in the orbital plane of the comet; as the earth passed through the orbital plane the reflection from the particles combined to give a luminous spike.

areography. The geography of *Mars. Surface coordinates are defined by *areographic longitude* and *latitude*. Longitude is measured in degrees westwards from a prime meridian chosen to pass through the dark-albedo feature Sinus Meridiani near the Martian equator. The exact position of the meridian is now fixed by a 0.5 km crater pit, Airy-0, within the crater Airy, 300 km south of the equator.

Argo or **Argo Navis (Ship).** A former very extensive unwieldy constellation in the southern hemisphere subdivided in the mid-18th century into the constellations *Carina, *Puppis, *Vela, and *Pyxis. Its stars are still identified by a single set of Bayer letters (*see* stellar nomenclature), the brightest being *Canopus or Alpha Carinae.

argument of the perihelion. *See* orbital elements.

Argyre Planitia. A 1000 km diameter basin in Mars' southern hemisphere, visible as a bright area to earth-based observers. Like *Hellas Planitia it is probably an impact feature. *See* Mars, surface features.

Ariane. The three-stage *launch vehicle built for the *European Space Agency (ESA) and designed to launch all ESA's major satellite missions, including geostationary communications satellites. The building programme was officially approved in Aug. 1973 and is financed mainly by France. Operational launches are scheduled, at present, from 1981, with four test launches 1979/80. There is little new technology in the heavy launcher: the first two stages use Viking engines that burn a hypergolic fuel (spontaneous ignition on admixture); the third stage is a liquid hydrogen/liquid oxygen engine.

Ariel. A satellite of Uranus, discovered in 1851. *See* Table 2, backmatter.

Ariel V. A British *x-ray astronomy satellite launched into a 550 km equatorial orbit in October 1974 from the San Marco platform off the Kenyan coast. Its six experiments (five UK, one US) were designed to detect, accurately locate, and measure spectral and temporal features of cosmic x-ray sources. Ariel V is spin-stabilized with the spin axis and period controlled by gas jets. Tracking and command and data reception is carried out from NASA ground stations at Quito and on Ascension Island. The UK control station is at the Appleton Laboratory with data links to the experiment groups. The satellite and all experiments remained fully operational until mid-1977 when the control gas was exhausted. Subsequent operation with magnetometer attitude control was extended into 1979.

Ariel V's major scientific achievements were the production of a new atlas of *x-ray sources, the establishment of *Seyfert galaxies as powerful x-ray emitters, and the discovery of slow *x-ray pulsators* (believed to be *neutron stars spinning with periods of minutes) and of abundant iron (by detection of its characteristic x-radiation) in *supernova remnants and *clusters of galaxies. Technically it was also the first small scientific satellite to be operated in near real-time.

Aries (Ram). A zodiac constellation in the northern hemisphere near Auriga, the brightest stars being the 2nd-magnitude *Hamal (α) and 3rd-magnitude Sheratan (β). Gamma Arietis is a wide visual binary with components equal in magnitude (4.4) and both A stars. Abbrev.: Ari; genitive form: Arietis; approx. position: RA 2.5h, dec +21°; area: 441 sq deg. *See also* first point of Aries.

Aries, first point of. *See* first point of Aries.

Arizona meteorite crater. The archetypal impact explosion crater, 1280 metres across and 180 m deep, formed about 6000 years ago by a 250 000 tonne *meteorite, 70 m in diameter, impacting at about 16 km s^{-1}. The rim is composed of folded and inverted rock layers that have been pushed out from a central point. The meteorite almost vaporized completely, only a few hundred tonnes being scattered over 80 km² around the crater in the form of tiny iron spherules.

armillary sphere. A device, dating back to antiquity, composed of a set of graduated rings representing circles on the *celestial sphere, such as the ecliptic, celestial equator, and colures. The whole globe often revolved about an axis—the polar axis—within horizon and meridian circles. Movable sighting adjustments enabled a star to be observed and its coordinates to be read off the relevant circles.

arm population. Young population I stars occurring in the spiral arms of the Galaxy. *See* populations I and II.

Arneb. *See* Lepus.

array. A system used in radio astronomy in which a number of discrete elements are connected together to form a composite *antenna. An element may for example be a single *dipole or a *dish. The ratio of the aperture area occupied by all the elements to the area that would be occupied by a continuous aperture of the same overall length is called the *filling factor*; it is a measure of the fraction of the aperture that is actually there. The *effective area* for an array operating at a wavelength λ with a *directivity D is given by
$$A_e = D\lambda^2/4\pi$$
It is usually less than the physical area, even when the filling factor is equal to unity, because of small irregularities in the aperture distribution. For a perfect array with a uniform aperture distribution the effective and physical areas are indeed equal.

An array may produce a *grating response* in one or more directions away from the main lobe (*see* antenna) which, if the elements of the array were all nondirectional, would be as strong as the main lobe. However, the *antenna patterns of the elements together with geometrical effects usually reduce the power in the grating response to an acceptable level. *See also* Butler matrix; feeder; aperture synthesis.

Arrow. Common name for *Sagitta.

Arsia Mons. A 19 km high 400 km wide volcano at the southern edge of the *Tharsis Ridge of Mars. It appears to be older than the other Tharsis volcanoes and has the largest summit caldera, 140 km across. *See* Mars, volcanoes.

artificial satellite. *See* satellite, artificial.

ascending node. *See* nodes.

Ascraeus Mons. A 19 km high volcano on the *Tharsis Ridge of Mars. It is 400 km wide at its base and has a summit caldera 50 km across. *See* Mars, volcanoes.

ashen light. 1. The faint glow claimed to be occasionally observed on the unlit area of Venus in its crescent phase. Its cause is unknown but it might result from bombardment of atmospheric atoms and molecules by energetic particles and radiation, as with terrestrial *airglow.
2. *See* earthshine.

aspect (configuration). The apparent position of any of the planets or the moon relative to the sun, as seen from earth. Specific aspects include *conjunction, *opposition, and *quadrature, which differ in the *elongation of the object concerned.

aspheric surface. Any surface that does not form part of a sphere. A paraboloid reflector, used in both optical and radio telescopes, is aspheric. Although production costs are high, aspheric surfaces can greatly improve lens and mirror performance.

association. A loose group of young stars of similar *spectral type. *O-associations* (or *OB-associations*) are groups of between 10 and 1000 highly luminous main-sequence stars of spectral types O to B2. They occur in regions rich in gas and dust in the spiral arms of the Galaxy. Over 50 are known with dimensions ranging from a few parsecs to several hundred parsecs. Often an *open cluster is found near the centre of an association, e.g. the Zeta Persei association surrounds *h and Chi Persei. *T-associations* are similar groups of *T Tauri stars. About 25 are known, most containing less than 30 stars though some contain as many as 400. They are often found in the vicinity of young open clusters (*see* O-cluster).

Associations are generally too sparsely populated to be gravitationally bound systems and there is strong evidence that they represent the after-effects of comparatively recent multiple star births. In some cases the stars appear to be expanding from a common centre: by extrapolating back their present velocities an estimate of the age of the system can be derived. For instance, the association II Persei shows an expansion age of slightly over one million years.

A stars. Stars of *spectral type A that are blue-white or white and have a surface temperature (T_{eff}) of about 7500 to 10 000 kelvin. Balmer absorption lines of hydrogen dominate the spectrum, reaching maximum strength for A0 to A3 stars; lines of ionized and neutral metals slowly strengthen with lines of ionized calcium, CaII (H, K), appearing in later subdivisions. *Colour indices for A0 stars are zero. Vega, Sirius, Deneb, and Altair are A stars.

asterism. A group of stars, such as the *Plough, in a constellation or, formerly, a constellation itself.

asteroid. *See* minor planets.

asteroid belt. The zone near to the plane of the *ecliptic and between the orbits of Mars and Jupiter that contains the majority of *minor planets. The inner edge is well-defined and the orbits strongly concentrated in the plane of the ecliptic. The orbital *inclination becomes more scattered on moving through the belt. The outer edge is rather diffuse. All the bodies are in *direct motion. *See also* Kirkwood gaps.

asthenosphere. *See* earth.

astigmatism. An *aberration of a lens or mirror system that occurs when light falls obliquely on the system and is focused not as a single point image but as two perpendicular and separated lines. In the reflecting system shown (see illustration) rays from points A and B on the mirror converge to the vertical line image ab; rays from C and D converge to the horizontal line image cd. The pencil of reflected rays, elliptical in cross section, cannot produce a sharp image anywhere along its path; the plane of optimum focus occurs between ab and cd where the pencil has its smallest cross section. Astigmatism is not as severe an aberration as *coma.

ASTP. *Abbrev. for* Apollo-Soyuz test project.

Astraea. The fifth *minor planet to be discovered, in 1845. *See* Table 3, backmatter.

astroblemes. Scars, usually ancient, left in the earth's crust by meteorite impact. Criteria

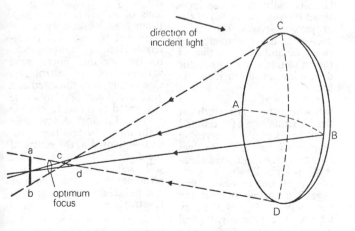

Astigmatism due to concave mirror

used for recognition are shatter cones, high-pressure silica polymorphs, shock microdeformations of quartz, impactite glasses, traces of nickel, and nickel-iron associations. All are of a geologically youthful age, none older than 10 million years. The largest are Vredefort, South Africa (40 km diameter), Nordlinger Ries, Germany (25 km), Deep Bay, Saskatchewan, Canada (14 km), Lake Bosumtwi, Ghana (10 km), and Serpent Mound, Ohio (64 km).

The velocity of a large meteorite, i.e. one over thousands of tonnes, is dissipated very little by atmospheric braking effects and it hits the ground at its cosmic velocity of 11 to 74 km s^{-1} and explodes. Most of the kinetic energy goes to accelerate and excavate the material from the volume that eventually becomes the crater. Usually only an exceedingly small fraction (<0.01%) of the projectile survives. The resulting crater has a diameter between 20 and 60 times that of the impacting body, this factor depending on the incident velocity.

astrochemistry. The study of the chemistry of celestial bodies and of the intervening regions of space. It involves the detection and identification, principally by spectroscopy, of the inorganic and organic chemicals present and the study of the reactions by which these neutral and charged atoms and molecules could have been produced and of future chemical processes. *See also* molecular-line radio astronomy.

astrodynamics. The application of *celestial mechanics, the ballistics of high-speed solids through gases, propulsion theory, and other allied fields to the planning and production of the trajectories of spacecraft.

astrograph. A multiple-lens refracting telescope with considerable light-gathering power, used to photograph a large area of the sky in a single exposure. Diameters range from about 20 to 40 cm.

Astrographic Catalogue. *See* Carte du Ciel.

astrolabe. An instrument, dating back to antiquity, used to measure the altitude of a celestial body and to solve problems of spherical astronomy. From the 15th century it was employed by mariners to determine latitude, until replaced by the sextant. There were various types of astrolabes. One simple form consisted of a graduated disc that could be suspended by a ring to hang in a vertical plane. A movable sighting device—the alidade—pivoted at the centre of the disc. Modern versions are still used to determine stellar positions and hence local time and latitude. *See also* prismatic astrolabe.

astrology. The prediction of the personality of individuals and groups of individuals and of the course of events in their lives by reference to the actual and relative positions of the sun, moon, and planets on the *zodiac, especially at the moment of birth. Astrology played an important role in many ancient civilizations. Accurate records of the motions of celestial bodies had to be kept and from these observations the separate science of astronomy developed. There is no apparent scientific basis to astrological predictions.

astrometric binary. *See* binary star.

astrometry (positional astronomy; spherical astronomy). The branch of astronomy concerned with the measurement of the positions of celestial bodies on the *celestial sphere, with the conditions, such as *precession, *nutation, and *proper motion, that cause the positions to change with time, and with the concepts and observational methods involved in making the measurements. *See also* radio astrometry; long-focus photographic astrometry.

astronautics. The science and technology of spaceflight.

Astronomer Royal. An eminent British astronomer appointed to the post of Astronomer Royal of England and, until the retirement of Sir Richard Woolley in 1971, Director of the Royal Greenwich Observatory. The two titles

ASTRONOMERS ROYAL

John Flamsteed	1675–1719
Edmund Halley	1720–42
James Bradley	1742–62
Nathaniel Bliss	1762–64
Nevil Maskelyne	1765–1811
John Pond	1811–35
Sir George Biddell Airy	1835–81
Sir William H.M. Christie	1881–1910
Sir Frank Watson Dyson	1910–33
Sir Harold Spencer Jones	1933–55
Sir Richard Woolley	1955–71
Sir Martin Ryle	1972–

were separated in 1972 when Sir Martin Ryle became the present Astronomer Royal. Astronomers Royal are listed in the table. The first Astronomer Royal, John Flamsteed, was appointed in 1675 following the foundation of the Royal Greenwich Observatory.

Astronomical Ephemeris, The. The annual *ephemeris published by H.M. Nautical Almanac Office, Royal Greenwich Observatory. With the exception of the introductory pages it is identical in content to *The American Ephemeris and Nautical Almanac,* issued by the Nautical Almanac Office, United States Naval Observatory. The two were unified in 1960. Prior to 1960 the British edition was known as *The Nautical Almanac and Astronomical Ephemeris,* usually abbreviated to *The Nautical Almanac,* which was first introduced in 1767. The title *The Nautical Almanac* is now used in both the UK and USA for the abridged edition of the Ephemeris for surface navigation.

Astronomical Netherlands Satellite (ANS). A joint Netherlands/US satellite that was launched in Aug. 1974 and contains a 20 cm ultraviolet telescope and two x-ray telescopes.

astronomical triangle. A *spherical triangle on the *celestial sphere formed by the intersection of the great circles joining a celestial body, S, the observer's zenith, Z, and the north celestial pole, P (see illustration). The relationships between the angles and sides of a spherical triangle are used for transformation between *equatorial and *horizontal coordinate systems: the angle at S (q) is the *parallactic angle;* that at P (t) is the *hour angle of S; that at Z (A) is the azimuth of S. The sides are equal to the *zenith distance (ζ), the complement of the terrestrial latitude (90°-ϕ), and the complement of the declination (90°-δ). The parallactic angle, for example, is given by the sine rule:

$$\sin q = (\cos \phi \sin t)/\sin \xi$$
$$= (\cos \phi \sin A)/\cos \delta$$

astronomical twilight. The period from sunset until the time when the sun is 18° below the horizon, i.e. until it has a *zenith distance of 108°, or

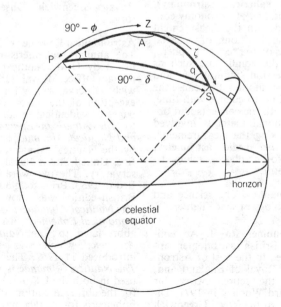

Astronomical triangle

the corresponding period from a zenith distance of 108° until sunrise. When the zenith distance becomes 108° the indirect illumination from the sun is about equal to the illumination of the night sky.

Other twilights are in use for navigational purposes: during *civil twilight* the sun's altitude changes from 0° to -6° or vice versa; during *nautical twilight* the sun's altitude changes from -6° to -12° or vice versa.

astronomical unit (AU or A.U.). A unit of length that is used for distances within the solar system and is equal to the mean distance between the earth and the sun. One astronomical unit is equal to 149 597 870 kilometres or 499.012 light seconds.

astronomy. The observational and theoretical study of celestial bodies, of the intervening regions of space, and of the universe as a whole. Astronomy is one of the oldest sciences and has developed in step with advances in instrumentation and other technology and with advances in physics, chemistry, and mathematics. The main branches are *astrometry, *celestial mechanics, and the major modern field of *astrophysics with its subsections of *cosmology, *radio astonomy, and *x-ray, *gamma-ray, *infrared, and *ultraviolet astronomy. *Stellar statistics is another modern field.

astrophysics. The study of the physical and chemical processes and characteristics associated with celestial bodies and with the intervening regions of space. It is concerned in particular with the production and expenditure of energy in systems such as the stars and galaxies and in the universe as a whole and with how this affects the evolution of such systems. Astrophysics developed in the 19th century and is closely related to atomic and nuclear physics, plasma physics, thermodynamics, spectroscopy, solid-state physics, and relativity. *Cosmology, *radio astronomy, *x-ray, *gamma-ray, *infrared, and *ultraviolet astronomy are usually considered subsections of astrophysics.

AT. *Abbrev. for* atomic time.

ataxite. *See* iron meteorite.

Aten. A small member of the *Apollo group of minor planets. Discovered in 1976, it was the first minor planet to be found with a semimajor axis of less than one astronomical unit and therefore an orbital period of less than one year. *See* Table 3, backmatter.

ATM. 1. *Abbrev. for* Apollo Telescope Mount.
   2. *Abbrev. for* Amateur Telescope Making, a textbook for amateur astronomers.

atmosphere. The gaseous envelope that surrounds a star, planet, or some other celestial body. The ability of a body to maintain an appreciable atmosphere depends on the *escape velocity at its surface and on its temperature. Small bodies, such as the moon, have low escape velocities that may be exceeded comparatively easily by gas molecules travelling with the thermal speeds appropriate to their temperature and mass. The speed of a gas molecule increases with temperature and decreases with molecular weight; the lighter molecules, such as hydrogen, helium, methane, and ammonia, can therefore escape into space more readily than heavier ones such as nitrogen, oxygen, and carbon dioxide. Some bodies, such as Pluto, may be so cold that most potential atmospheric gases lie frozen on their surfaces. A body may gain an atmosphere by any of several processes, including a temporary acquirement by collision with an object containing frozen gases, such as a cometary nucleus.

The study of planetary atmospheres—their contents, meteorology, and evolution—has developed rapidly with the advent of planetary space probes in the 1960s. Four planets—*Earth, *Mars, *Venus, and *Jupiter—have been examined at close range by probes and much has been learned about their atmospheres. The deep hydrogen, helium, methane, and ammonia atmospheres of Jupiter and the other *giant planets probably derive directly from the nebula and icy planetesimals from which the planets are now thought to have coalesced by accretion (see nebular hypothesis). The *terrestrial planets, however, were too small to capture the warm gases of the early nebula; the present atmospheres of Venus, Earth, and Mars are thought to have escaped from volcanoes after being liberated from chemical combination in subsurface rocks by radioactive heating.

The weather systems have also been studied by probes. The internal energy supply and rapid rotation of Jupiter (and possibly Saturn) cause their weather systems to differ greatly from those of the terrestrial planets, which are driven by the sun.

atmosphere, composition. The percentage composition of clean dry air in the earth's atmosphere is given in the table. The concentrations of CO_2, CO, SO_2, NO_2, and NO are greater in industrial areas. The amount of ozone increases in the ozonosphere (see atmospheric layers) and depends on meteorological and geographical variations. The fraction of water vapour in normal air varies from 0.1-2.8% by volume and 0.06-1.7% by mass, with meteorological and geographical variations and a decrease with altitude.

atmospheric extinction. See extinction.

atmospheric layers. The gaseous layers into which the

Composition of earth's lower atmosphere

gas		percentage composition of clean dry air	
		by volume	by mass
nitrogen	N_2	78.084	75.523
oxygen	O_2	20.947	23.142
argon	Ar	0.934	1.290
carbon dioxide	CO_2	0.032	0.050
neon	Ne	0.0018	0.0013
helium	He	0.0005	0.0001
methane	CH_4	0.0002	0.0001
krypton	Kr	0.0001	0.0003
carbon monoxide	CO	0.0001	0.0001
sulphur dioxide	SO_2	0.0001	0.0002

With smaller traces of hydrogen (H_2), nitrous oxide (N_2O), ozone (O_3), xenon (Xe), nitrogen dioxide (NO_2), radon (Rn), and nitric oxide (NO).

earth's atmosphere is divided, mainly on the basis of the variation of temperature with altitude. The *troposphere* is the lowest layer, extending from sea level to the *tropopause* at an altitude of between 8 km at the poles and 18 km within the tropics. It contains three-quarters of the atmosphere by mass and is the layer of clouds and weather systems. It is a region, heated from the ground by infrared radiation and convection, in which the temperature falls with increasing height to reach a minimum of approximately -55°C at the midlatitude and polar tropopause and maybe -80°C at the equatorial tropopause.

The *stratosphere*, lying above the tropopause, is an atmospheric layer in which the temperature is at first steady with altitude before increasing to a maximum of 0°C at about 50 km, which marks the *stratopause*. This temperature variation inhibits vertical air movements and makes the stratosphere stable. The heating arises from the absorption of high-energy solar ultraviolet radiation by ozone (O_3) molecules. These molecules, themselves formed by the action of ultraviolet radiation on oxygen molecules, constitute the *ozonosphere*: this atmospheric region lies between about 15 and 50 km altitude and thus is essentially synonymous with the stratosphere. The maximum concentration of ozone occurs at about 25 km.

Above the stratopause is the *mesosphere,* in which the temperature falls with height to reach about -90°C at the *mesopause* at an altitude of about 85 km. Within the mesosphere the heat contribution from ultraviolet absorption by ozone decreases as the ozone becomes less plentiful. The *ther-

mosphere, above the mesopause, has a temperature that rises with height as the sun's far-ultraviolet radiation is absorbed by oxygen and nitrogen. This process gives rise to ionized atoms and molecules and leads to the formation of the *ionosphere layers between altitudes of about 60 and 500 km. This is also the domain of *meteors and *aurorae.

Although the temperature climbs to about 500°C at 150 km and about 1300°C at 500 km, these are the kinetic temperatures (i.e. measures of the average random motions) of the few atoms and molecules found at such heights: their combined heating ability is negligible. The atmospheric density decreases to about 0.25 of its sea-level value at the tropopause, 0.0009 at the stratopause, and 0.000 007 at the mesopause. The density at 500 km averages less than one million millionth of its sea level value but is prone to large variations resulting from solar heating and disturbances during periods of enhanced *solar activity.

The *exosphere,* above 500 km, is the region in which atmospheric constituents lose collisional contact with each other because of their rarity and can leak away into space. It contains the *Van Allen radiation belts and the *geocorona and extends to the magnetopause where it meets the interplanetary medium.

atmospheric pressure. The force per unit area (i.e. pressure) exerted by a column of atmosphere extending vertically upwards to the limit of the atmosphere. It is usually expressed in millibars (mb), atmospheres (atm), or newtons per square metre (N m^{-2}). The standard atmospheric pressure at the earth's surface at sea level is 1013 mb or 101 325 N m^{-2}. On the surface of Venus and Mars the atmospheric pressure is about 90 000 mb and 7.5 mb, respectively.

atmospheric refraction. *Refraction of light entering the earth's atmosphere from the near-vacuum of space. Light coming into the atmosphere is bent towards the earth, with the result that stars and other celestial bodies appear to be displaced very slightly towards the *zenith. The amount of displacement, i.e. the resulting increase in altitude, depends on the body's distance from the observer's zenith and is greatest (about 35 arc minutes) for objects on the horizon. *See also* zenith distance.

atmospheric windows. Regions of the *electromagnetic spectrum that can be transmitted through the earth's atmosphere without significant absorption or reflection by atmospheric constituents. All spectral regions are at least partially absorbed but there are two nearly transparent ranges: the *radio window* and the *optical window* (see illustration). The radio window extends from a wavelength of a few millimetres to about 30

Atmospheric windows

metres (i.e. it includes frequencies from about 100 gigahertz to 10 megahertz); it thus allows high-energy radio waves from celestial *radio sources to be received by ground-based radio telescopes. Lower energies are reflected by the *ionosphere while above 100 GHz molecular absorption increases. Some far-infrared windows lie close to the radio window, centred on wavelengths of 0.4, 0.8 and 1.2 millimetres.

The optical window ranges from wavelengths of about 300 to 900 nanometres (nm) and thus includes some of the near-ultraviolet and near-infrared regions as well as the whole visible spectrum. In addition there are several narrow-band near-infrared windows: 0.95-1.1 micrometres (μm), 1.15-1.3 μm, 1.5-1.75 μm, 2-2.4 μm, 3.4-4.2 μm, 4.6-4.8 μm, 8-13 μm, and 16-18 μm. Radiation between the radio and optical windows is absorbed mainly by water vapour and carbon dioxide in the atmosphere. Observatories studying infrared sources are therefore best sited in very dry/or mountainous regions where the effect of water vapour is reduced and/or the atmosphere is thinner.

Beyond the low-wavelength end of the optical window no radiation can penetrate the atmosphere: ultraviolet radiation with wavelengths between about 230-300 nm is absorbed by atmospheric ozone; shorter wavelengths down to about 100 nm are blocked by oxygen and nitrogen molecules; the shortest wavelengths are absorbed by atoms of oxygen, nitrogen, etc., in the upper atmosphere. The spectral regions from space rendered inaccessible by atmospheric absorption must be studied by means of instruments carried in satellites, rockets, etc. Even then, high-energy ultraviolet and low-energy x-rays from distant stars can be absorbed by *geocoronal hydrogen and helium.

atom. The smallest part of a chemical *element that can retain its chemical identity, i.e. can take part in a chemical reaction without being destroyed. An atom can however be transformed into one or more other atoms by nuclear reactions. Nearly all the mass of an atom is concentrated in the small central *nucleus, which is made up of protons and neutrons. Electrons occupy the remaining space in the atom and may be thought of as orbiting the nucleus. The number of protons in a particular nucleus, i.e. the atomic number of the atom, is equal to the number of electrons in a neutral (nonionized) atom. The arrangement and behaviour of these electrons determine the interactions of the atom with other atoms and thus govern chemical properties and most physical properties of matter. *See also* energy levels; isotopes; ion.

atomic clock. The most accurate type of clock to date, in which the periodic process used as the timing mechanism is an atomic or molecular event with which is associated a constant and very accurately known *frequency. Various atoms and molecules have been employed in atomic clocks, the frequency involved in the timing mechanism being a microwave frequency; such frequencies can be produced and manipulated by sophisticated electronic techniques. The accuracy is at present better than one part in one thousand million million. *See also* atomic time.

atomic number. Symbol: Z. The number of protons in the nucleus of an atom. This is equal to the number of electrons orbiting the nucleus in a neutral atom. The *isotopes of an element have the same atomic number but different *mass numbers.

atomic time (AT). Symbol: t_A. A time system that operates at a constant rate and that is measured by means of an *atomic clock. The definition of the *second has been in terms of atomic time rather than *ephemeris time since 1964. At the time of definition the SI atomic second was equal to the ephemeris second.

International atomic time (TAI; IAT), evaluated by the Bureau Internationale de l'Heure, was adopted for all timing on Jan. 1 1972. It is a more precise time scale than can be determined from astronomical observations and thus replaced ephemeris time: the two rates agree to within two parts in one thousand million and for practical purposes ET is approximately equal to TAI + 32.2 seconds. Coordinated universal time (*see* universal time) is based on TAI and is used for all civil timekeeping.

attenuation. The reduction in magnitude of a quantity as a function of distance from source or of some other parameter. The attenuation of a beam of radiation, i.e. the reduction in intensity or flux density, results from absorption and/or scattering of the waves or particles as they pass through matter.

AU or **A.U.** *Abbrev. for* astronomical unit.

audiofrequency. Any frequency at which a sound wave would be audible, i.e. a frequency in the range 15 hertz to 20 000 hertz.

Auger shower. *Another name for* air shower. *See* cosmic rays.

augmentation. The difference at a particular time between the topocentric and geocentric semidiameter of the moon. *See also* diurnal libration.

Auriga (Charioteer). A conspicuous constellation in the northern hemisphere near Orion, lying in the Milky Way, the brightest stars being *Capella (α) and the 2nd-magnitude *Algol variable Menkalinan (β), *Epsilon Aurigae and *Zeta Aurigae are eclipsing binaries. The area contains the bright open clusters M36, M37, and M38. Abbrev.: Aur; genitive form: Aurigae; approx. position: RA 5.5h, dec +40°; area: 657 sq deg.

aurora. A display of diffuse changing coloured light seen high in the earth's atmosphere in polar regions. Aurorae show varying colours from whitish green to deep red and often have spectacular formations of streamers or drapery. They are formed usually within about 20° of the magnetic poles and occur at altitudes of approximately 100 km. They are caused by charged particles from the *solar wind or from *flares that become trapped by the earth's magnetic field and interact with gases in the upper atmosphere as they spiral around the magnetic field lines between the poles.

Australian National Radio Astronomy Observatory. The Australian C.S.I.R.O. observatory near Parkes, New South Wales, at which the chief instrument is a fully steerable 64 metre (210 ft) radio dish.

australite. *See* tektites.

autocorrelation function. A mathematical function that for a real function f(t) is given by
$$\int f(u)f(u-t)du$$
It describes, for example, the time or distance over which a signal is *coherent. The *cross-correlation function* describes how two different signals compare as they are displaced relative to one another; for two real functions f(t) and g(t), it is given by

$$\int g(u-t)f(u)du$$

autumnal equinox. *See* equinoxes.

axial period. The period of time during which a body makes one complete *rotation on its axis. For planets it is usually referred to the direction of a fixed star and is thus equivalent to the *sidereal day.

axis. 1. An imaginary line that usually passes through the centre of a body or system and about which the body is often symmetrical or has some form

of symmetry. It is the imaginary line about which a rotating body turns or about which an object, such as the celestial sphere, appears to rotate.
2. A reference line on a graph.

azimuth. *See* horizontal coordinate system.

azimuthal mounting. *See* altazimuth mounting.

B

Baade's star. *Another name for* Crab pulsar. *See* Crab nebula.

background radiation. *See* cosmic background radiation; microwave background radiation.

background temperature. *See* antenna temperature.

Baily's beads. A string of brilliant points of sunlight sometimes seen, very briefly, at the moon's edge just before or just after totality in a solar *eclipse. They are the result of sunlight shining through the valleys and mountains on the moon's limb as the sun just disappears or emerges from behind the moon. They were first described by Francis Baily after the eclipse of 1836. *See also* diamond-ring effect.

Balance. Common name for *Libra.

ballistic trajectory. The *trajectory of an object acted upon only by gravitational forces and the resistance of the medium through which it is travelling. The object is not propelled or guided except during the initial (nonballistic) phase of its flight.

Balmer series, lines. *See* hydrogen spectrum.

balun. *See* feeder.

band spectrum. A *spectrum consisting of one or more bands of emitted or absorbed radiation. The bands consist of numbers of closely spaced lines and result from transitions between *energy levels in molecules, allowing different levels of vibration or rotation.

bandwidth. 1. The range of frequencies over which an instrument, such as a *radio telescope, makes an appreciable response to input signals.
2. (coherence bandwidth). The range of frequencies over which electromagnetic or other emissions from a source maintain *coherence.

Barlow lens. An *achromatic *negative lens placed behind the eyepiece of a telescope, just inside the primary focal plane, in order to increase the effective focal length of the objective or primary mirror. This increases the magnification of the telescope so that a long-focus eyepiece may be used to give the higher powers needed to separate optical doubles, or for planetary observation under the best seeing conditions.

Barnard's loop or **ring**. *See* Orion.

Barnard's star (BD 4° 3561; Munich 15040). A red dwarf star in the constellation Ophiuchus that was discovered in 1916 by E.E. Barnard. The fourth-nearest star to the sun, it has the largest known *proper motion (10".3 per year) and thus moves a distance equivalent to the moon's diameter in 180 years. Observations of the star's position over many years show slight oscillating variations in right ascension and declination with a period of about 24 years. It is thought that this wobbling motion could be due to one or more planets, of about Jupiter's mass, orbiting the star. m_v: 9.53; M_V: 13.21; spectral type: M5 V; distance: 1.83 pc.

barred spiral galaxy. *See* galaxies; Hubble classification.

barycentre. *See* centre of mass.

baseline. The straight line or arc of a great circle between two observational points, as between the elements of an interferometer (*see* radio telescope). *See also* very long baseline interferometry.

basins, lunar. Huge multiringed *crater structures on the moon that are several hundred kilometres in diameter and were all produced during the first 700 million years of lunar history by the impact of asteroidal bodies. The youngest basins (*Orientale, *Imbrium, *Crisium, Nectaris, and *Serenitatis) may have been excavated during a cataclysmic period around 3900 million years ago. The *ejecta blankets of basins are extensive and provide the basis for *highland stratigraphy. Basins are so defined by their systems of concentric ring structures, most of which are now in the form of ridges and mountain arcs. Basins on the moon's nearside were filled with lava at some time up to 3000 million years ago to produce *maria: the resulting mare has the same name as the basin (e.g. Mare Imbrium). Several unfilled basins (*thassaloids*) exist on the farside, as yet unnamed, the largest of which was revealed as a vast depression by the *Apollo 15 laser altimeter. At least 28 basins are now known to exist. *See also* rille; mascons.

Bayer letters. *See* stellar nomenclature.

BC. *Abbrev. for* bolometric correction. *See* magnitude.

BD. *Abbrev. for* Bonner Durchmusterung.

beam. 1. *See* antenna.
  2. A well-defined elongated region of space down which energy is passing. In the beam model of double radio sources (*see* radio-source structure) the central galaxy shoots out two beams in opposite directions in which the energy may be transported by *relativistic electrons, low-frequency waves, or some other mechanism. Hot

beam efficiency. *See* antenna.

beamwidth. *See* antenna.

Becklin's star (Becklin-Neugebauer object; BN object). An intense unresolved infrared source discovered by Becklin and Neugebauer in 1966. It lies in the *Orion nebula and is believed to be a collapsing *protostar of mass about 10 times that of the sun. It has an effective temperature of about 600 kelvin and luminosity 1000 times that of the sun.

Beehive. *See* Praesepe.

Bellatrix. (γ **Ori**). A remote luminous blue-white giant that is the third brightest star in the constellation *Orion. m_V: 1.63; M_V: -4.1; spectral type: B2 III; distance: 140 pc.

Benatnasch. *See* Alcaid.

bent double radio source. *See* radio-source structure.

bent transit. *Another name for* broken-back transit. *See* meridian circle.

Berenice's Hair. Common name for *Coma Berenices.

Besselian year. The year starting when the *mean sun reaches a *right ascension of 18h 40m. This occurs sometime on Jan. 1, the precise time varying from year to year. The beginning of the Besselian year 1980, say, is written 1980.0.

Be stars. *Irregular variables of stellar type B in which bright emission lines of hydrogen are superimposed on the normal absorption spectrum. The stars are rotating very rapidly and are slowly losing mass to an expanding shell that surrounds the star and is drawn into a disc around the equator due to the rotation. Be stars are now known to be identical to *shell stars. Like *Ae stars they are thought to be similar to *T Tauri stars except that they have larger masses.

Beta (β). The second letter of the Greek alphabet used in *stellar nomenclature usually to designate the second-brightest star in a constellation or sometimes to indicate a star's position in a group.

Beta Canis Majoris stars. *See* Beta Cephei stars.

Beta Centauri (Hadar; β Cen). A luminous remote but still conspicuous blue-white giant that is the second-brightest star in the constellation Centaurus. It is a visual binary, separation 1″.2. m_V: 0.63; M_V: -5.0; spectral type: B1 III; distance: 130 pc.

Beta Cephei stars (Beta Canis Majoris stars). A small group of *pulsating variables, the prototypes being β Cep and β CaM. They are hot massive luminous stars of spectral types O9 to B3 with short periods of rotation (about 3-15 hours) and a very small variation in visual brightness (about 0.01-0.25 magnitudes), although the range is much greater in ultraviolet radiation. Some of the stars have two or even more periods of *radial-velocity variation. The pulsation mechanism is as yet unknown. Their position on the Hertzsprung-Russell diagram for *pulsating variables is just above the main sequence, beginning at type B3.

Beta Crucis (Mimosa; β Cru). A remote luminous blue-white subgiant that is the second-brightest star in the constellation Crux. It is a variable star (period 0.25 days) and may also be a double star. m_V: 1.29; M_V: -4.3; spectral type: B0.5 IV; distance: 130 pc.

beta decay. A type of *radioactive decay in which an atomic nucleus spontaneously transforms into a daughter nucleus and either an electron plus antineutrino or a positron plus *neutrino. The daughter nucleus has the same *mass number as the parent nucleus but differs in *atomic number by one. The electrons or positrons ejected by beta decay have a spread of energies, extra energy being taken up by the antineutrinos or neutrinos, respectively.

Beta Lyrae stars. A class of *eclipsing binaries that have periods of greater than one day and whose components are sufficiently close so that they are distorted by tidal forces into ellipsoidal shapes. The prototype, Beta Lyrae, is in a state of rapid *mass transfer as are other members of the class (*see also* equipotential surfaces). The components are much larger than the sun and usually of spectral types B or A. The *light curve shows continuous light variations, the secondary minimum being less than the primary. (The light curve is very similar to that of *RV Tauri- stars. The continuous variation is a result of the ellipsoidal shape and variable surface intensity. These binaries are intermediate between *Algol variables and *W Ursae Majoris stars.

beta particle. An energetic electron or positron ejected by *beta decay.

Beta Persei stars. *See* Algol variables.

Betelgeuse (α Ori). A remote luminous red supergiant that is the second-brightest star in the constellation *Orion. It is a semiregular variable with a period of about 5.8 years; the normal magnitude range is 0.3 to 0.9 but the magnitude has reached 0.15 and been as low as 1.3. It has a 14th-magnitude companion at a separation of 40″ and is a strong source of infrared radiation. An image of the surface of Betelgeuse has been produced by means of *speckle interferometry. M_V: -5.9; spectral type: M2 Iab; diameter: about 800 times solar diameter; distance: 180 pc.

Bethe cycle (Bethe-Weizsäcker cycle). *Other names for* carbon cycle.

biconcave, biconvex. *See* lens.

Biela's comet (1852 III). A comet, discovered in 1826, having a period of 6.62 years and a perihelion distance of 0.86 astronomical units. At the 1846 return it appeared to be distinctly elongated into a pearshaped form and actually divided into two separate comets some 10 days later, these comets travelling in practically the same orbit, one preceding the other by 280 000 km. The brightness of the two parts fluctuated drastically. At the next return in 1852 their separation had increased eightfold, the two periods differing by about 15 days. Neither comet has been seen since. Other comets have been observed to break up and to have portions detach from the main *nucleus; these include Brooks' comet (1889 V), Swift 1860 III, and 1882 III.

Bielids (Andromedids). A *meteor shower with its *radiant in Andromeda. It is named after *Biela's comet (1852 III), which passed within 30 000 km of the earth's orbit in 1832 and broke up in the 1850s. The associated *meteor stream is in the same orbit as the comet and produced a meteor storm on earth on Nov. 27, 1872 and 1885, the latter having an observed rate of over 75 000 meteors per hour. Due to the rapid regression of the *nodes of the orbit very little has been seen since.

big bang theory. A *cosmological model in which all matter and radiation in the universe originated in an explosion at a finite time in the past. This theory has been remarkably successful in explaining the expansion of the universe, the *microwave background radiation, and the cosmic abundance of *helium. An evolving universe was first discussed in the 1920s by Georges Lemaître and others. A *hot big bang,* in which the temperature of matter and radiation decreased with time, was suggested by George Gamow in the 1940s in an attempt to explain the observed abundances of the elements by cosmological *nucleosynthesis. A neglected prediction—that of primeval-fireball radiation from the hot early phases following the big bang—was verified by the discovery of the microwave background radiation in 1965.

The uncertainty principle of quantum mechanics prevents our speculating on times shorter than 10^{-43} seconds after the big bang. The next 10^{-4} seconds were characterized by the presence of all *elementary particles and their *antiparticles, which were spontaneously created at the

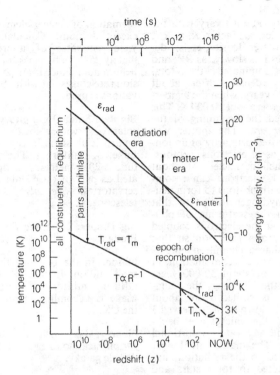

Big-bang theory of evolution of universe

enormous energy densities (energy per unit volume) then present. As the temperature dropped below 10^{12} kelvin (K) with the adiabatic expansion of the universe (i.e. expansion with no heat leaving or entering the system), the production rates of first protons and heavy particles and then electrons and light particles rapidly decreased. At about one second only that small fraction of matter destined to be the present material universe remained: the majority of particles had annihilated with their antiparticles to produce photons of electromagnetic radiation. These photons then dominated the energy density, ε of the early universe (see illustration). This was the *radiation era*.

Radiation loses energy in the expansion however—by the *cosmological redshift—and together with the decrease in photon number density this causes the radiation energy density, ε_{rad}, to vary with the *cosmic scale factor, R, as R^{-4}. The matter density, ε_{matter}, decreases less slowly, as R^{-3}, and thus dominated the total energy density after about 10 000 years, when the temperature was about 10 000 K. This marked the beginning of the *matter era*. The matter was composed primarily of an ionized gas of electrons, protons, and helium nuclei. At about 3000 K electrons can recombine with protons to form neutral hydrogen. Prior to that epoch the scattering of photons on the electrons coupled matter and radiation so that they shared a common temperature. Since recombination, at a time of about 30 000 years after the big bang, the radiation has cooled from about 3000 K down to the observed 3 K of the microwave background at a rate proportional to R. The matter, no longer coupled to the radiation, has interacted to form stars and galaxies.

Deuterium and helium were synthesized in the dense hot phase at about 100 seconds after the big bang when the temperature was about 10^9 kelvin: primordial *neutrons combined with protons to form deuterium before the neutrons, with a half-life of 12 minutes, were able to decay. The deuterium nuclei could then combine to give helium. Most of the helium in the universe today was formed at this time. The lack of any stable nuclei with *atomic numbers of five and eight prevented formation of heavier elements, except lithium. Calculations yield estimates of approximately 25-30% by mass for the helium abundance if an expansion rate consistent with observations is used.

Big Bear Solar Observatory. A solar observatory located on an artificial island in Big Bear Lake, S. California. The altitude is 2053 metres. It is operated as part of the *Hale Observatories. *See also* solar telescope.

Big Dipper. *US name for* the Plough.

billion. In the US, one thousand million. In the UK, one million million. American usage is becoming current in the UK.

binary galaxy. *Another name for* physical double galaxy. *See* double galaxy.

binary pulsar. The pulsar PSR 1913 + 16 (period 59.03 milliseconds) that is in orbit about an unseen companion. The short period of revolution (0.323 days) and other data relating to the orbit have allowed estimates to be made of masses for the two components. Current best values are around 1.4 solar masses for both the pulsar and its companion, indicating that the latter is likely to be either a *white dwarf or a *neutron star.

It has been found that the orbital period is decreasing by 10^{-4} seconds per year. The binary system is thus losing *angular momentum either by loss of mass or energy. In such a system, where gravitational effects are very considerable, energy could be carried away by *gravitational waves: observational data agrees very well with figures predicted for the gravitational-wave mechanism by general *relativity.

binary star (binary). A pair of stars that are revolving about a common *centre of mass under the influence of their mutual gravitational attraction. In most cases the stars may be considered to be moving in elliptical orbits described by *Kepler's laws. Binary and *multiple stars are very common in the Galaxy: in a recent survey of 123 nearby sunlike stars over half (57%) were found to have one or more companions. Studies of the orbital motions in binaries are important because they provide the only direct means of obtaining stellar masses. Optical determination of orbits is only possible if the components are sufficiently far apart to be distinguished (*see* visual binary). In an *astrometric binary* one component is too faint to be observed directly; the presence of this *unseen component* is inferred from perturbations in the motion of the visible component (*see* Sirius).

Orbital motion in *spectroscopic binaries is revealed by variations in radial velocity. Most spectroscopic binaries are *close binaries,* in which the components are too close together to be seen separately. Stars in a close binary are often distorted into nonspherical shapes and material may pass from one component to the other at various stages during their evolution (*see* equipotential surfaces). Consequently their orbits may deviate from simple two-body orbits.

The orbital planes of binaries appear to be randomly oriented and only a minority of systems are *eclipsing binaries, most of which are also spectroscopic binaries. *See also* x-ray binary.

bipolar group. *See* sunspots.

Bird of Paradise. Common name for *Apus.

birefringent filter. *See* Lyot filter.

birth of stars. *See* stellar birth.

black body. A body that absorbs all the radiation falling on it, i.e. a body that has no reflecting power. It is also a perfect emitter of radiation. The concept of a black body is a hypothetical ideal. The radiation from stars, and their *effective and *colour temperatures, can however be described by assuming that they are black bodies.

Black-body radiation is the *thermal radiation that would be emitted from a black body at a particular temperature. It has a continuous distribution of wavelengths. The graph of the energy, or intensity, of the

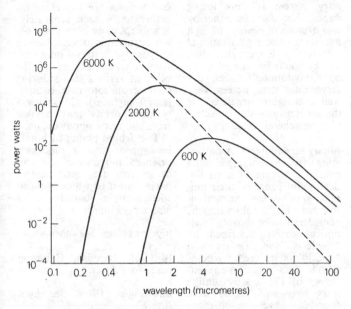

Black body radiation curves for different temperatures

radiation has a characteristic shape (see illustration) with a maximum value at a given wavelength, λ_{max}. At lower temperatures the black-body radiation is mainly in the infrared region of the spectrum. As the temperature increases the maximum of the curve moves to shorter wavelengths. Curves at different temperatures follow the relationship

$\lambda_{max} T = 2.898 \times 10^3$

where the wavelength is measured in micrometres. This is known as the *Wien displacement law*. The radiation from a hot black body is greater at every wavelength than the radiation from a cooler black body. The total radiation flux thus increases rapidly with increasing temperature, as described by *Stefan's law.

There have been a number of formulae deduced to explain the experimental curves of wavelength distribution. Wien's formula gives good results at short wavelengths while the Rayleigh-Jeans formula agrees with experiment at long wavelengths. In fact, the radiation emitted from a black body follows *Planck's radiation law* over the whole wavelength range. This law was derived by Max Planck from his *quantum theory, propounded in 1900. If $E(\lambda, T)$ is the energy emitted per unit wavelength interval at wavelength λ, per second, per unit area, into unit solid angle, by a body at thermodynamic temperature T, then

$E(\lambda, T) = (2hc^2/\lambda^5)[e^{hc/\lambda kT}-1]^{-1}$

h is the Planck constant, k the Boltzmann constant, and c the speed of light.

Planck's radiation law only describes the continuous spectrum emitted by a black body. The continuous radiation from a star, as opposed to a black body, does not usually follow Planck's law exactly, although over a broad region of the spectrum the law is a good approximation.

black drop. An optical phenomenon seen near the beginning and end of *transits of Venus across the sun's disc. Caused by irradiation and refraction of sunlight in the planet's atmosphere, it takes the form of an apparent link between the sun's limb and the silhouette of the planet against the solar surface. It reduced the accuracy of early attempts to determine the *solar parallax by observing transits from different parts of the earth.

black dwarf. *See* white dwarf.

black hole. An object so collapsed that its *escape velocity exceeds the velocity of light. Although a black hole has yet to be unambiguously detected, there are strong theoretical grounds for believing that they exist. A collapsing object becomes a black hole when its radius has shrunk to its *Schwarzschild radius and light can no longer escape. Although gravity will make the object shrink beyond this limit, the 'surface' having this critical value of radius—the *event horizon—marks the boundary inside which all information is trapped. Calculations, however, indicate that space and time become highly distorted inside the event horizon, and that the collapsing object's ultimate fate is to be compressed to an infinitely dense *singularity at the centre of the black hole.

The most promising candidate for black holes are massive stars that explode as *supernovae, leaving a core in excess of three solar masses. This core must undergo complete *gravitational collapse because it is above the stable limit for both *white dwarfs and *neutron stars. Once formed, a black hole can only be detected by its gravity. Finding black holes only a few kilometres across (the size of the event horizon for a single-star black hole) is exceedingly difficult, but chances are increased if the black hole is a member of a close binary system. If the components of a binary system are close enough, *mass transfer can occur between the primary star and its more compact companion. Matter will not fall directly on to the companion, however, for it has too much angular momentum; instead it forms a rapidly spinning disc—an *accretion disc*— around the compact object. If the latter is a black hole considerable energy can be produced, predominantly at x-ray wavelengths, as matter in the accretion disc loses momentum and spirals in.

One candidate for a black hole in a binary system has already been detected. The x-ray source *Cygnus X-1 comprises a 20 solar-mass B0 supergiant accompanied by an invisible companion with a mass about 10 times that of the sun. It is possible that this massive nonluminous object is a black hole emitting x-rays from its accretion disc.

Black holes of all masses have been proposed. At one extreme are the *supermassive black holes* of 10^6 to 10^9 solar masses that may lie in the centres of galaxies and give rise to the *quasar phenomenon and the phenomena of other *active galaxies. If a huge black hole is able to form and to capture sufficient gas and/or stars from its surroundings, the rest-mass energy of infalling material can be converted into radiation or energetic particles. At the other mass extreme are *mini black holes*, weighing only 10^{11} kg and with radii of 10^{-10} metres. These could have formed in the highly turbulent conditions existing after the big bang. They create such intense localized gravitational fields that quantum-mechanical effects are thought to make them 'evaporate', exploding in a final burst of gamma rays.

Blazhko effect. *See* RR Lyrae stars.

blink comparator. *See* comparator.

BL Lacertae objects (BL Lac objects). Extremely compact violently variable extragalactic objects that resemble *quasars but lack both emission and absorption lines in their spectra. A few dozen are known, the first to be identified being BL Lacertae, which was mistakenly classified as a variable star and much later (1968) shown to be the optical counterpart of a peculiar radio source. To date, all identified BL Lac objects have been found to be radio sources but, like quasars, the peak of their emission lies in the infrared.

Light and radio waves from these objects are strongly polarized, indicating intense magnetic fields, and there are rapid variations in both strength and direction. BL Lac objects are also violently variable in luminosity at all wavelengths, sometimes flaring up five magnitudes (a factor of 100) in only a few months. This indicates that the size of their energy-producing regions cannot be more than a few light-months across. Distance measurements to BL Lac objects have been hampered by their absence of spectral lines but some (BL Lac, AP Lib) are embedded in a faint 'fuzz', believed to be the surrounding galaxy. This contains weak absorption lines, from which it is possible to measure a *redshift. Such measurements place the BL Lac objects at the distance of the nearest quasars, and with a comparable energy output.

bloomed lens. *See* coating of lenses.

blueshift. An overall shift of the spectral lines in a spectrum towards shorter wavelengths. It is observed in the spectra of celestial objects approaching the earth. *See* redshift.

BN object. *See* Becklin's star.

Bode's law (Titius-Bode law). A relationship between the distances of the planets from the sun. Take the sequence 0, 3, 6, 12, 24, ..., where each number (except the 3) is twice the previous one, add 4 to each, and divide by 10. The resulting progression (0.4, 0.7, 1.0, 1.6, 2.8, 5.2, ...) is in good agreement with the actual distances in *astronomical units (AU) of most planets, provided that the minor planets are included and considered as one entity at a mean distance of 2.8 AU. The law fails to predict the correct distances for Neptune and Pluto, but may have some significance with respect to the formation of the solar system. Named after Johann Bode, who published it in 1772, it was formulated by Johann Titius of Wittenberg in 1766.

Bok globules. Small dark globules seen against a background of stars or of an *HII region; they are named after the American astronomer Bart J. Bok. They are believed to represent a phase in the contraction of *protostars when the material has become sufficiently dense to be opaque. Bok has discovered that most globules have diameters between 10 and 25 thousand astronomical units, and absorb between one and five magnitudes of light. During the globule stage it is thought that the protostars collapse more slowly than before but the internal temperature and pressure build up to become sufficient for nuclear reactions to commence. It is possible that the onset of nuclear burning is marked by the star becoming a *Herbig-Haro object.

bolide. A brilliant *meteor that appears to explode, i.e. a detonating *fireball. The brighter ones are caused by ablating *meteorites that subsequently fall to earth. About 5000 bolides occur in the earth's atmosphere each year.

bolometer. A radiation-sensing instrument that is used in astronomy to measure the total energy flux carried by all wavelengths entering the earth's atmosphere, and is especially useful in the infrared and microwave regions of the spectrum. It is es-

bolometric correction

sentially a small resistive element capable of absorbing electromagnetic radiation. The resulting temperature rise is a measure of the power absorbed. There is a large variety of bolometers, including semiconductor and supercooled devices.

bolometric correction (BC). *See* magnitude.

bolometric magnitude. *See* magnitude.

Boltzmann constant. Symbol: k. A constant given by the ratio of the universal gas constant to the Avogadro number and equal to $1.380\ 622 \times 10^{-23}$ joules per kelvin. It is the constant by which the mean kinetic energy of a gas particle can be related to its thermodynamic temperature.

Boltzmann equation. An equation, derived by the Austrian physicist Ludwig Boltzmann in the 1870s, that shows how the distribution of atoms or ions in their various *excitation levels depends on the temperature of the system; the system is in thermal equilibrium, with *excitation balanced by de-excitation. It gives the ratio of the number density (number per unit volume) of atoms or ions, N_2, in one energy level to the number density, N_1, in another lower energy level as

$$N_2/N_1 = (g_2/g_1)e^{-E/kT}$$

g_1 and g_2 are the degeneracies of the two levels, i.e. the multiplicity of energy levels with the same energy, E is the energy required to excite the atom or ion from the lower to the higher energy level, and T is the thermodynamic temperature. Thus as T increases a greater number of atoms or ions will become excited.

The Boltzmann equation together with the *Saha ionization equation are widely used to interpret the absorption and emission spectra of stars and to determine stellar temperatures and densities.

Bonner Durchmusterung (BD). A general star catalogue prepared over a period of 25 years by the Prussian astronomer F.W.A. Argelander and published in Bonn 1859-62. The original catalogue contained 324 189 stars—i.e. all those visible in the three inch Bonn refractor—lying between *declinations of +90° to -2°. The limiting magnitude was about 9.5. Accompanying star charts were published in 1863. It was extended by E. Schönfeld in 1886 to a declination of -23°and then included about 458 000 stars.

Cataloguing was extended to the south polar regions through an analogous work, *The Córdoba Durchmusterung (CD)*, compiled at the Córdoba Observatory, Argentina, and finally completed in 1930. This catalogue contains about 614 000 stars, brighter than 10th magnitude, from declinations -23° to -90°.

Boötes (Herdsman). A large constellation in the northern hemisphere near Ursa Major.

The brightest stars are *Arcturus (the brightest star in the northern sky) and the 2nd-magnitude yellow-blue visual binary Epsilon Boötis, one of several double stars. The area also contains one of the most distant *radio galaxies, 3C-295. Abbrev.: Boö; genitive form: Boötis; approx. position: RA 14.5h, dec +30°; area: 907 sq deg.

Boss General Catalogue. A whole-sky star catalogue listing the positions and *proper motions of 33 342 stars for the epoch 1950.0, down to a limiting magnitude of 7. It was prepared initially by the American astronomer Lewis Boss, with a preliminary publication in 1910, and was completed by his son Benjamin Boss in 1937.

bound-free absorption. A process of absorption that occurs when an electron bound to an atom is given sufficient energy by an absorbed photon of radiation to free itself from the atom. If however the photon energy can only excite the electron to a higher energy level so that it remains bound to the atom, *bound-bound absorption* takes place.

bow shock. *See* interplanetary medium.

Brackett series. *See* hydrogen spectrum.

Brans-Dicke theory. A relativistic theory of gravitation put forward in the 1960s by Carl Brans and Robert Dicke as a variant of Einstein's general theory of *relativity. It is considered by many astronomers to be the most serious alternative to general relativity. It is a scalar-tensor theory, i.e. a theory in which the gravitational force on an object is due partly to the interaction with a scalar field and partly to a tensor interaction.

bremsstrahlung. (literally: 'braking radiation'). Electromagnetic radiation arising from the rapid deceleration of electrons. It has a continuous spectrum. *See* thermal emission; synchrotron emission.

bridge. *See* radio-source structure.

brightest stars. The stars with the greatest apparent visual *magnitude. Many are very much more luminous than the sun and others appear bright because of their proximity. The most luminous lie in the supergiant and giant regions of the *Hertzsprung-Russell diagram. *See* Table 6, backmatter.

bright nebula. An *emission nebula or *reflection nebula as opposed to a *dark nebula. *See also* nebula.

brightness. The intensity of light or other radiation emitted from—absolute or intrinsic brightness—or received from —apparent brightness—a celestial body, the latter decreasing as the distance from the body increases. Intrinsic brightness is directly related to *luminosity. Apparent brightness is a subjective evaluation and is therefore considered in terms of apparent *magnitude: a star one magnitude less than another is about 2.5 times brighter. If two stars belong to the *main sequence then the brighter star is the hotter of the two.

brightness temperature. Symbol: T_b. The apparent temperature of radiation from a distant emitting region, i.e. the temperature that a *black body—a perfect radiator—would have to radiate the same observed *flux density per unit solid angle. It may be equal to the physical temperature of, say, an interstellar cloud if the radiation is by *thermal emission and the cloud is optically thick (*see* optical depth). If the cloud is optically thin, the brightness temperature is reduced. *See also* antenna temperature.

bright rays. *See* rays, lunar.

broken-back transit (bent transit; prism transit). *See* meridian circle.

Brooks 2. *See* Jupiter's comet family.

B stars. Stars of *spectral type B that are massive hot blue stars with a surface temperature (T_{eff}) of about 10 000 to 28 000 kelvin. Absorption lines of neutral helium (He I) dominate the spectrum, reaching maximum intensity for B2 stars. Balmer lines of hydrogen strengthen from B0 to B9; lines of ionized magnesium and silicon are also present. B0, B1, and B2 stars are found in OB *associations in the spiral arms of galaxies. Spica, Rigel, Bellatrix, and Alpha Crucis are B stars.

Bull. Common name for *Taurus.

burst. 1. A brief flux of intense radiation with a sudden onset and rapid decay, as is observed from Jupiter and from the sun at radio wavelengths. *See also* x-ray burst sources; gamma-ray bursts.
 2. (flare). *See* train.

burst source (burster). *See* x-ray burst sources.

Butler matrix. An electronic phasing device by which the signals from 2^n separate elements of an *array are combined together with their correct relative *phases to produce 2^n outputs, each of which corresponds to a main lobe (*see* antenna) pointing in a different direction.

butterfly diagram. *See* sunspot cycle.

B-V. *See* colour index.

Bw stars. Blue stars that, unlike normal *B stars, have only weak helium lines.

C

Caelum (Chisel). A small inconspicuous constellation in the southern hemisphere near Orion, the brightest stars being of 4th magnitude. Abbrev.: Cae; genitive form: Caeli; approx. position: RA 4.5h, dec -40°; area: 125 sq deg.

calcium star. *Obsolete name for F star.*

calendar. Any of various present-day, past, or proposed systems for the reckoning of time over extended periods: days are grouped into various periods that are suitable for regulating civil life, fixing religious observances, and meeting scientific needs. All calendars are based either on the motion of the sun (solar calendar), or of the moon (lunar calendar), or both (lunisolar calendar) so that the length of the year corresponds approximately to either the *tropical (solar) year or the *lunar year. Their complexity results primarily from the incommensurability of the natural periods of *day, *month, and *year: the month is not a simple fraction of the year and the day is not a simple fraction of the month or the year. Present-day calendars include the *Gregorian calendar, which is in use throughout most of the world, and the Moslem, Jewish, and Chinese calendars.

calendar year. The interval of time that is the basis of a *calendar. The *Gregorian calendar year contains an average of 365.2425 days.

calibration. A procedure carried out on a measuring instrument, such as a *radio telescope, by means of which the magnitude of its response is determined as a function of the magnitude of the input signal. The calibration of a radio telescope provides an absolute scale of the output deflection against *antenna temperature.

California nebula (IC 1499). An *emission nebula—an *HII region—that lies in the constellation Perseus and is ionized by the star xi Persei.

Callisto. The fourth and faintest *Galilean satellite of Jupiter, though it is in fact the second largest with a diameter of 4900 km—almost identical to that of Mercury. *See* Jupiter's satellites; Table 2, backmatter.

Caloris Basin. *See* Mercury.

Camelopardalis (Giraffe). A large inconspicuous constellation in the northern hemisphere near Ursa Major, the brightest stars being of 4th magnitude. It contains the prototype of the Z Camelopardalis *dwarf novae stars. Abbrev.: Cam; genitive form: Camelopardalis; approx. position: RA 6h, dec +70°; area: 757 sq deg.

camera tube. *See* television camera.

canals. Linear markings on *Mars that were first observed by Giovanni Schiaparelli in 1877 but later charted by many observers, most notably by Percival Lowell who advocated that they were irrigation ditches dug by Martians to distribute their planet's scarce water resources. *Mariner and *Viking spacecraft observations show that the canals do not exist.

Cancer (Crab). An inconspicuous zodiac constellation in the northern hemisphere near Ursa Major, the brightest stars being of 4th magnitude. There are many double and variable stars, including the multiple star Zeta Cancri. The area also contains the open clusters *Praesepe and the fainter M67 and the strong radio source NGC 2623. Abbrev.: Cnc; genitive form; Cancri; approx. position: RA 9h, dec +20°.

Canes Venatici (Hunting Dogs). An inconspicuous constellation in the northern hemisphere near Ursa Major, the brightest star being the 3rd-magnitude binary *Cor Caroli. It contains the bright globular cluster M3 and several galaxies including M94 and the *Whirlpool galaxy, M51. Abbrev.: CVn; genitive form: Canum Venaticorum; approx. position: RA 13h, dec +40°; area: 465 sq deg.

Canis Major (Great Dog). A conspicuous constellation in the southern hemisphere, lying partly in the Milky Way. The brightest stars are *Sirius (the brightest and one of the nearest stars in the sky), the 1st-magnitude giant *Adhara (ε), and the giant Mirzam (β) and supergiant Wezen (δ), both of 2nd magnitude and very remote and luminous. The area contains the open cluster M41. Abbrev.: CMa; genitive form: Canis Majoris; approx. position: RA 7h, dec -20°; area: 380 sq deg.

Canis Minor (Little Dog). A small constellation in the northern hemisphere near Orion, lying partly in the Milky Way. The brightest star is the visual binary *Procyon. Abbrev.: CMi; genitive form: Canis Minoris; approx. position: RA 7.5h, dec +5°; area: 183 sq deg.

Canopus (α Car). A conspicuous and luminous cream supergiant that is the brightest star in the constellation Carina and the second-brightest star in the sky. m_v: -0.72; M_v: -3.1; spectral type: F0 Ib; distance: 30 pc.

canyon. *See* valley, lunar; Mars, surface features.

Capella (α Aur). A conspicuous yellow giant that is the brightest star in the constellation Auriga. It is a spectroscopic triple star (period 104 days) and has a high lithium content. m_v: 0.09; M_v: -0.6; spectral type: F8-G0 III, G5 III, M5 V; distance: 13.7 pc.

Cape Photographic Durchmusterung. A general star catalogue of the southern sky compiled by J.C. Kapteyn from photographic plates produced by David Gill in Cape Town. It was published 1896-1900. It lists positions and *magnitudes of some 455 000 stars, brighter than 10th magnitude, from declinations -19° to -90°.

Caph. *See* Cassiopeia.

Capricornids. A minor *meteor shower, radiant: RA 315°, dec -15°, that maximizes on July 25. Another minor shower, the *Alpha Capricornids,* radiant: RA 309°, dec -10°, maximizes on Aug. 1.

Capricornus (Sea Goat). A zodiac constellation in the southern hemisphere near Sagittarius, the brightest stars being of 3rd magnitude. Alpha Capricorni is a naked-eye double star (magnitudes 3.7, 4.5): the brighter component is a binary and the fainter component an optical double. The area contains the globular cluster M30. Abbrev.: Cap; genitive form: Capricorni; approx. position: RA 21h, dec -20°; area: 414 sq deg.

captured rotation. *See* synchronous rotation.

capture theory. *See* encounter theories; moon.

carbonaceous chondrite. An uncommon class of *meteorites but very important because of their mineralogical and chemical composition, especially as regards the presence of hydrated minerals and organic (carbon) compounds. They are very easily crumbled and contain water-soluble compounds and must therefore be collected soon after fall. Although all meteorites were formed very early in the solar system's history, carbonaceous chondrites are possibly the most primitive form of matter in the solar system.

carbon cycle. A chain of *nuclear fusion reactions by which energy may be generated in stars. The overall effect of the cycle is the transformation of hydrogen nuclei into helium nuclei with emission of gamma-ray photons (γ), positrons (e^+), and neutrinos (ν). The major sequence of reactions is as follows:

$$^{12}C + {}^{1}H \rightarrow {}^{13}N + \gamma$$
$$^{13}N \rightarrow {}^{13}C + e^+ + \nu$$
$$^{1}H + {}^{13}C \rightarrow {}^{14}N + \gamma$$
$$^{1}H + {}^{14}N \rightarrow {}^{15}O + \gamma$$
$$^{15}O \rightarrow {}^{15}N + e^+ + \nu$$
$$^{1}H + {}^{15}N \rightarrow {}^{12}C + {}^{4}He$$

The carbon nucleus, ^{12}C, reappears at the end of the cycle and can be regarded as a catalyst for the reaction:

$$4{}^{1}H \rightarrow {}^{4}He + 2e^+ + 3\nu + 4\gamma$$

because nitrogen (N) and oxygen (O) intermediates are involved, the cycle is often termed the *carbon-nitrogen cycle* or *CNO cycle*.

The carbon cycle is very strongly temperature dependent and becomes the dominant energy-producing mechanism at core temperatures ex-

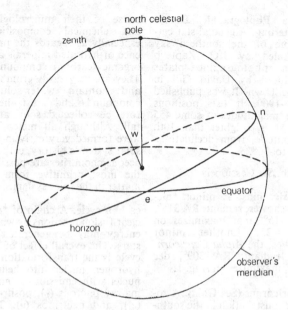

Cardinal points

ceeding about 1.6×10^7 kelvin. It is therefore thought to be the major source of energy in hot massive stars of spectral types O, B, and A. *See also* proton-proton chain reaction.

carbon-nitrogen cycle. *See* carbon cycle.

carbon stars (C stars). Rare red giant stars of low temperature that have an over-abundance of carbon relative to oxygen and also an unusually high abundance of lithium. WZ Cassiopeia is an example. The spectra show strong bands of carbon compounds, including C_2, CN, and CH. In the earlier Harvard classification (*see* spectral types) these stars were divided into *R stars* and *N stars:* N stars are similar to *M stars, being cooler and much redder than the K-type R stars. *See also* S stars.

cardinal points. The four principal points on the *horizon (see illustration). The *north point* (n) lies at the intersection of the horizon with the *celestial meridian nearest the north *celestial pole; the *south point* (s), diametrically opposite, is the equivalent intersection point closest to the south celestial pole. The *east point* (e) and *west point* (w) lie at the intersections of the horizon with

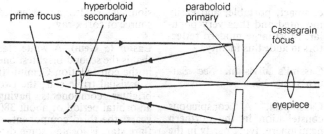

Ray path in Cassegrain telescope

the *celestial equator, the east point being 90° and the west point 270° clockwise from the north point.

Carina (Keel). A constellation in the southern hemisphere near Crux, lying partly in the Milky Way. It was once part of the constellation *Argo. The brightest stars are the supergiant *Canopus and three 2nd-magnitude stars. The irregular variable *Eta Carina is associated with extensive nebulosity. The area also contains the large globular cluster NGC 2808. Abbrev.: Car; genitive form: Carinae; approx. position: RA 9h, dec -60°; area: 494 sq deg.

Carme. A small satellite of Jupiter, discovered in 1938. *See* Jupiter's satellites; Table 2, backmatter.

Carte du Ciel. A projected whole-sky photographic atlas proposed in Paris in 1887 and planned to include stars of magnitude 14 or brighter. This ambitious scheme has involved 18 observatories in photographing areas of the sky only 2° square. It is as yet incomplete in its original form, technical advances now making it possible to photograph much greater areas with very much less work. The associated *Astrographic Catalogue,* listing the positions of stars down to 12th magnitude, was essentially complete by 1958.

cascade. *See* cosmic rays.

Cassegrain telescope. A compound *reflecting telescope designed by the Frenchman Guillaume Cassegrain in 1672. It is similar to the *Gregorian telescope but has a secondary mirror that has a convex hyperboloid shape (see illustration). The secondary is mounted inside the focal plane of the primary mirror and acts in a way analogous to a *Barlow lens, increasing the magnification. The magnification at the Cassegrain focus can be altered by changing the secondary mirror. The telescope is

Cassini's division. *See* Saturn's rings.

Cassiopeia. A conspicuous constellation in the northern hemisphere, lying partly in the Milky Way. The five brightest stars form a W-shape, which includes the variable *Schedar (α), the blue-white visual binary Cih (γ), which is an irregular variable, and the cream-coloured Caph (β), all of 2nd magnitude. Eta, close to Alpha, is also a binary star. The area contains some fine open clusters (M52, M103, and NGC 457) and globular clusters and the remnants of two recent supernovae—*Tycho's star and the radio source *Cassiopeia A. Abbrev.: Cas; genitive form: Cassiopeiae; approx. position: RA 1h, dec +608; area: 598 sq deg.

Cassiopeia A. An intense *radio source about three kiloparsecs distant in the constellation Cassiopeia. It has the largest *flux density at low frequencies of all the discrete radio sources (apart from the sun), the value being 11 000 *jansky at 178 megahertz. It is a *supernova remnant—probably of a supernova explosion (unrecorded) in the late 17th century—and has a ring-like structure about four arc minutes in diameter. The radio radiation is *synchrotron emission. It is also an extended source of soft x-rays.

Castor (α Gem). A white star that is the second-brightest one in the constellation Gemini. It is a visual triple star, the two brighter components having an orbital period of about 380 years; the third component, a flare star, is located some distance away. All three are spectroscopic binaries. m_v: 1.56; M_V: 0.8; spectral type: A1 V (A), A5 (B); distance: 14 pc.

cataclysmic variable (eruptive variable). Any star that displays a sudden large and unpredictable increase in brightness and then fades. *Novae, *recurrent novae, *dwarf novae, and *flare stars are cataclysmic variables. All are thought to be close binary systems. *Super-novae are somewhat different in that they are irreversibly altered by the cataclysmic explosion.

catadioptric telescope. A telescope that uses both refraction and reflection to form the image at the prime focus. By introducing a full-aperture *correcting plate in front of the primary mirror, the telescope designer can correct many *aberrations over an exceptionally wide field of view, combining the best features of both *refracting and *reflecting telescopes. *See* Schmidt and Maksutov telescopes.

catastrophic theories. *See* encounter theories.

catoptric system. An optical system in which the principal optical elements are mirrors. *Compare* catadioptric telescope; dioptric system.

Cayley formation. *See* highland light plains.

CCD camera. A highly sensitive solid-state electronic device that in astronomy can be used to generate, extremely rapidly, a pictorial representation of an area of the sky and, using a computer, to analyse the signal producing the picture and print out required information. It is a *television camera: the target area consists of many thousands of tiny charge-coupled devices (CCDs). When light or near-ultraviolet radiation falls on the target, the photons generate electron-hole pairs in the semiconductor substrate. The number of pairs depends on the radiation intensity. The accumulated charges are transferred out of the array in a controlled manner, forming a digital video signal. This output signal can be displayed on a television screen and can also be fed into a computer for analysis. A picture can be produced in a few minutes: almost every photon can be detected, so the *quantum efficiency of the device is very high.

cD galaxy. *See* clusters of galaxies.

Celescope experiment. *See* Orbiting Astronomical Observatory.

celestial axis. The line joining the north and south *celestial poles and passing through the centre of the *celestial sphere; the extension of the earth's axis to the celestial sphere.

celestial equator. The great circle in which the extension of the earth's equatorial plane cuts the *celestial sphere. The plane of the celestial (and hence terrestrial) equator is perpendicular to the celestial axis and is the reference plane for the equatorial coordinates *right ascension and *declination. The orientation of the celestial equator is slowly changing as a result of the *precession of the earth's axis.

celestial latitude (ecliptic latitude). *See* ecliptic coordinate system.

celestial longitude (ecliptic longitude). *See* ecliptic coordinate system.

celestial maser. *See* maser source.

celestial mechanics. The study of the motions and equilibria of celestial bodies, subjected to mutual gravitational forces, by the application of Newton's law of *gravitation and the general laws of mechanics, based on *Newton's laws of motion. Satellite and planetary motions, tides, precession of the earth's axis, and lunar libration are all described by these laws within the limits of accuracy of measurements. Newtonian mechanics is much simpler to use than the more accurate general theory of *relativity. Despite fundamental differences, general relativity reduces in a first approximation to Newtonian mechanics, observational and predicted Newtonian values being very close except in the case of the *advance of the perihelion of Mercury.

celestial meridian. The great circle on the *celestial sphere that passes through the north and south *celestial poles and the observer's *zenith and intersects the observer's horizon at the north and south points (*see* cardinal points). It is the projection of the observer's terrestrial meridian onto the celestial sphere. *See also* culmination.

celestial poles. The two points at which the extension of the earth's polar axis cuts the *celestial sphere. As a result of the *precession of the earth's axis the positions of the poles are not fixed but trace out a circle on the celestial sphere in a period of about 25 800 years. The *north celestial pole* revolves about a point in the constellation Draco and at present lies in Ursa Minor close to the star *Polaris. The *south celestial pole* lies in the constellation Octans. *See also* pole star.

celestial sphere. An imaginary sphere of indeterminate but immense radius that provides a convenient surface on which to draw and study the directions of celestial bodies and other points in the sky as seen from an observation point on earth at the sphere's centre. Celestial bodies, etc., may thus be thought of as points projected onto the surface of the celestial sphere. The earth's radius is negligible compared with that of the celestial sphere so that every observer may be considered as located at the sphere's centre. The earth's west-to-east rotation causes an apparent rotation of the celestial sphere about the same (extended) axis once every 24 hours but in an east-to-west direction.

The measurements that may be made with the celestial sphere, using the four main *coordinate systems of positional astronomy, concern not the distances but only the directions of celestial bodies. Angular relationships are established by the methods of spherical trigonometry (*see* spherical triangle). The principal reference circles on the celestial sphere are the *celestial equator, the *ecliptic, and an observer's *horizon, which are all great circles. The principal points are the poles of these circles and the points of intersection—the *equinoxes and *cardinal points—of the circles (see illustration).

cell (mirror cell). The enclosure that holds the *primary mirror in a reflecting telescope. It must hold the mirror so that the *collimation of the optical elements is maintained as the direction of observation

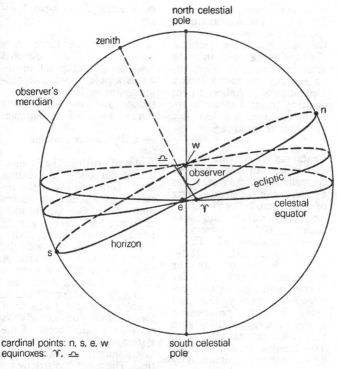

cardinal points: n, s, e, w
equinoxes: ♈, ♎

Celestial sphere

changes. It must also support the mirror so that it does not sag to an unacceptable degree under its own weight. Without carefully designed support over the whole rear face the mirror of the 508 cm Hale telescope would sag by 0.0625 mm, which is about 500 times the allowable quarter wave limit. The cell is usually provided with a closure or lid that protects the optical surface when the telescope is not in use.

Centaurus (Centaur). An extensive conspicuous constellation in the southern hemisphere almost surrounding Crux and lying partly in the Milky Way. The brightest stars are the triple star *Alpha Centauri, the remote 1st-magnitude *Beta Centauri, and several stars of 2nd and

3rd magnitude. *Proxima Centauri is the nearest star to the sun. The area also contains the fine naked-eye globular cluster Omega Centauri, the strong radio source *Centaurus A, and the *x-ray binary Centaurus X-3. Abbrev.: Cen; genitive form: Centauri; approx. position: RA 13h, dec -50°; area: 1060 sq deg.

Centaurus A. An intense *radio source in the southern constellation Centaurus. It has a *flux density at 86 megahertz of 8700 *jansky by *synchrotron emission from a huge area spanning 600 kiloparsecs of space. It is identified with the peculiar galaxy NGC 5128 lying at a distance of only four megaparsecs from the solar system, which makes it a very large *radio galaxy. The giant galaxy is also a pronounced source of x-rays and infrared radiation; gamma rays, too, have been detected. It is an elliptical star mass, cut across by broad belts of gas and dust.

Centaurus X-3. An *x-ray binary, discovered by the *Uhuru satellite in 1971. The binary nature was determined from the observation of regular x-ray eclipses with a period of 2.09 days; in 1974 the optical companion star was identified as a 13th magnitude supergiant of spectral type O6.5. Rapid pulsation of the x-ray source at a 4.8 second period is widely interpreted as being associated with the rotation of a magnetized *neutron star. It may thus be regarded as an x-ray *pulsar.

central force. A force on a moving body that is directed towards a fixed point or towards a point moving according to known laws. The gravitational attraction between the sun and a planet is an example.

central peak. *See* craters.

centre of gravity. The point in a material body at which a single force, which is the resultant of all external forces on the body, may be considered to act. The centre of gravity of a body in a uniform gravitational field coincides with the body's *centre of mass. A body does not possess a centre of gravity however if the external forces are not equivalent to a single resultant force or if the resultant does not always act on the same point of the system. This situation occurs in a nonuniform gravitational field. The moon has strictly no centre of gravity although the resultant force of the earth's gravitational attraction always passes within a few metres of the centre of mass.

centre of inertia. *See* centre of mass.

centre of mass (centre of inertia; barycentre). The point in a material system at which the total mass of the system may be regarded as being concentrated and that moves as if all external forces on the system could be reduced to a single force acting at that point. When a body has a centre of gravity, which it does in a uniform gravitational field, this point coincides with the centre of mass.

Two bodies, moving under the influence of their mutual gravitation, will orbit around their centre of mass, which lies on the line between them. The distances, r_1 and r_2, of the bodies from this point depend on their masses, m_1 and m_2, the more massive body lying closer. For circular orbits:
$$r_1/r_2 = m_2/m_1$$

centrifugal force. *See* centripetal force.

centripetal acceleration. *See* centripetal force.

centripetal force. A force, such as gravitation, that causes a body to deviate from motion in a straight line to motion along a curved path, the force being directed towards the centre of curvature of the body's motion. The force reacting against this constraint, i.e. the force equal in magnitude but opposite in direction, is the *centrifugal force*. The centrifugal force results from the inertia of all solid bodies, i.e. their resistance to acceleration, and unlike gravitational or electrical forces, cannot be considered a real force. The centripetal force is equal to the product of the mass of the body and its *centripetal acceleration*. The latter is the acceleration towards the centre and for a body moving in a circle at a constant angular velocity ω it is given by $ω^2r$, where r is the radius.

Cepheid. *Short for* Cepheid variable.

Cepheid instability strip. *See* pulsating variables.

Cepheid variables. A large and important group of very luminous yellow or orange supergiants that are *pulsating variables with periods ranging from about 1-50 days. Over 700 are known in our Galaxy and several thousand in the Local Group. There are two categories, which have each followed different evolutionary paths: *classical Cepheids* are massive young population I objects found in spiral arms on the galactic plane; *W Virginis stars* are much older and less massive population II objects found in the galactic centre and halo, especially in *globular clusters, and are thus similar in distribution to *RR Lyrae stars. The classical Cepheids are about 1.5-2 magnitudes more luminous than W Virginis stars of the same period. The luminosity variations of both categories are continuous and extremely regular so that the periods can be measured very accurately. Characteristic periods are 5-10 days (classical) and 12-30 days (W Virginis stars); the *amplitudes are typically 0.5-1 in magnitude.

The prototype of the classical Cepheids is *Delta Cephei*,

discovered 1784. The changes in brightness were found in the 1890s to be accompanied by and principally caused by changes in stellar temperature and also by changes in radius (see illustration). This was explained in terms of pulsations in the outer layers of the stars that appear at a late evolutionary state (*see* pulsating variables). The relation between period of pulsation and light variation of Cepheid variables was discovered during 1908-12 by Henrietta Leavitt. Miss Leavitt could only measure the apparent rather than the absolute magnitude of the Cepheids under study. An independent determination of the distance of a Cepheid of known period would lead to the graphical relation between period and luminosity, or absolute magnitude (*see* distance modulus). It was quickly realized by Shapley that this *period-luminosity relation would be an invaluable tool for measurements of distance out to the nearest galaxies and thus for studying the structure of our own Galaxy and of the universe. Much work has been done to establish the true values on the graph of period versus absolute magnitude. Baade and Kukarkin were consequently able to demonstrate the existence of the two distinct categories of population I and population II Cepheids, having separate but parallel period-luminosity relations. The prototype of the population II Cepheids is the much studied star W Virginis.

A *period-colour relation* was also discovered whereby the *colour index of a Cepheid increases, i.e. becomes more positive, as the period increases: a relatively long period implies a redder and cooler star. It was thus also possible to derive a *period-spectrum relation* between *spectral type and period. Both Cepheid categories follow these relations although the W Virginis stars (spectral types typically F6-G8) tend to be bluer than the classical Cepheids (typically F1-G6), even though they are less luminous.

Cepheus. A constellation in the northern hemisphere near Cassiopeia, lying partly in the Milky Way, the brightest star, Alderamin (α), being of 2nd magnitude. It contains several pulsating variable stars, including the prototype *Cepheid variable Delta Cephei, the red irregular variable Mu Cephei or *Garnet star, and the prototype of the *Beta Cephei stars. Abbrev.: Cep.; genitive form; Cephei; approx. position: RA 22h, dec +70°; area: 588 sq. deg.

Cerenkov counter. An instrument for detecting and measuring the velocity of very energetic particles: such as cosmic rays: light—*Cerenkov radiation*—is emitted by the particles as they pass through a transparent nonconducting (dielectric) material at a speed greater than the speed of light in that material. The radiation is emitted at a fixed angle, α,

Variation in brightness, temperature, radius, and radial velocity of expanding and contracting layers of Delta Cephei

to the direction of motion of the particle, where
$$\cos\alpha = c/nv$$
v is the particle velocity, n the refractive index of the transparent medium, and c the speed of light in a vacuum. This cone of light is focused onto a *photomultiplier so that an amplified electric pulse is registered. The velocity of the particle is determined from the angle of the cone.

Ceres. The first *minor planet to be discovered, found by Piazzi in 1801 following the predicted existence by *Bode's law of a planet at about 2.8 AU from the sun. It is the largest minor planet (diameter 1003 km) and may contain as much as half the total mass of the minor-planet system. It has a low *albedo of 0.05 and a spectrum similar to that of the *carbonaceous chondrite meteorites. Unlike the rather smaller minor planet *Vesta it can never be seen by the unaided eye. See Table 3, backmatter.

Cerro Las Campanas Observatory. See Las Campanas Observatory.

Cerro Tololo Interamerican Observatory. An observatory near La Serena, Chile, at an altitude of 2400 metres. It is run by a group of American universities—the Association of Universities for Research in Astronomy. Its principal instruments are a 4 metre (158 inch) reflecting telescope, which began operating in 1974, and a 1.5 m (60 inch) reflector.

Cetus (Whale). An extensive equatorial constellation near Orion, the brightest star being the 2nd-magnitude orange giant Diphda (β). It contains several *Mira stars (including the prototype *Mira Ceti), the *flare star UV Ceti, and the *Seyfert galaxy M77. Abbrev.: Cet; genitive form: Ceti; approx. position: RA 1.5h, dec -10°; area: 1231 sq deg.

chain craters. See crater chain.

Chameleon. A small inconspicuous constellation in the southern hemisphere near Crux, the brightest stars being of 4th magnitude. Abbrev.: Cha; genitive form: Chameleontis; approx. position: RA 11h, dec -80°; area: 132 sq deg.

Chandler wobble (wandering of the poles). A small continuous variation in the location of the geographic poles on the earth's surface. It leads to a *variation of latitude* of points on the earth since latitude is measured from the equator midway between the poles. The variation in polar location is resolved into two almost circular components: one (diameter: 6 metres; period: 12 months) results from seasonal changes in ice, snow, and atmospheric mass distribution; the second (diameter: 3-15 metres; period: about 14 months) is believed to arise from movements of material within the earth. *See also* universal time.

Chandrasekhar limit. The limiting mass for a *white dwarf star: 1.44 solar masses. A star whose mass exceeds this limit will be forced to undergo further *gravitational collapse to become a *neutron star or even a *black hole, because its material will be unable to support itself against the force of gravity. *Compare* Schönberg-Chandrasekhar limit.

channels. Meandering valleys, some over 1000 km long, that look like dried watercourses on the Martian surface. They are not related to the mythical Martian canals. *See* Mars, surface features.

charge. A property of certain *elementary particles that causes them to attract or repulse each other. Charged particles have associated electric and magnetic fields that allow them to interact with each other and with external electric and magnetic fields. Charge is conventionally 'negative' or 'positive': like charges repel, unlike charges attract. The *electron possesses the natural unit of negative charge, equal to 1.6022×10^{-19} coulombs. The *proton carries a positive charge of the same magnitude. If matter is charged, it is due to an excess or deficit of electrons with respect to protons.

Charioteer. Common name for *Auriga.

Charon. *See* Pluto.

chemosphere. An alternative name for the *atmospheric layers, at altitudes between about 25 and 200 km, in which reactions involving solar radiation and atmospheric atoms and molecules occur.

Chi Persei. *See* h and Chi Persei.

Chiron. *Minor planet number 2060 that was discovered in 1977 but was subsequently identified on photographs taken as early as 1895. It has a 50.68 year orbit that differs from orbits of all other known minor planets in lying almost entirely beyond that of Saturn. With an uncertain diameter of 300 km, it is possible that Chiron is one of the brighter members of a distant swarm of minor planets. *See* Table 3, backmatter.

Chisel. Common name for *Caelum.

chondrite. A type of *stony meteorite that contains *chondrules. They are the most abundant class of meteorite in the solar system (about 86%). *Compare* achondrite.

chondrules. Near-spherical bodies composed chiefly of silicates, with sizes between 0.2 mm and 4 mm, found embedded in *chondrites. They are usually aggregates of olivine, $(MgFe)_2SiO_4$, and pyroxene, $(Mg, Fe)SiO_3$. They may also be single crystals, wholly glass, or crystal and glass, in a wide range of proportions. They appear to have been free fluid drops made spherical by surface tension and then solidified and crystallized. There is a possibility that these silicate drops were produced by lightning discharges in the dusty primitive solar nebula or that they are crystallized droplets of impact melt produced when two minor planets collided. Chondrule-like bodies have been found in lunar soils.

christmas tree. *See* feeder.

chromatic aberration. An *aberration of a lens—but not a mirror—whereby light composed of different wavelengths, i.e. ordinary white light, is brought to a focus at different distances from the lens (see illustration). It arises from the variation with wavelength of the *refractive index of the lens material: red light is refracted (bent) less than blue light (*see* dispersion). False colours therefore arise in the image. Chromatic aberration can be reduced by using an *achromatic lens. Before the introduction of achromats, objective lenses of very long focal length were used in telescopes to reduce the aberration; this led to very cumbersome instruments.

chromosphere. The stratum of the solar atmosphere immediately above the *photosphere, in which the temperature rises over a few thousand kilometres from 4000 kelvin at the temperature minimum to around 50 000 kelvin where it reaches the transition region (*see* sun). Above the transition region lies the *corona. The rise in temperature (which continues in the transition region and inner corona) is thought to be produced by the following process: low-frequency sound waves are generated by turbulence in the *convective zone; these are accelerated as a result of the exponentially decreasing density, and magnetohydrodynamic shock waves form; as the shock waves propagate through to the corona, they dissipate, transferring energy

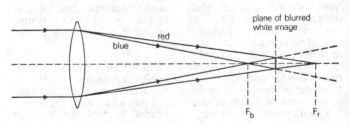

Chromatic aberration of convex lens

to the medium. The temperature of the medium thus increases with height.

The chromosphere is visible under natural circumstances only when the photosphere is totally eclipsed by the moon (*see* eclipse). It is then seen in profile at the sun's limb. It may, however, be imaged on a *spectroheliogram or filtergram at times other than totality. *See also* flash spectrum; chromospheric network; spicules.

chromospheric network. A large-scale cellular pattern in the solar *chromosphere. It is most clearly defined at the centre of the sun's disc in the light of the K *Fraunhofer line of singly ionized calcium, when it appears as a mosaic of cells outlined (at least partially) by bright mottles. Its appearance is similar though less pronounced in the core of the Hα line of neutral hydrogen, but in the wings (i.e. on either side of the core) the mottles appear dark instead of bright. Both bright and dark mottles have been tentatively identified with *spicules, visible beyond the sun's limb. They seem to differ only in their height above the *photosphere, bright mottles being situated in the lower chromosphere, dark mottles at heights of up to about 10 000 km.

The chromospheric network coincides with the underlying *supergranulation, the horizontal flow within each supergranular cell concentrating magnetic flux at its boundary; at the boundary an enhanced vertical magnetic field is produced, which is responsible for the localized excess heating that manifests itself as the bright mottles.

Chryse Planitia. A relatively smooth Martian plain, 1600 km across and 2.5 km below the average planetary surface. It appears to have suffered water erosion in the past and was chosen as the site of the *Viking 1 landing in 1976. *See* Mars, surface features.

Cih. *See* Cassiopeia.

Circinus (Compasses). A small constellation in the southern hemisphere near Centaurus, lying in the Milky Way. The brightest star is of 3rd magnitude. Abbrev.: Cir; genitive form: Circini; approx. position: RA 15h, dec -60°; area: 93 sq deg.

circle of longitude. Any great circle that passes through the poles of the *ecliptic and is thus at right angles to the plane of the ecliptic. *Circles of latitude* lie parallel to the ecliptic plane.

circle of position. An imaginary small circle on the earth's surface on which an observer is located and that is centred on the *substellar point, F, of a particular star. Its radius is equal to the *zenith distance, ζ, of the star (see illustration). The star will thus have the

Circle of position

same *zenith distance from every point on the circle of position.

circulator. A three-or four-port electronic device in which signals entering port number 1 appear only at port number 2, while signals entering port number 2 emerge only at port number 3, and so on. It is used, for example, in a *radio telescope employing a negative-resistance preamplifier, such as a *parametric amplifier or a *maser amplifier; the circulator isolates the preamplifier from the *antenna, which, if connected directly, might oscillate.

circumpolar stars. Stars that are permanently above the horizon from a given observational point on earth, i.e. they never set. For stars to be circumpolar from a given geographical latitude ϕ, their declination (angular distance above or below celestial equator) must be greater than $90°\text{-}\phi$. For an observer at the north pole all stars in the northern hemisphere of the celestial sphere will be circumpolar, while at the equator there are no circumpolar stars although almost all stars will be seen at some time during the year.

cislunar. Of or in the region of space between the earth and the moon.

civil time. Clock time, i.e. *mean solar time, commonly divided into two 12-hour periods beginning at midnight and noon of each civil day.

civil twilight. *See* astronomical twilight.

civil year. The length of the year as reckoned for ordinary purposes. Any civil year must contain an exact number of days, which in the *Gregorian calendar usually amounts to 365 but is equal to 366 in *leap years.

classical Cepheids. *See* Cepheid variables.

cleft. *See* rille.

Clock. Common name for *Horologium.

clock star. A bright star, usually situated near the celestial equator, whose position and proper motion are very accurately known so that it can be used for the exact determination of time, for the determination of the error of observatory clocks, and for correction of the positional observations of other stars.

close binary. *See* binary star.

closed universe. *See* universe.

cluster. A group of stars whose members are sufficiently close to each other to be physically associated. Clusters range from dense congregations of many thousands of stars to loose groups of only a few stars. Current theories of stellar evolution suggest that a majority of stars are formed in clusters rather than singly; this view is supported by the relatively well-defined *Hertzsprung-Russell diagrams of clusters, which indicate that cluster members are of essentially the same composition and age. (*See also* turnoff point). Since the stars in a cluster are all at approximately the same distance, the observed differences in appearance are believed to be due mainly to differences in mass. An estimate of distance is obtained from the technique of *main-sequence fitting. *See* globular cluster; open cluster.

clusters of galaxies. Groups of galaxies that may contain up to a few thousand members. The majority of galaxies appear to occur in clusters or in smaller groups such as doubles or triples. Our own Galaxy is a member of a small irregular cluster—the *Local Group. The nearest large cluster is the *Virgo cluster. The densest clusters, which typically contain a thousand or more members, are apparently roughly spherical and consist almost entirely of elliptical and S0 galaxies. Irregularly shaped clusters may be large, as with the Virgo cluster, or small but tend to be less dense and to contain all types of galaxies.

Large clusters with an unusually high concentration of galaxies in the centre are called *rich clusters*. Several thousand have been catalogued, examples being the *Coma cluster and *Perseus cluster. It has been shown that the mass required to keep the galaxies in rich clusters gravitationally bound is, on average, about 10 times greater than the mass observed in the main body of the galaxies (*see* missing mass). A second aspect, found by x-ray and radio measurements, is the large amount of hot gas that fills the space between the galaxies; the mass of the gas is not however sufficient to explain the missing mass. Another feature typical of most rich clusters is a centrally located supergiant elliptical *galaxy that has a greatly extended halo of faint stars and is usually a strong radio source: these are termed *cD galaxies*. It is thought by many astronomers that these observations could be explained in terms of a central gravitational maelstrom into which the more massive galaxies inevitably fall.

There is some evidence that clusters of galaxies are in turn grouped into *superclusters,* possibly as much as 50 or 100 megaparsecs in diameter. This evidence is not however accepted by all astronomers.

cluster variables. *See* RR Lyrae stars.

C-M diagram. *Abbrev. for* colour-magnitude diagram. *See* Hertzsprung-Russell diagram.

CNO cycle. *See* carbon cycle.

Coalsack (Southern Coalsack). A prominent *dark nebula about 170 parsecs away in the southern constellation Crux. It has an angular diameter of 4° and can be seen with the naked eye, projected against the Milky Way.

coaltitude. *See* zenith distance.

coating of lenses. The deposition of one or more thin uniform transparent coatings on lens surfaces, usually by vacuum evaporation or other electronic techniques; this reduces the amount of scattered light resulting from reflection at each refracting surface, say the interfaces between air and lens and between lens and air. A *bloomed lens* is coated with a single film of suitable refractive index and thickness: destructive *interference can then occur between light reflected from the coating-air surface and that reflected from the lens-coating surface. *Multilayer coating* depends on thin-film interference. Using up to 30 layers of controlled thickness, extremely low levels of reflected light can be achieved. Multilayer coating can also be used to produce very high levels of reflection from a *mirror.

coelostat. A flat mirror that can be driven by a clock mechanism so as to rotate from east to west about an axis parallel to the earth's rotational axis, thus compensating for the apparent west to east rotation of the celestial sphere. The mirror may thus continuously reflect light from the same area of the sky into the field of view of an instrument that is fixed in position, usually by means of an additional optical system. *See also* siderostat; heliostat; solar telescope.

coherence. The degree to which an oscillating quantity maintains a constant *phase and *amplitude relationship at points displaced in space or time. Hence the *coherence length* of a train of waves describes the distance over which the oscillations are appreciably correlated while the *coherence time* defines the time for which the character of the wavetrain remains more or less unchanged. *See also* autocorrelation function.

coherence bandwidth. *See* bandwidth.

colatitude. The complement of the celestial latitude (β) of a

celestial body, i.e. the angle (90°-β). *See* ecliptic coordinate system.

collapsar. A collapsed star. *See* black hole.

collimation. The alignment of (or act of aligning) the optical elements of a telescope. For a simple refractor the only adjustment required is to position the object lens at right angles to the optical axis by adjusting the screws on its supporting cell. In a simple Newtonian reflector the orientation of both the primary and diagonal mirrors must be adjusted, and the latter positioned correctly opposite the *draw tube. Final small adjustments are then made to obtain bright star images as free as possible from *coma.

collimator. A device used to produce a parallel or near parallel beam of light or other radiation in an instrument. One example, used in spectroscopes, is a converging lens or mirror at whose focal point is a narrow slit upon which light is focused from behind.

colour. 1. The colour of a star is measured and quoted in terms of its *colour index but basically depends on surface temperature: the star will radiate predominantly blue, white, yellow, orange, or red light in descending order of temperature (*see* spectral types). Only the brightest stars have recognizable colours. Two stars of the same colour may have very different *luminosities.
2. *Short for* colour index.

colour difference photography. A method for enhancing subtle colour differences on the moon by combining photographs exposed through infrared and ultraviolet filters.

colour excess. *See* colour index.

colour index. The difference between the apparent *magnitude of a star measured at one standard wavelength and the apparent magnitude at another (always longer) standard wavelength. Its value depends on the spectral distribution of the starlight, i.e. whether it is predominantly blue, red, etc., and it is therefore an indication of the *colour (i.e. temperature) of the star. It is independent of distance. Prior to the UBV system the *international colour index* was mainly used; this is the difference between photographic and photovisual *magnitudes (m_{pg}-m_{pv}). In the now widely used UBV system (*see* magnitude) colour index is usually expressed as the difference B-V, where B and V are the magnitudes measured with blue starlight (at a wavelength of 420 nanometres) and greenish-yellow starlight (540 nm), respectively. The colour index U-B is also used, where U is the apparent magnitude measured with ultraviolet radiation (360 nm) from a star.

Stars are classified into

Colour index for different luminosity classes of spectral types

*spectral types, which are further subdivided into luminosity classes; each has a characteristic *intrinsic colour index,* given as $(B-V)_0$ and $(U-B)_0$ (see graph). $(B-V)_0$ and $(U-B)_0$ defined as zero for A0 main-sequence stars and are therefore negative for hotter stars, i.e. those emitting more ultraviolet (O and B stars), and positive for cooler ones (A1 to M stars). Since colour index is easily measured, it is usually used on graphs in preference to spectral type or temperature. Any excess of the measured value of colour index of a star over the expected intrinsic value indicates that the starlight has become reddened by passage through interstellar dust (*see* extinction). The difference between the values is the *colour excess, E,* of the star:

$E = (B-V) - (B-V)_0$

The value of E gives the amount of reddening.

There are also colour indices relating to *magnitudes measured at red and infrared wavelengths.

colour-luminosity diagram. *See* Hertzsprung-Russell diagram.

colour-magnitude diagram (C-M diagram). *See* Hertzsprung-Russell diagram.

colour temperature. Symbol: T_c. The surface temperature of

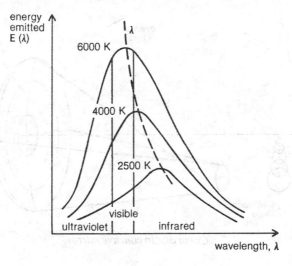

Planck radiation curves for various colour temperatures

a star expressed as the temperature of a *black body (i.e. a perfect radiator) whose energy distribution over a range of wavelengths corresponds to that of the star. It can thus be found by matching the energy distribution in the star's continuous spectrum to that of a black body (given by Planck's radiation law). With increasing temperature the star emits a higher proportion of blue and ultraviolet radiation and the position (wavelength) of maximum radiated energy on the energy distribution curve shifts accordingly; the basic shape of the curve remains unchanged (see illustration). Colour temperature is related to *colour index $(B-V)$ by an approximation of Planck's radiation law:

$T_c = 7300/[(B-V) + 0.6]$ kelvin

The colour temperature of a star always exceeds its *effective temperature, the difference being greatest for hot (O and B) stars. The sun's colour temperature is 6500 kelvin.

Columba (Dove). A constellation in the southern hemisphere near Canis Major, the two brightest stars being of 3rd magnitude. Abbrev.: Col; genitive form: Columbae; approx. position: RA 6h, dec -35°; area: 270 sq deg.

colures. The two great circles passing through the celestial poles and intersecting the ecliptic at either the *equinoxes—*equinoctial colure*—or the *solstices—*solstitial colure*.

Coma due to concave mirror

coma. 1. An *aberration of a lens or mirror that occurs in a telescope when the optical elements are misaligned so that light falls on the objective or primary mirror at an oblique angle. The light is not imaged as a point in the focal plane but as a fan-shaped area: each zone of the lens or mirror produces an off-axis image in the form of a circular patch of light, the diameter and position of the circle centre varying steadily from zone to zone (see illustration). The dimensions of the resulting combination of zone images, i.e. the fan-shaped image, depends on the obliquity of the light falling on the lens or mirror.

2. The principal part of most *comets, consisting of a diffuse luminous nebulous cloud of gas and dust that surrounds the *nucleus. The visual boundary of the coma merges with the sky background. The size varies with the comet's distance from the sun and with comet age but is typically 10^4 to 10^5 km in diameter. The gas and dust originate in the nucleus; the gas density is between 10^4 and 10^6 molecules per cm^3. The luminosity is mainly produced by *fluorescence from a variety of carbon, nitrogen, hydrogen, and oxygen radicals. The shape seems roughly circular but fan-shaped comas have been reported. It is transparent: even faint stars shine through. *See also* head.

Coma Berenices (Berenice's Hair). An inconspicuous constellation in the northern hemisphere near Boötes, the brightest stars being of 4th magnitude. The north galactic pole lies in this constellation, which contains a huge cluster of faint galaxies, the *Coma cluster. Abbrev.: Com; genitive form: Comae Berenices; approx. position: RA 12.5h, dec +20°; area: 386 sq deg.

Coma cluster. A remote spherical rich *cluster of galaxies, near the north galactic pole, in the constellation Coma Berenices that contains at least 1000 bright elliptical and S0 galaxies. The core is dominated by two immense galaxies that are both thought to be radio sources: an elliptical, NGC 4889, and an S0 galaxy, NGC 4874. The cluster is about 120 megaparsecs distant and is moving away from us at approximately 6700 km s^{-1} (*see* recession of the galaxies).

Comas Solá. A *comet that had its orbit severely disturbed when it approached within 0.19 astronomical units (AU) of Jupiter in 1912. The major changes were in period (from 9.43 years to 8.53 years), *perihelion distance (from 2.15 AU to 1.77 AU), and *inclination (from 18.1° to 13.7°).

combined magnitude. The apparent brightness of two or more stars so close that they are observed as a single star. It is not equal to the sum of the individual magnitudes but is a logarithmic function of the sum of the individual brightnesses (*see* magnitude).

comet. A minor member of the solar system that travels around the sun in an orbit that generally is much more eccentric than the orbits of planets. Typical comets have three parts: the *nucleus, *coma, and *comet tail. The gas and dust tails of a comet only appear when the comet is near the sun and always point away from the sun. The nucleus is the permanent solid portion of a comet and is thought by the majority of astrophysicists to be a kilometric-sized *dirty snowball;* this model was first suggested by Fred L. Whipple. The solar system contains only a few observable comets that have nuclei in the tens of kilometres size range. The numbers of actual (as opposed to observed) comets increase enormously as one goes to smaller and smaller sizes, but these get much less easy to see. No nuclei have been observed as they invariably subtend much too small an angle to the earth-bound observer.

Cometary orbits fall into two classes. *Short-period comets* have periods of less than 150 years and orbits lying completely or almost completely inside the planetary system. They seem to have been captured into the solar system by the gravitational attraction of the major planets. At any epoch about 50 of these are bright enough to be detected as they come to *perihelion. They have a mean *eccentricity of 0.56 and a mean *inclination of 11°. Nearly all of them (Halley's comet is an exception) are moving in direct orbits, i.e. in the same direction as the planets. The second class have near-parabolic orbits with periods in excess of 150 years. These orbits are orientated at random, indicating that these comets do not come from any specific direction in space. Many of them have been seen only once (for example Kohoutek has a period in excess of 70 000 years).

The majority of comets observed from earth have perihelions near the earth. This is simply due to observational selection. The brightness of a comet is proportional to $1/r^n\Delta^2$, where r is the distance between the comet and the sun and Δ is the comet-earth distance. The power n is on average 4.2 but can vary widely around this value. On average a comet passes perihelion about 200 times before decaying away. Cometary decay produces *meteor streams.

Comets are named after their discoverers. In any year, say 1978, comets are designated 1978a, 1978b, etc., in order of discovery. Permanent designations 1978 I, 1978 II, etc., are given later, these being in order of increasing perihelion distance.

There are two prevalent theories of cometary origin. In the first, comets were produced out of icy *planetisimals, at the same time as the origin of the solar system, and were stored in the *Oort cloud. In the second they are produced by gravitational accretion every time the sun passes through an interstellar dust cloud. This latter theory predicts that the solar system's cometary reservoir is periodically being topped up. The first theory predicts a cometary population that decreases with time.

cometary group. A group of comets, occurring in the solar system, which, apart from the time of perihelion passage, have very nearly the same orbital elements. 1668, 1843 I, 1880 I, 1882 II, 1887 I and 1948 VII is a typical group.

cometary nebula. A fan-shaped *reflection nebula whose illuminating star lies at the vertex of the fan. The *Hubble nebula is an example.

comet family. A distinct group of comets, the members of which have *aphelion distances that coincide approximately with points on the orbit of a particular major planet: this grouping is seen on a plot of comet numbers as a function of aphelion distance, in which the distribution peaks at the orbits of the planets Jupiter, Saturn, Uranus, and Neptune. The planet has captured the comet from a long-period orbit. Two frequency concentrations (at 53 AU and more strongly at 83 AU) have been found, indicating possible aphelia of transplutonian planets. *See* Jupiter's comet family.

comet tails. Tails of gas and dust that point away from the cometary *nucleus and only appear when a *comet is near the sun, i.e. closer than about two astronomical units (see illustration). Not all comets have tails. The luminosity is

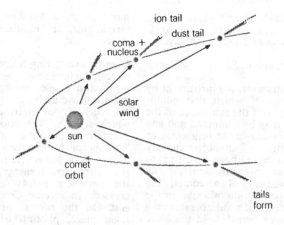

Comet tails

due to both molecular and atomic emission and to reflection of sunlight. The gas and dust is forced away by *radiation pressure and by *solar-wind interactions. The *ion tail* is usually long and narrow, showing many diverse visible and rapidly changing structures. It consists exclusively of ionized molecules, moving at velocities between 10 and 100 km s^{-1}, the repulsive force being 20 to 100 times greater than gravity. The tail can be a million km wide and lengths of 10 million km are not uncommon. *Dust tails* are more strongly curved than ion tails. They contain solid particles, which simply reflect sunlight. Particles are ejected from the *nucleus along curved paths known as *syndynames*. The tail is an envelope over the reflection regions of particles of similar mass. The closer the dust tail is to the ion tail in curvature the smaller are the responsible particles.

command and service module (CSM). *See* Apollo.

command module (CM). *See* Apollo.

common proper motion stars. *See* visual binary.

compact galaxy. A member of a class of *peculiar galaxies, first catalogued by Fritz Zwicky in the 1960s, that are so intensely bright that they are almost indistinguishable from stars on normal photographs. They show a variety of properties and include several distinct types of galaxy. Some appear to be very young objects with a high proportion of their mass still in the form of gas. Others are characterized by a bright starlike nucleus and include *Seyfert galaxies, *N galaxies, *BL Lacertae objects, and *quasars. *See also* active galaxy.

companion. *See* visual binary.

comparator. An instrument by means of which two photographs of the same area of the sky may be compared and any difference in the brightness or position of an object quickly detected without resorting to exact measurements. The movement of a comet, the proper motion of a star, or the presence of a variable star may thus be revealed. In the *blink comparator* the two photographs are viewed in rapid succession through an eyepiece: changes in position produce an apparent movement of an image while variable stars are detected as a pulsation in brightness. In the *stereocomparator* the two photographs are viewed simultaneously by binocular vision: objects that have changed in position or brightness appear to stand out of the plane of the picture.

comparison spectrum. The spectrum of one or more known substances under terrestrial conditions that is used in astronomy as a standard of comparison for investigating spectra of celestial objects; it provides, for example, a zero Doppler shift.

Compasses. Common name for *Circinus.

components. Stars, or possibly planets, that are constituent members of a *binary star, *triple star, or *multiple-star system.

Compton scattering (Compton effect). An interaction between a *photon of electromagnetic radiation and a charged particle, such as an electron, in which some of the photon's energy is given to the particle. The photon is therefore reradiated at a lower frequency (i.e. with a lower energy) and the particle's energy is increased. In *inverse Compton emission* the reverse process takes place: photons of low frequency are scattered by moving charged particles and reradiated at a higher frequency.

concave. Curving inwards. A concave mirror converges light. A biconcave or planoconcave lens, thinner in the middle than at the edges, has a diverging action.

concentration. The diameter of the telescopic image of a star, etc. The image is increased from point size by unavoidable optical (diffraction) effects in the telescopic itself (*see* resolving power) and is further increased by distortion due to atmospheric conditions at the time of observation, i.e. by the *seeing.

configuration. *See* aspect.

confocal lenses. *See* afocal system.

confusion. The running to-

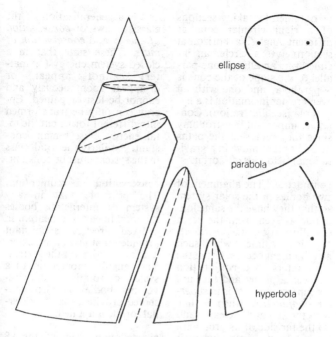

Conic sections

gether of the traces from different *radio sources in the output of a *radio telescope. Confusion becomes important at the value of *flux density when there is, on average, more than about one source in the beam at once. The radio telescope becomes *confusion limited* when this flux density is appreciably higher than its *sensitivity. Further increase in sensitivity will not then result in fainter sources being detected.

conic sections (conics). A family of curves that are the locus of a point that moves so that its distance from a fixed point (the focus) is a constant fraction of its distance from a fixed line (the directrix). The fraction, e, is the eccentricity of the conic. The value of the eccentricity determines the form of the conic. If e is less than one the conic is an ellipse. A circle is a special case of this with $e = 0$. If $e = 1$ the conic is a parabola and if e exceeds one the conic is a hyperbola.

These curves are known as conic sections because they can

be obtained by taking sections of a right circular cone at different angles: a horizontal section gives a circle, an inclined one an ellipse, one parallel to the slope of the cone is a parabola, and one with an even greater inclination is a hyperbola (see illustration). Conics are important in astronomy since they represent the paths of bodies that move in a gravitational field. *See also* orbit.

conjunction. The alignment of two bodies in the solar system so that they have the same longitude as seen from the earth (see illustration at *elongation*). It is also the time at which such an alignment occurs. Often the term refers to a conjunction between a planet and the sun. An *inferior planet is in *superior conjunction* when in line with the sun and the earth and on the far side of its orbit with respect to the latter; when it lies between the sun and the earth it is said to be in *inferior conjunction*. Conjunction may also occur between two planets, or between the moon and a planet or the moon and the sun—the latter occurring at every new moon. *See also* synodic period.

conservation of mass-energy. The principle that in any system of interacting bodies the total mass plus total energy is constant, where energy and mass are to be considered interconvertible according to the equation $E = mc^2$. The principle is a consequence of the theory of special *relativity and is a generalization of the classical laws of *conservation of energy* and *conservation of mass*. These state that in a closed system energy (or matter) cannot appear or disappear spontaneously and cannot be lost or gained. Energy can only be transformed into different forms but the total energy must remain constant. Similarly the total mass in the system must be constant.

conservation of momentum. The principle that in any system of interacting bodies the total linear *momentum in a fixed direction is constant provided that there is no external force acting on the system. The *angular momentum of a system of rotating and/or revolving bodies is also conserved provided that no external torque is applied.

constellation. Any of the 88 areas into which the whole of the northern and southern hemisphere of the sky (or *celestial sphere) is now divided. Every star, galaxy, or other celestial body lies within, or sometimes overlaps, the boundaries of one of the constellations. These boundaries were established unambiguously by the International Astronomical Union in 1930 along arcs of *right ascension and *declination for the *epoch 1875, Jan. 1.

Originally the constellations had no fixed limits but were groups of stars considered by early civilizations to lie within the imagined outlines of mythological heros, creatures, and other forms. In the *Almagest, Ptolemy, in about AD 140, listed 48 constellations that were visible from the Mediterranean region. During the 16th, 17th, and 18th centuries many new groupings were formed (and several later discarded), especially in the previously uncharted regions of the southern hemisphere. Since then further additions have proved unacceptable.

The constellations seen on any clear night depend on the latitude of the observer, and change with the time of year and time of night. If, for a particular latitude, ϕ, the stars are *circumpolar (never set), they will be visible on every night of the year. Other stars never rise above the horizon: these have *declinations $\phi - 90°$ or $\phi + 90°$ for the northern and southern hemispheres, respectively. The remaining stars can be seen only when they are above the horizon during the night, a star rising earlier, on average, by about two hours per month.

Each constellation bears a Latin name, as with Ursa Major, the genitive form of which is used with the appropriate letters or number for the star names (*see* stellar nomenclature), as with Alpha Ursae Majoris. For simplification the three-letter abbreviations for the constellations are more usually used, as with UMa. *See* Table 5, backmatter.

contact. *See* eclipse.

contact binary. *See* equipotential surfaces.

continental drift. *See* earth.

continuous spectrum. A *spectrum consisting of a continuous region of emitted or absorbed radiation in which no discrete lines are resolvable. The emission from hot fairly dense matter will produce a continuous spectrum, as with the continuous emission of ultraviolet, visible, and infrared radiation from the sun's *photosphere. *Synchrotron emission is another example of continuous emission.

continuum. The *continuous spectrum that would be measured for a body if no absorption or emission lines were present.

contour map. *See* radio-source structure.

convection. A process of heat transfer in which energy is transported from one region of a fluid to another by the flow of hot matter in bulk into cooler region. *See also* energy transport; convective zone.

convective zone. The layer directly beneath the surface of a *main-sequence star, in which the temperature is sufficiently low to enable the nuclei of hydrogen and heavier elements to recombine with free electrons to form atoms and negative ions. The presence of these atoms and ions and their ability to absorb photons increases the *opacity of the medium to the passage of radiation from below and results in a steeper temperature gradient, thereby triggering turbulent convection. Thus convection supersedes radiation as the primary mode of *energy transport. Smaller quantities of energy are transported by low-frequency sound waves, generated by the turbulence; these waves propagate through the *photosphere to the outer layers of the star, where they play a major role in determining the local conditions (*see* chromosphere). *See also* granulation; stellar structure.

convex. Bulging outwards. A convex mirror diverges light. A biconvex or planoconvex lens, fatter in the centre than at the edges, has a converging action.

convolution. A mathematical operation that is performed on two functions and expresses how the shape of one is 'smeared' by the other. Mathematically, the convolution of the functions f(x) and g(x) is given by

$$\int_{-\infty}^{\infty} f(u)g(x-u)du$$

It finds wide application in physics; it describes, for example, how the transfer function of an instrument affects the response to an input signal. *See also* radio-source structure.

coordinated universal time (UTC). *See* universal time.

coordinate system. A system by which the direction of a celestial body or a point in the sky can be defined and determined by two spherical coordinates, referred to a fundamental great circle lying on the *celestial sphere and a point on the fundamental circle (see illustration). One coordinate (*a*) is the angular distance of the celestial body measured perpendicular to the fundamental circle along an auxiliary great circle passing through the body and the poles of the fundamental circle. The other coordinate (*b*) is the angular distance measured along the fundamental circle from a selected zero point to the intersection of the auxiliary circle.

There are four main coordinate systems: the *equatorial, *horizontal, *ecliptic, and *galactic coordinate systems (see table). They are all centred on the earth. Transformations can be made from one system to another by means of the relationships between the angles and sides of the relevant *spherical triangles. The *astronomical triangle, for example, relates equatorial and horizontal coordinates; the triangle formed by the celestial body and the poles of the equator and ecliptic relates equatorial and ecliptic coordinates.

Copernican system. A *heliocentric system of the solar system that was proposed by Nicolaus Copernicus and eventually published in 1543 in his book *De Revolutionibus*. It uses some of the basic ideas of

Coordinate system

the *Ptolemaic system, including circular orbits and epicycles, and was no more accurate in its predictions. Copernicus, however, maintained that the planets move around the sun (in the relative positions accepted today), the sun's position, S, being offset from the centre of the orbits. The apparent motions of celestial bodies such as the sun were explained in terms of the rotation of the earth about its axis and also the earth's orbital motion.

The planetary motion can be represented by two uniform

circular motions: one is an epicyclic motion of the planet P about a point D (see illustration) and the other, unlike that of the Ptolemaic system, is a uniform circular motion of D about the centre, C, of the orbit. This requires that the rate of motion of D about C is exactly half that of the epicyclic rate of motion with respect to a fixed direction.

There was a strong and prolonged reaction—especially by the Church—to the Copernican system, which effectively displaced the earth as the centre of the universe. There was also a sudden revival in astronomical observation in order to test the theory, notably by Tycho Brahe. Tycho's detailed observations, which showed the inadequacies of the Copernican system, were used in the formulation of *Kepler's laws of planetary motion.

Copernican System. *See* Copernicus.

Copernicus. A young lunar crater to the south of Mare Imbrium that is 90 km in diameter and has an extensive system of *rays and *secondary craters. Possible ray material was collected by the crew of Apollo 12 and dated at 850 million years; this provides a calibration point for the *Copernican System*, which is the most recent era in lunar history.

Copernicus satellite. The *Orbiting Astronomical Satellite OAO-3, launched in Aug.

Copernican system

1972. It was used primarily for measurements of ultraviolet radiation but also carried x-ray equipment. *See* ultraviolet astronomy; x-ray satellites.

Coprates canyon. *See* Valles Marineris.

Cor Caroli (α CVn). A bluish-white star that is the brightest one in the constellation Canes Venatici and was named in honour of Charles II. It is a visual binary (separation 20″), the brighter component (A or α^2 CVn) being the prototype of the *spectrum variable stars. It has a period of 5.469 days and its spectrum shows strong profuse lines of rare earths, iron-peak elements, and silicon. m_v: 2.89 (A), 5.60 (B); spectral type: B9.5 pv (A), F0 V (B).

Córdoba Durchmusterung. *See* Bonner Durchmusterung.

core. 1. The central region of a star, such as the sun, in which energy is generated by thermonuclear reactions.

2. The central region of a differentiated planet or satellite, such as the *earth or *moon.

corona. The outermost part of the solar atmosphere. It has two main components: the *K-corona (or inner corona) consists of rapidly moving free electrons, exhibits a *continuous spectrum, and attains an extremely high temperature of around 2 000 000 kelvin at a height of about 75 000 km; the *F-corona (or outer corona) consists of relatively slow-moving particles of interplanetary dust, exhibits an *absorption spectrum, and extends for many million kilometres into the *interplanetary medium. A third component, the *E-corona, consists of relatively slow-moving ions and exhibits an *emission spectrum superimposed on the continuum of the K-corona.

The *white-light corona* comprises the overlapping K-corona and F-corona. It is visible under natural circumstances only in profile beyond the sun's limb on the rare occasions when the *photosphere is totally eclipsed by the moon. It may however be observed out to distances of several solar radii at times other than totality, with the aid of a balloon- or satellite-borne externally occulted *coronagraph. Similarly the E-corona may be observed at specific wavelengths with the aid of the Lyot coronagraph, used at certain high-altitude observatories. The corona may also be observed against the sun's disc at extreme-ultraviolet and x-ray wavelengths, using rocket- or satellite-borne instrumentation.

The overall shape of the corona changes with the phase of the *sunspot cycle. At sunspot minimum it is roughly symmetrical, with long equatorial *streamers,* and has *plumes*—orientated in the direction of the sun's general magnetic field—around the poles. At sunspot maximum it is less symmetrical, although more evenly distributed about the sun's disc as a whole. Its changing shape is due principally to the presence of individual streamers above *active regions, the mean *heliographic latitude of which progresses towards the equator as the sunspot cycle proceeds. Solar radio emission at metre wavelengths originates in the corona and may exhibit an intense burst or series of bursts at the time of a large *flare.

Corona Australids. A minor *meteor shower that is observable in the southern hemisphere, radiant: RA 245°, dec -48°, and that maximizes on March 16.

Corona Australis (Southern Crown). A small inconspicuous constellation in the southern hemisphere near Scorpius, lying partly in the Milky Way. The brightest stars are of 4th magnitude. It contains the just-visible globular cluster NGC 6541. Abbrev.: CrA; genitive form: Coronae Australis; approx. position: RA 19h, dec -40°; area: 128 sq deg.

Corona Borealis (Northern Crown). A small constellation in the northern hemisphere near Ursa Major in which there is a semicircle of stars, the brightest being the 2nd-magnitude *Alphecca. It also contains the prototype of the variable *R Coronae Borealis stars and the *recurrent nova T Coronae Borealis. Abbrev.: CrB; genitive form: Coronae Borealis; approx. position: RA 15.5h, dec +30°; area: 179 sq deg.

coronagraph. An optical instrument designed in 1930 by the French astronomer Bernard Lyot for observing and photographing the sun's *corona, prominences, etc., at times other than at solar eclipses. The image of the bright solar surface is artificially eclipsed by means of a blackened occulting disc set at the centre of the focal plane of the objective lens of a telescope. A lens is used since scattering of light from its surface is much less than with a mirror. Strong sunlight scattered from the elements of the instrument must be eliminated so that only the faint coronal light is focused on the photographic plate or some other detector. To reduce atmospheric scattering, high-altitude sites are used. Narrow-band optical filters are usually placed in front of the detector so that the emission lines of the *E-corona can be studied.

coronal holes. Regions of exceptionally low density and temperature compared with the surrounding solar *corona. They overlie areas of the sun's disc that are characterized by relatively weak divergent (and therefore primarily unipolar) magnetic fields. They are thought to be the primary source of the long-lived high-speed streams in the *solar wind.

coronal lines. Bright emission lines observed in the spectrum of the sun's *E-corona.

coronal rain. *See* prominences.

correcting plate (correcting lens; corrector). The lens placed at the front of a *catadioptric telescope to correct the *aberration of its spherical primary mirror. In this way a very wide field can be obtained, sensibly free from aberrations, even at *focal ratios well below one. *See also* Schmidt and Maksutov telescopes.

correlation receiver. A radio-astronomy *receiver in which either the *noise power from a single *antenna is split and multiplied with itself, or the noise power from the two arms of an interferometer (*see* radio telescope) are multiplied together. Signals that are uncorrelated (completely unrelated) give an output whose average value is zero when multiplied together. Correlated signals, however, produce components near zero frequency that give a steady deflection in the output of the multiplier. Correlation receivers are used to reduce the effect of manmade *interference and of changes in the *gain and other parameters of the system. *See also* phase-switching interferometer.

Corvus (Crow). A small fairly conspicuous constellation in the southern hemisphere near the star Spica, the four brightest stars being of 2nd and 3rd magnitude. It contains the associated radio sources NGC 4038 and 4039. Abbrev.: CrV; genitive form: Corvi; approx. position: RA 12.5h, dec -20°; area: 184 sq deg.

COS-B. A satellite of the European Space Agency that was launched in Aug. 1975 to investigate celestial objects emitting gamma rays and to measure time variations of such emission. *See* gamma-ray astronomy.

cosine rule. *See* spherical triangle.

cosmic abundance. The relative proportion of each *element found in the universe. The standard cosmic abundance is based on that of the solar system, determined from observations of the relative line strengths in the spectrum of solar radiation, from geological surveys of the earth, and from analysis of *meteorites (see table). The solar abundance in terms of numbers of atoms gives 94% hydrogen, 5.9% helium, about 0.06% oxygen, 0.04% carbon, and trace numbers of other atoms. Abundances deduced for most stars from their spectra usually agree quite well with the standard, although old stars tend to have rather less of the heavier elements (*see* populations I and II) and there are a number of odd stars with abundances quite unlike the rest. The abundances deduced from the absorption and emission lines of the *interstellar medium do not agree so well, principally in that there is an apparent shortage of refractory elements such as iron. It is possible that these elements are present, however, but are bound up with the cosmic dust.

Cosmic abundance of elements, using solar system as standard

element	atomic number	approx. composition by mass
hydrogen (H)	1	73%
helium (He)	2	25%
oxygen (O)	8	0.8%
carbon (C)	6	0.3%
neon (Ne)	10	0.1%
nitrogen (N)	7	0.1%
silicon (Si)	14	0.07%
magnesium (Mg)	12	0.05%
sulphur (S)	16	0.04%

cosmic background radiation. Unresolved radiation from space. The cumulative effect of many unresolved—and individually weak—discrete

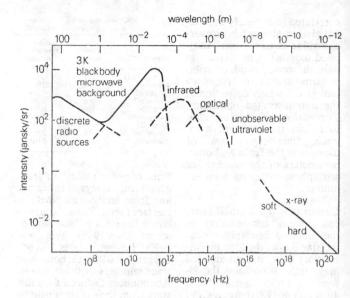

Cosmic background radiation

sources provides the background at radio, x-ray, and possibly other wavelengths (see illustration p. 72).

One important radiation is the *microwave background radiation, which peaks at about 2.5 mm wavelength (i.e. a frequency of 1.2×10^{10} hertz). This is considered to be due to the hot *big bang. *See also* x-ray astronomy; gamma-ray astronomy; radio-source structure.

cosmic censorship, principle of. *See* singularity.

cosmic dust. Extraterrestrial particles that are found in all regions of space and range in size from less than a micrometre to large meteoritic pieces. The particles occur in craters on lunar and planetary surfaces; they are revealed in the earth's environment by the scattering of sunlight to produce *zodiacal light and in the *interstellar medium by the reddening of starlight. The particles can also clump together to form such objects as massive clouds, seen as *dark nebulae.

cosmic noise (Jansky noise). *See* antenna temperature.

cosmic rays. Highly energetic

particles that move through space at close to the speed of light and that continuously bombard the earth's atmosphere from all directions. They were discovered by V.F. Hess during a balloon flight in 1912. The energies of the individual particles are immense, ranging from about 10^8 to over 10^{20} electronvolts (eV). The intensity (or particle flux) of cosmic rays is very low; it is greatest at low energies—several 1000 particles per square metre per second—dropping with increasing energy but apparently levelling off at high energies.

Cosmic rays are primarily nuclei of the most abundant elements (atomic weights up to 56) although all known nuclei are represented. *Protons (hydrogen nuclei) form the highest proportion; heavy nuclei are very rare. Also present are a small number of *electrons, positrons, *neutrinos, and *gamma-ray photons. All these particles are known collectively as *primary cosmic rays*. On entering the earth's atmosphere the great majority collide violently with atomic nuclei in the atmosphere producing *secondary cosmic rays;* these consist mainly of *elementary particles.

The large number of particles produced from one such collision form an *extensive air shower* (or simply an *air shower*), in which there is a very rapid, highly complex, but well-defined sequence of reactions. The initial products are principally charged and neutral *pions. The neutral pions decay (disintegrate) almost immediately into gamma rays, which give rise to *cascades* of electrons and positrons by *pair production and to more photons by *bremsstrahlung radiation from the decelerating electrons and positrons. The charged pions decay into *muons, which subsequently decay into neutrinos and electrons. The original primary cosmic ray nucleus, together with many of the secondary cosmic rays, undergo more collisions to create further generations of particles. Following the initial collision the number of particles in the shower increases to a maximum at some point high in the atmosphere, where the creation of new particles is balanced by absorption; the number then decreases. Very few primary cosmic rays reach the earth's surface, the final products including electrons, muons, neutrinos, gamma rays, and some initial products of the primary collision. The maximum number of particles in an air shower depends on the energy of the primary cosmic ray: half a million particles can be produced at maximum development by a primary nucleus of 10^{15} eV. It is however a major problem to distinguish air showers produced by different primary nuclei of the same energy.

Lower-energy primary cosmic rays ($<10^{13}$ eV), i.e. those with the largest intensity, can be studied directly by sending particle detectors, such as scintillation counters, above the earth's atmosphere in satellites, spaceprobes, rockets, and balloons. High-energy particles ($>10^{16}$ eV) have too low a flux for direct measurements and can only be studied through their extensive air showers using arrays of detectors at ground level; particles with energies exceeding 10^{20} eV have been detected with giant arrays of instruments covering areas greater than one km². Intermediate energies require sensitive detector arrays on mountaintops or large detectors in satellites.

Most of the lighter primary cosmic-ray nuclei are considered products of collisions of heavier nuclei that occur during their journey from source. These parent nuclei, with other heavy nuclei and most of the primary electrons, are created in some as yet unknown *nucleosynthesis process of catastrophic origin. Medium and lower energy particles probably originate in *supernova explosions: both the Crab nebula and the supernova remnant of the Vela pulsar are thought to be sources. The sun produces very low energy cosmic rays, the strongest solar *flares generating energies up to 10^{10} eV. The extraordinary acceleration processes that could produce the highest energies are not yet identified. There is no evidence as yet for a maximum energy for cosmic rays, although this has been predicted.

It is generally believed that almost all cosmic rays with energies less than about 10^{18} eV are generated by sources within the Galaxy. It is also thought that these particles are confined to the Galaxy for probably tens of millions of years by the complex *galactic magnetic field, which has an estimated value of about 3-6 $\times 10^{-10}$ tesla. Cosmic-ray particles, being charged, are affected by a magnetic field: they are deflected from their initial paths and become trapped in the galactic field, the lowest energies being most deflected. During their long confinement their directions of travel become almost uniformly scattered. As cosmic rays enter the solar system they can be further scattered by magnetic irregularities in interplanetary space: the total intensity at the earth's orbit is twice as great at sunspot minimum than at sunspot maximum. Although lower-energy cosmic rays are confined very effectively, higher-energy particles tend to 'leak out' of the Galaxy; above a critical energy at which the effect of the magnetic field becomes negligible, they can all escape. Thus the cosmic-ray intensity should decrease with energy.

The intensity however levels off at high energies and this has been taken as evidence of other sources of cosmic rays outside the Galaxy. Since high-energy particles travel in approximately straight lines (being unaffected by magnetic fields) their arrival direction should indicate their source direction: measurements made with the giant detector arrays point to sources well beyond the galactic plane, i.e. to possible extragalactic sources. Gamma rays can arise in space from the interaction of primary cosmic rays with interstellar matter. It is hoped that measurements of the distribution of gamma rays in space, especially those of high energies ($>10^8$ eV), will reflect the distribution of cosmic rays and thus reveal the galactic or extragalactic nature of their sources.

cosmic rising (or setting). *See* acronical rising (or setting).

cosmic scale factor. Symbol: R. A measure of the size of the universe as a function of time. R is related to the *redshift parameter, z, by
$$1 + z = R(t_0)/R(t_1)$$
where t_0 is the present time and t_1 is the time at emission of the radiation. The *proper distance* between the origin and a point at a comoving coordinate r (i.e. in a system expanding with the universe) is given by the product of $R(t)$ and $\sin^{-1}r$, r, or $\sinh^{-1}r$ depending on whether the *universe is closed, Euclidean, or open, respectively. R is also related to the *Hubble constant, which equals $(dR/dt)/R$.

cosmic year (galactic year). The period of revolution of the sun about the centre of the Galaxy, equal to about 220 million years.

cosmogeny. The study of the origin and evolution of cosmic objects and in particular the solar system. *See* solar system, origin; galaxy evolution; stellar evolution.

cosmological constant. *See* static universe; cosmological models.

cosmological models (world models). Possible representations of the universe in simple terms. Models are an essential link between observation and theory and act as the basis for prediction. Complications are added only when necessary (Occam's razor). A simple model for a two-dimensional universe is the surface of an expanding balloon, on which *Hubble's law and the isotropy of the *microwave background radiation may be demonstrated.

Most standard cosmological (*Friedmann*) models of the universe are mathematical; they assume the *homogeneity and *isotropy of an expanding (or contracting) universe in which the only force that need be considered is *gravitation. The *big bang theory is such a model. These models result from considerations of Einstein's field equations of general *relativity. When the pressure is negligible the equations reduce to
$$(dR/dt)^2/R^2 + kc^2/R^2 = (8\pi/3)G\varrho$$
for energy conservation and
$$\varrho R^3 = \text{constant}$$
for mass conservation. R is the *cosmic scale factor, ϱ the *mean density of matter, G the gravitational constant, and c

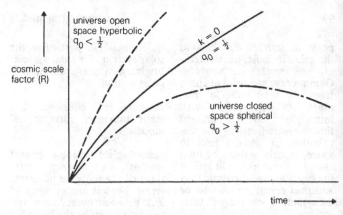

Models of universe with different deceleration parameters, q_0

the speed of light; k is the curvature index of space of value +1 (closed *universe), -1 (open universe), or 0 (flat or *Einstein-de Sitter universe*). See illustration.

Other models involving the *cosmological constant,* Λ, have been proposed, such as the *de Sitter model,* in which no mass is present, the *Lamaître model,* which exhibits a coasting phase during which R is roughly constant, the *steady-state theory, and those in which the gravitational constant, G, varies with time. *See also* static universe.

cosmological principle. A principle stating that all positions in the universe are equivalent. This is the basis for much of modern *cosmology and is an extreme case of the Copernican view that the earth is not the centre of the universe. Neglecting local irregularities, measurements of the limited regions of the universe available to earth-based observers are then valid samples of the whole universe. *See also* steady-state theory.

cosmological redshift. The *redshift resulting from the expansion of the universe rather than from gravitational effects of intervening matter (*see* gravitational redshift) or from the motion of an intergalactic object away from the solar system. *See also* cosmic scale factor.

cosmology. The study of the origin, evolution, and structure of the *universe. *Cosmological models describing the behaviour of the *cosmic scale factor with time are constructed from gravitational theory. These are then tested by comparison with, for example, *source counts, the *Hubble diagram, and the cosmic *helium abundance. The angular size of cosmic objects does not necessarily vary linearly with *redshift and can be used as a further test—the *angular-size redshift test*. The nonlinearity arises because such objects were closer to us in the past and because radiation is bent by the gravitational attraction of intervening matter.

Cosmos. A series of over 1000 Soviet satellites, the first of which was launched in March 1962 and the 1000th in March 1978. The many and varied applications have included astronomical, ionospheric, atmospheric, geomagnetic, geodetic, and biological studies. The satellites have also apparently been used for navigation, reconnaissance, ocean surveillance, and military communications.

coudé telescope. Any telescope in which the light emerges along the *polar axis so that its direction remains fixed while the object observed goes through its normal diurnal motion. This arrangement is essential when the light is subjected to analysis by bulky and delicate instruments, such as the *spectrograph, which can then be mounted in a fixed position at the *coudé focus*.

There are two main ways in which the fixed coudé focus can be secured. With an equatorially mounted telescope, one or more extra mirrors can be placed so as to reflect the emerging light along the polar

a *Light path in telescope with coudé facility*

b *Coudé telescope: equatorial Springfield*

axis (see illustration *a*). This is the practice in large reflecting telescopes where the coudé focus is provided in addition to the prime and Cassegrain foci. It is also the basis for the amateurs' *Springfield mount* (see illustration *b*) in which the diagonal of a Newtonian telescope sends the light along the declination axis to an additional diagonal, which reflects it up the polar axis to the observer's eye.

In the other alternative the whole telescope is mounted in a fixed position along the polar axis; a large moveable plane mirror is driven to reflect the observed object into the telescope. This is the basis of many *solar telescopes.

count. *Short for* source count. *See also* radio source catalogue.

counterglow. *See* gegenschein.

Crab. Common name for *Cancer.

Crab nebula (MI; NGC 1952). A turbulent expanding mass of gas and dust lying about 2000 parsecs away in the constellation Taurus. Luminous filaments of ionized gas give it its crablike appearance. It is the *supernova remnant of a type I *supernova that was almost certainly observed by Chinese and Japanese astronomers in 1054 and was sufficiently bright (magnitude about -5) to be visible in daylight for over three weeks. At present it is about four parsecs in diameter. The Crab nebula emits *synchrotron radiation of all wavelengths (with the possible exception of gamma rays) and is a particularly strong source of x-rays (Taurus X-1) and radio waves (*Taurus A).

The star whose explosion produced the Crab nebula is now a young *optical pulsar, (the *Crab pulsar*, NP 0532) identified as such in 1967. Its pulsations are also observed at radio, infrared, x-ray, and gamma-ray wavelengths. These pulsations have the shortest period (0.0331 seconds) of all known pulsars. The pulsar is the power house for the Crab nebula: energy is being lost by the pulsar in the form of highly energetic electrons, causing the pulsar to slow down very gradually; the electrons interact with the intense magnetic field extending into the nebula and radiate synchrotron emission. The energy loss of the pulsar equals the total energy radiated by the nebula (10^{30} to 10^{31} J s^{-1}). The ultraviolet component of this radiation ionizes the gas in the filaments, causing the atoms to fluoresce. The Crab nebula can thus be considered an *emission nebula.

Crab pulsar. *See* Crab nebula.

Crane. Common name for *Grus.

Crater (Cup). An inconspicuous constellation in the southern hemisphere near Leo, the brightest stars being of 4th magnitude. Abbrev.: Crt; genitive form: Crateris; approx. position: RA 11.5, dec -15°; area: 282 sq deg.

crater chain (chain craters). Any linear association of *craters. Some lunar crater chains are incipient sinuous *rilles, i.e. lava tubes (in *maria) that have suffered only partial collapse. Others are not confined to the maria and consist of overlapping *secondary craters. Chains of volcanic craters (as in fissure eruptions) have not been unequivocally recognized but *see* wrinkle ridges.

crater cluster. A group of *secondary craters. Crater clusters were visited by the crews of Apollo 15 and 17.

crater counting. *See* craters.

craters. Circular rocky formations observed on the moon and several other bodies in the solar system. They have diameters ranging from less than one metre to more than 1000 km, although the larger structures are more usually termed *basins or *maria. The largest lunar craters are considered to be the 200 km *walled plains*. Some of the more interesting lunar craters are listed in the table. Craters appear all over the moon's surface and are most prominent when near the terminator. Small craters are everywhere more numerous than larger

Lunar Craters

Crater	Characteristics*	Coordinates
Aitken	mare-filled, farside (L)	173°E 17°S
Alphonsus	dark halo crater, rilles (L)	3°W 13°S
Archimedes	flat-floored, Mare Imbrium (M)	4°W 29°N
Aristarchus	young, bright, TLP source (S)	47°W 24°N
Aristillus	young, close to Apollo 15 site (S)	2°E 34°N
Arzachel	central peak, S of Alphonsus (M)	2°W 18°S
Autolycus	young, close to Apollo 15 site (S)	2°E 31°N
Bailly	extremely large (L)	68°W 66°S
Clavius	very large (L)	14°W 58°S
Cleomedes	large, N of Crisium (L)	56°E 27°N
Copernicus	young, rayed, secondaries (M)	20°W 10°N
Cyrillus	overlaps with Theophilus (L)	24°E 3°S
Eratosthenes	type crater for Eratosthenian System (M)	12°W 14°N
Gassendi	N Mare Humorum, TLP source (L)	39°W 17°S
Giordano Bruno	extremely young, historical? (S)	119°E 46°N
Grimaldi	very dark floored, TLP source (L)	68°W 5°S
Jules Verne	farside crater (L)	148°E 35°S
Kepler	very young, rayed, Oc. Procellarum (S)	38°W 8°N
Langrenus	large, E Mare Fecunditatis (L)	62°E 9°S
Plato	very dark floored, N Mare Imbrium (L)	9°W 52°N
Proclus	partially rayed, W Mare Crisium (S)	47°E 16°N
Ptolemaeus	large, flat-floored (L)	2°W 9°S
Reiner	very deep, Oc. Procellarum (S)	55°W 7°N
Schickard	large, S Mare Humorum (L)	55°W 44°S
Schiller	elongated, secondary? (M)	39°W 52°S
Stevinus	young, bright, S Mare Fecunditatis (M)	54°E 33°S
Theophilus	young, unrayed, N Mare Nectaris (L)	27°E 12°S
Tsiolkovsky	mare-filled, farside (L)	130°E 21°S
Tycho	very young, prominent ray system (M)	11°W 43°S
Wargentin	flooded with Orientale eject (M)	60°W 49°S
Van de Graaff	double, farside, magnetic anomaly (L)	174°E 27°S

*diameter ranges: L, greater than 100 km; M, 50-100 km; S, smaller than 50 km

ones and crater-size distribution curves provide the basis for the estimation of relative ages by *crater counting*. The surface of *Mercury is also now known to be heavily cratered while on Mars the craters predominate in the southern hemisphere (*see* Mars, surface features).

The vast majority of craters are now believed to have originated by meteoritic impact, although in some cases their present appearance results from later modification by erosion or volcanic processes. A few craters in the lunar maria, particularly those associated with sinuous *rilles, may be volcanic in origin. Some such as those at the summits of the volcanoes of the earth or Mars, are solely volcanic. Other small craters result from subsidence or surface collapse.

In profile, a typical crater has a floor slightly below the level of the surrounding surface and a raised rim that slopes gently outwards but falls more steeply towards the crater floor. The inner slope of the rims of large craters may be *terraced*, while the terrain beyond the rim may be hummocky and broken with numerous very small craters. Some large craters also have mountainous *central peaks*.

The impact events that created most of the surviving craters on solar-system bodies are thought to have occurred between 3000 and 4000 million years ago as the bodies collided with numerous *planetesimals. Possibly the planetesimals were themselves the debris from collisions between larger objects that accreted at the same time as the planets, about 4600 million years ago, between the orbits of Mars and Jupiter. Most of the present-day survivors of the planetesimals, the minor planets, orbit between Mars and Jupiter but the *Apollo and *Amor objects do stray within the orbits of the earth or Mars and, together with cometary nuclei, still produce occasional impacts, though at a much reduced rate.

During a crater-forming impact, the kinetic energy of the planetesimal, which may be several kilometres across and moving at many kilometres per second, goes towards vaporizing and compressing much of itself and the rock on which it impinges. It is the subsequent explosion of these gases that excavates the crater, throwing material (*ejecta) away from the crater centre to form the rim and the crater's hummocky *ejecta blanket. Chunks within the ejecta give rise to secondary impact craters nearby. In a large crater the steep inner rim can then slump to form the observed terraces, while the central peaks may arise from a rebound of the rocks immediately below the site of the explosion. Lunar craters also initially have *ray systems.

Later impacts may destroy the crater completely, form a new overlapping crater, or partially fill the crater with ejecta. Lava may well up through cracks in the floor of the largest impact features to produce lava planes, such as the lunar maria. There may be rim obliteration and erosion and the occurrence of isostatic uplift. As a result the depth-to-diameter ratio, which for lunar craters may be as high as 1:5, decreases. On the earth, erosion by wind, water, ice, and tectonic forces has removed nearly all the evidence that our planet was subject to the same ancient bombardment as the moon; only a few recent impact craters, such as the *Arizona meteorite crater, remain relatively intact.

crepe ring. *See* Saturn's rings.

Crimean Astrophysical Observatory. An observatory of the Academy of Sciences in the USSR, the main site of which is near Nauchnyy in the Crimea at an altitude of 640 metres. The chief instrument is a 260 cm (102 inch) reflecting telescope, which has been operating since 1961 and is Europe's largest reflector. An earlier site at Simeis, restored

Crisium, Mare. in 1945 after its destruction in World War II, has both optical and radio telescopes.

Crisium, Mare. A lava-filled *basin close to the moon's eastern limb from which soil was returned by *Luna 24.

critical density. *See* mean density of matter.

cross-correlation function. *See* auto-correlation function.

Crow. Common name for *Corvus.

Crown. Common name for *Corona Borealis or *Corona Australis.

crust. *See* earth.

Crux or **Crux Australis (Southern Cross).** A very small but conspicuous constellation in the southern hemisphere, lying in the Milky Way. The four brightest stars, *Alpha, *Beta, Gamma, and Delta Crucis form a cross (*see* illustration). The 1st-magnitude Alpha and the 2nd-magnitude red giant Gamma form the longer arm that points approximately to the south pole. The 1st-magnitude Beta and 3rd magnitude Delta form the transverse arm. The 4th-magnitude Epsilon, between Alpha and Delta, interferes with the figure's regularity, making it more kite-shaped. The area also contains the brilliant cluster the *Jewel Box and the dark *Coalsack nebula. Abbrev.: Cru; genitive form: Crucis; approx. position: RA 12.5h, dec -60°; area: 68 sq deg.

C stars. *See* carbon stars.

Culgoora. Site of the C.S.I.R.O. Solar Observatory in New South Wales, Australia. The principal instrument is a large *radio heliograph, which provides a near instantaneous moving display of radio activity in the sun's atmosphere: 96 dishes, 14 metres in diameter, are equally spaced around the circumference of a three km diameter circle. The resolving power is thus that of a three km dish. The received signals are transmitted to a central receiver and processed electronically to produce a television image of the radio sun.

culmination (transit). The passage of a star across an observer's *celestial meridian. As a result of the earth's rotation this occurs twice daily although both culminations can be observed only for *circumpolar stars. The star will reach its greatest altitude above the horizon at *upper* (or *superior*) *culmination;* this occurs when the star crosses the meridian between the north celestial pole (P) and the south point (s) of the horizon (northern hemisphere) or between the south pole (P') and the north point (n) (southern hemisphere). The other meridian passage is the *lower* (or *inferior*) *culmination* when the altitude above or below the hori-

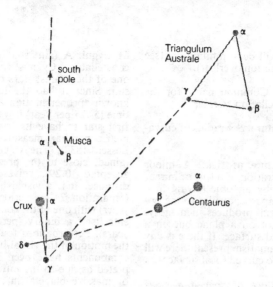

Crux and nearby bright constellations

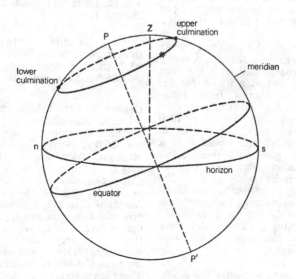

Culminations of a circumpolar star

zon will be at a minimum (see illustration overleaf).

Cup. Common name for the constellation *Crater.

curvature. *See* radius of curvature.

curvature of field. A minor *aberration of a lens or mirror whereby an object lying in a plane perpendicular to the optical axis produces an image lying not on a plane but on a curved surface. If there is any *astigmatism present there will be two curved focal surfaces.

curvature of space-time. *See* space-time.

cusp. Either of the tapering points of the crescent phase of the moon, Venus, or Mercury.

Cyclops. An immense *array of radio dishes, covering maybe 60 km², that has been proposed as a means for detecting signals from extraterrestrial civilizations. The design calls for over 1000 dishes, each 100 metres in diameter, and would have a very great effective signal-collecting area. Although an extremely costly project, the array could also be used for radio astronomy, for radar studies in the solar system, and for tracking space vehicles to very great distances. *See also* exobiology.

cyclotron radiation. *See* synchrotron emission.

61 Cygni. A faint star in the constellation Cygnus that is one of the nearest stars to the sun. Since it had the largest known *proper motion at the time (5″.25 per year) it was the first star to have its trigonometric parallax measured (F. Bessel, 1838). The value obtained, close to the presently accepted 0.29″, gives the distance. It is a visual binary (separation 24″.6, period 720 years) with one of the components itself a binary (period 5 years). Perturbations found in the motion of the visual-binary components have been interpreted as due to the influence of massive planets. m_v: 5.20 (A), 6.03 (B); M_v: 7.53 (A), 8.36 (B); spectral type: K5 V (A), K7 V (B); mass: 0.59 (A), 0.50 (B) times solar mass; distance: 3.42 pc.

Cygnids. A minor *meteor shower, radiant: RA 290°, dec +55° (near Kappa Cygnus), that maximizes on Aug. 20.

Cygnus (Swan). A large conspicuous constellation in the northern hemisphere, lying in the Milky Way. The five brightest stars—Alpha (1st mag.), Gamma and Epsilon (2nd mag.), and Beta and Delta (3rd mag.)—form the *Northern Cross*: α, γ, β comprise the long arm and ε, γ, δ the transverse arm. Alpha is the extremely luminous supergiant *Deneb and Beta the beautiful binary *Albireo. There are several other noticeable binaries and some variables, including the *Mira star Chi Cygni, the brightest *dwarf nova SS Cygni, and the prototype of the *P Cygni stars. There is a large open cluster, M39, some complex emission and dark nebulosity, and the supernova remnant the *Cygnus Loop, comprising several bright nebulae. The constellation also contains the star 61 Cygni, the powerful radio source *Cygnus A, the x-ray source *Cygnus X-1, and the *x-ray binary Cygnus X-3. Abbrev.: Cyg; genitive form: Cygni; approx. position: RA 20.5h, dec +40°; area: 804 sq deg.

Cygnus A. An intense *radio source in Cygnus that is the third strongest radio source after the sun and Cassiopeia A, having a *flux density at 178 megahertz of 8100 *jansky. It is a typical double radio source, a *radio galaxy with a *redshift of 0.057 indicating a distance from the solar system of around 170 megaparsecs. The bright outer lobes are about 50 kpc from the central 16-magnitude galaxy with which the radio source is identified and over which there is some weak radio emission. Cygnus A is also a source of low-energy x-rays.

Cygnus Loop. An *emission nebula in the constellation Cygnus: the *Veil nebula* forms part of the Cygnus Loop. It is the *supernova remnant of an old supernova that exploded probably some 50 000 years ago. The remnant now has an area of about 3° and is about 800 parsecs away. Its rate of expansion has gradually slowed to its present 100 km s^{-1} as the expanding shell collides with and sweeps up any surrounding interstellar matter. This creates a shock wave that compresses, heats, and ionizes the gas atoms, which then become visible as the emission nebula.

Cygnus X-1. An intense galactic x-ray source considered to represent the best observational evidence for the existence of a *black hole of stellar mass. Identification of Cygnus X-1 with the luminous supergiant star HDE 226868 was followed by the discovery of a 5.6 day binary period. Analysis of the optical observations showed the mass of the 'unseen' companion star (the x-ray source) to lie in the range 6-15 solar masses, well above the limit allowed for a *neutron star (≤ 3 solar masses) or *white dwarf (≤ 1.4 solar masses). Support for the view that Cygnus X-1 involves gas accretion from the optical supergiant onto a black hole is provided by the rapid (millisecond) and irregular variability of the x-ray emission.

D

dark halo crater. A small lunar crater, originally believed to be volcanic (e.g. a cinder cone or fumarole) but now considered to be produced when an im-

dark nebula

pacting meteorite excavates dark surface materials.

dark nebula (absorption nebula). A cloud of interstellar gas and dust that is sufficiently dense to obscure partially or completely the light from stars and other objects lying behind it and sufficiently large and suitably located to produce a noticeable effect. These *nebulae can be observed as dark extrusions in front of *bright nebulae or as blank regions or regions with a greatly diminished number of stars in an otherwise bright area of sky. They are situated along the spiral arms of the Galaxy and are composed predominantly of molecular hydrogen; small clouds of molecular hydrogen are called *globules* and can sometimes be seen in large numbers superimposed on bright nebulae. Although having no optical features, dark nebulae can be studied through their radio and infrared emissions. The *Coalsack, *Horsehead nebula, and the *Great Rift are dark nebulae.

Davy Chain on the moon. A prominent *crater chain on the moon that was produced by secondary *ejecta.

Dawes limit. *See* resolving power.

day. The period of the earth's rotation on its axis with reference to some specified point in the sky. The true rotation period (*see* sidereal day) is referred to the background of fixed stars and is equal to 23h 56m 4.1s of *mean solar time. The four-minute difference between the sidereal day and the ordinary 24 hour day—the mean solar day—results from the superimposition of the earth's orbital motion round the sun on the rotational motion.

dec. *Short for* declination.

deceleration parameter. Symbol: q_0. The rate at which the expansion of the universe changes with time as a result of self-gravitation. It is dimensionless and can be given in terms of the *cosmic scale factor, R:

$q_0 = -R(d^2R/dt^2)/(dR/dt)^2$

Values of q_0 less than 0.5 are associated with an ever-expanding open universe; values exceeding 0.5 imply that the universe is closed and will collapse at a finite time in the future. If q_0 is equal to 0.5 then the universe is infinite and space is Euclidean. *See also* mean density of matter; universe.

declination (dec). Symbol: δ. A coordinate used with *right ascension in the *equatorial coordinate system. The declination of a celestial body, etc., is its angular distance (from 0° to 90°) north (counted positive) or south (counted negative) of the celestial equator. It is measured along the hour circle passing through the body.

declination axis. The axis at right angles to the *polar axis about which a telescope on an *equatorial mounting is turned in order to make adjustments in declination. The optical axis then follows a particular hour circle across the sky.

declination circle. A *setting circle on the declination axis that enables an equatorially mounted telescope to be set at the declination of the object to be observed. Its scales are graduated from 0°, when the telescope is aligned with the celestial equator, to ±90°, when it is aligned with the north or south poles.

deferent. *See* Ptolemaic system.

degeneracy pressure. *See* degenerate matter.

degenerate matter. Matter in a highly dense form that can exert a pressure as a result of quantum mechanical effects. Degenerate matter occurs in *white dwarfs and *neutron stars. During the *gravitational collapse of a dying star, the electrons are stripped from their atomic nuclei, and nuclei and electrons exist in a closely packed, highly dense mass. As the density increases, the number of electrons per unit volume increases to a point when the electrons can exert a considerable pressure, called *degeneracy pressure*. This pressure is a result of the uncertainty principle of quantum mechanics. Unlike normal pressure, degeneracy pressure is essentially independent of temperature, depending primarily on density. At the immense densities typical of white dwarfs (10^7 kg m^{-3} or more) it becomes sufficiently large to counteract the gravitational force and thus prevents the star from collapsing further. The gross properties of a white dwarf are therefore described in terms of a gas of degenerate electrons.

Above a certain stellar mass (1.44 solar masses) equilibrium cannot be attained by a balance of gravitational force and electron degeneracy pressure. The star must collapse further to become a neutron star. It is then the degeneracy pressure exerted by the tightly packed neutrons that balances the gravitational force. *See also* black hole.

degenerate star. A star composed primarily of degenerate matter, i.e. a *white dwarf or a *neutron star.

Deimos. A satellite of Mars, lying further from the planet than the slightly larger satellite *Phobos. It is an irregular body measuring 11 × 12 × 15 km, and orbits above Mars' equator at a distance of 23 490 km from the centre of the planet in a period of 1.26 days. Like Phobos, it may be a captured *minor planet of a type similar to the *carbonaceous chondrite meteorites. However, it appears to have none of the grooves and crater chains found on Phobos while many of its craters are less distinct, probably because of infilling by crater ejecta that may lie tens of metres thick on the surface. The larger craters, such as *Voltaire* and *Swift,* measure only about two km across. *See* Table 2, backmatter.

Delphinus (Dolphin). A small constellation in the northern hemisphere near Cygnus, the brightest stars being of 3rd magnitude. Gamma Delphinus is a 4th-magnitude double. Abbrev.: Del; genitive form: Delphini; approx. position: RA 20.5h, dec +10°; area: 189 sq deg.

Delta (δ). The fourth letter of the Greek alphabet, used in *stellar nomenclature usually to designate the fourth-brightest star in a constellation or sometimes to indicate a star's position in a group.

Delta Aquarids. *See* Aquarids.

Delta Cephei. *See* Cepheid variables.

Delta Scuti stars. A class of relatively young *pulsating variables, typified by Delta Scuti, that are about 1.5-3 solar masses and range in spectral type from about A3 to F6. They were originally grouped with the *dwarf Cepheids. They are numerous but tend to be inconspicuous because of the very small variation in brightness, usually about 0.01-0.4 magnitudes. Like dwarf Cepheids they have very short periods ranging from about 0.8-5 hours. A typical feature is a variation in the amplitude and shape of the light curve; irregularities in the light-curve contour result from the superimposition of two or more harmonic modes in which the star is pulsating.

Demon Star. *See* Algol.

Deneb (α Cyg). A very luminous remote white supergiant that is the brightest star in the constellation Cygnus and lies at one end of the long arm of the Northern Cross. It has very low proper motion (0″.003 per year). m_V: 1.25; M_V: -7.2; spectral type: A2 Ia; distance: about 500 pc.

Denebola (β Leo). A white star that is the third-brightest in the constellation Leo. m_V: 2.13; M_V: 1.6; spectral type: A3 V; distance: 13 pc.

density. 1. Symbol: ϱ. The mass per unit volume of a body or material. The mean density of a celestial body is its total mass divided by its total volume. A wide variation in densities is found in the universe, ranging from about 10^{-20} kg m^{-3} for interstellar gas to over 10^{17} kg m^{-3} for neutron stars. The *mean density of matter in the universe is of the order of 10^{-28} kg m^{-3}.

2. The number of electrons, ions, or other particles per unit volume.

density parameter. *See* mean density of matter.

density-wave theory. A theory

that attempts to explain how the structure of a spiral galaxy is sustained by the flow of material in and out of the regions of the spiral arms under the influence of gravity. It was proposed in 1964 by C.C. Lin and F.M. Shu and based on work by B. Lindblad. The spiral structure is not permanent: it is regarded as a wave pattern in which the spiral arms are regions of low gravitational potential where matter concentrates. The spiral arms therefore comprise the loci of wave crests of a density wave. The wave pattern is not tied to the matter but moves through it at an angular velocity that can differ appreciably from that of the matter; the relative motion is on average about 30 km s^{-1}. It is this motion that somehow maintains the overall appearance of the galaxy.

Stars move into the travelling gravitational troughs of the spiral arms and can remain there for a considerable time until their orbits take them out again. Interstellar gas also comes under the influence of the density wave. It has been suggested that when the gas encounters the moving wave it undergoes a sudden compression, which in turn could be a possible trigger for interstellar cloud collapse and star formation. Young stars should then be found preferentially in spiral arms, which is indeed the case.

Although the density-wave theory explains the spiral shape of galaxies and the presence of young stars in spiral arms, it does not show how the density wave originated. In addition it predicts the dissipation of the spiral arms after only a few galactic rotations, i.e. after several hundred million years, which is much shorter than the lifetime of galaxies.

Descartes formation. A range of lunar hills to the NW of Mare Nectaris that have an unusually high *albedo. They were once thought to consist of acid volcanics but are now considered to have been impact-generated.

descending node. *See* nodes.

de Sitter model. *See* cosmological models.

detached binary. *See* equipotential surfaces.

deuterium. *See* hydrogen.

diagonal mirror. *See* Newtonian telescope.

Diamond of Virgo. *See* Spica.

diamond-ring effect. A bright flash of sunlight from a conspicuous *Baily's bead sometimes very briefly observed, together with the sun's inner corona, at the moon's limb just before or after totality in a solar eclipse.

diaphragm. *See* stop.

dichotomy. The moment of exact half-phase of the moon, Mercury, or Venus.

Dicke-switched receiver. A radioastronomy *receiver in which the input is continuously switched between the *antenna and a reference noise source of constant noise power. The switching, at a rate of between 10 and 1000 hertz, is carried out by a *Dicke switch,* which is typically a semiconductor-diode switch or a ferrite switch. The switching reduces the fluctuations in the output of the receiver that arise from *gain changes in the amplification stages and seriously limit the performance of total-power *receivers. The amplitude of the component at the switching frequency in the output of the first detector is proportional to the difference in powers between the noise source and the antenna; it is detected using a *phase-sensitive detector. The effect of a change in gain is much reduced provided that the change takes place much more slowly than in one switch cycle.

differential rotation. Rotation of different parts of a system at different speeds. It occurs in a system, such as a star, composed primarily of gas. A solid body, like the earth, rotates at a constant rate for all positions. The *sun and the giant planets, such as Jupiter, rotate differentially such that equatorial regions move somewhat quicker than those closer to the poles. The outer regions of a galaxy also rotate differentially (*see* galactic rotation).

differentiation. The process by which layers of different density are formed from originally homogeneous molten rock. Melting and differentiation occurred very early in the history of the earth and began with the formation of a dense iron core. The heat released by the infalling molten iron caused a large fraction of the planet to melt. The lightest material, with low melting points, could then float upwards, cool, and form a primitive crust. Differentiation also occurred early in the moon's history. It is uncertain whether other terrestrial planets have differentiated.

diffraction. The apparent bending of light or other electromagnetic radiation at the edges of obstacles, with the resulting formation of light and dark bands or rings (diffraction patterns) at the edges of the shadow. An example of a diffraction pattern is the *Airy disc observed in telescope images of distant objects. A small aperture or slit produces similar annular or banded patterns. Diffraction results from the wave nature of electromagnetic radiation. *See also* interference.

diffraction grating. An optical device that is used for the production and study of *spectra, usually by means of a *spectrograph. Its action depends on the *diffraction of light or other radiation by a very large number of very close and exactly equidistant linear slits.

Diffraction by a plane transmission grating

The slits are produced by ruling very fine closely spaced scratches on glass or polished metal, forming either a *transmission* or *reflection grating*. Reflection gratings can be plane or concave; the latter can act as a focusing element for incident light.

The diffracted light, once focused, produces a series of sharp spectral lines for each resolvable wavelength present in the incident light. For a plane wave of a single wavelength λ, incident on a transmission grating at angle i, the successive wave trains passing through the slits will travel different pathlengths. If the path difference between two adjacent slits is a whole number of wavelengths, the wavetrains will be brought to a focus as a bright image of the radiation source at a particular angle of refraction, d (see illustration), where

$$\sin d = n\lambda/s - \sin i$$

s is the spacing between adjacent slits and n is an integer. Bright images will in fact be produced for each of the angles d corresponding to $n = 1, 2, 3 \ldots$ These numbers denote the *orders* of the image. If, for a particular order, d is made equal to i, then

$$\sin d = n\lambda/2s$$

If the incident light is composed of various wavelengths, the image of any particular order, $n = 1, 2, 3 \ldots$, will appear at different points since d varies with wavelength. Hence a spectrum is produced. The angular dispersion of the spectral lines, i.e. $dd/d\lambda$, will be high when the slits are very fine and closely spaced (i.e. s is small). For the resolving power—separation of images of very nearly equal wavelength—to be high, the total number of slits must be large: several thousand slits per centimetre are common for the visible and ultraviolet regions of the spectrum.

diffraction pattern. The pattern of alternating light and dark rings or bands produced when light is diffracted by a circular opening or object (rings) or a rectangular slit or an edge (bands). *See also* Airy disc.

Dione. A satellite of Saturn, discovered in 1684, and probably having an icy surface. *See* Table 2, backmatter.

dioptric system. An optical system, such as a refracting telescope, in which all the optical elements are lenses.

dipole. A simple linear balanced *antenna that is fed from its centre. It usually consists of two equal-length straight opposing elements, though these may sometimes be folded back on themselves and their far ends joined together in the middle to form a *folded dipole*. The elements may be of any length but the overall length of the dipole is usually a multiple of half a wavelength of the received (or transmitted) radio wave; the reactive component of the *impedance seen at its feed point is then small.

A *half-wave dipole* has two elements of length one quarter of a wavelength each and has a *radiation resistance of about 70 ohms. A *full-wave dipole* is twice as long as a half-wave dipole and has a radiation resistance of several thousand ohms. The *gain of a half-wave dipole over an isotropic radiator is about two decibels.

directivity. A measure of the ability of an *antenna to transmit or receive radiation in one direction only. It is defined as the ratio of the maximum radiation intensity in the wanted direction to the average radiation intensity over all directions. If an antenna is operating at a wavelength λ and has an effective area A_e (*see* array), then the directivity, D, is given by
$$D = 4\pi A_e/\lambda^2$$
See also gain.

direct motion. 1. The apparent west to east motion of a planet as seen from earth against the background of stars. *Retrograde motion* is an apparent motion from east to west. When a superior planet near *opposition is overtaken by the earth, moving with higher relative velocity, its normal direct motion can become temporarily retrograde and it appears to undergo a loop or zigzag in the sky (see overleaf); the turning points between these motions, when the planet appears motionless in the sky, are known as *stationary points*.
2. The anticlockwise orbital motion, as seen from the north pole of the ecliptic, of a planet or comet around the sun or of a satellite around its primary. A clockwise orbital motion is *retrograde*. The planets all have direct motion.
3. The anticlockwise rota-

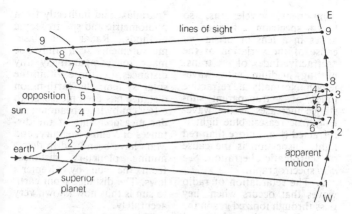

Direct and retrograde motion of superior planet

tion of a planet, as seen from its north pole. Venus and Uranus have a clockwise *retrograde* rotation.

dirty-snowball theory. *See* comet; nucleus.

disc. 1. The two-dimensional projection of the surface of a star or planet.
2. *See* Galaxy.

disc population stars. *See* populations I and II.

dish. A type of *antenna consisting of a large spherical or parabolic metal reflector, usually circular in outline, by means of which radio waves are brought to a focus; they are collected at the focus by a secondary antenna, called a *feed*. At low radio frequencies the feed may be a *dipole mounted in front of a small reflector; at high frequencies it is usually a horn antenna (*see* waveguide).

The dish is often mounted so that it can be steered to point in different directions and can be made to track a moving object. Some dishes however, like the 300-foot dish at Green Bank, USA, can only move in one coordinate (*see* coordinate system) and rely on the earth's rotation to provide coverage in the other coordinate. The 1000-foot dish mounted in a natural hollow in the ground at Arecibo, Puerto Rico, achieves partial beam steering by moving the feed.

dispersion. 1. The separation of a beam of light into its component colours, i.e. into its component wavelengths, so that a spectrum is formed. It arises in a lens or prism because of the variation of the *refractive index of the transmitting medium with wavelength. Normally the refractive index of a transparent substance increases as the wavelength decreases: blue light is refracted (bent) more than red light. Dispersion is the cause of *chromatic aberration. *See also* spectrograph.

2. The retardation of radio waves that occurs when they pass through ionized gas in the interstellar medium, the speed of propagation depending on frequency: the lower the frequency the greater the delay. This delay is not observable in a continuous radio signal but is detectable in a pulsed one, such as that from a pulsar. The amount of dispersion gives an indication of distance to a pulsar. *See also* dispersion measure.

dispersion measure (DM). The number of electrons per unit volume, n_e, integrated along the line of sight to a source:
$$DM = \int n_e dl$$
where dl is an element of the path. It is the factor that determines, for example, the time delay (i.e. *dispersion) between the arrival of radio pulses from a *pulsar at two different frequencies.

distance determination. The distances of celestial objects can be determined directly by means of radar or laser ranging and from measurements of *parallax, and indirectly from photometric and spectroscopic methods. Radar and laser measurements can only be made for comparatively tiny distances, i.e. those within the solar system: the earth-moon distance has been measured to a few cm by laser beaming; radar measurements of the distances of Venus, Mercury, and Mars have been used in determining planetary distances from the sun by *Kepler's laws. The distance from earth to sun is thus now known very accurately.

This earth-sun distance is used as the baseline in the determination of the *annual parallax of stars, from which stellar distances out to about 30 parsecs can be determined trigonometrically. At greater distances the error in measuring parallax becomes as large as the parallax itself. These distances have therefore to be found indirectly from measurements of *luminosity, *magnitude, and other stellar properties: distances are determined from relationships connecting these properties, including *distance modulus, *moving-cluster parallax, and the *period-luminosity relation for *Cepheid variables.

A distance scale can be established whereby the distances found for one set of distance indicators, such as open clusters (using *main-sequence fitting, for example), can be used to calibrate the next most distant indicators—Cepheid variables—occurring in such clusters. Cepheid var-

Distance scale ladder

iables can in turn be used for further calibration (see illustration). The P-L relation for Cepheids, plus observations of novae and supernovae, allows the distance to the nearer galaxies to be found. Estimates of distance for more remote galaxies rely on measurements of *redshift and *Hubble constant.

distance modulus. The difference between the apparent magnitude, *m*, and the absolute magnitude, *M*, of a star and therefore a measure of distance (*see* magnitude):

$m - M = 5 \log(d/10) = 5 \log d - 5$

where *d* is the distance in parsecs. Distance modulus is used to determine the distances of stars and stellar clusters, especially those over about 30 parsecs away.

distortion. An *aberration of a lens or mirror in which the image has a distorted shape as a result of nonuniform lateral magnification over the field of view. Common forms of lens distortion are shown in the illustration. In *barrel distortion* the magnification decreases towards the edge of the field; in

no distortion   barrel distortion   pincushion distortion

diurnal circle. *See* diurnal motion.

diurnal libration. Daily geometrical *libration brought about by the change in the position of the observer with respect to the earth-moon line as the earth rotates on its axis. This libration is most significant near the equator, where it amounts to 57′2″.6; this is equal to the moon's mean equatorial horizontal parallax (*see* diurnal parallax). An observer closer to the poles will however experience an equivalent geometrical libration because of his displacement from the *ecliptic. *See also* physical libration; augmentation.

diurnal motion. The apparent daily motion of celestial bodies across the sky from east to west. It results from the earth's rotation from west to east, the axis of the diurnal motion coinciding with the earth's axis. The apparent path of a celestial body arising from diurnal motion is a circle parallel to the celestial equator and is termed a *diurnal circle*.

diurnal parallax (geocentric parallax). The *parallax of a celestial body that results from the change in position of an observational point during the earth's daily rotation. It is the angle measured at a given celestial body between the direction to the earth's centre and that to an observer on the earth's surface. It is thus equal to the difference between the topocentric and geocentric *zenith distances, η and η_0, respectively (see illustration). It is only appreciable for members of the solar system. As the observer's position changes during the day, the diurnal parallax of the body varies from a minimum, when the body is on the observer's meridian, to a

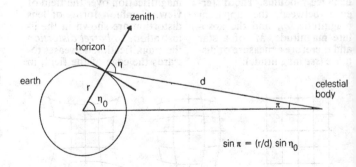

$$\sin \pi = (r/d) \sin \eta_0$$

Diurnal parallax

maximum, when it is on the horizon. This maximum value is the object's *horizontal parallax*, π. It is given by:

$$\sin \pi = r/d$$

where r is the earth's radius and d is the distance to the celestial object.

The earth's nonspherical shape causes the length of the baseline to vary with latitude. If this variation is significant, as it is with nearby bodies like the moon, then for the horizontal parallax to be a maximum it must be measured from the equator. This gives the *equatorial horizontal parallax:* the average value for the moon—the *lunar parallax*—is 3422″.45; the value for the sun—the *solar parallax—averages 8″.794.

D-layer (D-region). *See* ionosphere.

D lines. A close prominent doublet of lines in the yellow region of the visible spectrum of sodium. They occur at wavelengths of 589.6 nm and 589.0 nm and, being easily recognizable, are useful as standards in spectroscopy. The D lines are strong *Fraunhofer lines.

docking. The locking together of two spacecraft in space following their rendezvous.

Dog Star. *See* Sirius.

Dolphin. Common name for *Delphinus.

dome. 1. A movable hemispherical roof of an observatory that protects the telescope and associated equipment from bad weather and from the heating effect of the sun. The telescope has access to the sky through a *shutter* in the dome, which is closed when the telescope is not in use. Rotation of the dome on a circular track allows different areas of the sky to be brought into view.

2. A low lunar hill with a gentle convex-upward profile, of probable volcanic origin. Domes are confined to the *maria and are generally less than 15 km in diameter.

Donati's comet (1858 VI). A spectacular *comet famous for its *coma with multiple *haloes:* parabolic envelopes with vertices towards the sun and foci near the apparent *nucleus. These haloes were regular and sharp and are thought to have been produced by repetitive ejection of material from a single active area exposed successively to solar radiation as the solid cometary nucleus rotated every 4.6 hours.

Doppler broadening. Broadening of spectral lines due to the random motion of emitting or absorbing atoms. As a result of the *Doppler effect, atoms moving away from the observer show lines with a slight shift to longer wavelengths; atoms moving towards the observer show a slight shift to shorter wavelengths. The overall effect is that the line is broader than the natural width (determined by quantum mechanical uncertainty).

The motion of the atoms may be due to thermal motion, in which case the effect is larger for lighter atoms. Turbulence of stellar material, rapid rotation of a star, or an expanding stellar atmosphere can also produce Doppler broadening. *See also* line broadening.

Doppler effect. A change in apparent frequency—and hence wavelength—of a wave motion as a result of relative motion of the source and observer. For *electromagnetic radiation emitted from a moving source, the magnitude of this change is known as the *Doppler shift*. For a source moving away from the observer, the observed wavelength is longer than it would be if source and observer had no relative motion along the line joining them. This change to longer wavelengths (i.e. towards the red end of the visible spectrum for visible wavelengths) is called a *redshift. Conversely, if the source is moving towards the observer, there is a change to shorter wavelengths, i.e. a *blueshift.

When there is no relative motion between source and observer, the wavelength of a spectral line can be given by λ. If the relative velocity along the line of sight of source and observer is v, the change in wavelength ($\Delta\lambda$) of the spectral line is given by

$$\Delta\lambda/\lambda = v/c$$

where c is the speed of light. For values of v comparable with c, a relativistic expression is used:

$$\Delta\lambda/\lambda = [(c + v)/(c - v)]^{1/2} - 1$$

v is taken to be positive for a receding object and negative for an approaching one. Doppler shifts can be observed in all regions of the electromagnetic spectrum.

The Doppler effect was first suggested by the Austrian physicist C.J. Doppler in 1842. It has proved invaluable in astronomy for measuring the *radial velocities of celestial objects and is also used for determining the orbital motions (and in some cases the masses) of *spectroscopic binaries and the rotational motions of bodies. *See also* Doppler broadening.

Doppler shift. *See* Doppler effect.

Dorado (Swordfish). A small constellation in the southern hemisphere, the brightest star being of 3rd magnitude. It contains a large portion of the Large Magellanic Cloud in which lies the extremely luminous star S Doradûs and the complex Great Looped nebula, 30 Doradûs. Abbrev.: Dor; genitive form: Doradûs; approx. position: RA 5.5h, dec -60°; area: 179 sq deg.

30 Doradûs (Great Looped nebula; Tarantual nebula; NGC 2070). A very extensive very luminous *emission nebula in the Large *Magellanic Cloud (LMC) that is a grouping of ionized hydrogen (*HII) regions characterized by rapid complex motions. The largest HII region has a diameter of about 230 parsecs and the complex as a whole could cover an area of three by one kiloparsecs. It has a dense concentration of *Wolf-Rayet stars. It is the brightest object (absolute magnitude -19) in the LMC and is visible to the naked eye. It is also the strongest radio source in the LMC.

double cluster. A pair of *clusters that lie relatively close together. A well-known example is the pair *h and Chi Persei, in which the individual clusters appear to be of similar age.

Double Cluster in Perseus. *See* h and Chi Persei.

double galaxy. Two galaxies observed in close proximity. *Physical doubles* are spatially close pairs assumed to be in orbit about a common centre of mass. *Optical doubles* are widely separated pairs that only appear close because of a chance alignment along the line of sight. Although orbital motions cannot be directly observed, the relative radial velocities of galaxies in a gravitationally bound pair can be used to estimate their total mass.

double-lined spectroscopic binary. *See* spectroscopic binary.

double radio source. *See* radio-source structure.

double star. A pair of stars that appear close together in the sky. There are two types: *optical double stars* and *physical double stars*. Stars in an optical double, such as *Deneb, appear close only because they happen to lie in very nearly the same direction from earth; they are actually too far apart to be members of a dynamically linked system. Stars in a physical double are sufficiently close together for their motions to be affected by their mutual gravitational attraction. They are more usually known as *visual binaries.

Dove. Common name for *Columba.

Draco (Dragon). An extensive straggling constellation in the northern hemisphere near Ursa Major. The brightest stars are the 2nd-magnitude yellowish giant Eltanin (γ) and several of third magnitude including *Thuban (α), the *pole star in ancient times. The area contains several binaries, such as Psi Draconis, and the bright planetary nebula NGC 6543. Abbrev.: Dra; genitive form: Draconis; approx. position: RA 9.5 -20·5h, dec 50°-85°; area: 1083 sq deg.

draconic month (nodical month). The time interval of 27.2122 days between two successive passages of the moon through the ascending *node (or descending node) of its orbit.

Draco system. A member of the *Local Group of galaxies.

Dragon. Common name for *Draco.

Draper classification. *Another name for* Harvard classification. *See* spectral types.

draw tube. A simple tube of 31.75 mm bore into which fits the telescopic *eyepiece in use. It is equipped with a mechanism, such as a rack and pinion, allowing fine adjustment of the eyepiece position in order to obtain a sharp focus. Some draw tubes are threaded to accommodate eyepieces with an 'RAS thread' but these are not very common.

drift scan. A profile of a *radio source made with a radio telescope by aiming the *antenna at a point beside the source and allowing the rotation of the earth to move the source through the antenna beam.

drive clock. *See* equatorial mounting.

Dubhe (α UMa). An orange-yellow giant that is one of the seven brightest stars in the *Plough and is one of the two *Pointers. It is a close visual binary (period 44 years). m_v: 1.81; M_v: -0.6; spectral type: G9 III; distance: 30 pc.

Dumbbell nebula (M27; NGC 6853). A *planetary nebula about 220 parsecs away in the constellation Vulpecula. It has an hourglass shape with a large angular diameter (330 arc seconds) and low surface brightness.

durchmusterung. A star catalogue.

dust tail. *See* comet tails.

dwarf. *Short for* dwarf star.

dwarf Cepheids (AI Velorum stars; RRs stars). A group of *pulsating variables that until recently were classed as *RR Lyrae stars but have been shown to have a lower luminosity in addition to a shorter period of about 1.3 to 5 hours. The variation in luminosity (i.e. *amplitude) ranges from 0.2 or 0.3 up to 1.2 but can be higher. At the lower amplitudes they are difficult to differentiate from *Delta Scuti stars but are apparently older. Instabilities of period and light-curve shape are characteristic of these stars and many exhibit the Blazhko effect found in some *RR Lyrae stars. The pulsations are probably mixtures of oscillations of a fundamental mode and those at higher harmonic frequencies (*see* pulsating variables).

dwarf galaxy. A *galaxy that is unusually faint either because of its very small size, its very low surface brightness, or both. Since galaxies exist in a continuous range of sizes from the giant ellipticals downwards, the dividing line between average and dwarf is somewhat arbitrary. Since no spiral or S0 galaxies have been observed with total magnitudes below -16, this is often used as a convenient demarcation line. Dwarf galaxies can only be seen when comparatively nearby so that their true abundance and distribution are uncertain. Also the lower limit to galactic size is not known: dwarf galaxies in the *Local Group have diameters ranging down to 300 parsecs but there may well be smaller and fainter systems still to be detected.

dwarf novae. A small group of intrinsically faint stars that are characterized by sudden increases in brightness occurring at intervals of a few weeks or months, the maximum brightness lasting only a few days. The change in brightness (i.e. *amplitude) is between 2-6 magnitudes. The first to be discovered, U Geminorum, is typical of the majority, which are therefore classified as *U Geminorum stars*. This subgroup displays a fairly smooth decline in brightness from the maximum, unlike the much smaller subgroup of *Z Camelopardalis stars* that can undergo *standstills,* i.e. periods of nearly constant intermediate brightness, before dropping to minimum brightness. Both the occurrence of the standstill and its duration—a few days to many months—are quite unpredictable. There are also periods of erratic light variations. The brightest and most carefully observed dwarf nova is *SS Cygni:* its period varies widely from the mean of 51 days; the maxima vary in shape, brightness, and duration and have been divided into long, short, and anomalous maxima. SS Cygni is now known to be an *x-ray binary.

Most if not all dwarf novae are close *binary stars in which the primary is usually a *white dwarf. The secondary is a cooler main-sequence star, spectral type K or G. The components have similar masses (about 0.7 to 1.2 solar masses); the orbital periods are between about 3 to 15 hours. The secondary is undergoing irregular expansion and in doing so can fill one Roche lobe of the *equipotential surfaces of the system. Hydrogen-rich gas then streams from the secondary and takes up a disc-shaped orbit around the primary. The increases in brightness could arise as 'hot spots' on the accretion disc or could occur when mass ejection from the secondary reveals lower hotter levels of the star. *W Ursae Majoris stars are thought by many astronomers to be the progenitors of dwarf novae.

dwarf star (dwarf). Any star lying on the *main-sequence of the *Hertzsprung-Russell diagram. It is therefore of luminosity class V; its *spectral type is sometimes prefixed by *d,* as in *d*G2.

dynamic parallax. An iterative method used to determine the distance and the mass of a *visual binary star, usually from measurements of the orbital period (P years) and apparent mean orbital size (l'') and an estimate of the mass of the two stars ($M_1 + M_2$ solar masses). The approximate semimajor axis (a AU) of the orbit is calculated from *Kepler's third law:

$$(M_1 + M_2)P^2 = a^3$$

The ratio of a/l gives an estimate of the distance of the binary. If the apparent magnitude is known, the absolute magnitude can be deduced from the *distance modulus and a more accurate value of total mass is hence obtained using the *mass-luminosity relation. The process is repeated as necessary using the corrected mass estimates, an inaccurate estimate in mass leading to a much smaller error in distance.

E

Eagle. Common name for *Aquila.

Eagle nebula (M16). A large *emission nebula—an HII region—that is located about 2.5 kiloparsecs away in the direction of the constellation Serpens. It contains some hot young stars that are part of an open cluster.

early-type star. Hot stars of *spectral type O, B, and A. They were originally thought, wrongly, to be at an earlier stage of evolution than *late-type stars.

earth or **Earth.** The third planet of the solar system and the only one known to possess life. It is the largest of the inner planets (equatorial radius: 6378 km) and has one natural satellite, the moon. A summary of the earth's orbital and physical characteristics are given in Table 1, backmatter. The earth has a substantial *atmosphere, mainly of nitrogen and oxygen, and a *magnetosphere linked to a magnetic field. (*See* atmosphere, composition; atmospheric layers; geomagnetism; Van Allen radiation belts.) Two-thirds of the planet is covered by water at an average ocean depth of 3900 metres; the average land elevation is 860 metres.

Studies of the refraction and reflection of seismic waves propagating through the earth show that it consists of three main internal layers: the crust, mantle, and core. The *crust,* with a thickness of about 32 km under the continents and 10 km under the oceans, has a density of about three times that of water. It consists largely of sedimentary rocks, such as limestone and sandstone, resting on a base of igneous rocks, such as granite (under the continents) and basalt (under both the continents and oceans). The *mantle* extends to a depth of 2900 km where its density reaches 5.5 times that of water. Its composition is thought to include a high proportion of silicate rocks. The *core* increases in density from 10 times that of water at its junction with the mantle to 13 times at the centre. It may be similar in composition to the nickle-iron *meteorites, which contain 6% nickle. The inner core, from 5150 km to the centre, appears solid but the outer core is liquid and is regarded as the seat of the earth's magnetic field (*see* geomagnetism).

It is believed that the internal temperature rises from 1200 kelvin (K) at a depth of 100 km, to 3000 K at 1000 km, and 6400 K at the centre. The heat required to maintain these temperatures is derived from the natural radioactivity of the earth's constituent rocks, but would have been greatly augmented soon after formation by gravitational compression and the impact of meteoritic material. This led to the widespread melting and *differentiation that produced the present layered structure.

According to the theory of *plate tectonics,* the crust and the upper 50-100 km of the mantle, together forming the *lithosphere,* comprise a layer of hard rock broken into about a dozen separate plates covering the earth's surface. It is thought that convection within the *asthenosphere*—a hotter and more plastic layer of the mantle below the lithosphere—causes the plates and their associated land masses to move in relation to each other: the phenomenon of *continental drift*. Friction between plates gives rise to enhanced earthquake and volcanic activity at their borders. Where plates are moving apart new material wells up from below to create new lithosphere. Where plates collide, one of them may dip below the other, possibly depositing its crustal load in an episode of mountain building.

See also age of the earth; solar system; solar system, origin.

earthgrazers. *Informal* Members of the *Amor and *Apollo groups of *minor planets, such as *Adonis, *Atan, *Hermes, and *Icarus, all having orbits that can bring them within about 0.1 astronomical units (15 million km) of the earth.

earth-rotation synthesis (supersynthesis). *See* aperture synthesis.

earthshine (ashen light). Sunlight reflected from earth. Close to new moon, earthshine reflected by the moon back to the earth enables the whole lunar disc to become visible— the old moon in the new moon's arms. An observer on the moon sees the earth illuminated by earthshine and at new moon can measure the earth's *albedo.

eastern elongation. *See* elongation.

east point. *See* cardinal points.

eccentric anomaly. *See* anomaly.

eccentricity. Symbol: e. A measure of the extent to which an elliptical orbit departs from circularity. It is given by the ratio $c/2a$, where c is the distance between the focal points of the ellipse and $2a$ is the length of the major axis. For a circular orbit $e = 0$. The major planets and most of their satellites have an eccentricity range of 0–0.25 (see table). Many comets and some of the minor planets and planetary satellites have very eccentric orbits. *See also* conic sections.

planet	eccentricity
Mercury	0.205 628
Venus	0.006 787
Earth	0.016 722
Mars	0.093 377
Jupiter	0.048 45
Saturn	0.055 65
Uranus	0.047 24
Neptune	0.008 58
Pluto	0.249

echelle grating. A *diffraction grating with very fine lines ruled much further apart than in other gratings. Such a grating has very high resolution but only over a fairly narrow wavelength band. The resolving power at a given wavelength depends primarily on the total ruled width and the angles of incidence and refraction; a high resolution can be obtained with coarse ruling if the grooves are properly shaped to reflect light, etc., in a narrow bundle. The echelle grating of the *echelle spectrograph* is used at high-order diffraction to give larger angular dispersion. Since the high orders overlap, cross dispersion with a second grating or prism must be used so that the spectral images of the different orders can be separated and stacked above one another at the detector.

eclipse. The total or partial obscuration of light from a celestial body as it passes through the shadow of another body. A planetary satellite is eclipsed when it passes through the shadow of its primary or another satellite. An eclipse of the sun is strictly an *occultation.

An eclipse of the sun—a *solar eclipse*—or of the moon—a *lunar eclipse*—occurs when the sun, moon, and earth lie in or nearly in a straight line (see illustration, *a*). If the plane of the moon's orbit lay exactly in the plane of the *ecliptic a solar eclipse would take place at each new moon and a lunar eclipse at each full moon. The two planes are however inclined at an angle of about 5°, intersecting at the *nodes of the moon's orbit. Eclipses are only observed when the sun is at or near a node and the moon is near the same node (solar eclipse) or the opposite one (lunar eclipse). The *ecliptic limits* are the maximum angular distances of the new or full moon from its node for an eclipse to take place.

Although the moon is 400 times smaller than the sun, it is

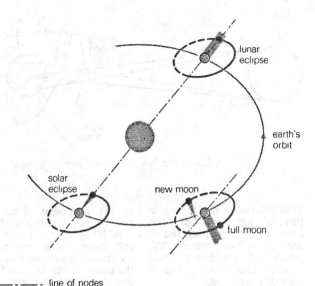

a Solar and lunar eclipses

also about 400 times nearer the earth. As a result, sun and moon have almost exactly the same angular size (about ½°), so that it is possible for the moon to obscure the sun. The earth and moon both cast shadows in sunlight, the shadow having a dark cone-shaped inner region—the *umbra—and an outer lighter penumbral region. A solar eclipse occurs, between sunrise and sunset at new moon, when the moon passes directly in front of the sun so that the earth lies in the moon's shadow (see illustration, *b*). When the moon is sufficiently close to earth so that its apparent diameter exceeds that of the sun, then the umbra of the moon's shadow can just reach the earth's surface. It moves rapidly in a general west to east trend over a very narrow curved zone of the surface, known as the *path of totality,* which can be up to 250 km wide but averages about 160 km. An observer at a point where only the penumbra will move past sees a *partial eclipse,* in which only part of the sun is obscured. An observer in the path of totality will experience a *total eclipse,* in which the sun is completely obscured. If the moon is far enough away to appear smaller

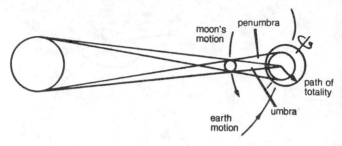

b *Lunar shadow in solar eclipse*

than the sun, a rim (or annulus) of light will be seen around the eclipsed sun and an *annular eclipse* occurs. The period of annularity never exceeds 12.5 minutes and is normally much less.

In a total solar eclipse, *first contact* occurs when the moon just appears to touch the sun's western limb. As the moon gradually covers the sun, the landscape darkens and animals become disturbed. *Totality* begins at *second contact* when the sun disappears from sight. The maximum duration of totality is 7m40s but is usually much less. Totality ends at *third contact*, just as the crescent sun emerges, and at *fourth contact* the whole disc of the sun is once more seen. The time between first and last contact can approach four hours. During totality the *chromosphere, *corona, and associated phenomena can be observed and studied. There are between two and five solar eclipses each year. Total eclipses are however very rare at any particular place: the next UK occurrence is in 1999.

A lunar eclipse occurs, at full moon, when the moon passes into the shadow cone of the earth. It can be seen from any place at which the moon is visible above the horizon. A *total eclipse* occurs when the moon enters completely into the umbra of the earth's shadow. If only part of the moon enters the umbra the eclipse is *partial*. When the moon only enters the penumbral region, a *penumbral eclipse* takes place in which a slight, usually quite unappreciable darkening of the moon's surface occurs. The maximum duration of totality is 1h42m. The moon can usually be seen throughout totality, being illuminated by sunlight refracted by the earth's atmosphere into the shadow area. Since the bluer wavelengths are removed by scattering, the moon has a coppery-red colour. There are up to either two or three lunar eclipses each year. Up to seven eclipses can occur in one year, either five solar and two lunar or four solar and three lunar.

See also Saros.

eclipse year. The period of time between two successive passages of the sun through the same *node of the moon's orbit. It is equal to 346.620 days, being less than the *sidereal year because of regression of the moon's nodes. There are almost exactly 19 eclipse years in one *Saros.

eclipsing binary. A *binary star whose orbital plane is orientated such that each component is totally or partially eclipsed by the other during each orbital period. The effect observed is a periodic decrease in the light from the system (see illustration). The deeper minimum corresponds to the eclipse of the brighter star. The light curve gives the period of rotation and from the depths and shapes of the minima it is possible to estimate the inclination of the orbital plane. The duration of the eclipses compared to the time between eclipses indicates the radii of the stars in terms of the distance between them. If the system is also a doublelined *spectroscopic binary, the individual masses and radii of the stars can be calculated.

Eclipsing binaries tend to be composed of large stars with small orbits and therefore the majority are close binaries (*see* binary star). There are three major classes of eclipsing binaries: *Algol variables have spherical or near spherical components; *Beta Lyrae stars

Schematic of orbits and light curve of eclipsing binary

and *W Ursae Majoris stars have components that are distorted into ellipsoids by their proximity to each other and suffer *mass transfer.

ecliptic. The great circle in which the plane of the earth's orbit intersects the *celestial sphere. Although the orbital plane is in fact defined by the motion of the centre of mass of the earth-moon system around the sun, the ecliptic usually refers to the earth alone, the resulting errors being negligible. The ecliptic may thus be generally taken as coincident with the sun's apparent annual path across the sky. The orbits of the moon and planets, apart from Pluto, when projected on to the celestial sphere also lie very near the ecliptic.

The planes of the ecliptic and celestial equator are inclined at an angle equal to the tilt of the earth's axis. This angle, known as the *obliquity of the ecliptic, is about 23°27'. The *equinoxes lie at the two points of intersection of ecliptic and celestial equator. The *poles of the ecliptic* lie at 90° from all points on the ecliptic at the positions RA 18h, dec +66°.5 and RA 6h, dec -66°.5.

ecliptic coordinate system. A *coordinate system in which the fundamental reference circle is the *ecliptic and the zero point is the vernal *equinox (♈). The coordinates are *celestial* (or *ecliptic*) *latitude* and *celestial* (or *ecliptic*) *longitude* (see illustration).

The celestial latitude (β) of a star, etc., is its angular distance (from 0° to 90°) north (counted positive) or south (counted negative) of the ecliptic; it is measured along the great circle through the body and the poles of the ecliptic. The celestial longitude (λ) of a body is its angular distance (from 0° to 360°) from the vernal equinox, measured eastwards along the ecliptic to the intersection of the body's circle of longitude; it is measured in the same direction as the sun's apparent annual motion. Although observations are taken from the earth's surface the coordinates should strictly be geocentric and, tabulated as such, are universally applicable. A slight correction is therefore applied to convert surface (topocentric) observations to geocentric values.

The ecliptic system is older but less used than the *equatorial and *horizontal coordinate systems. It is sometimes used to give the positions of the sun, moon, and planets. *See also* heliocentric coordinate systems.

ecliptic limits. *See* eclipse.

E-corona. The part of the solar *corona that consists of relatively slow-moving ions and exhibits an *emission spectrum superimposed on the *continuum of the *K-corona. The emission lines have shown that the ions of the E-corona are very highly ionized, some species having lost up to 15 electrons. At visible wave-

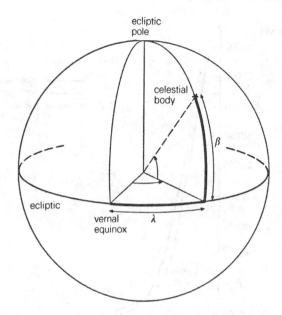

Ecliptic coordinate system

lengths, only *forbidden lines are present; the most prominent of these are the 'red' line at 637.4 nanometres (due to the presence of Fe X), the 'green' line at 530.3 nm (due to Fe XIV), and the 'yellow' line at 569.4 nm (due to Ca XV).

ecosphere. The zone of space around the sun or some other star in which a planet would experience external conditions that are not incompatible with the existence of life. Prevailing conditions, as of atmospheric content or pressure, might however prove too hostile.

effective area. *See* array.

effective temperature. Symbol: T_{eff}, T_e. The surface temperature of a star, expressed as the temperature of a *black body (i.e. a perfect radiator) having the same radius as the star and radiating the same total amount of energy (E) per unit area per second. A star is thus being considered a black body and its effective temperature is thus given by *Stefan's law:

$$T_{eff}^4 = E/\sigma = L/4\pi R^2 \sigma$$

σ is Stefan's constant and L and R are the star's luminosity and radius. Although accurate determinations of effective temperature are difficult, each *spectral type has a character-

Effective temperature of different spectral types

istic T_{eff}, which ranges from about 2500 to 40 000 kelvin (see graph overleaf). The sun's effective temperature is 5800 K.

Radiation temperature, T_r, is equivalent to the effective temperature measured not over the whole spectral range but over a narrow portion of the spectrum, for example at visible wavelengths, which gives the *optical temperature*.

Einstein-de Sitter universe. *See* cosmological models.

Einstein Observatory. *See* HEAO satellites.

Einstein shift. *See* gravitational redshift.

ejecta. Material excavated during the formation of a *crater or *basin. *See also* ejecta blanket; secondary craters.

ejecta blanket. A deposit of *ejecta from a *crater or *basin, the thickness of which decreases with distance from the crater or basin rim. The surface may exhibit transverse *dunes,* radial *furrows,* and *secondary craters.

Elara. A small satellite of Jupiter, discovered in 1905. *See* Jupiter's satellites; Table 2, backmatter.

E-layer (E-region). *See* ionosphere.

ELDO. *Abbrev. for* European Launcher Development Organization. *See* European Space Agency.

electromagnetic radiation. A flow of energy that is produced when electrically charged bodies, such as electrons, are accelerated. Light is the most familiar form of such radiation. It can equally be considered as a wave motion or as a stream of particles.

Electromagnetic waves consist of oscillating electric and magnetic fields that travel through space or through air, etc., carrying energy. The two fields vibrate in planes at right angles to each other and to the direction of wave motion, and are self-propagating. An electromagnetic wave thus requires no medium for its propagation (unlike sound waves). It travels through a vacuum at the *speed of light, c, which is a fundamental constant equal to about 3×10^5 km s^{-1}. The speed is slightly reduced on entering a medium, such as air or glass. Like other periodic waves, electromagnetic waves have a *wavelength λ and a *frequency ν, which are related by $\lambda\nu = c$.

Electromagnetic radiation can manifest itself as both waves and particles, or as either, and thus has a dual nature. Although *reflection, *interference, and *polarization can be explained in terms of wave motion, a particle-like nature must often be invoked, as when radiation interacts with atoms or molecules. A particulate nature was originally proposed by Newton but in its present form the concept is part of *quantum theory. Light and other kinds of radiation are considered as streams of particles travelling at the speed of light, c, and having zero mass; the particles are tiny packets of energy, called *photons. Energy is absorbed or emitted by matter in discrete amounts, i.e. in the form of photons. The photon is thus a quantum of electromagnetic radiation. The energy, E, of the photons is related to the frequency ν, of the radiation by $E = h\nu$, where h is the *Planck constant.

The range of frequencies (or wavelengths) of electromagnetic radiation is known as the *electromagnetic spectrum* (see illustration). The spectrum can be divided into various regions, which are not sharply delineated. These regions range from low-frequency low-energy *radio waves through *infrared radiation,

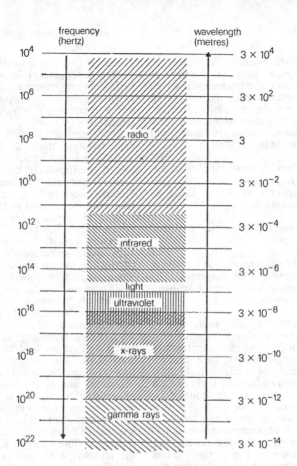

Electromagnetic spectrum

*light, *ultraviolet radiation, x-rays, to high-frequency high-energy *gamma rays.

electromagnetic spectrum. *See* electromagnetic radiation.

electromagnetic wave. *See* electromagnetic radiation.

electron. Symbol: e. A stable *elementary particle (a lepton) that has a negative *charge of 1.6022×10^{-19} coulomb, a mass of 9.1096×10^{-28} gram, and *spin ½. Electrons are constituents of all *atoms, moving around the central nucleus. They can also exist independently. The *antiparticle of the electron is the *positron*, e⁺ or ē.

electronographic camera. A highly sensitive electronic device that can be used for recording astronomical images. It is an *image tube in which the beam of electrons liberated from the photocathode is accelerated by an electric field and focused by a magnetic field onto a film coated with an *electronographic emulsion*. This is a very fine grain high-resolution 'nuclear' photographic emulsion in which electrons are recorded directly: every high-energy electron leaves a developable track so that an image can be produced by a process similar to that for a photograph. The resulting *electronograph* looks similar to a photograph with the important difference that the density at any point on the electronograph is proportional to the intensity at the corresponding point in the optical image through almost the whole intensity range. The device thus has a highly linear response. Photometric information can be obtained directly by measuring the density of the image. The device also has a high *quantum efficiency.

Several methods have been devised to prevent gases evolved from the emulsion from poisoning the photocathode. In a recent device photocathode and film are separated by a mica window a few micrometres thick, which allows passage of about 75% of the high-energy electrons but is impervious to gases. The film is held in close contact with the mica.

electronvolt. Symbol: eV. A unit of energy equal to the energy acquired by an electron falling freely through a potential difference of one volt. It is equal to 1.6022×10^{-19} joule. High-energy *electromagnetic radiation is often referred to in terms of the energy of its photons: a photon energy of 100 eV is equivalent to a radiation frequency of 2.418×10^{16} hertz. The energies of *elementary particles are usually quoted in eV.

element. Any of a large number of substances, including *hydrogen, *helium, carbon, nitrogen, oxygen, and iron, that consist entirely of *atoms of the same *atomic number, i.e. with the same number of protons in their nuclei. The atoms are not all identical however: *isotopes of an element can occur with different numbers of neutrons in their nuclei. When arranged in order of increasing atomic number along a series of horizontal rows, elements with similar chemical and physical properties fall into groups. The similarities in properties arise from similarities in the electron arrangements within the atom. This table of elements is called the periodic table. Over 100 elements are now known, more than 90 of which have stable forms and occur naturally on earth.

The creation of the elements

and the means by which they become distributed throughout the universe are major areas of study in astronomy (*see* nucleosynthesis). The abundance of each element in the universe, i.e. its *cosmic abundance, depends on the method of its synthesis—by nuclear reactions in stars, by *cosmic-ray collisions, etc.—and on its lifetime in its immediate surroundings and its long-term stability.

elementary particles: The simplest forms of matter. Each particle is characterized by a set of fundamental properties including *rest mass, *charge, and *spin. Only the *electron, *proton, *neutrino, and *photon are completely stable particles; the *neutron is stable when bound in a nucleus. Other particles, such as the *muon and the *pions, are produced (along with the stable particles) by a variety of high-energy reactions and decay into the stable particles.

Elementary particles are usually classified by the types of interaction in which they can take part. The *hadrons,* which include the proton, neutron, and the pions, can participate in strong interactions, weak interactions, and, if charged, in electromagnetic interactions; the *leptons,* i.e. the electron, muon, tau lepton, and their associated neutrinos, cannot participate in strong interactions but are restricted either to the other two or, for neutrinos, to the weak interaction only. *See also* antiparticles; quarks.

elements, abundance. *See* cosmic abundance.

elements of an orbit. *See* orbital elements.

ellipse. A type of *conic section with an eccentricity less than one. The longest line that can be drawn through the centre of an ellipse is the major axis while the shortest line is the minor axis. The two axes are at right angles. There are two foci, which lie on the major axis and are symmetrically positioned on opposite sides of the centre. The sum of the distances from the foci to a point moving round the ellipse is constant and equal to the length of the major axis. An orbiting body moves in an ellipse with the primary at one of the foci. *See also* orbit; Kepler's laws.

ellipsoid. A surface or solid whose plane sections are circles or ellipses. An ellipse rotated about its major or minor axis is a particular type of ellipsoid, called a *prolate spheroid* (major axis) or an *oblate spheroid* (minor axis).

ellipsoidal reflector. A reflector whose surface is part of an ellipsoid. An object situated at one focus will be imaged, after reflection, at the other focus.

elliptical galaxy. *See* galaxies; Hubble classification.

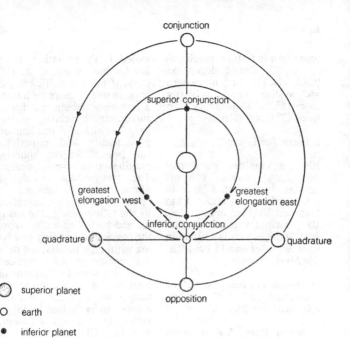

Elongations of superior and inferior planet

elliptically polarized wave. *See* polarization.

ellipticity. *See* oblateness.

El Nath (β Tau). A bluish-white giant that is the second-brightest star in the constellation Taurus. It was originally part of Auriga, as Gamma Aurigae. m_v: 1.65; M_v: -2.9; spectral type: B7 III; distance: 80 pc.

elongation. The angular distance between the sun and a planet or the moon, i.e. the angle sun-earth-planet/moon. It is the difference in celestial longitude (*see* ecliptic coordinate system) between the sun and the object. An elongation of 0° is called *conjunction, one of 180° is *opposition, and one of 90° is *quadrature (see illustration). When an inferior planet follows the sun in its daily motion, appearing east of the sun in the evening, it is in *eastern elongation*. When it precedes the sun, appearing west of the sun in the morning, it is in *western elongation*. The inferior planets, which cannot

Eltanin

come to quadrature, reach positions of greatest elongation (GE). The GE for both eastern and western elongation varies from 18° to 28° (Mercury) and from 47° to 48° (Venus).

Eltanin. *See* Draco.

Elysium Planitia. Mars' second main centre of volcanic activity, located on a bulge in the Martian crust about 3 km high and 2000 km in diameter. Its principal volcano is *Elysium Mons,* which is 250 km in base diameter and 15 km high. *See* Mars, volcanoes.

emersion. The reappearance of a celestial body following its eclipse or occultation.

emission lines. *See* emission spectrum.

emission measure. *See* thermal measure.

emission nebula. A region of hot interstellar gas and dust that shines by its own light. The gas atoms in the cloud are ionized—stripped of one or more electrons. The electrons liberated by the ionization process are free to move about the nebula. The light, and other radiation, emitted by the nebula results from subsequent interactions between ions and free electrons.

There are three classes of emission nebula, all of extremely low gas density. In *HII (ionized hydrogen) regions, such as the *Orion nebula, the gas is ionized by ultraviolet (UV) radiation from nearby hot young O and B stars; if there is sufficient gas close to one or more such stars a luminous nebula will form, surrounded by cooler neutral gas. The size is dependent on gas density and ultraviolet flux (*see* Strömgren sphere), ranging up to several parsecs. The temperature is always about 10 000 kelvin. In *planetary nebulae, typified by the *Ring nebula, the gas is ionized by UV radiation from the remnant of a dying star lying within the nebula; the nebula comprises gas expelled from this star. In *supernova remnants the ionization mechanisms are very complex: the atoms can be ionized by ultraviolet *synchrotron radiation, as in the *Crab nebula, by interactions following the collision of an expanding remnant with surrounding interstellar gas atoms, as in the *Cygnus Loop, or possibly by other processes.

Emission nebulae contain primarily free electrons and hydrogen ions, together with some helium ions and trace amounts of oxygen, carbon, nitrogen and other elements. These components can interact in various ways. Firstly, an electron can be recaptured by an ion, usually a hydrogen ion. This *recombination leads to the emission of radio, infrared, and optical radiation. Secondly, an electron can collide with an ion, such as ionized oxygen, without being captured. Instead, energy is transferred from electron to ion, which becomes excited into an *energy level just above the ground state. The excited ion returns to its normal energy state by emitting light: at higher densities these transitions would be 'forbidden' (see forbidden lines). Thirdly, free electrons can be decelerated when they pass near an ion, energy being radiated away as weak *bremsstrahlung—mainly radio waves.

The spectacular colours, predominantly red and green, of an emission nebula are due to recombination and excitation. The most intense visible radiation from hydrogen is the red hydrogen alpha emission; red light is also emitted during a forbidden transition of singly ionized nitrogen. The strongest emission, however, results from a forbidden transition of doubly ionized oxygen (O^{++} or O III), the emission being green; the O III ions can only be produced by highly energetic UV photons. See also nebula.

emission spectrum. A *spectrum formed by emission of *electromagnetic radiation by matter. Energy has first to be supplied, either as heat (high temperature) or by some other mechanism such as absorption of electromagnetic radiation by the matter or by impact of electrons. The energy raises the atoms or molecules to higher *energy levels. In a simple example, the electron in a hydrogen atom may 'jump' from its normal orbit to one further from the nucleus (see hydrogen spectrum). The emission spectrum of hydrogen, or any other emitting material, is formed by transitions from such excited states to lower energy states, the excess energy appearing in the form of photons with characteristic frequencies.

The emission spectrum therefore consists of a specific pattern of narrow peaks—*emission lines*—which occur at these frequencies. These lines may be superimposed on a *continuum, but not necessarily so. Emission lines can occur in stellar spectra if, for example, the star is surrounded by a hot shell of gas. See also excitation; absorption spectrum.

emissivity. Symbol: ε. A measure of a body's ability to radiate *electromagnetic radiation as compared to that of a perfect radiator—a *black body—at the same temperature.

Enceladus. A satellite of Saturn, discovered in 1789. See Table 2, backmatter.

Encke's comet. The most observed comet, having been seen at nearly 50 apparitions. The comet has a period of 3.30 years—one of the shortest—and is thought to be the parent body of the Taurid *meteor stream. The period of Encke is decreasing by 2.5 hours per orbital revolution. This is caused by a jet effect from its rotating nucleus, which is nonthermally symmetrical. The brightness of Encke has hardly decreased during the 165 years of observation. Since the orbit of Encke is extremely well known and a considerable bank of scientific data has been established, Encke is one of the prime targets for a spacecraft investigation of a comet.

Encke's division. *See* Saturn's rings.

encounter theories (catastrophic theories). Theories that propose that the planets were formed from material ejected from the sun or a companion star during an encounter with another object. Although the first such theory, by G.L.L. de Buffon in 1745, postulated that a comet collided with the sun, most later theories have invoked an approach or collision involving another star.

The *planetesimal theory* (1901-05) of T.C. Chamberlin and F.R. Moulton suggested that a passing star caused the sun to eject filaments of material. These condensed into *planetesimals from which the planets formed by *accretion.

Tidal theories (1916-18) by Harold Jeffreys and James Jeans suggested that a star grazed the sun, drawing out into solar orbit a cigar-shaped filament of material that fragmented to form the planets: the larger planets—Jupiter and Saturn—condensed from the thicker central regions of the filament.

Variations on these ideas postulate that the encounter took place between a passing star and a binary companion star to the sun. This overcomes one objection to most encounter theories: that material ejected from the sun could not enter near-circular orbits at the observed planetary distances. However a remaining major problem with all these ideas is that most of the material drawn out from the sun or its postulated companion would have fallen back; any material ejected at stellar temperatures would expand violently and dissipate into space before it could form planets.

A modification of the Jeans-Jeffreys theory, the *capture theory* (1964) of M.M. Woolfson, attempts to overcome these problems. It proposes that a filament of matter was drawn off a relatively cool tenuous protostar moving past and tidally distorted by the condensed and more massive sun. Condensations of protoplanets formed from the fragmented filament, six of which were captured by the sun. A gradual rounding-off of the initial elliptical orbits and a major collision between the two innermost bodies led to the present distribution of solar-system bodies.

See also nebular hypothesis; solar system, origin.

energy. Symbol: E. A measure of the capacity of a body or system for doing work, i.e. for changing the state of another body or system. The SI unit of energy is the joule; the erg (10^{-7} joule) is also used.

There are various forms of energy including heat, mechanical, electrical, nuclear, and radiant energy, all of which are interconvertible in the presence of matter. Mass is also regarded as a form of energy. In any closed system, the total energy, including mass, is always constant. *See also* conservation of mass-energy; nuclear fusion; nuclear fission.

energy level. One of a number of discrete energies that, according to quantum mechanics, is associated with an atomic or molecular system. For instance, a hydrogen atom, which consists of a proton and an orbiting electron, has an energy due to electrostatic interaction between the electron and proton and to the motion of the electron. This energy can only have certain fixed values, which correspond to different orbits of the electron, and these constitute the *electronic energy levels* of the atom. The state of lowest energy is the *ground state* of the atom and higher states (in which the electron is further from the nucleus) are known as *excited states*. Neither the hydrogen atom nor any other type of atom can take up energy continuously but can only 'jump' from one energy state to another: the energy is said to be *quantized*. This behaviour is quite general, applying to energies of vibration and rotation in molecules and to energy difference due to interaction of magnetic moments of the electrons.

energy transport. The flow of heat energy from a hot to a cooler region by means of radiation, convection, or conduction. In stars the energy flows from the hot interior to the surface, where it is radiated away into space. The most important means of energy transport in most stars is *radiative transport,* in which high-energy photons lose energy to the hot plasma through scattering, mainly by free electrons, and by absorption through photoionization of ions; this occurs in the *radiative zone* of the star. (The equation of radiative transport is given at *stellar structure*.) Convection is another mode of energy transport in which collections of hot fluid in the *convective zone of a star rise towards the surface, release their heat energy, and sink again to pick up more energy. Convection is a highly efficient process but the conditions must be right for its onset. Conduction is negligible in most stars but becomes important in very dense degenerate stars, such as white dwarfs.

English mounting. *Another name for* yoke mounting. *See* equatorial mounting.

ephemeris. 1. A work published regularly, usually annually, in which are tabulated the daily predicted positions of the sun, moon, and planets for that period, together with information relating to certain stars, times of eclipses, etc. It is used in astronomical observation and navigation. *See also* Astronomical Ephemeris.

2. (pl. form: **ephemerides**). A series of predicted geocentric positions of a celestial object at constant intervals of time.

ephemeris second. *See* ephemeris time.

ephemeris time (ET). Symbol: t_E. A time system for which a constant rate has been defined and that depends for its determination on the laws of dynamics. In comparison *sidereal time and *universal time do not have an invariable rate: they depend on the earth's rotational period, which shows small irregular and unpredictable variations. ET is reckoned from the instant, near the beginning of 1900, when the sun's mean longitude was 279° 41′ 48″.04, at which instant the measure of ephemeris time was 1900 Jan 0d 12h precisely. The primary unit of ET is the *tropical year at this epoch of 1900 Jan 0.5 ET, which contains a calculable number of seconds: the *ephemeris second* is the fraction
1/31 556 925.9747
of the tropical year 1900 Jan 0.5 ET. The ephemeris second was adopted in 1957 as the fundamental invariable unit of time until replaced in 1964 by the second of *atomic time. The ephemeris day contains 86 400 ephemeris seconds.

ET can be found by comparison of the observed positions of sun, planets, and primarily the moon with their corresponding ephemerides (predicted geocentric positions of a body on the celestial sphere with constant time separation). In practice the observed position is recorded in *universal time and a correction, ΔT, is added to give ET. ΔT is not constant but is increasing in an irregular manner; it can now be determined over a short period with reasonable accuracy. Between 1970-79, ΔT has increased from 40 seconds to about 50 seconds.

ET was adopted in 1958 as the fundamental invariable measure of time but was replaced in 1972 by international *atomic time (TAI): ET is equal to TAI + 32.2s. The two rates agree to within two parts in one thousand million.

epicycle. *See* Ptolemaic system.

epoch. A precise instant that is used as a fixed reference datum, especially for stellar coordinates and *orbital elements. For example, with coordinate systems based on the vernal *equinox stellar coordinates are continuously changing, primarily as a result of the *precession of the equinoxes. Coordinates must therefore refer to a particular epoch, which can be the time of an observation, the beginning of the year in which a series of observations of an object was made, or the beginning of a half century.

Epsilon (ε). The fifth letter of the Greek alphabet, used in *stellar nomenclature usually to designate the fifth-brightest star in a constellation.

Epsilon Aurigae. An *eclipsing binary star about one kiloparsec distant in the constellation Auriga. It consists of an extremely luminous A8 Ia supergiant about 30 times the mass of the sun with an invisible secondary star. The binary period is about 27 years. The invisible secondary is likely to be a collapsed star, possibly a *black hole.

Epsilon Eridani. One of the nearest stars to the sun. There is evidence of one or more associated planets but the orbit(s) have not been determined. *See also* Table 7, backmatter; Ozma project.

Epsilon Indi. *See* Table 7 (nearest stars), backmatter.

equation of light. *See* light-time.

equation of the centre. A major *inequality in the moon's motion that is a direct result of the large *eccentricity (0.0549) of the moon's orbit and hence of its varying orbital velocity (*see* Kepler's laws).

equation of the equinoxes. *See* sidereal time.

equation of time. The amount that must be added to the *mean solar time to obtain the *apparent solar time, i.e. it is the difference in time as measured by a sundial and by a clock. The equation of time varies through the year; it has two maxima and two minima

Yearly variation in equation of time

and is zero on four dates: April 15/16, June 14/15, Sept 1/2 and Dec 25/26 (see illustration). A positive value indicates that apparent time is ahead of mean time; the greatest positive value is 16.4 minutes, the greatest negative value being 14.2 minutes. The curve is the sum of two components, each reflecting a nonuniformity in the apparent motion of the sun: one component arises from the ellipticity of the earth's orbit, the other from the inclination of the ecliptic to the celestial equator (*see* mean sun).

equator. 1. The great circle on the surface of a planet or star that lies in a plane—the *equatorial plane*—that passes through the centre of the body and is perpendicular to the axis of rotation.
2. *Short for* celestial equator.

equatorial. *Short for* equatorial mounting.

equatorial bulge. *See* geoid.

equatorial coordinate system. The most widely used astronomical *coordinate system in which the fundamental reference circle is the celestial equator and the zero point is the vernal *equinox, ♈ (see illustration). The coordinates are *right ascension (α) and *declination (δ), which are measured along directions equivalent to those of terrestrial longitude and latitude, respectively. *Hour angle (t) is sometimes used instead of right ascension and *north polar distance instead of declination. Because of the slow westerly drift of the vernal equinox around the ecliptic, the coordinates right ascension and declination are normally referred to the *mean equinox for a standard epoch, which in present use is usually 1950.0.

equatorial horizontal parallax. *See* diurnal parallax.

equatorial mounting (equatorial). A telescope *mounting in which one axis (the polar axis) is parallel to the earth's axis, while the second axis (the declination axis) is at right angles to it. Its great advantage is that when the telescope is clamped in declination, and the polar axis is driven to turn once in 24 hours in the opposite direction to the earth's rotation, then any star will remain stationary in the *field of view. Almost every large telescope is equatorially mounted.

There are several types of equatorial mountings, some being shown in the illustration. In the *fork mounting* the telescope swings in declination about an axis carried on two prongs of a fork; the fork itself rotates about a shaft that is the polar axis. In the *yoke* (or *English*) *mounting* the polar axis is in the form of a long frame with a bearing at each end. The telescope swings in declination about an axis between the sides of the frame. This mounting is simple, very rigid, and needs no counter-

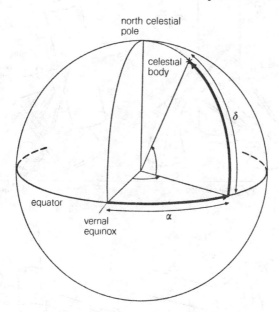

Equatorial coordinate system

poise weights; the polar region is however inaccessible. The *horseshoe mounting* is a modification of the fork and yoke mountings: the upper end of the polar-axis frame is made into a horseshoe shape to accommodate the telescope tube; the polar region may then be observed. Many of the giant reflectors use this exceptionally stable mounting. In the *German mounting* the declination axis is carried as a tee on the top end of the polar axis. The telescope is carried on one end of the declination axis with a counterpoise on the other end.

The drive mechanism for the telescope is usually called a *drive clock;* in modern telescopes it is often an electric motor controlled by a variable frequency generator. A telescope is usually driven, about its polar axis, at the *sidereal rate* so that one rotation is completed in 23h 56m 4s. Small corrections are required in drive rates because of atmospheric refraction and when following the moon or planets, which move relative to the stars. These are obtained by varying the frequency supplied to the drive motor. *Compare* altazimuth mounting.

Equatorial mountings

equatorial plane. See equator.

equator of illumination. The plane of symmetry of a lunar or planetary *phase.

equinoctial colure. See colures.

equinoctial points. See equinoxes.

equinoxes. 1. (equinoctial points). The two points on the celestial sphere at which the *ecliptic intersects the *celestial equator. They are thus the two points at which the sun in its apparent annual motion crosses the celestial equator. The sun crosses from south to north of the equator at the *vernal equinox* (or *spring equinox*), symbol: ♈ which lies at present in the constellation Pisces. The sun crosses from north to south of the equator at the *autumnal equinox*, symbol: ♎, which at present lies in Virgo. The vernal equinox is the zero point for both the *equatorial and *ecliptic coordinate systems. The equinoxes are not fixed in position but are moving westwards around the ecliptic as a result of *precession of the earth's axis; the advance is about 50 arc seconds per year. *See also* first point of Aries; true equinox; mean equinox.

2. The two instants at which the centre of the sun in its apparent annual motion crosses the celestial equator. In the northern hemisphere the *vernal equinox* occurs on about March 21 and the *autumnal equinox* on about Sept. 23. The dates are reversed in the southern hemisphere. On the days of the equinoxes the hours of daylight and of darkness are equal.

Compare solstices.

equipotential surfaces. Imaginary surfaces surrounding a star or star system over which the gravitational field is constant. For a single star the surfaces are spherical and may be considered as the contours of the potential well of the star. In a close *binary star the equipotential surfaces of the

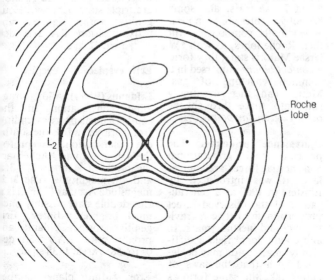

Equipotential surfaces of close binary star

components interact to become hourglass-shaped (see illustration). The surfaces 'meet' at the *inner Lagrangian point*, L_1, where the net gravitational force of each star on a small body vanishes; the contour line through this point defines the two *Roche lobes*.

When both components of the binary star are contained well within their Roche lobes they form a *detached binary system*. If one star has expanded so as to fill its Roche lobe it can only continue to expand by the escape of matter through the inner Lagrangian point. This stream of gas will then enter an orbit about or collide with the smaller component. The system is then a *semidetached binary*: *dwarf novae, *Beta Lyrae stars, and some *Algol variables are examples. When both components fill their Roche lobes, as with *W Ursae Majoris stars, they form a *contact binary* immersed in a common envelope of gas. Matter can then eventually spill into space through the *outer Lagrangian point*, L_2.

equivalence principle. The principle that gravitational and inertial forces are equivalent. It was introduced by Einstein in 1907 as a generalization of the observed direct proportionality between gravitational and inertial *mass. In any gravitational field a sufficiently small region may be defined such that the laws of nature take the same form as in an accelerated system in the absence of gravity. *See* relativity, general theory.

equivalent focal length (e.f.l.) The property of a compound lens, such as an eyepiece, that can be used in the same way as the focal length of a single thin lens in, say, the determination of magnifying power. For a two-lens system the e.f.l. is given by

$$f_1 f_2 / (f_1 + f_2 - d)$$

where f_1 and f_2 are the focal lengths of the components and d their separation.

Equuleus (Little Horse). A very small inconspicuous constellation in the northern hemisphere near Pegasus, the brightest stars being of 4th magnitude. Delta and Epsilon are triple stars. Abbrev.: Equ; genitive form: Equulei; approx. position: RA 21h, dec +7°; area: 72 sq deg.

Erfle eyepiece. *See* eyepiece.

Eridanus (River). A long straggling constellation in the southern hemisphere extending from Orion towards the southern polar region. Except for the 1st-magnitude *Achernar the brightest stars are of 3rd and 4th magnitude. The 3rd magnitude Acamar (Theta) is a fine double star, possibly once much brighter. Abbrev.: Eri; genitive form: Eridani; approx. position: RA 1.5-5h, dec 0° to -58°; area: 1138 sq deg.

Eros. *Minor planet number 433 that was discovered in

1898 and is a member of the *Amor group. It passed within 23 million km of the earth in 1975. It appears to be a stony spindle-shaped object measuring 7 × 16 × 35 km. *See* Table 3, backmatter.

eruptive variable. *See* cataclysmic variable.

ESA. *Abbrev. for* European Space Agency.

escape velocity. The minimum velocity required for an object to enter a parabolic trajectory around a massive body and hence escape from its vicinity. The object will keep moving away from the body although it travels at decreasingly lower speeds as the massive body's gravitational attraction continues to slow it down. If the object does not attain escape velocity it will enter an elliptical orbit around the body. If escape velocity is exceeded the object will follow a hyperbolic path.

The escape velocity at a distance r from the centre of a massive body, mass m, is given by $\sqrt{(2Gm/r)}$, where G is the gravitational constant. The escape velocity from the earth's surface is about 11.2 km/s, from the moon 2.4 km/s, and from the sun 617.7 km/s. For a massive body to retain an atmosphere the average velocity of the gas molecules must be well below escape velocity.

ESO. *Abbrev. for* European Southern Observatory.

ESO/SRC Southern Sky Survey. *See* Southern Sky Survey.

ESRO. *Abbrev. for* European Space Research Organization. *See* European Space Agency.

ET. *Abbrev. for* ephemeris time.

Eta Aquarids. *See* Aquarids.

Eta Carinae (η Car). A very remote cool (250 K) star in the constellation Carina. It is an irregular nova-like variable with great variation in magnitude at very irregular periods. From 1835-45 it outshone every star except Sirius, reaching a magnitude of -0.8 in 1843; it has now faded to 7.7. It is however the brightest infrared source at a wavelength of 20 micrometres. It is thought that it could be a slow supernova of large mass or a *Hubble-Sandage variable. It is surrounded by a dense expanding cloud of gas and dust—the bright *Eta Carinae nebula* (NGC 3372)—on which is superimposed the dark *Keyhole nebula*. Distance: about 2000 pc.

ether. An extremely elastic medium of negligible density that was thought in the 19th century to permeate all space and thus provided a medium through which light and other electromagnetic waves from celestial bodies could travel. (At that time all waves were considered to require a medium for their propagation).

The speed of travel of the electromagnetic waves, i.e. the speed of light, was assumed to be fixed with respect to the ether. The earth, however, should move with respect to the ether and so, from considerations of relative motion, the speed of light should vary very slightly when measured in different directions.

Experiments were set up in the late 19th century to detect and measure the motion of the earth relative to the ether. The most notable experiment was carried out in 1887 by A.A. Michelson and E.W. Morley (the *Michelson-Morley experiment*). No such motion was found, indicating both that the ether did not exist and that electromagnetic waves did not require any medium for their propagation. The concept of the ether was thus discarded. The result of these experiments was later explained by Einstein's special theory of *relativity (1905) in which it was stated that the speed of light was independent of the velocity of the observer and was therefore invariant.

Europa. The second and smallest of the *Galilean satellites of Jupiter, discovered in 1610. *See* Jupiter's satellites; Table 2, backmatter.

European Southern Observatory (ESO). An observatory at La Silla, near La Serena, Chile, at an altitude of 2400 metres; even in normal conditions the *seeing is exceptional due to the clear dry stable climate. It is a European intergovernmental organization (members: France, Germany, Netherlands, Sweden, Belgium, and Denmark) with its headquarters at Darching, near Munich, where all activities have been concentrated since 1979. The chief instrument at the Observatory is the 3.6 metre (142 inch) reflecting telescope that commenced operation in 1976. It works at focal ratios of f/3, f/8, and f/30 at the prime, Cassegrain, and coudé foci, respectively. The limiting magnitude should be 24 magnitudes or more. The observatory also has a one metre *Schmidt telescope, which has been involved in the preparation of the *Southern Sky Survey.

European Space Agency (ESA). An organization for coordinating, promoting, and funding Europe's space programme. It was formed in 1975 from the merging of the *European Space Research Organization* (*ESRO*) and the *European Launcher Development Organization* (*ELDO*), both formed in 1962. There are at present 11 full-member countries, including the UK. The head office is in Paris and the technical centre—the European Space Research and Technology Centre (ESTEC)—is at Noordwijk in the Netherlands; the European Space Operations Centre (ESOC) is at Darmstadt in Germany. The ESA has a network of ground stations for data reception from satellites at Odenwald in West Germany, Redu in Belgium, Fucino in Italy, and Villafranca de Castillo in Spain.

Several ESA satellites have been launched (by NASA), including *COS-B and *GEOS. ESA is building its own launch vehicle, *Ariane, in addition to its contribution—*Spacelab—to the joint ESA/NASA Spacelab/Shuttle programme. It is also in the process of testing equipment on OTS-2 (Orbital Test Satellite), which was launched, May 1978, as a forerunner of the European communications satellite (ECS) system.

EUV (extreme ultraviolet). *Another name for* XUV. *See* XUV astronomy.

EVA. *Abbrev. for* extravehicular activity. Operations, including repair and maintenance, undertaken by an astronaut or cosmonaut outside his spacecraft. EVA was first achieved in 1965 by the Russians. *See* Voskhod; Gemini project.

evection. The periodic (31.807 day) *inequality in the motion of the moon that may amount to a displacement in longitude of up to $1°16'20''.4$. It arises from changes in the *eccentricity of the moon's orbit (0.0432 to 0.0666) that are brought about by solar attraction. Evection was discovered by Hipparchus. *See also* variation; parallactic inequality; equation of the centre; annual equation.

evening star. An informal name for *Venus when it lies east of the sun and is a brilliant object in the western sky after sunset. Such periods of visibility follow superior *conjunction and precede inferior conjunction. The term is sometimes applied loosely to evening visibility of Mercury or other planets.

event horizon. The boundary of a *black hole: a spherical 'surface' having a radius equal to the *Schwarzschild radius for a body of a given mass. The event horizon marks the critical limit where the *escape velocity of a collapsing body becomes equal to the speed of light and hence no information can reach an external observer. Calculations predict that a body shrinking within its event horizon—or indeed any body that falls inside the event horizon of another—inevitably contracts to a *singularity at the precise centre of the black hole.

Evershed effect. A radial flow of material in the penumbra of a *sunspot, outward from the edge of the umbra, with a velocity of about 2 km s^{-1}. It forms part of a pattern of circulation determined by the configuration of the intense localized magnetic field. The gas ascends at or just beyond the outer edge of the penumbra, doubles back on itself in the *chromosphere, and then descends into the umbra (for the cycle to be repeated).

evolution, stellar. *See* stellar evolution.

excitation. A process in which an electron bound to an atom is given sufficient energy to transfer from a lower to a higher orbit but not to escape from the atom. The atom is then in an *excited state*. The atom may be excited in two ways. In collisional excitation a particle, such as a free electron, collides with the atom and transfers some of its energy to it. If this energy corresponds exactly to the energy difference between two *energy levels, the atom becomes excited. The atom rapidly returns to a lower energy level, usually its ground state, possibly passing through several intermediate levels on the way. Photons are emitted during these transitions, so that an emission line spectrum results. The photon energies are equal to the energy differences between the levels involved.

In radiative excitation, a photon of radiation is absorbed by the atom. The photon energy must correspond exactly to the energy difference between two energy levels. This process produces an absorption line in a spectrum. The atom rapidly returns to its ground state, emitting photons in the process (as before). Since the photons can be emitted in any direction and can have different (lower) energies than the absorbed photon, the absorption line dominates the spectrum. *See also* Boltzmann equation; ionization.

excited state. *See* excitation; energy level.

exit pupil (Ramsden disc; Ramsden circle). The image of the objective or primary mirror of a telescope formed on the eye side of the *eyepiece. It is seen as a bright disc when the telescope is pointed at any extended bright surface. A convenient way to determine the magnification is to divide the telescope *aperture by the diameter of the Ramsden disc.

exobiology. The theories concerning 'living' systems that may exist on other planets or their satellites in the solar system or on planetary systems of other stars, the methods of detecting them, and their possible study. The probability of this *extraterrestrial life* occurring at some time on some planetary system of some star is now considered fairly high: it depends on the fraction of the possible 10^{20} stars in the universe—10^{11} in our Galaxy—that have planetary systems and on the fraction of these planetary systems that could support, on one or more members, these 'living' organisms.

The degree to which exobiology may differ from terrestrial biology is a matter of dispute. There are no universal laws of biology as there are assumed to be for physics; organisms could evolve on planets or satellites moderately different from earth, in an unrecognizable form. There is no absolute definition of life but it is believed that any living material is built up from the common elements carbon, hydrogen, oxygen, and nitrogen and that the presence of water is probably necessary for sustaining life. In addition the environment should have an atmosphere to act as a shield from high-energy cosmic and stellar radiation, and to aid respiration where necessary, and the temperature should not be too adverse.

Stars are being formed in the Galaxy at a mean rate of about 10 per year. The course and rate of evolution of living organisms probably depends on the type of star and its stability, as well as the planetary environment. The conditions may not be appropriate for the origin and evolution of living organisms in this present era. If they can form, however, many scientists believe that these organisms will inevitably become more developed. The number of life forms that might at this moment be 'intelligent' in a human sense and also able to communicate this intelligence is very uncertain: one group of scientists has suggested that at present there could be as many as one million advanced civilizations in our Galaxy; other estimates have been very much lower. Communications, however, are severely limited by distance: signals from a star 100 light-years (30.7 parsecs) away take 100 years to reach earth.

There is to date no direct evidence for extraterrestrial life. Searches for microorganisms in the soils of the moon and Mars have been unsuccessful, although the *Viking results on the Martian soil are still under debate. Serious searches for intelligent life beyond the earth have also been made. Wavelengths ranging from radio to ultraviolet have been used with targets varying from nearby stars to other galaxies. The pioneering programme was the *Ozma Project of 1960 in which Frank Drake listened fruitlessly to the radio waves from the two nearest and most sunlike stars—Tau Ceti and Epsilon Eridani. Later experiments, using more sophisticated equipment and targeting more stars, have also proved unsuccessful. Most have been conducted at the 21 cm hydrogen-line frequency, although the 1.35 cm line of water has also been used. Radio telescopes are now being employed to listen to galaxies in the *Local Group and more distant galaxies. The rapidly increasing level of man-made radio interference will make it progressively harder to detect weak signals from distant sources. Searches for extraterrestrial intelligence (SETI) are thus deemed by many scientists to be both feasible and timely.

EXOSAT. A proposed orbital

observatory for measurements of x-ray sources that is being developed for the European Space Agency. ESA hopes to launch the satellite in 1981 with its *Ariane launcher. It will be placed in a near-polar highly eccentric orbit (apogee: 200 000 km, perigee: 300 km) and will be able to observe x-ray sources by *lunar occultation techniques as well as in the arbitrary pointing mode.

exosphere. See atmospheric layers.

expanding universe. The observation that radiation from distant galaxies is redshifted leads to the conclusion that all galaxies (beyond the Local Group) are receding from us or, more precisely, that the distance between the galaxies is continuously increasing. The *universe as a whole is therefore expanding. This expansion was proposed by Edwin Hubble in 1929. It can be explained by the *big bang theory. The universe may eventually contract if the *deceleration parameter exceeds 0.5. See redshift; Hubble constant; Hubble's law.

Explorers. A large group of US scientific satellites, the first of which—Explorer 1—was America's first satellite and was launched on Jan. 31 1958; it weighed 14 kg. It was able to confirm the existence of one of the *Van Allen radiation belts (already predicted) by means of the shielded *Geiger counter that formed part of its 8.2 kg payload. Explorers have been used in studies of the upper atmosphere and the interplanetary medium and of micrometeoroids, and as radio, x-ray, and gamma-ray astronomy satellites.

extended source. See radio-source structure.

extensive air shower (air shower). See cosmic rays.

extinction. The reduction in the amount of light or other radiation received from a celestial body as a result of absorption and scattering of the radiation by intervening interstellar dust and by the earth's atmosphere. The extinction decreases with wavelength of the radiation and increases with the pathlength through the absorbing medium and with the density of the medium.

The starlight is also reddened since the extinction of blue light by interstellar dust is greater than that of red light. The reddening may be given in terms of the colour excess, E,

$$E = (B\text{-}V) - (B\text{-}V)_0$$

where $(B\text{-}V)$ and $(B\text{-}V)_0$ are the observed and intrinsic *colour indices of the star. Most stars are reddened by a few tenths of a magnitude although values of up to two magnitudes are not uncommon. Stars lying behind extremely dense matter may be reddened by maybe ten magnitudes and might only be detectable at radio or far-infrared wavelengths.

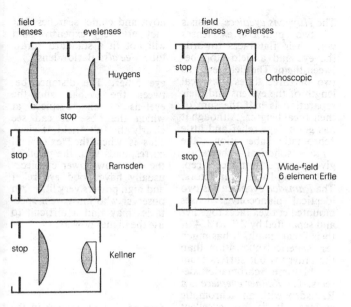

Types of eyepieces

extragalactic nebula. *Obsolete term for* galaxy.

extraterrestrial life. *See* exobiology.

extravehicular activity. *See* EVA.

extreme population I stars. *See* populations I and II.

extreme ultraviolet (XUV; EUV). The region of the electromagnetic spectrum bridging the gap between ultraviolet radiation and x-rays. *See* XUV astronomy.

extrinsic variable. *See* variable star.

eyelens. *See* eyepiece.

eyepiece. The magnifying system of lenses through which a telescopic image is viewed. It normally comprises a *field lens,* which receives the light rays, a stop, which defines the *field of view, and an *eyelens,* which directs the light into the eye. The magnification of a telescope is normally varied by using a short focal length eyepiece to give high powers, and one of long focal length for low powers.

Some eyepieces are shown overleaf in the illustration.

The *Huygens eyepiece* consists of two planoconvex lenses with their flat faces towards the eye and a field stop between them. The field lens has two or three times the focal length of the eyelens and their separation is half the sum of their focal lengths. Although it has good *eye relief and little *chromatic aberration, the large *spherical aberration gives poor performance except with long-focus refractors. The *Ramsden eyepiece* has two identical planoconvex lenses mounted convex faces together and separated by 2/3 to 3/4 of their focal length. It has much less spherical aberration than the Huygens but suffers from troublesome chromatic defects. The *Kellner eyepiece* is a Ramsden with an achromatic eyelens. It is an excellent eyepiece, widely used in binoculars and field glasses. For best results the lenses must be coated to minimize ghost images formed by internal reflections. The *orthoscopic eyepiece*, in its commonest form, has a planoconvex eyelens and a cemented triplet field lens. Although expensive this eyepiece gives excellent results and has a long eye relief, which suits the bespectacled user. By the use of extra lens elements and specially figured (aspheric) lenses, *wide-field eyepieces* may be obtained; these have fields subtending up to 90° at the eye; the *Erfle* has a field of 68° and the *Bertele* up to 80°. Such eyepieces are useful for observation work requiring exceptionally wide fields, such as nova and comet searches, but they are very expensive and will not fit a standard *draw tube. *See also* Barlow lens.

eye relief. The distance between the eyeball and the eyelens of an *eyepiece at which the observer can see clearly the full *field of view. This is when the *exit pupil corresponds with the pupil of the eye. Low-power eyepieces usually have good eye relief and high powers very little. An observer who must wear spectacles may find it difficult to use the higher powers.

F

faculae. Bright patches in the upper part of the solar *photosphere that have a higher temperature than their surroundings and occur in areas where there is an enhancement of the relatively weak vertical magnetic field. With the exception of polar faculae, which consist of isolated granules and appear in high *heliographic latitudes around the minimum of the *sunspot cycle, they are intimately related to *sunspots: they form shortly before the spots—in the same vicinity—and persist for several weeks after their disappearance. Faculae are best seen when near the sun's limb, where *limb darkening renders them more readily visible. They are approximately coincident, albeit at a lower level, with the *plages visible in monochromatic light.

fall. *See* meteorite.

false-colour photography. The use of colours in a photographic image to indicate the distribution of temperature, intensity of emitted radiation, etc., of an object, such as the sun, or over a region of the earth, sun, etc. The colours do not occur in the original object.

False Cross. A cross-shaped group of four 2nd-magnitude stars in the former constellation Argo. The group is now divided between Carina (Iota and Epsilon) and Vela (Kappa and Delta).

fan beam. *See* antenna.

Faraday rotation. The rotation of the plane of polarization experienced by a beam of plane-polarized radiation when it passes through a region where there are free electrons and a magnetic field. The plane of polarization is rotated through an angle equal to $\lambda^2 RM$, where λ denotes the radiation wavelength. RM is the *rotation measure* and is proportional to the integral along the line of sight of the product of the electron density, n_e, and the component of the magnetic flux density, B, parallel to the line of sight.

farside. *See* nearside.

fast ejection. *See* flares.

fast nova. *See* nova.

fault. A vertical or near vertical crustal displacement on the moon, Mars, or some other solid body. Lunar faults are usually double but single-sided faults also occur. *See also* wrinkle ridges; rille; Mars, surface features.

Faye's comet. A comet with a *coma that increased in size as it approached the sun. Most comets, Encke being a typical example, have comas that decrease with heliocentric distance. Between 1843 and 1955 Faye's brightness decreased from 4.2 to 11.1 magnitudes.

F-corona. The outer part of the solar *corona, responsible for the greater part of its intensity beyond a distance of about two solar radii. It consists of relatively slow-moving particles of interplanetary dust and exhibits an *absorption spectrum due to the scattering, by the particles, of light from the *photosphere. It extends for many million kilometres into the *interplanetary medium and merges, in the plane of the *ecliptic, with the particles that give rise to the *zodiacal light.

feed. *See* dish.

feeder. A system of wires or *waveguides used to connect a distant radio *antenna with its transmitter or *receiver so that radio-frequency power is transferred with minimum loss (energy dissipation). The wires are arranged to form a *transmission line,* which carries the power in the form of electromagnetic waves. When a transmission line is terminated by a resistance equal to its characteristic impedance all the power travelling down the line is absorbed by the load and none is reflected. The line is then said to be *matched* to its load. Transmission lines are of two types: balanced, as with the open-wire feeder, and unbalanced, as with coaxial cable. It is important that a balanced load, such as a *dipole, is fed by a balanced feeder. Where an unbalanced feeder is required to drive a balanced load, a balanced-to-unbalanced transformer, or *balun,* must be used.

In an *array many elements must be connected together to one receiver. One method of doing this is to join adjacent pairs of elements, and then pairs of pairs and so on so that the total length of feeder from the receiver to any element is the same. Such an arrangement is sometimes called a *christmas tree. See also* Butler matrix.

fibrils. Predominantly horizontal strands of gas in the sun's lower *chromosphere— typically 11 000 km long and 725-2200 km wide—that are visible (in the monochromatic light of certain strong *Fraunhofer lines) as dark absorption features against the brighter disc. They form an ordered pattern in and around *active regions, where they tend to be aligned in the direction of the localized magnetic field. The overall configuration is stable over a period of several hours, though an individual fibril has a lifetime of only 10-20 minutes.

field. 1. The region in which a physical agency exerts an influence, measured in terms of the force experienced by an object in that region. Thus a massive body has an associated gravitational field within which an object will feel an attractive force. There are several types of field, including gravitational, magnetic, electric, and nuclear fields, each of which has its own characteristics and exerts its influence, strong or weak, over a particular range.

2. *Short for* field of view.

field curvature. *See* curvature of field.

field equations. *See* relativity, general theory.

field galaxy. Any galaxy that is not a member of a *cluster of galaxies. The ratio of field galaxies to cluster galaxies has been estimated as 1:23 or less.

field lens. *See* eyepiece.

field of view (field). The area over which the image is rendered by the optical system of an instrument such as a telescope.

field pattern. *Another name for* antenna pattern. *See* antenna.

figuring. The process of grinding and polishing the surface of a lens or mirror base to the exact shape desired.

filaments (Formerly sometimes termed **dark flocculi**). Clouds of gas in the sun's upper *chromosphere/inner *corona, with a higher density and at a lower temperature than their surroundings, that are visible (in the monochromatic light of certain strong *Fraunhofer lines) as dark absorption features against the brighter disc. When viewed beyond the limb they appear as bright projections and are termed prominences. *See* prominences.

filar micrometer (position micrometer). A measuring device incorporated into the eyepiece of an instrument and used in astronomy to measure the angular separation and orientation of a visual binary star or to determine the position of an object relative to a nearby star whose coordinates are known. It consists essentially of two parallel wires that lie in the focal plane of the eyepiece and can be rotated about the optical axis of the telescope. A third wire lies perpendicular to the parallel wires. Two stars are aligned along this perpendicular wire. The position of one of the parallel wires can be controlled by turning a screw with a finely graduated micrometer head. The parallel wires are adjusted so as to coincide with the images of the two stars and the separation of the images can then be accurately measured.

filling factor. *See* array.

filter. A device that transmits part of a received signal and rejects the rest. The signal may be in the form of a beam of light or other radiation or may be an electrical signal. Optical, ultraviolet, and infrared filters are dyed plastic or gelatin, glass or glass-like substances, or confined liquids, all of which absorb incident radiation except for a relatively narrow band of wavelengths. Such filters are used in astronomical *photometry, especially in measurements of *magnitudes in the UBV system; transmitting bands typically 100 nanometres wide, they are termed broadband filters. Much narrower wavelength bands, of maybe one nm, can be obtained with *interference filters.

Electrical filters are transfer devices whose *attenuation varies with frequency. Filters that allow low or high frequencies to pass without serious attenuation are called *low-pass* and *high-pass filters*, respectively. A filter that allows only a limited range of frequencies through is a *band-pass filter* while its converse is a *band-stop filter*. *See also* bandwidth.

filtergram. *See* spectroheliogram.

find. *See* meteorite.

finder. A low-power telescope with a wide *field of view that has its optical axis aligned with that of the main telescope. Since the field of view of the average amateur astronomer's telescope used at its lowest power is only about half a degree, some means of pointing it in the correct direction is needed. *Setting circles enable this to be done with a permanent *equatorial mounting but for a portable or *alt-azimuth mounting a finder is essential. It should have a field of view of at least four to eight degrees and be provided with illuminated cross wires or a graticule.

fireball. A bright visual *meteor of *magnitude greater than -10. About 50 000 to 100 000 occur in the earth's atmosphere each year.

first point of Aries. Another name for the vernal *equinox, which over 2000 years ago, when Hipparchus first used the term, lay in the constellation Aries. Due to the westerly *precession of the equinoxes the vernal equinox now lies in Pisces and will subsequently move into Aquarius. Likewise the *first point of Libra,* or the autumnal equinox, once lay in Libra but is now located in Virgo.

first point of Libra. *See* first point of Aries.

first quarter. *See* phases, lunar.

Fishes. Common name for *Pisces.

fission. *See* nuclear fission.

FitzGerald contraction. *See* Lorentz contraction.

Five-Kilometre telescope. *See* aperture synthesis.

fixed stars. Originally stars in general, which, until the early 18th century and the discovery of *proper motion, were thought to have no relative motion and thus remained fixed in position in the sky. In comparison the planets were described as *wandering stars.*

FK4. *See* fundamental catalogue.

Flamsteed number. *See* stellar nomenclature.

flares. Sudden short-lived brightenings of small areas of the sun's upper *chromosphere/inner *corona that are usually visible only in the monochromatic light of certain strong *Fraunhofer lines. They represent an explosive release of energy—in the form of particles and radiation—that causes a temporary heating of the surrounding medium and may accelerate electrons, protons, and heavier ions to high velocities.

A typical flare attains its maximum brilliance in a few minutes and then slowly fades over a period of up to an hour. Although attempts have been made to classify them according to their structure, the most objective system considers the maximum area covered by the flare and to a lesser extent its intensity. The overwhelming majority of flares are small, covering areas of less than several hundred million square kilometres, but those that are larger are sometimes associated with a number of interesting phenomena.

During the 'flash phase' of a large flare, when it suddenly increases in brightness and rapidly expands to its maximum extent, material may be ejected in the form of a *surge, *spray or (for a particularly energetic event) *fast ejection*—in which a compact portion of the flare is expelled without fragmentation, with a velocity of about 1000 km s^{-1}. Large flares may initiate a *Moreton wave,* which is a magnetohydrodynamic shock wave that spreads out from the centre of disturbance in a sector of about 90°, as it travels with a velocity of the order of 1000 km s^{-1} across the vertical magnetic fields of the inner corona. This causes a depression and subsequent relaxation of the underlying *chromospheric network and may induce distant *filaments, perhaps several hundred thousand kilometres from the flare, to undergo several damped vertical oscillations. Filaments in the vicinity of the flare may also be activated, before and/or during the flare, by changes in the configuration of the local magnetic field. This activation usually takes the form of increased internal motion or flow along the filament (small surges), sometimes accompanied by a gradual ascent of the filament itself. The aftermath of large flares may include coronal condensations of relatively dense material at temperatures of up to 4 000 000 kelvin.

The effects of energetic flares are by no means confined to the sun. Fade-out of short-wavelength radio signals is often experienced in the daylight hemisphere of the earth and is due to a temporary strengthening (by increased ionization) of the reflecting property of the D layer of the *ionosphere (50-90 km altitude), which suppresses the passage of the signals to the higher layers where they are normally reflected. This is accompanied by a sudden increase in the electrical conductivity of the E layer (90-140 km altitude) and by disturbances of the earth's magnetic field. Ultraviolet radiation from the flare is responsible for these effects. X-rays may also be emitted and solar radio emission (from the inner corona) frequently exhibits an intense burst or series of bursts at centimetre or metre wavelengths. Occasionally, within about half an hour of the flare, low-energy *cosmic rays reach the earth, and within about 26 hours, an average, less energetic charged particles may also arrive. These latter particles spiral around the earth's magnetic field lines, causing *geomagnetic storms and their luminous counterpart, *auroral displays.

The nature of flares and the physical mechanism responsible for them are not completely understood. They invariably occur in *active regions, close to the line of inversion (*see* sunspots), where the gradient of the horizontal component of the magnetic field is steepest. It is thought that the emergence of new magnetic flux in the active region or reorganization of the field for other reasons squeezes together the field lines of opposite polarity and that this produces the flare; but much remains to be done before a quantitative picture will emerge.

A phenomenon similar to solar flares, but far more energetic, is thought to be responsible for the rapid brightening of *flare stars.

flare stars (UV Ceti stars). Intrinsically faint cool red main-sequence stars that undergo intense outbursts of energy, possibly from localized areas of the surface, causing transient but appreciable increases in the brightness of the star. The brightness can change by two magnitudes or more in several seconds, decreasing to its normal minimum in about 10 to 20 minutes. Radio bursts are usually emitted almost simultaneously. Flare stars are *cataclysmic variables; they are of *spectral type M or sometimes K with spectral emission lines of hydrogen and ionized calcium. A large proportion are close binary stars. UV Ceti (M6e V) is a typical flare star; it is the fainter component of the nearby star Luyten 726-8.

flash spectrum. The *emission spectrum of the solar *chromosphere. It is observable only for a few seconds during a total solar eclipse, at the beginning and end of totality. The majority of the spectral lines have the same wavelengths as the *Fraunhofer lines of the photospheric spectrum. Their presence in successive spectrograms taken during an eclipse provides information on the distribution of elements in the chromosphere.

flattening of the earth. *See* geoid.

F-layer (F-region). *See* ionosphere.

flocculi. A term that has been applied to several different solar phenomena. In the older literature *plages and *filaments were often referred to as *bright* and *dark flocculi* respectively, while today the term is sometimes used to denote the bright mottles of the *chromospheric network.

fluorescence. The transformation of *photons of relatively high energy (i.e. high frequencies, especially ultraviolet frequencies) to lower-energy photons through interactions with atoms. It is also the lower-energy radiation that is produced by the process.

flux. A measure of the energy, number of particles, etc., emitted from or passing through a surface per unit time.

flux collector. *See* infrared telescope.

flux density. A measure of the signal power to be expected from a discrete source of emission. In radio astronomy, where emissions are *noise signals, the dimensions of flux density are energy per unit time per unit *bandwidth per unit area. The unit is the flux unit or *jansky.

Fly. Common name for *Musca.

flyby. A trajectory that takes a spaceprobe close to a planet or satellite but does not permit it to enter an orbit about the body or land on it.

Flying Fish. Common name for *Volans.

f-number. *See* focal ratio.

focal length. Symbol: f. The distance between the centre of a mirror or lens to the *focal point* or *focus* (see illustration). With a converging system, i.e. a concave mirror or thin convex lens, the focus, F, is the point to which a narrow beam of light from a distant object, i.e. a parallel beam closely aligned to the optical axis, is brought to a sharply defined or focused image. In a convex mirror or thin concave lens it is the point from which a parallel beam, made divergent by the mirror or lens, appears to diverge. If the two surfaces of a lens do not have identical curvatures the lens will have two different focal lengths and focal points, depending on which surface the light falls.

The *focal plane* is the plane through the focus, at right angles to the optical axis, in which the image of a distant

Focal length and focal plane

focal plane

object will be formed. In some cases, as in the *Schmidt telescope, the image is focused on a curved surface—the *focal surface*—rather than a plane. *See also* equivalent focal length.

focal plane. *See* focal length.

focal point (focus). *See* focal length.

focal ratio. The ratio f/d of the *focal length, f, of a lens or mirror to its effective diameter, d, i.e. to its aperture. The numerical value of the ratio, often called the *f-number*, is usually written f/4 or f:4 for a ratio of 4, say. The reciprocal of the focal ratio (d/f) is the *aperture ratio. The *limiting magnitude, or the faintest detectable star, depends on the focal ratio of a telescope: for telescopes used under the same observing conditions, the larger the ratio the fainter the limit; if photographs are being taken, however, the larger the ratio the longer the necessary exposure time.

focal surface. *See* focal length.

focus (focal point). *See* focal length.

folded dipole. *See* dipole.

following spot (f-spot). *See* sunspots.

Fomalhaut (α PsA). A white star that is the brightest one in the constellation Piscis Aus-

trinus. m_v: 1.16; M_v: 1.9; spectral type: A3 V; distance: 7.0 pc.

forbidden lines. Lines that are not found in spectra under normal terrestrial conditions but are observed in certain astronomical spectra. In *emission nebulae, for example, atoms can be excited by impact of low-energy electrons. Under normal laboratory conditions such atoms would be de-excited by collisions with other atoms, etc., before they had time to radiate. Such collisions are very infrequent in nebulae and the *'forbidden' transitions* between the excited state and a state of lower energy can occur, producing the forbidden lines.

force. Symbol: F. Any physical agency that alters or attempts to alter a body's state of rest or of uniform motion. The force required to accelerate a body of mass m is given by ma, where a is the acceleration imparted. There are many kinds of forces, the weakest of which is the gravitational force. The SI unit of force is the newton.

fork mounting. *See* equatorial mounting.

Fornax (Furnace). An inconspicuous constellation in the southern hemisphere fairly near Orion, the brightest stars being of 4th magnitude. Abbrev.: For; genitive form: Fornacis; approx. position: RA 3h, dec -30°; area: 397 sq deg.

Fornax system. A member of the *Local Group of galaxies.

Foucault knife-edge test. A method for testing the accuracy of the shape of a mirror or lens surface. It was developed in the 1850s by J.B.L. Foucault. A light beam reflected from a spherical mirror forms an image at the mirror's radius of curvature. A knife edge moved across the reflected beam at the focal plane will cause the whole image to darken uniformly only if the surface curvature is accurately spherical. An annular zone, half light half shade, will occur if a depression or raised area is present on the surface.

Foucault's pendulum. A simple pendulum that demonstrates the rotation of the earth, first shown by J.B.L. Foucault in 1851. It consists of a massive ball that is suspended by a long wire and can swing freely with a minimum amount of friction. The pendulum swings steadily, tracing a straight line on the floor beneath it. The plane in which it swings, however, is observed to rotate during the day as a result of the earth's rotation. The period of rotation of the plane depends on the latitude of the place: at the poles the plane would appear to move through a complete circle in one sidereal day (23h 56m 4s) while on the equator it would not rotate at all. The pendulum is in fact swinging in a plane that is fixed relative to the stars.

Fourier transform. A mathematical operation by which a function expressed in terms of one variable, x, may be related to a function of a different variable, s, in a manner that finds wide application in physics. The Fourier transform, $F(s)$, of the function $f(x)$ is given by
$$F(s) = \int_{-\infty}^{\infty} f(x) \exp(-2\pi i x s) dx$$
and
$$f(x) = \int_{-\infty}^{\infty} F(s) \exp(2\pi i x s) ds$$
The variables x and s are often called *Fourier pairs*. Many such pairs are useful, for example, time and frequency: the Fourier transform of an electrical oscillation in time gives the spectrum, or the power contained in it at different frequencies.

Fox. Common name for *Vulpecula.

Fra Mauro formation. The *ejecta blanket associated with the *Imbrium Basin. It was the target for Apollo 14.

frame of reference. A set of coordinate axes to which position, motion, etc., in a system may be referred and by which they can be measured.

Franklin-Adams charts. A whole-sky photographic star atlas of 206 plates prepared by John Franklin-Adams and published in 1914 as the *Photographic Chart of the Sky*. It has a limiting magnitude of 16.

Fraunhofer lines. Absorption lines in the solar (photospheric) spectrum, first studied in detail by Joseph von Fraunhofer in 1814. He catalogued over 500 lines and labelled the more striking ones with letters. Over 25 000 lines have now been identified. The most prominent lines at visible wavelengths are due to the presence of singly ionized calcium, neutral hydrogen, sodium, and magnesium, while many weaker lines are due to iron. The strength of a particular line depends not only on the quantity of the element present but also on the degree of *ionization and level of *excitation of its atoms. By isolating an individual strong line it is possible, by virtue of its residual intensity due to the reemission of radiation in the *chromosphere, to observe the chromosphere in the light of the element concerned. *See* spectroheliogram. *See also* H and K lines; D lines; photosphere.

free-bound emission. The radiation emitted when a free electron is captured by an ion. *See* recombination-line emission.

free fall. Motion of a body in a gravitational field when no resistance is offered to the *acceleration of gravity. *See also* weightlessness.

free-free emission. *See* thermal emission.

frequency. Symbol: f, ν. The number of oscillations per unit time of a vibrating system. The frequency of an electromagnetic wave is the number of wave crests passing a point per unit time. It is related to wavelength λ by $\nu = c/\lambda$, where c is the speed of light. The frequency of electromagnetic waves is measured in hertz.

frequency changer. *Another name for* mixer. *See* receiver.

Friedmann universe. *See* cosmological models.

fringe. *Short for* interference fringe. *See* interference; radio telescope.

f-spot. *Short for* following spot. *See* sunspots.

F stars. Stars of *spectral type F that are white with a surface temperature (T_{eff}) of about 6000 to 7500 kelvin. The hydrogen lines in the spectrum weaken rapidly and the lines of ionized calcium (Ca II, H and K) strengthen from F0 to F9; there are many lines of neutral and singly ionized metals and heavy atoms. Canopus, Polaris, and Procyon are F stars.

full moon. *See* phases, lunar.

full-wave dipole. *See* dipole.

fundamental catalogue. A catalogue of *fundamental stars in which the precise positions and proper motions are listed for a given epoch. Other star catalogues can be compiled by reference to a fundamental catalogue. The *Fourth Fundamental Catalogue (FK4)*, published in 1963 in Heidelberg, contains data for 1535 stars to a limiting magnitude of 7.

fundamental stars. A number of stars distributed over the whole celestial sphere whose positions and proper motions are known so accurately that they can be used to determine the positions of other stars.

FU Orionis. A very young star, thought to be a *T Tauri star, that is located in Orion in a cloud of gas and dust and is a strong source of infrared radiation. It flared up from about 16th to 10th magnitude in 1936 and has remained near that brightness. It has a high lithium content.

Furnace. Common name for *Fornax.

fusion. *See* nuclear fusion.

G

gain. 1. A measure of the amplification of an electronic device. If the power input to the device is P_1 and the power output is P_2, the gain expressed in decibels is given by

$$G = 10 \log_{10}(P_2/P_1)$$

Gains measured in this way can be added when amplifying stages are connected in cascade.

2. A measure of the directional advantage of using one radio *antenna as compared with another. It is usual to express the gain, G, of a particular antenna over an isotropic radiator. For a lossless antenna it is given by

$$G = 4\pi A_e/\lambda^2$$

where A_e is the effective area (*see* array) and λ is the wavelength; the gain is equal to the *directivity in this case. Sometimes the comparison is with a *dipole, which itself has a gain over an isotropic radiator of two decibels.

galactic centre. *See* galactic coordinate system; Galaxy.

galactic circle. *See* galactic equator.

galactic cluster. 1. *See* open cluster.
2. *See* clusters of galaxies.

galactic concentration. Counts of stars made at different galactic latitudes show that the apparent density of stars (the number per unit area of sky) increases towards the plane of the *Milky Way. Moreover, comparisons of counts taken to different limiting magnitudes show that faint stars are relatively more concentrated than bright ones. Since the fainter stars are, on average, the most distant, this implies that the *Galaxy is a flattened system that extends to much greater depths along the galactic plane than at right angles to it. *See also* star gauges.

galactic coordinate system. A

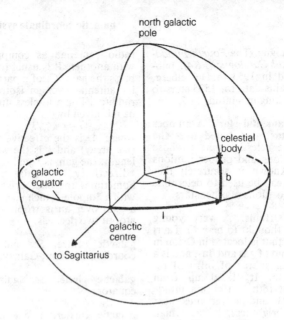

Galactic coordinate system

*coordinate system used to study the structure, surroundings, and contents of the *Galaxy. The fundamental circle is the *galactic equator and the zero point lies in the direction of the *galactic centre* (in the constellation Sagittarius) as seen from earth (see illustration). The coordinates are *galactic latitude* and *longitude*.

The galactic latitude (b) of a celestial body is its angular distance (from 0° to 90°) north (counted positive) or south (counted negative) of the galactic equator; it is measured along the great circle passing through the body and the galactic poles. The galactic longitude (l) of a celestial body is its angular distance (from 0° to 360°) from the galactic centre measured eastwards along the galactic equator to the intersection of the great circle passing through the body.

The position of zero galactic longitude, i.e. the presumed galactic centre, was agreed on (1959) by the International Astronomical Union (IAU); it lies at RA 17h 42.4m, dec -28° 55′ (1950). This zero position is used in the new IAU system in which the coordinates l^{II}, b^{II} are used to distinguish them from the old IAU system coordinates l^{I}, b^{I}, which referred to a slightly different galactic equator and a zero point defined by the ascending *node of galactic equator on celestial equator.

galactic disc. *See* Galaxy.

galactic equator (galactic circle). The great circle on the *celestial sphere that represents the path of the *Galaxy. The galactic centre lies on the galactic equator in the direction of the constellation Sagittarius and is the zero point for the *galactic coordinate system. The *galactic plane*, which contains the galactic equator, is the plane that passes most nearly through the central plane of the spiral disc of the Galaxy. The galactic plane and celestial equator are inclined at an angle of about 62°.

galactic halo. *See* Galaxy.

galactic latitude, longitude. *See* galactic coordinate system.

galactic magnetic field. A very weak and complex system of magnetic fields, known to exist in the interstellar regions of the Galaxy because of the *Faraday rotation of radiation from pulsars and other radio sources, from the *Zeeman splitting of the 21 cm hydrogen line arising in interstellar clouds, and from *interstellar polarization. The magnetic flux density lies in the range 10^{-10} to 10^{-9} tesla, with $3-6 \times 10^{-10}$ tesla being the best estimate at present. The direction of the field seems to be in the plane of the Galaxy, along the spiral arms. *See also* cosmic rays.

galactic plane. *See* galactic equator; Galaxy.

galactic poles. The two points on the celestial sphere that are 90° north and 90° south of every point on the *galactic equator. The *north galactic pole* lies in the constellation Coma Berenices at RA 12h 49.0m, dec +27°24′ (1950). The *south galactic pole*, diametrically opposite, lies in Sculptor.

galactic radio background radiation. *See* interstellar medium; radio source.

galactic rotation. The rotation of the *Galaxy about its centre, all its components sharing in this rotation to varying degrees. The predominant motion in the galactic disc is circular and parallel to the galactic plane. As expected in a system with a strong central concentration of mass, the general rotation pattern is differential: the average speed of rotation decreases with distance from the galactic centre so that stars in the inner regions of the disc complete their orbits more quickly than those further away (*see* Oort's formulae). The rotation velocity of the *local standard of rest in the galactic plane is about 250 km s^{-1}. Objects in the galactic halo display much more random motions than those in the disc and the system as a whole has only a small residual rotation with respect to the galactic centre.

galactic year. *See* cosmic year.

galaxies. Giant assemblies of stars, gas, and dust into which most of the visible matter in the universe is concentrated. Each galaxy exists as a separate, though not always entirely independent, system held together and organized largely by the gravitational interactions between its various components. When capitalized, the term denotes specifically our own system, the Milky Way *Galaxy. Owing to their cloudlike appearance when viewed by eye through simple telescopes, galaxies other than our own were once known as 'nebulae'. This term is now strictly applied only to clouds of interstellar gas and dust.

The majority of the galaxies close enough to be observed in any detail can be divided into three broad categories: elliptical, spiral, and irregular. *Elliptical galaxies* are spheroidal systems with no clearly defined internal structures. They contain little or no interstellar matter and the stars within them (mainly *spectral types K and M) are predominantly old and well advanced in their evolution. Most ellipticals are highly concentrated objects, particularly near the centre. On photographs they look somewhat like fuzzy elliptical patches of light, usually diminishing smoothly in brightness from the centre outwards. The shape of the outline varies from almost circular through to narrow ellipses about three times as long as they are wide. The latter are presumably the most oblate systems seen edge-on. In all other cases the true oblateness is uncertain because the orientation of the galaxies to our line of sight is not known. However, there are too many galaxies with circular outlines for them all to be flattened systems seen pole-on, so some must actually be spherical.

Spiral galaxies are flattened, disc-shaped systems containing prominent *spiral arms* of interstellar matter and bright young stars that wind outwards from a dense central nucleus. Although two-armed spirals are the most common, systems with one arm or even, very occasionally, three arms have also been observed. In *normal spirals* the arms emerge directly from the nucleus, usually from opposite sides; in *barred spirals* the arms emanate from each end of a bright central bar that extends across the nucleus. Both types exist in a wide range of forms. At one extreme there are galaxies with large dominant nuclei and thin tightly wound spiral arms; at the other extreme there are galaxies with inconspicuous nuclei and prominent loosely wound spiral arms.

Spiral galaxies are rich in gas and dust, most of which is distributed in clouds along the spiral arms. Stars in the nuclei of spirals appear to be predominantly well advanced in their evolution and the brightest individual stars observed there are *red giants of *population II. A fairly smooth axially symmetric distribution of old stars also extends beyond the nucleus. However, owing to the intrinsic faintness of most of its constituent stars, this system is much less conspicuous than the spiral arms embedded within it.

Although star formation is undoubtedly still taking place in the spiral arms, what triggers this process is still uncertain, as is the origin of the arms themselves. It has been proposed, in the *density-wave theory, that the spiral structure is maintained over a long period by gravitational effects. Alternatively, it has been suggested that the spiral structure arises and persists as a result of local regenerative effects involving supernovae. All spiral galaxies are in rapid differential rotation, stars in the outer parts completing their orbits, on average, more slowly than those nearer the centre. As far as can be ascertained, the spiral arms always trail away from the direction of rotation.

Irregular galaxies have no discernable symmetry in shape or structure. They vary enormously in appearance but all are below average in size and contain large amounts of interstellar matter. In the early Hubble classification this term was applied to any galaxy that could not be fitted into the elliptical or spiral classes. Subsequent research has uncovered a very diverse assortment of objects within this category, including exploding and *interacting galaxies, in addition to the small or dwarf systems presently labelled as irregular.

In addition, astronomers now recognize a class of galaxies that appear to be intermediate between the spirals and ellipticals. Called *S0 galaxies,* these possess a central nucleus and the beginnings of a disc but show little or no evidence of spiral structure. The stars within them are predominantly old and very little gas is apparent, although dark dust clouds are sometimes seen in the disc. The term *lenticular galaxy* is used either to denote specifically these S0 galaxies, or to denote any galaxy that superficially resembles a biconvex lens.

The colours of galaxies vary according to the age of the stars responsible for most of the light. In general, ellipticals and S0s are the most red and irregulars, because of their high content of very young stars, are the most blue. Differences in the dominant *spectral type are also reflected in the mass to luminosity ratio, spirals and irregulars being, on average, brighter than ellipticals or S0s of similar mass.

The brightest (and also rarest) galaxies are *supergiant ellipticals* with absolute magnitudes of about -22.5 and masses in excess of 10^{12} solar masses. Ellipticals also display the greatest variation in mass, ranging down to extreme dwarfs (about 10^6 times the sun's mass) that are no brighter than the most luminous *globular clusters. Spirals appear to exist only as large or *giant* systems, with masses typically of the order of 10^{10} or 10^{11} times that of the sun. No irregulars are as bright as the giant spirals and some are extreme dwarfs.

Of the 1000 brightest galaxies, about 75% are spiral, 20% elliptical, and 5% irregular. However, when allowance is made for the many *dwarf galaxies, the true proportions turn out to be nearer to 30:60:10. Few galaxies exist in total isolation. Double and multiple systems are common and many galaxies are also members of larger groups known as *clusters of galaxies.

How and why galaxies have evolved to their present shapes is still uncertain although it appears that spirals contain much more *angular momentum than ellipticals. Turbulence and vorticity in the early universe may play a role. However, it is generally conceded now that the principal types of galaxy represent separate species rather than one species seen at different stages in its evolution. There is, in fact, no direct evidence that any other galaxy is significantly older or younger than our own Galaxy (which is thought to be about 12 thousand million years old), and all galaxies appear to contain a mixture of stellar *populations. *See also* galaxies, evolution; interacting galaxies; active galaxy; compact galaxy; ring galaxy; Hubble classification; recession of the galaxies.

galaxies, classification. Galaxies may be classified according to a variety of criteria, including morphology (shape, concentration, and structure), spectroscopy (integrated spectrum, etc.), colorimetry (integrated colour, etc.), and photometry (luminosity, etc.). The *Hubble classification is based on morphology while *Morgan's classification is basically a quantitative spectroscopic classification.

galaxies, evolution. The formation and evolution of galaxies is an outstanding problem. The collapse of mass condensations under their own self-gravity is undoubtedly important but would have been inhibited by *radiation pressure before the epoch of recombination (*see* big bang theory). Thereafter each mass condensation behaves as a separate closed universe, expanding and then collapsing. The virial theorem predicts a final size about one half that of the maximum, if no energy is dissipated. The *mean density of matter in galaxies can then be combined with the dynamic behaviour of a closed 'mini-universe' to obtain a maximum age. For our Galaxy this age is not in disagreement with age estimates from *globular clusters and suggests that galaxies formed at redshifts of between 10 and 100 (see cosmic scale factor; big bang theory).

Galaxy (Milky Way System). The giant star system to which the sun belongs. The Galaxy has a spiral structure and, like other spiral galaxies, is highly flattened. It is estimated to contain of the order of 100 billion (10^{11}) stars, the bulk of which are organized into a relatively thin *disc* with an ellipsoidal bulge, or *nucleus*, at its centre. This system is embedded in an approximately spherical *halo* of stars and *globular clusters (see illustration *a*). The diameter of the disc is approximately 30 kiloparsecs and its maximum thickness (at the centre) is about four kiloparsecs. The halo is more sparsely populated than the disc and its full extent is uncertain, although its diameter is known to be greater than that of the disc. The characteristic *spiral arms*, which contain many of the brightest stars in the Galaxy, wind outwards from the nuclear region, in or close to the central plane of the disc.

a *Schematic views of Galaxy, (i) from above, (ii) from the side* b *Named spiral arms of Galaxy from optical observations*

The sun is situated only a few parsecs north of the central plane, near the inner edge of one of the spiral arms. Our distance from the centre is between 8 and 11 kiloparsecs, with 10 kiloparsecs being accepted by most astronomers as the most likely value.

The entire Galaxy is rotating about an axis through the centre, the disc rotating fairly rapidly, the halo more slowly (*see* galactic rotation). At the sun's distance from the centre, the systematic rotation of the disc stars is approximately 250 km s^{-1}, whereas the halo system is rotating with a speed of only about 50 km s^{-1}.

All the objects in the halo are old and belong to *population II. They increase in density towards the centre of the Galaxy but show little concentration towards the galactic plane. Stars in this system are believed to have condensed early in the life of the Galaxy, maybe 12 billion years ago, when the gas cloud from which it formed was still almost spherical. The bulk of the stars in the disc, and probably in the nucleus also, are stars of intermediate age (3-5 billion years) belonging to the disc *population. The young *population I stars are mainly confined to a layer about 500 parsecs thick along the centre of the disc.

Most of the interstellar gas and dust detected in the Galaxy lies in or close to the galactic plane, and a large part of it is concentrated into clouds distributed along the spiral arms. The youngest stars in the Galaxy, notably the *T Tauri stars and the very bright, short-lived *O and *B stars, are also largely confined to the spiral arms and it is almost certain that stars are still being formed from the interstellar clouds there. Tracing of the spiral structure is complicated by our position in the disc. Three relatively nearby sections of arms have been traced optically— principally by mapping the distribution of O-B stars and associated *emission nebulae. These are the *Orion arm,* in which the sun is located; the *Perseus arm,* located about two kiloparsecs further out along the plane; and the *Sagittarius arm,* which lies about two kiloparsecs nearer the centre (see illustration *b*). *Interstellar extinction makes optical tracing impossible beyond the first few kiloparsecs of the sun in most directions, and more distant spiral features have been traced mainly by mapping the distribution of neutral hydrogen in *HI regions by radioastronomy techniques.

Interstellar extinction also blots out most of the central nucleus at optical wavelengths. Information about this region has been derived mainly from radio, x-ray, and infrared observations. The nucleus appears to consist almost entirely of stars, with interstellar matter contributing only a very small proportion of the mass. The star density appears to increase rapidly towards the centre: at 0.1 parsecs it is estimated to be about 10^8 times that in the solar neighbourhood. Radio observations indicate that there is a thin, rapidly rotating disc of hydrogen extending out to a radius of about 750 parsecs and also considerable quantities of gas moving rapidly away from the centre. In particular, there are two expanding arms of gas: one at three kiloparsecs this side of the centre, moving towards us at 50 km s^{-1}, and another on the far side receding at 135 km s^{-1}. A powerful nonthermal radio source, *Sagittarius A, appears to lie right at the centre.

The total mass of the Galaxy is estimated to be a little less than 2×10^{11} solar masses, of which about 10% exists in the form of interstellar matter. Where our Galaxy fits in the *Hubble classification is still open to argument. Some observations suggest a type Sb structure, others a type Sc; there is even tentative evidence of a central bar, although this evidence is not accepted by all astronomers.

Galilean satellites. The satellites Io, Europa, Ganymede, and Callisto of Jupiter, discovered in 1610 by Galileo and, independently, by Simon Marius. They are bright enough to be seen with the aid of binoculars. *See* Io; Jupiter's satellites; Table 2, backmatter.

Galilean telescope. The first type of astronomical telescope, developed in 1609 by Galileo from Hans Lippershey's 'magic tubes'. It is made up from a single long-focus object lens and a powerful negative lens used as an eyepiece (see illustration). This optical system, which gives an upright image, survives in the modern opera glass. Galileo's best telescopes had a magnification of about 30 times; although very imperfect they led to the great achievements of 17th-century astronomy.

Ray path in Galilean refractor

Galileo spacecraft. A US spacecraft that is scheduled to be launched towards Jupiter in 1983. It will study 11 of Jupiter's satellites and will also send a probe into the atmosphere of Jupiter itself.

Gamma (γ). The third letter of the Greek alphabet, used in *stellar nomenclature usually to designate the third-brightest star in a constellation or sometimes to indicate a star's position in a group.

gamma-ray astronomy. The study of radiation from space at the extreme short-wavelength end of the electromagnetic spectrum (less than 0.01 nanometres) and with the largest photon energies (usually exceeding 500 000 electronvolts [eV]). The first cosmic gamma-rays were detected with a scintillation counter on the OSO-3 satellite in 1968. More detailed observations confirmed their origin in a narrow band along the galactic plane, being particularly strong within 30° of the galactic centre; these observations were provided by the NASA SAS-2 satellite launched in November 1972 and suggested that the gamma rays came from interactions in interstellar gas. The spark-chamber detector on SAS-2, sensitive in the extreme energy band 20-100 MeV, also found discrete γ-ray sources coincident with the two most rapid pulsars—the Crab and Vela pulsars.

The European Space Agency's COS-B satellite, launched in August 1975, also carried a large spark chamber and operated successfully for several years, providing improved exposure and angular resolution of 20-200 MeV γ-rays. In addition to the Crab and Vela pulsars some 20 other galactic sources, plus the bright *quasar 3C 273, were detected by COS-B up to mid-1978. A diffuse galactic and extragalactic *gamma-ray background* was also indicated: this is possibly related to *cosmic ray interactions with interstellar matter, perhaps by production and decay of neutral *pions (π_0). The great penetration of γ-rays makes their observation uniquely interesting in the study of energetic events in astronomy. For example, nuclear γ-rays at 0.5-3 MeV could bring the first direct observational proof of *nucleosynthesis of the heavy elements in supernovae.

gamma-ray bursts. Intense flashes of hard x-rays or gamma rays, detected at energies up to one million electronvolts. They were discovered by US Air Force satellites in 1967 but not declassified until 1973, crude positions being obtained by *light-time differences between a pair of Vela satellites. A characteristic gamma-ray burst has a rise time of less than one second and a duration of several seconds. These bursts are quite rare, occurring at a rate of about 10 per year. Scarcity and poor positional accuracy has so far prevented optical identification of gamma-ray burst sources, and their origin and nature remains unknown.

gamma rays (γ-rays). Very high energy *electromagnetic radiation, i.e. radiation with the shortest wavelengths and the highest frequencies. Gamma-ray wavelengths are less than 10^{-11} metres and are sometimes taken to be less than 10^{-12} metres or even 10^{-13} metres. There is no sharp cutoff between the gamma-ray region and the adjacent x-ray region of the electromagnetic spectrum. Gamma rays, like x-rays, are usually described in terms of photon energy, $h\nu$, where ν is the frequency of the radiation and h is the Planck constant. The energies of gamma rays generally exceed 500 000 electronvolts.

Ganymede. The third and brightest of the *Galilean satellites of Jupiter. With a diameter of 5100 km, it is larger than the planet Mercury. Photographs from *Voyager 1 show the surface to be heavily cratered and to be covered with ridges and grooves. *See* Jupiter's satellites; Table 2, backmatter.

Gargantuan Basin. *See* Procellarum, Oceanus.

Garnet star (μ Cep). A red supergiant in the constellation Cepheus. It is a *semiregular variable with a magnitude that ranges from 3.6 to 5.1 although it is usually about 4.5. It is also a triple star. Spectral type: M2 Ia.

gaseous nebula. Any *nebula whose light is generated by the gas component rather than the dust component of the cloud, i.e. an *emission nebula.

gauss. Symbol: G. The c.g.s. (electromagnetic) unit of *magnetic flux density. One gauss is equal to 10^{-4} tesla.

Gaussian gravitational constant. Symbol: k. A defining constant equal to
  0.017 202 098 950 radian
i.e. to
  3548.187 607 arc seconds
It is given by the relationship, derived from Newton's form of *Kepler's third law:
$$n^2a^3 = k^2(1+m)$$
where n is the mean daily motion of a planet in radians, a is the semimajor axis of its orbit in astronomical units, and m is its mass in terms of solar mass. It is also given by the ratio $2\pi/T$, where T is the earth's sidereal year in ephemeris days.

GE. *Abbrev. for* greatest elongation. *See* elongation.

gegenschein (counterglow). A very faint patch of light that can be seen on a clear moonless night on the ecliptic at a point 180° from the sun's position at the time of observation. It is part of the *zodiacal light.

Geiger counter (Geiger-Müller counter). A gas-filled electron tube used to detect ionizing radiation, such as x-rays and alpha particles, and to count particles. A high electric potential difference (about 1 keV) maintained across the electrodes of the device is not quite sufficient to produce a gas discharge. Radiation entering the device ionizes the gas along its path. The resulting electrons are accelerated towards the anode by the electric field and produce more ions and electrons along their path. The resulting discharge is rapidly quenched by the gas and the electrons, collected on the anode, generate a current pulse, which is amplified and registered by a detector or counting device. The output is independent of the nature and energy of the ionizing radiation.

Gemini (Twins). A conspicuous zodiac constellation in the northern hemisphere near Orion, lying partly in the Milky Way. The brightest stars are the 1st-magnitude *Pollux (β) and the somewhat fainter multiple star *Castor (α), the 2nd-magnitude spectroscopic binary Alhena (γ), and several of 3rd magnitude. There are many interesting objects including variable stars, such as the *Cepheid Zeta Geminorum and the prototype of the U Geminorum *dwarf novae stars, binary stars, such as Eta Geminorum, the open cluster M35 visible to the naked eye, and the planetary nebula NGC 2392. Abbrev.: Gem; genitive form: Geminorum; approx. position: RA 7h, dec +25°; area: 514 sq deg.

Geminids. An important and active winter *meteor shower that maximizes on Dec. 13, meteors being visible between Dec. 7-15. The shower has a radiant of RA 113°, dec +32°, and a *zenithal hourly rate of 58; the meteoroids hit the atmosphere with a velocity of about 36 km s^{-1}. The *meteorstream orbit has a low semimajor axis (1·5 AU) and there is no known cometary association. The shower has yielded a fairly constant rate during the last century.

Gemini project. A series of US space missions that extended the knowledge gained from the *Mercury project and preceded the *Apollo programme. The project demonstrated that man could function effectively over long periods of weightlessness, both inside and outside a spacecraft, and that two spacecraft could be made to rendezvous and dock while in orbit. Gemini 3 was the first manned flight, launched in March 1965, with two astronauts on board. America's first spacewalk was made by Ed White in Gemini 4, launched in June 1965. Geminis 6 and 7 rendezvoused in Dec. 1965, and in Mar. 1966 Gemini 8 made the first successful docking with an unmanned target. Gemini 7 remained in orbit for a record 330.6 hours. Gemini 12, launched Nov. 1966, completed the project.

Gemma. *See* Alphecca.

general precession. *See* precession.

general relativity. *See* relativity, general theory.

geocentric coordinate system. A *coordinate system, such as the *ecliptic coordinate system, in which the position of a celestial body is referred to the centre of the earth. *Compare* heliocentric coordinate system; topocentric coordinates.

geocentric parallax. *See* diurnal parallax.

geocentric system. A model of the solar system or of the universe, such as the *Ptolemaic system, that has the earth at its centre.

geocentric zenith. *See* zenith.

geocorona. A tenuous cloud of hydrogen and some helium extending more than 50 000 km from the earth to the *magnetosphere and forming the outermost *atmospheric layer. It scatters Lyman-alpha radiation (*see* hydrogen spectrum) from the sun, producing a glow that is visible on long-distance far-ultraviolet photographs of the earth. It also interferes with far-ultraviolet astronomical observations from earth orbit.

geodesy. The study and measurement of the size, shape, and gravitational field of the earth, using both the measurements of the precise form of the *geoid and observations of artificial satellites, such as the *GEOS satellites.

geoid. The form of the earth obtained by taking the average sea level surface and extending it across the continents. The geoid differs from a sphere in that the earth's rotation has produced an *equatorial bulge:* the equatorial diameter (12 756.28 km) is greater than the diameter through the poles (12 713.55 km). The *flattening of the earth*—the *oblateness—corresponds to the difference between these diameters as a fraction of the equatorial diameter and has a value of 1/298.53.

Studies of the *perturbations affecting the orbits of artificial earth satellites have shown that the geoid departs from an oblate spheroid, or ellipsoid. The true geoid is elevated by 81 metres to the north of Australia and by 60 metres south of South Africa and west of the British Isles; in addition there are 50-metre depressions south of New Zealand and both east and west of the USA. The greatest departure is a 113 metre depression to the south of India.

geology. The study of the history, structure, and composition of the earth's crust. It overlaps with *geophysics, the study of the physics of the whole planet. The principle divisions of geology include *physical geology, historical geology,* and *economic geology.*

Physical geology includes mineralogy, which is the study of minerals; petrology, the study of the formation, composition, and structure of different types of rock; structural geology, concerned with how these rocks combine to form the crust; and geomorphology (or physiography), the study of the relation between geographical features and subsurface geological structures.

Historical geology includes stratigraphy, which is concerned with the chronology of rock strata; palaeogeography, the study of the distribution of geographical features (seas, deserts, mountains) during early periods of the earth's history; and palaeontology, dealing with the development of life as revealed by the fossil record on different geological strata.

Economic geology is concerned with the study of valuable mineral deposits, such as ores, coal, and oil.

geomagnetism. The earth's magnetic field (or its study), which approximates that of a bar magnet at the centre of the earth with its axis inclined by 11.4° to the earth's rotation axis. The magnetic flux density varies from 0.6 gauss near the magnetic poles to 0.3 gauss near the equator, i.e. from 60-30 microtesla, but can depart by up to 20% from the average dipole flux density without any correlation with major surface features. The dipole field changes only slowly with time but there are larger local variations in strength and direction, amounting, for example, to a compass swing of 35° in 240 years at London. Violent short-term fluctuations occur during *geomagnetic storms. Studies of magnetized rocks show that the entire magnetic field has reversed in direction at intervals of a few hundred thousand years. The source of the field is believed to lie in a complex dynamo action involving electric currents flowing in the rotating and convective outer core of the earth, which is composed of partly molten nickle-iron. *See also* ring current; magnetosphere.

geomagnetic storms. Sudden alterations in and subsequent recovery of the earth's magnetic field due to the effects of solar *flares. The variations are complex in the auroral zone and polar regions but at middle latitudes the horizontal component of the field shows four distinct phases.

The first is *Storm Sudden Commencement* (SSC), when a sharp rise in field strength (over 2.5 to 5 minutes) is caused by compression of the *magnetosphere by a flare-generated shock wave. The second is the *Initial Phase* (IP), lasting from 30 minutes to several hours, when the earth is surrounded by the high-speed post-shock plasma and field, and is effectively isolated from the interplanetary magnetic field. The surface field strength is higher than the pre-SSC value. In the *Main Phase* (MP) an increase in particle population, or particle acceleration due to reconnection of the geomagnetic and interplanetary fields, or in magnetospheric fluctuations produces a *ring current at three to five earth radii. This generates a magnetic field opposed to the earth's and causes a decrease of the surface field strength of 50-400 nanoteslas, which lasts from a few hours to more than a day. During the fourth phase, *Recovery Phase* (RP), which is typically longer than the MP, the ring current decays by diffusion of the trapped particles and plasma instabilities. The surface field strength may return to, or just below, the pre-SSC value.

Storms are usually accompanied by ionospheric and auroral activity, and some may recur after 27 days due to the persistence of a particular solar active region.

geometrical libration (optical libration). *See* libration.

geophysics. The physics of the earth. It includes the study of the history, motion, and constitution of the planet; movements within it, such as those associated with continental drift, mountain-building, earthquakes, glaciers, tides, and atmospheric circulation; geomagnetism; and the interaction between the earth and its interplanetary environment. It overlaps with geology—the study of the earth's crust and surface.

GEOS. The first all-scientific satellite in geostationary orbit, built by the European Space Agency for the purpose of studying the earth's *magnetosphere. GEOS-1, launched in April 1977, failed to reach a geostationary orbit and was eventually placed in an elliptical orbit; it was replaced by GEOS-2, which was successfully launched on July 14 1978 and carries seven experiments.

geostationary orbit. *See* geosynchronous orbit.

geosynchronous orbit. An earth orbit made by an artificial satellite (moving west to east) that has a period—23 hours 56 minutes 4.1 seconds—that is equal to the earth's period of rotation on its axis. If the orbit is inclined to the equatorial plane the satellite will appear from earth to trace out a figure-of-eight, once per day, between latitudes corresponding to the angle of orbital inclination to the equator. If the orbit lies in the equatorial plane, and is circular, the satellite will appear from earth to be almost stationary; the orbit and orbiting body are then termed *geostationary*. A geostationary orbit has an altitude of 35 900 km.

A geosynchronous or geostationary orbit is very difficult to achieve, requiring a very high orbital velocity. Satellites in such orbits are used for communications and navigation and also for certain types of earth observations. Most communications satellites are now geostationary, with groups of three or more, spaced around the orbit, giving global coverage.

German mounting. *See* equatorial mounting.

GHA. *Abbrev. for* Greenwich Hour Angle. *See* Greenwich Mean Time.

ghost crater. A lunar *crater that has been almost totally buried by *mare lavas. Ghost craters survive in the shallow irregular *maria where they can be made use of to calculate mare depths.

giant. A large highly luminous star that lies above the *main sequence on the *Hertzsprung-Russell diagram. Giants are grouped in luminosity classes II and III (*see* spectral types) and generally have absolute *magnitudes brighter than 0. Despite their great size, they are not necessarily more massive than typical main-sequence stars; their atmospheres are consequently very tenuous—a feature that shows in their spectra. Giants represent a late phase in *stellar evolution, when the central hydrogen supplies have been exhausted and the star has to turn to other forms of energy production. Capella and Arcturus are typical examples of giant stars. *See also* red giant; supergiant.

giant galaxy. *See* galaxies.

giant planets. The planets Jupiter, Saturn, Uranus, and Neptune, which have diameters between 3.8 and 11.2 times that of the earth and masses of between 14 and 318 earth masses. They orbit the sun at mean distances ranging from 5.20 astronomical units (AU) for Jupiter to 30.06 AU for Neptune in periods of from 11.86 to 164.79 years. All have low densities—from 0.7 to 1.7 times that of water—and are probably composed largely of hydrogen in its molecular or metallic state. Their visible surfaces are thought to be clouds of ammonia or methane. Jupiter, Saturn, and Uranus have ring systems, while all four giant planets share 31 satellites between them (*see* Table 2, backmatter).

giant radio galaxy. *See* radio galaxy.

giant star. *See* giant.

gibbous moon. *See* phases, lunar.

giga-. Symbol: G. A prefix to a unit, indicating a multiple of 10^9 of that unit. For example, one gigahertz is 10^9 hertz.

Giraffe. Common name for *Camelopardalis.

glitch. A discontinuity in the *phase of the pulse train from a *pulsar.

globular cluster. A spherically symmetrical compact *cluster of stars, containing from several tens of thousands to maybe a million stars that are thought to share a common origin. An example is the *Great Cluster in Hercules. A few globular clusters, such as Omega Centauri, appear to be slightly flattened. The concen-

Colour-magnitude diagram for typical globular cluster

tration of stars increases greatly towards the centre of the cluster, where the density may be as much as 1000 stars per cubic parsec. About 125 globular clusters are known. They appear to move in giant and highly eccentric elliptical orbits about the galactic centre. Unlike open clusters, they are not concentrated towards the galactic plane; instead they show a roughly spherical distribution in the galactic halo (*see* Galaxy).

Globular clusters are *population II systems: all the stars within them are relatively old (older than the sun) and have a very low metal content; their brightest occupants are *red giants. The distribution and other characteristics of globular clusters suggest that they were formed early in the life of the Galaxy, possibly some 12 to 13 thousand million years ago, before the main body of the galactic disc had evolved. Since most of the member stars will have evolved away from the main sequence, the *Hertzsprung-Russell diagram for stars of a globular cluster

globule

differs greatly from the conventional H-R diagram (see illustration). The *turnoff point from the main sequence gives a measure of the age of a cluster. Distances to globular clusters are usually calculated from the apparent magnitudes of the *RR Lyrae stars within them. *See also* open cluster.

globule. *See* dark nebula; Bok globule.

GMAT. *Abbrev. for* Greenwich Mean Astronomical Time.

GMST. *Abbrev. for* Greenwich Mean Sidereal Time. *See* Greenwich Sidereal Time.

GMT. *Abbrev. for* Greenwich Mean Time.

gnomon. A device used in ancient times to measure the altitude of the sun and hence to determine the time of day and time of year. It consisted of a vertical shaft of known height that cast a shadow of measurable length and direction on a graduated horizontal base. The ratio of height to shadow length gave the tangent of the altitude angle. The term is also applied to the metal projection on a sundial, used for the same purpose.

Goldstone. Site of the *Jet Propulsion Laboratory, California.

Gould Belt (Local System). A local formation of stars and clouds of gas and dust that appears to be a spur attached to the lower edge of the Orion arm of the *Galaxy. The system is about 700 parsecs wide and 70 parsecs thick and includes the II Persei, Scorpio-Centaurus, and *Orion O-associations (*see* association). The sun lies approximately 12 parsecs north of the Belt's equatorial plane and about 100 parsecs from its centre.

grading function. *See* aperture synthesis.

gram. A unit of mass equal to one thousandth of a *kilogram.

Grand Tour. A series of NASA space missions designed to take advantage of the *planetary alignment in the period 1977-82, whereby one or more probes could visit the outer planets in a relatively short time: the gravitational field of a giant planet would be used to swing the passing probe into the correct path towards the next target planet, where the process would be repeated, hence reducing the fuel requirements of the mission. The planets' proximity would reduce the total flight time. The Grand Tour was cancelled because it was too costly but the basic slingshot action will be used in the *Voyager project to the outer planets.

granulation. A network of convective cells in the solar *photosphere. It consists of bright irregularly shaped (often polygonal) *granules,* separated by dark narrow intergranular lanes. The individual granules are about 1000 km in diameter, have an upward velocity of about 0.5 km s^{-1}, and exhibit a horizontal flow of material—outward from the centre—with a velocity of about 0.25 km s^{-1}; their average lifetime is about eight minutes. They represent the changing tops of currents from the *convective zone bringing hot gases to the photosphere, where the gases cool and then return via the intergranular lanes.

Granulation is readily visible under favourable conditions, when it gives the photosphere a mottled appearance. It may be best seen near the centre of the sun's disc, where foreshortening is not significant. *See also* supergranulation.

graticule. A glass plate or cell bearing a grid, cross wires, or graduated scale that is set in the focal plane of the eyepiece of a telescope and is used for positioning or measuring.

grating. *Short for* diffraction grating.

grating response. *See* array.

grating ring. *See* aperture synthesis.

grating spectrograph. A *spectrograph in which a *diffraction grating is used to produce the spectra.

gravitation. The ability of all material bodies to attract each other. This mutual attraction, or *gravitational force,* is the most familiar force of nature and was first expressed in mathematical form by Isaac Newton. *Newton's law of universal gravitation* was published in 1687 in *Principia*. It states that the force of attraction, F, between two bodies is proportional to the product of their masses, m_1 and m_2, and inversely proportional to the square of their distance apart, d:

$$F = Gm_1m_2/d^2$$

The constant of proportionality, G, is the *gravitational constant. The force experienced by the mass m_1 is equal to but in the opposite direction to that felt by the mass m_2: the two masses are attracted towards each other. Newton showed that any body behaves, gravitationally, as if its mass were concentrated at its centre. Thus gravitational force acts along a line joining the centres of two bodies. In a system in equilibrium, such as the solar system or a star like the sun, gravitational force is balanced by an equal force acting in the opposite direction.

The region of space surrounding a massive body and in which the gravitational force is appreciable is the *gravitational field* of that body. The magnitude of the field at a particular point is the gravitational acceleration (in the direction of the massive body) that would be experienced by any object at that point (*see* acceleration of gravity).

Gravitation is the weakest force known. It is many orders of magnitude smaller than the other fundamental forces of nature—the strong, electromagnetic, and weak interactions. It is, however, the only means by which bodies with no overall electric charge can interact over the immense distances in the solar system, in the Galaxy, between the galaxies, and in the universe as a whole.

Newton's theory of gravitation has proved adequate in most circumstances but was challenged by the more complex general theory of *relativity put forward by Albert Einstein in 1916. According to general relativity, gravitational fields change the geometry of *space-time: both space and time are curved or warped around a massive body. Matter tells space and time how to curve and space-time tells matter how to move. *See also* Kepler's laws; quantum gravitation.

gravitational collapse. Contraction of a body arising from the mutual gravitational pull of all its constituents. Although there are several examples of such contraction processes in astronomy, 'gravitational collapse' usually refers to the sudden collapse of the core of a massive star at the end of nuclear burning, when its internal gas pressure can no longer support its weight. This may initially result in a *supernova explosion, removing much of the star's mass. The eventual degree of gravitational collapse is determined by the mass that remains. The three most likely end-products (in order of increasing mass) are *white dwarfs, *neutron stars, and *black holes.

gravitational constant. Symbol: G. A universal constant that appears in Newton's law of *gravitation and is the force of attraction between two bodies of unit mass separated by unit distance. It is equal to 6.672×10^{-11} N m^{2}kg^{-2}. Predictions that G is decreasing very slightly with time (by less than one part in 10^{10}) as a result of the expansion of the universe have not yet been definitely disproved or proved by experimental evidence.

gravitational field. *See* gravitation.

gravitational force. *See* gravitation.

gravitational mass. *See* mass.

gravitational radiation. *See* gravitational waves.

gravitational redshift (Einstein shift). The *redshift of spectral lines that occurs when radiation, including light, is emitted from a massive body. In order to 'climb out' of the body's gravitational field, the radiation must lose energy. The radiation frequency must therefore decrease and its wavelength, λ, shift by $\Delta\lambda$ towards a greater value. The redshift is given by

$$\Delta\lambda/\lambda = Gm/c^2r$$

G is the gravitational constant, m and r the mass and radius of the massive body, and c is the speed of light. Gravitational redshift was predicted by Einstein's general theory of *relativity and although extremely small has been detected in the spectra of the sun and several *white dwarfs. The redshift of the earth's gravitational field has been determined very accurately using beams of radiation travelling upwards through a tall building. Predicted and measured values agree very closely.

gravitational waves (gravitational radiation). Extremely weak wavelike disturbances that were predicted by Einstein's general theory of *relativity (1916). They are produced when massive bodies are accelerated or otherwis disturbed. They are ripples in the curvature of *space-time that travel at the speed of light, with a wide range of frequencies, and carry energy away from the source. They should affect all matter: gravitational waves hitting a suspended body, for example, should make it vibrate slightly. The interactions are very small, however. No conclusive evidence of the existence of gravitational waves has yet been forthcoming from the various highly sensitive experiments designed to detect them.

These waves should be emitted during supernova explosions or the collapse of massive stars into black holes. They should also be emitted by two massive stars in close orbit. Recent observations of the only known *binary pulsar show that its orbital period is decreasing by 10^{-4} seconds per year. This corresponds almost exactly with the decrease predicted to result from the emission of gravitational waves and is therefore possibly indirect evidence for their existence.

graviton. See quantum gravitation.

gravity. *Another name for* gravitation.

grazing incidence. X-rays reflect from surfaces for the same physical reason that light reflects, mainly due to coherent scattering in the surface material. However, since x-ray photon energies are large, the *refractive index, n, is slightly less than unity and reflection occurs only up to a critical angle, θ, where $\cos\theta = n$. The angle θ is usually small and thus x-ray reflection is only high for rays that graze the surface; hence 'grazing incidence'. For example θ is about 1° for a gold surface and x-rays of wavelengths of about 0.3 nm (4 keV). θ is proportional to x-ray wavelength and also increases with the *atomic number of the reflecting surface. *X-ray mirrors* are finding increasing application in *x-ray astronomy.

grazing occultation. A lunar

*occultation in which the astronomical body is only momentarily occulted by mountain peaks along the moon's limb.

Great Bear. Common name for *Ursa Major.

great circle. The circular intersection on the surface of a sphere of any plane passing through the centre of the sphere. A sphere and its great circles are concentric and of equal radius. *Compare* small circle.

Great Cluster in Hercules (M13, NGC 6205). The finest *globular cluster to be seen from the northern hemisphere, discovered in Hercules by Edmund Halley in 1714. It has a dense broad centre that is just visible to the naked eye; a 7.5 cm telescope will show some of the outer stars.

Great Dog. Common name for *Canis Major.

greatest elongation (GE). *See* elongation.

Great Looped nebula. *See* 30 Doradûs.

Great Nebula in Orion. *See* Orion nebula.

Great Red Spot. An immense oval feature centred 22° south of Jupiter's equator; it has a north-south width of 14 000 km and a variable east-west width of 24 000 to 40 000 km. Often visible in small telescopes, it has been observed for more than 100 years and may be identical with a feature recorded in the 17th century. Its prominence has varied with colour fluctuations between pale pink and brick red. Early explanations involved a solid island adrift in Jupiter's atmosphere or an atmospheric disturbance above a Jovian mountain or basin; the latter was disputed by evidence that it drifts erratically in longitude.

Observations by *Pioneers 10 and 11 reveal the Great Red Spot as a vortex of cold anticlockwise-rotating clouds elevated 8 km above the surrounding cloud deck by upward convection of warm gases from below. The red colour may come from condensation of phosphorus at the cloud tops or from contamination by organic molecules, such as nitriles, synthesized by lightning discharges deep in Jupiter's atmosphere. Probably it should be regarded as an outsized and long-lived, but not permanent, meteorological feature akin to smaller temporary spots observed on the planet.

Great Rift. A chainlike complex of large *dark nebulae that more or less obscures the light from a narrow but extensive band of the Milky Way between the constellations Cygnus and Sagittarius.

Great Square of Pegasus. *See* Pegasus.

Greek alphabet. *See* stellar nomenclature; Table 13, backmatter.

Green Bank. Site of the *National Radio Astronomy Observatory, West Virginia.

greenhouse effect. A phenomenon whereby an environment is heated by the trapping of infrared radiation. Although the effect may be of secondary importance in the heating of greenhouses, it is responsible for the high surface temperature of Venus and operates also on earth and in the deep atmospheres of the giant planets. Sunlight that filters to the Venusian surface through the dense clouds is absorbed and re-radiated at (longer) infrared wavelengths. The mainly carbon dioxide atmosphere is relatively opaque to infrared radiation, which it absorbs and re-emits. By the time the influx of sunlight to the surface is balanced by the escape of infrared radiation from the atmosphere, enough infrared radiation is being exchanged between the atmosphere and the surface to maintain both at a high temperature.

Greenwich Hour Angle (GHA). *See* Greenwich Mean Time.

Greenwich Mean Astronomical Time (GMAT). *Greenwich Mean Time beginning at noon. It was the time system in use before Jan. 1, 1925. GMAT is converted to *universal time, i.e. present-day GMT, by the addition of 12 hours.

Greenwich Mean Sidereal Time (GMST). *See* Greenwich Sidereal Time.

Greenwich Mean Time (GMT). The basis of standard time throughout the world. It is the *mean solar time in the time zone that is centred on the *Greenwich meridian,* i.e. the meridian of zero longitude passing through Greenwich. Since 1925 GMT has been counted from midnight rather than from noon and is therefore equal to the *hour angle of the *mean sun—the *Greenwich Hour Angle (GHA)*—plus 12 hours. It is thus now the same as *universal time. *See also* standard time.

Greenwich meridian. *See* prime meridian; Greenwich Mean Time.

Greenwich Observatory. *See* Royal Greenwich Observatory.

Greenwich sidereal date (GSD). A concept analogous to that of the *Julian date. It is the number of *sidereal days that have elapsed at Greenwich since the beginning of the sidereal day that was in progress on Julian date 0·0. The integral part of the date is the *Greenwich sidereal day number.* The nonintegral part is the *Greenwich Sidereal Time expressed either as a fraction of the sidereal day or in hours, minutes, and seconds.

Greenwich Sidereal Time (GST). The *sidereal time at Greenwich, i.e. the *hour angle at Greenwich of the vernal *equinox. GST at 0 hours GMT on various days of the year can be found from astronomical tables. To determine the GST at some other precise time, 3.94 minutes must be added for every complete day after a tabulated date together with a correction for part of a day. A further correction must be made to obtain *local sidereal time. *Greenwich Mean Sidereal Time* (*GMST*) is the hour angle at Greenwich of the *mean equinox (*see also* sidereal time).

Gregorian calendar. The calendar that is now in use throughout most of the world and that was instituted in 1582 by Pope Gregory XIII as the revised version of the *Julian calendar. The simple Julian four-year rule for leap years was modified so that when considering century years only one out of four, i.e. only those divisible by 400, were to be leap years: 1700, 1800, and 1900 were not leap years. There are therefore 365.2425 days per year averaged over 400 years. This greatly reduced the discrepancy between the year of 365.25 days used in the Julian calendar and the 365.2422 days of the *tropical year, which had resulted in the accumulation of 14 days over the centuries.

The revision came into effect in Roman Catholic countries in 1582, the year being brought back into accord with the seasons by eliminating 10 days from October: Thursday Oct. 4 was followed by Friday Oct. 15. The vernal *equinox, which should have occurred on March 11 and which had originally fallen on March 25 in Julius Caesar's time was thus adjusted to March 21. Gregory also stipulated that the New Year should begin on Jan. 1. Non-Catholic countries were slow to accept the advantages of the Gregorian reform. Britain and her colonies switched in 1752 when an additional day had accumulated between old and new calendars: Sept. 2, 1752 was followed by Sept. 14 and New Year's Day was changed from March 25 to Jan. 1, beginning with the year 1752. The very slight discrepancy between the Gregorian year and the tropical year amounts to about three days in 10 000 years.

Gregorian telescope. The first compound *reflecting telescope to be devised, designed by the Scottish mathematician James Gregory. It has a small concave secondary mirror that is mounted beyond the focal plane and reflects the light back through a central hole in the primary mirror (see illustration). The design, published in 1663, required a secondary mirror of ellipsoid figure. By the time this could be made in the 18th century the design was obsolete.

Grigg-Skjellerup comet. A comet that, after Encke, has the second shortest cometary period, 4.91 years.

GRIST. *Abbrev. for* Grazing Incidence Solar Telescope. A large high-resolution solar x-ray and ultraviolet telescope being constructed for the European Space Agency and planned to fly in several missions in *Spacelab after 1983. It will examine wavelengths in the spectral region between 10 and 170 nanometres, which covers a wide selection of solar phenomena, with a spatial resolution of one arc second.

groove. *See* rille.

ground state. *See* energy levels.

Grus (Crane). A fairly conspicuous but isolated constellation in the southern hemisphere, the two brightest stars, Alpha and Beta Gruis (a red giant), being of 2nd magnitude. Abbrev.: Gru; genitive form: Gruis; approx. position: RA 22.5 h, dec -50°; area: 366 sq deg.

GST. *Abbrev. for* Greenwich Sidereal Time.

G stars. Stars of *spectral type G that are yellow stars with a surface temperature (T_{eff}) of about 5000 to 6000 kelvin. The lines of ionized calcium (Ca II, H and K) dominate the spectrum and there are a large number of strong neutral metal lines (strengthening) and ionized metal lines (weakening) from G0 to G9. Molecular bands of CH and CN appear. The sun and Capella are G stars.

guide telescope. A telescope carried upon the same *equatorial mounting as an astronomical camera and used to ensure that the camera is accurately guided for the whole of an exposure, which may be of many hours' duration. For this duty a high magnification and an illuminated graticule

Ray path in Gregorian telescope

are required. The telescope drives must also be provided with slow motions in both declination and right ascension so that corrections can be applied manually as the selected *guide star* begins to drift from the cross wires. The camera will often be a larger telescope with a photographic plate exposed at its prime focus.

In a modern well-equipped observatory the tedious duty of manual guidance is now taken over by automatic devices that monitor the position of the guide star image and adjust the drive to hold it steady.

Gum nebula. An immense *emission nebula that dominates the Milky Way in the southern constellations Puppis and Vela. It is an expanding shell of ionized gas some 35° (250 parsecs) in diameter and about 400 parsecs distant. It has been suggested that it is a fossil *Strömgren sphere produced by a supernova that exploded maybe up to a million years ago and gave off sufficient ultraviolet radiation to ionize interstellar gas out to an immense distance. Although this original source no longer remains, several very hot stars within the shell might be ionizing the gas thus preventing it from returning to a neutral state. The *Vela pulsar and its supernova remnant lie in the same direction as the Gum nebula, possibly just beyond it.

gyrofrequency. *See* synchrotron emission.

H

HA. *Abbrev. for* hour angle.

Hadar. *See* Beta Centauri.

Hadley Rille. A sinuous *rille in Palus Putredinis, adjacent to the Apennine (third) ring of the *Imbrium Basin. It was sampled by the crew of *Apollo 15.

hadron. *See* elementary particles.

Hale Observatories. The group of observatories comprising the *Mount Wilson Observatory, Palomar Observatory* and *Big Bear Solar Observatory*, all in California and *Las Campanas Observatory,* Chile. The group is named after Gregory Ellery Hale, who played a great part in the development of the first two. The observatories are sponsored by the California Institute of Technology and the Carnegie Institute of Washington.

The Mount Wilson Observatory was built at an altitude of 1742 metres. Its first major instrument was a 60 inch (152 cm) reflector, completed in 1908, which was followed by a 100 inch (254 cm) reflector. The 100 inch, named the *Hooker telescope* after its benefactor, began regular operation in 1919 as the world's largest telescope. Its great success led Hale to consider an even larger telescope 200 inches (508 cm) in diameter. This project was underway by 1928. A site was chosen on Palomar Mountain, near Pasadena, for the observatory, at an altitude of 1706 metres. The mirror, ready for use in late 1947, almost ten years after Hale's death, had been cast in 1934 after considerable design problems. The first photographs were taken in 1948.

The 200 inch mirror of the *Hale telescope* is of low-expansion Pyrex glass, with a reflecting surface of aluminium; it weighs 13.15 tonnes. The chosen *focal ratio of f/3.3 meant that a much shorter tube could be used than with previous telescopes, which traditionally had an f/5 ratio. The prime focus is 16.5 metres from the prime mirror; there is an observing cage at the focus where much of the photographic work is done. A hyperboloid secondary mirror in the prime-focus cage can be used to provide a Cassegrain facility of focal ratio f/16 and two other similar mirrors with additional optics produce coudé systems of f/30. The Hale telescope was the world's largest telescope until the Soviet *Academy of Sciences' six metre instrument was built near Zelenchukskaya.

Additional to the 200 inch there is a 48 inch *Schmidt telescope. This has an aperture of 48 inches (122 cm) and a field of semiangle of three degrees, which produces a photograhic plate about 35.5 cm square. It was used in the production of the *Palomar Sky Survey. It is one of the world's finest Schmidts.

Hale telescope. *See* Hale Observatories.

half-life. The time taken for the number of atoms of a radioactive isotope to be reduced, by radioactive decay, to one half. The *mean life* is the average time before decay of a large number of similar elementary particles or atoms of a radioisotope. Mean life is equal to $1.44 \times$ half-life.

half-wave dipole. *See* dipole.

half-width. Half the width of a spectral line, measured at half the height. In some branches of spectroscopy it is customary to use the term to mean the full width at half height. *See also* line profile.

Halley's comet. A comet that was first recorded in 240 BC and has a period of 76 years, one return coinciding with the battle of Hastings (1066) and leading to the comet's representation on the Bayeux Tapestry. The orbit is retrograde, the comet moving in the opposite direction to the planets. It is named after Edmund Halley who showed that comets move around the sun in accordance with Newton's theory of gravitation. He noticed that the orbits of the bright comets of 1531, 1607, and 1682 were very similar and concluded that they were actually one and the same comet. He calculated its orbit and predicted that this comet would return in 1758. Unfortunately Halley died 16 years before the comet returned. Halley's comet can be seen again in the early spring of 1986. It is a firm contender for a spacecraft visit.

halo. 1. *See* Galaxy.
  2. *See* Donati's comet.

halo population II stars (halo objects). *See* populations I and II.

Hamal (α Ari). An orange giant that is the brightest star in the constellation Aries. m_V: 2.00; M_V: 0.3; spectral type: K2 III; distance: 22 pc.

h and Chi Persei (Sword-Handle; Double Cluster in Perseus). A fine double cluster in the constellation Perseus. Both the constituent *open clusters, NGC 869 (h Persei) and NGC 884 (Chi Persei) are visible to the naked eye. They are relatively young clusters, about 10 million years old, and appear to be associated with the surrounding Zeta Persei O-association. *See* association; O-cluster.

H and K lines. Lines in the spectrum of the calcium ion (Ca II), occurring at the visible/ultraviolet wavelengths of 393.3 nm (H) and 396.8 nm (K). They are very prominent *Fraunhofer lines.

Hare. Common name for *Lepus.

harmonic law. *See* Kepler's laws.

Haro galaxies. Galaxies found in the southern sky by G. Haro, that are all blue objects with sharp emission lines.

Harvard classification (Henry Draper system). *See* spectral types.

harvest moon. The full moon that occurs closest to the autumnal *equinox. At this time the *retardation is small (so that the moon appears to rise at the same time on successive evenings) because of the low inclination of the plane of the moon's orbit to the horizon.

Hayashi line. A near-vertical downward-directed track on the *Hertzsprung-Russell diagram that is believed to be followed by *protostars as they contract towards the *main sequence. Details of pre-main-sequence evolution are unclear but young stars on the Hayashi track are thought to be convective throughout, exceedingly luminous, and possibly so rarified that they are transparent.

HD. A prefix used to designate a star as listed in the *Henry Draper Catalogue.

H-D system. *Short for* Henry Draper system. *See* spectral types.

head. The coma and nucleus of a *comet when seen together. In this context 'nucleus' means the diffuse starlike luminous condensation sometimes ob served in the coma. The head generally contracts as *perihelion is approached and expands again afterwards. Remarkable changes in size, luminosity distribution, and the number of observed 'nuclei' can take place inside the head in a few hours.

HEAO satellites (High Energy Astrophysical Observatory satellites). A series of three large (about 3000 kg) earth satellites built by NASA for second-generation studies of *x-ray sources and *cosmic rays. *HEAO-A* was launched into a 400 km, 22° inclination orbit in August 1977 and carried a payload of large x-ray detectors with which to conduct a more sensitive sky survey than that of *Uhuru and *Ariel V. HEAO-A, or HEAO-1 after launch, was intended as a forerunner of the large *grazing-incidence x-ray telescope that was orbited, from November 1978, on *HEAO-2*.

HEAO-2 has been renamed the *Einstein Observatory*. Its telescope has brought *x-ray astronomy to a par with optical astronomy, providing images of x-ray sources with a resolution of several arc seconds and a point source sensitivity 1000 times that of earlier experiments. An important early result of the Observatory is the discovery of intense x-ray emission from *quasars over three million parsecs from earth. This radiation may account for a significant proportion of the low-level x-ray background radiation (*see* x-ray astronomy). Other observations have revealed the distribution of hot gas in *supernovae remnants, as well as distant *clusters of galaxies and also x-ray emission from a wide variety of stellar sources.

HEAO-C, due for launch in 1979, carries a number of large cosmic-ray instruments to study the mass and energy distribution of primary cosmic rays, particularly those in the group heavier than iron in the periodic table.

heavy elements. All elements heavier than hydrogen and helium. The *cosmic abundance of the heavy elements is very much lower than that of hydrogen and helium.

heavy-metal stars. *See* S stars.

heliacal rising (or setting). The rising or setting of a star or planet at or just before sunrise so that it can be seen in the morning sky, often for the first time in a year.

heliocentric coordinate system. A *coordinate system referring usually to the plane of the *ecliptic but specifying position with respect to the centre of the sun rather than that of earth (*see* geocentric coordinate system). The coordinates, *heliocentric latitude* and *longitude,* are used to determine the relative positions of objects in the solar system.

heliocentric parallax. *See* annual parallax.

heliocentric system. A model of the solar system or of the universe that has the sun at its centre. Aristarchus of Samos proposed a heliocentric system in the third century BC in which the planets moved at uniform rates around the sun in circular orbits. The theory was abandoned due to large discrepancies between prediction and observation, and the geocentric *Ptolemaic system was subsequently favoured until the establishment of the heliocentric *Copernican system.

heliographic latitude. The angular distance from the sun's equator, measured north or south along the meridian. The position of a point on the sun's disc may be defined in terms of this and its *heliographic longitude.

heliographic longitude. The angular distance from a standard meridian on the sun (taken to be the meridian that passed through the ascending *node of the sun's equator on 1854 Jan. 1 at 12h 0m UT and calculated for the present day assuming a uniform *sidereal period of rotation of 25.38 days; see sun), measured from east to west (0° to 360°) along the equator. The position of a point on the sun's disc may be defined in terms of this and its *heliographic latitude.

heliometer. An obsolete refracting telescope with a split objective lens that was used to measure the angular distance between two stars and to measure position angles. The two sections of the objective were moved until the images of the stars produced in the two sections were made to coincide. The amount of movement gave a precise measure of the angular distance.

heliosphere, heliopause. *See* solar wind.

Helios probes. Two 370 kg interplanetary spacecraft designed to observe the sun and the *solar wind. Constructed in West Germany, they were launched by the US in Dec. 1974 and Jan. 1976 into orbits taking them within 45 million km of the sun: closer than any previous probes.

heliostat. A *coelostat mirror mounted on a synchronously controlled drive mechanism so that it may reflect sunlight in a fixed direction as the sun moves across the sky. Together with additional optical elements it is a basic component of *solar telescopes.

helium. Chemical symbol: He. A chemical *element with an *atomic number of two. In its most abundant form it consists of a nucleus of two protons and two neutrons, around which orbit two electrons. This nucleus, i.e. the positively charged helium ion, is exceptionally stable; it is called an alpha particle. A second far less abundant isotope, helium-3, has two protons and one neutron as its nucleus. Two radioactive (i.e. unstable) isotopes also exist.

Helium is the second most abundant element in the universe (after *hydrogen): about 25% by mass and 6% or more by numbers of atoms. All but about 1% of this approximate 25% cosmic abundance is now considered to have been synthesized in the first few minutes of the universe (*see* big bang theory). Helium is also synthesized, from hydrogen, by *nuclear fusion reactions in the centres of main-sequence stars. It is by these reactions—the *proton-proton chain reaction and the *carbon cycle—that the energy of the stars is generated. When the hydrogen supplies in the stellar cores are exhausted, the helium is itself consumed by further fusion processes to form carbon (*see* triple alpha process; helium flash).

Helium is not easy to detect. It was first discovered in 1868 by Norman Lockyer: a yellow emission line of a then unknown element was observed in the spectrum of the sun's chromosphere, recorded during a solar eclipse. High temperatures are required to excite helium to emit or absorb radiation. Its spectral lines do not appear, for example, in the sun's absorption spectra, which originates in the photosphere. Even greater temperatures are required to ionize helium, i.e. to remove one or both of its electrons. Singly ionized helium does however occur in hot *O stars and in *emission nebulae. Helium, although a minor component in the inner planets, is a major constituent in the atmospheres of Jupiter and Saturn and possibly of Uranus and Neptune.

helium burning. The *nuclear fusion of three helium nuclei to form carbon. *See* triple alpha process.

helium flash. The explosive onset of helium burning (by the *triple alpha process) in the degenerate core of an evolved star when the core temperature reaches about 10^8 kelvin. The sudden increase in energy production causes a rapid rise in temperature, in turn increasing the reaction rate, but this runaway process eventually removes the degeneracy and the reaction once again becomes sensitive to the gas pressure.

helium star. *Obsolete name for* B star.

Helix nebula (NGC 7293). A *planetary nebula lying in the southern hemisphere about 140 parsecs away in the direction of Aquarius. It has the largest apparent diameter (about one degree) of any planetary but is too faint to be seen with the naked eye.

Hellas Planitia. A 1600 km diameter and 3 km deep basin in Mars' southern hemisphere. It can be seen as a bright area from the earth and is probably an impact feature. *See* Mars, surface features.

Helmholtz-Kelvin contraction (Kelvin-Helmholtz contraction). *See* Kelvin timescale.

Henry Draper Catalogue. Nine volumes of star catalogues compiled by Annie J. Cannon and E.C. Pickering of the Harvard College Observatory and published 1918-24 as part of the *Harvard Annals*. The catalogue (a memorial to the astronomer Henry Draper) lists the stellar spectra of 225 300 stars, each allocated a *spectral type according to the Harvard classification.

Henry-Draper system (H-D system). *Another name for* Harvard classification. *See* spectral types.

Herbig-Haro objects. Small bright concentrations of gas and dust, believed to represent the final stages in the evolution of a *protostar. Some of these objects, particularly those in the *Orion nebula, have changed their shape over the last 50 years; this has led to speculation that we are seeing light from the nucleus of young stars reflected off their surrounding cocoons of dust.

Hercules. An extensive but rather faint constellation in the northern hemisphere near Ursa Major. The brightest stars, Beta and Zeta Herculis, are of 3rd magnitude with many others of 3rd and 4th magnitude. There are several binary stars, including Zeta and *Ras Algethi (α) and the *x-ray binary Hercules X-1; globular clusters include the *Great Cluster in Hercules (M13), visible to the naked eye, and the smaller slightly fainter M92. Abbrev.: Her; genitive form: Herculis; approx. position: RA 17h, dec +30°; area: 1225 sq deg.

Hercules X-1. An *x-ray binary star, the second to be established as such (after *Centaurus X-3), based on *Uhuru satellite observations. It exhibits a complex behaviour, with regular 1.24 second *x-ray pulsations,* eclipses (by the binary companion star) every 1.75 days, and longer-term modulation of the x-ray intensity over a 35 day cycle. Hercules X-1 was the first x-ray binary to be optically identified; the optical counterpart, HZ Her, varies through the binary period from *spectral type A/F to B due to x-ray heating. Its mass (about 1.5 times that of the sun) and the 1.24 second pulsations strongly suggest the x-ray component to be a rotating *neutron star. Such stars are sometimes termed x-ray pulsars. The x-ray pulsations are believed to arise from channelling of the accreting gas onto the magnetic poles of the neutron star. The discovery in 1976 of a hard x-ray emission line at 55 keV, attributed to cyclotron radiation from near the star's surface, provides support for this view and gives a direct measure of the intense polar magnetic field (about 10^8 tesla).

Hermes. A *minor planet discovered in 1937 when it approached within 0.006 astronomical units (780 000 km) of the earth, coming closer than any other *earthgrazer. A very small member of the *Apollo group, it was lost because of uncertain knowledge of its orbit. *See* Table 3, backmatter.

Herstmonceux. Site of the *Royal Greenwich Observatory.

hertz. Symbol: Hz. The SI unit of frequency, defined as the frequency of a periodic phenomenon that has a period of one second. The frequency range of *electromagnetic radiation is about 3000 Hz (very low frequency radio waves) to about 10^{22} Hz (high-frequency gamma rays).

Hertzsprung gap. A region on the *Hertzsprung-Russell diagram, to the right of the main sequence, in which few stars are found. This is because of a star's rapid evolution through this region, away from the main sequence. This occurs during the period when hydrogen is being burnt in a shell around the core of helium, before the onset of helium burning (*see* stellar evolution). The gap can be easily seen on H-R diagrams of open clusters.

Hertzsprung-Russell diagram (H-R diagram). A two-dimensional graph that demonstrates the correlation between *spectral type and *luminosity of stars, discovered independently by the Danish astronomer E. Hertzsprung in 1911 and the American astronomer H.N. Russell in 1913. Instead of a uniform distribution the stars were found to form well-defined groups or bands on the graph of spectral type plotted against absolute visual *magnitude: about 90% lie along a diagonal band known as the *main sequence; the somewhat brighter *giant stars, belonging to the later spectral types, form another sequence as do the very brightest and relatively rare *supergiants; *white dwarfs and other groupings can also be distinguished.

There are several forms of H-R diagram. In most observational applications the absolute visual magnitude, M_V, is the horizontal axis and the *colour index, $B-V$, the vertical axis, colour index being related to but more easily measured than spectral type. These are *colour-magnitude H-R diagrams*. Sometimes apparent rather than absolute magnitude is used. Other studies use bolometric *magnitude against *effective temperature—*theoretical H-R diagrams*—or luminosity against colour index—*colour-luminosity H-R diagrams*.

The H-R diagram is of great importance in studies of *stellar evolution. Diagrams obtained on the basis of theoretical calculations can be tested against observationally determined diagrams. They can be drawn for the brightest

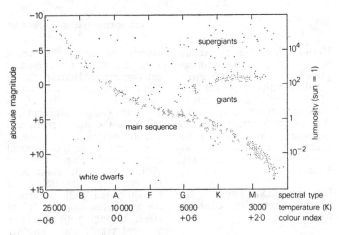

Hertzsprung-Russell diagram for bright stars

stars (see illustration), for stars in a particular locality such as the solar neighbourhood (mainly small cool main-sequence stars), for *pulsating variables, for star clusters, etc. The two broad stellar populations (*populations I and II) can be demonstrated by the H-R diagrams of a young *open cluster (no giants), a somewhat older open cluster (a few giants), and a much older *globular cluster (many giants and supergiants). The H-R diagram can also be used for distance determination by both *main-sequence fitting for stellar clusters and by spectroscopic *parallax for individual main-sequence stars, the star's spectral type fixing its position on the diagram and thus indicating its absolute magnitude and hence its *distance modulus.

Hesperus. The Greek name for *Venus when it was visible as an *evening star. *Compare* Phosphorus.

heterochromatic magnitude. *See* magnitude.

Hevelius formation. The *ejecta blanket of the *Orientale Basin. *See also* highland light plains; Fra Mauro formation.

hexahedrite. *See* iron meteorite.

Hidalgo. *Minor planet number 944 that was discovered in 1920 and has an eccentric 14.04 year orbit that carries it between 2.00 AU and 9.64 AU from the sun. Until the discovery of *Chiron in 1977, it had the furthest aphelion and longest period of any known minor

hierarchical universe. A *cosmological model, originally due to Charlier in 1908, in which inhomogeneities exist on increasingly larger scales: galaxies occur in *clusters, which in turn are clustered into superclusters, and so on. There is little observational evidence for clumping of galaxies on scales greater than about 20 megaparsecs.

highland light plains (Cayley formation). The widespread relatively level 'ponds' of light-coloured materials that have accumulated in depressions in the lunar *highlands. Originally thought to be highland volcanics they are now believed to be rubble derived from impacts, possibly from the *Orientale Basin.

highlands, lunar (terra). The topographically elevated and more rugged regions of the *moon. They have a higher *albedo, and thus appear brighter, than the *maria and consist largely of anorthosite, a calcium-aluminium silicate rock. Lunar highland areas include most of the farside and southern part of the *nearside of the moon. *Craters are larger and more abundant here than in the maria; this reflects the greater age of the highland crust, parts of which date back 4600 million years to the formation of the moon. *See also* mountains, lunar; moonrocks.

high-velocity stars. Very old (*population II) stars in the galactic halo that are relatively near the sun but do not share in the common circular motion around the galactic centre of the sun and other stars in the sun's neighbourhood (*see* local standard of rest). They are actually moving slower than the sun in elliptical orbits that often carry them far out of the galactic plane. Their apparently high velocities (>65 km s^{-1}) relative to the sun result from this difference in relative motions.

HII region (H$^+$ region). A region of predominantly ionized hydrogen in interstellar space, existing mainly in discrete clouds. The ionization may be caused by photoionization by ultraviolet photons, especially in regions of star formation, and also by cosmic rays, x-rays, or shock waves in the medium. In comparison with the 21 cm radio emission of neutral hydrogen in *HI regions, the ionized hydrogen of HII regions emits radio waves by *thermal emission and *recombination-line emission. HII regions are roughly spherical with a sharply delineated boundary (*see* Strömgren sphere). Their size is usually less than 200 parsecs. By studying the apparent diameters of the HII regions in a distant galaxy, the distance to the galaxy can be estimated. *See also* interstellar medium; emission nebula.

Himalia. A satellite of Jupiter, discovered in 1904. *See* Jupiter's satellites; Table 2, backmatter.

Hind's nebula (NGC 1554-5). A *reflection nebula in the constellation Taurus, illuminated by the star T Tauri. The nebula has varied considerably in brightness in the past century.

Hirayama families (minor-planet families). Groupings of *minor planets whose members orbit at approximately the same distance from the sun and have very similar orbital characteristics (period, *eccentricity, *inclination). More than 40 families are recognized, some containing more than 70 members. They are named after the Japanese astronomer K. Hirayama who discovered the first nine families in 1928. It is believed that each family represents debris resulting from a collision between larger minor planets in the past. In contrast, the *Amor group and the *Apollo group each comprises minor planets with a range of orbital characteristics. A large number of minor planets are members of Hirayama families.

HI region (H⁰ region). A diffuse region of neutral, predominantly atomic, hydrogen in interstellar space. Although the temperature (about 100 kelvin) is too low for optical emission, neutral hydrogen emits radio radiation at the spot frequency of 1420.4 megahertz, which corresponds to a wavelength of about 21 cm. The emission, termed the *hydrogen line* or *21 cm line* emission, is associated with a *forbidden transition between two closely spaced *energy levels of the ground state, related to the relative electron and proton spin orientation in the hydrogen atom.

This radio emission has allowed the distribution and relative velocity of neutral hydrogen to be studied both in our own and in nearby spiral *galaxies, using, for example, electronic *line receivers. Motions within the galaxy cause the observed frequency (1420.4 MHz) to be displaced by the *Doppler effect. The velocity is determined from the frequency shift. *See also* interstellar medium; HII region.

H line. *See* H and K lines.

Hoba meteorite. The world's largest single meteorite mass, found in 1920 near Grootfontein, Namibia. It is an *iron meteorite weighing 60 tonnes and measures $2 \cdot 7 \times 2 \cdot 7 \times 1$ metres. It still lies at its original resting place and produced no crater, being only partly buried in the ground.

Hohmann transfer (or orbit). *See* transfer orbit.

H₂O maser. *See* maser source.

homogeneity. The property by which all positions in the universe are equivalent, in terms of the measurement of pressure, density, etc., at a given cosmic time. *See* cosmological principle; isotropy.

Hooker telescope. *See* Hale Observatories.

horizon. The great circle in which an observer's *horizontal plane* meets the *celestial sphere. The horizontal plane through an observer's position lies at right angles to the plumb line and is tangential to the earth's surface, which must be considered spherical. The *zenith and *nadir are the poles of the horizon. *See also* cardinal points.

horizontal branch. A horizontal strip on the Hertzsprung-Russell diagram of a *globular cluster, to the left of the red giant branch. It consists of low-mass stars that have probably lost mass during the red-giant phase; they have absolute magnitudes of about zero. A gap occurs in the horizontal branch where only variable stars (*RR Lyrae stars) are found.

horizontal (or horizon) coordinate system. A *coordinate system in which the fundamental reference circle is the *hori-

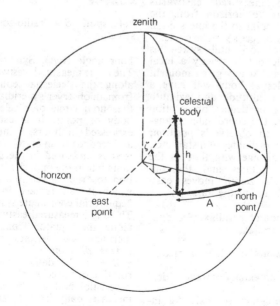

Horizontal coordinate system

zon and the zero point is the north point (*see* cardinal points). The coordinates are *altitude* and *azimuth* (see illustration).

The altitude (*h*) of a celestial body is its angular distance north (counted positive) or south (counted negative) of the horizon; it is measured along the vertical circle through the body and ranges from 0°, when the object rises or sets, to 90°, when the object is directly overhead at the zenith. The *zenith distance (ζ) is the complement of the altitude (90° - *h*) and is frequently used instead of altitude in the horizontal system. The azimuth (*A*) of a body is its angular distance measured eastwards along the horizon from the north point to the intersection of the object's vertical circle.

The horizontal system is simple but is strictly a local system. At any given moment a celestial body will have a unique altitude and azimuth for a particular observation point, the coordinates changing with observer's position. The coordinates of a star, etc., also change with time as the earth rotates and the observer's zenith moves eastwards among the stars.

horizontal parallax. *See* diurnal parallax.

horizontal plane. *See* horizon.

horn antenna. *See* waveguide.

Horologium (Clock). An inconspicuous constellation in the southern hemisphere near the star Achernar. The brightest stars are of 5th magnitude, apart from one of 4th magnitude. Abbrev.: Hor; genitive form: Horologii; approx. position: RA 3.5h, dec -50°; area 249 sq deg.

Horsehead nebula (NGC 2024). A *dark nebula in Orion located near the star Zeta Orionis. It is seen as a dark region against the bright background of an emission (HII) nebula.

horseshoe mounting. *See* equatorial mounting.

hot big bang. *See* big bang theory.

hot spot. *See* radio-source structure.

hour angle (HA). Symbol: *h*. The angle measured westward along the *celestial equator from an observer's meridian to the *hour circle of a celestial body or point. It is usually expressed in hours, minutes, and seconds from 0h to 24h. It is thus measured in the same units but in the opposite direction to *right ascension, RA, which it can replace in the *equatorial coordinate system. The angle measured eastward along the equator from the meridian is sometimes called the *meridian angle*.

Due to the daily apparent rotation of the celestial sphere, a celestial body's hour angle increases daily from 0h at the meridian; after six hours the hour angle is 6h and 24 hours later the celestial body again crosses the meridian. The hour angle of the vernal *equinox at any instant gives the *sidereal time. For a particular star then:

RA + HA = sidereal time

hour circle. Any of a series of great circles on the *celestial sphere along which *declination is measured. The hour circle of a particular celestial body or point passes through the body or point and the north and south celestial poles. *See also* hour angle.

H-R diagram. *Short for* Hertzsprung-Russell diagram.

Hubble classification. A classification scheme for galaxies introduced by Edwin Hubble in 1925. Although more complex and refined schemes have since been devised the simple but slightly revised Hubble scheme is still widely used. The scheme recognizes three main types of galaxy: elliptical, spiral, and barred spiral. Each is divided into subtypes according to gross observable characteristics, the principle one being shape. These subtypes are arranged into a morphological sequence (see illustration).

Elliptical galaxies are denoted by the letter E followed by a number from 0 to 7 that indicates the apparent degree of flattening; this number is the nearest integer to

$$10(1-b/a)$$

where a and b are respectively the semimajor and semiminor axes of the observed profile. Thus E0 galaxies are almost circular in outline and E7, the most elliptical, have an $a:b$ ratio of about 3:1. The spirals (type S) and barred spirals (type SB) are separated into subtypes a, b, and c along a sequence that follows a progressive decrease in importance of the nucleus and an increase in the openness and prominence of the spiral arms.

The less commonly observed irregular galaxies are designated Irr I if their contents are predominantly blue *population I objects and Irr II if they contain mostly population II objects. Irregular galaxies were not included in Hubble's original diagram; they are now generally placed to the right of the S or SB sequences. The S0 galaxies, placed at the junction of the elliptical and spiral sequences, resemble the spirals in general shape but lack spiral arms. This type was predicted by Hubble in 1925 but only identified later.

Hubble constant. *Symbol:* H_0. The rate at which the expansion velocity of the universe changes with distance; it is commonly measured in km s^{-1} megaparsec^{-1}. Current estimates place H_0 at about 55 km s^{-1} Mpc^{-1} but higher values up to about 100 km s^{-1} Mpc^{-1} are also used. The expansion velocity is measured from galaxy *redshifts but the distance estimates are more complicated. Most of the steps

Hubble diagram

Hubble's 'tuning fork' diagram

of the cosmic distance ladder (*see* distance determination) are involved, including *parallax, *Cepheid variables, *HII regions, and the brightest galaxies. Various corrections to this distance ladder over the years have resulted in drastic revisions to the estimates of H_0: it is now approximately 10 times smaller than when measured by Hubble.

The inverse of H_0 has the dimensions of time and is a measure of the *age of the universe—the *Hubble time*. In an evolving universe the Hubble 'constant' actually changes with time at a rate dependent upon the *deceleration parameter, q_0. It is only independent of time in a *steady-state universe where q_0 equals -1. In terms of the *cosmic scale factor, R, H_0 equals $(dR/dt)/R$.

Hubble diagram. A plot of the *redshift of galaxies against their distance (see illustration) or of redshift against apparent magnitude. Apparent magnitude is a crude measure of galactic distance, the value for the brightest or the third or tenth brightest member of a *cluster of galaxies being used. *See* Hubble's law; Hubble constant.

Hubble flow. *See* Hubble's law.

Hubble nebula (Hubble's variable nebula; NGC 2261). A fan-shaped *reflection nebula in the constellation Monoceros that is illuminated by the star R Monocerotis. R Mon is an A-F peculiar variable star and causes the brightness and outline of the nebula to vary.

Hubble-Sandage variable. The most massive and luminous stars (M_v>-9 and all spectral types) occurring in the nearby galaxies M31 and M33 that were found to be *variable stars. The variation in brightness, discovered by H

Hubble diagram compiled for several kinds of galaxies

Hubble and A. Sandage in 1953, were in most cases slow, small, and irregular although some displayed much greater variation. They are thought to be young unstable stars that cannot evolve normally owing to their huge mass, which could exceed 50 solar masses. *Eta Carinae and *P Cygni are possible examples in our Galaxy.

Hubble's law (velocity-distance relation). The law, first proposed by the American astronomer Edwin Hubble in 1929, stating that the recession velocity, V, of a distant extragalactic object (one outside the *Local Group) is directly proportional to its distance, D. The constant of proportionality is known as the *Hubble constant, H_0, thus

$$V = H_0 D$$

The law is a direct consequence of a uniformly expanding isotropic universe. The separation of points on a balloon, as measured along the surface of the balloon, behaves in just this manner. Variations from a linear relationship—or *Hubble flow*—due to the *velocity dispersion in the Local Group

Hubble time

and in nearby *clusters of galaxies are detectable.

Hubble time. *See* Hubble constant.

Hunting Dogs. Common name for *Canes Venatici.

Huygens eyepiece. *See* eyepiece.

Hyades. An *open cluster of over 200 stars in the constellation Taurus. It is about 46 parsecs away. Its brightest members form a V-shaped group that is visible to the naked eye. The Hyades is a *moving cluster. Its age is about 500 million years.

hybrid ring. A four-port device by which two input signals A and B are combined to give two output combinations $A + B$ and $A - B$, with a high degree of isolation between one input port and the other and between one output port and the other, and such that identical cables are matched (*see* matching) at all four ports.

Hydra (Sea Serpent or Monster). The largest constellation, straggling between Leo in the northern hemisphere and Centaurus in hemisphere and Centaurus in the south. Its stars are rather faint apart from the solitary 2nd-magnitude *Alphard. It contains several types of variable stars, the just-visible open cluster M48, the globular cluster M68, the bluish planetary nebula NGC 3242, and the spiral galaxy M83. Abbrev.: Hya; genitive form: Hydrae; approx. position: RA 8 -15h, dec +7° to 35°; area: 1303 sq deg.

hydrogen. Chemical symbol: H. The simplest chemical *element, with an *atomic number of one and the lowest density of all elements. In its most abundant form it consists of a *proton orbited by a single electron. Two other forms, i.e. isotopes, exist: *deuterium* has a proton plus a neutron as its nucleus; *tritium*, which is radioactive (i.e. unstable), has a proton plus two neutrons.

Hydrogen is the most abundant element in the universe: 94% by numbers of atoms, 73% by mass. It is generally agreed that all this hydrogen is primeval in origin, created in the earliest phase of the universe (*see* big bang theory). The amount of deuterium formed relative to the amount of hydrogen would have been very sensitive to the density of matter at the time of formation: deuterium readily combines with an additional neutron to form tritium, which rapidly decays into an isotope of helium. Thus if the universe was very dense in its first few minutes, most of the deuterium would have been converted to helium. Deuterium is not easy to detect but recent measurements of the ratio of deuterium to hydrogen give an upper limit of about 2 × 10⁻⁵ for interstellar gas.

Hydrogen can exist in atomic, molecular, and ionized forms. Ionized hydrogen, H⁺, more usually denoted HII, results when neutral atoms are stripped of their electrons. It occurs at high densities and temperatures, as in the centres of stars. *Nuclear fusion of hydrogen ions, i.e. protons, in the stellar core generates the energy of main-sequence stars by either the *proton-proton chain reaction or the *carbon cycle. Ionized hydrogen can also be produced by *photoionization, as in *emission nebulae, and may then be detected by its *recombination line emission (*see also* HII region).

In addition to the positive ion, the negative ion, H⁻, can occur when an electron attaches itself loosely to a neutral atom. A continuous spectrum of radiation is emitted in the process. The negative ion is not very stable and breaks up easily with the second electron escaping with any amount of energy: this results in a continuous absorption spectrum. These two processes, producing continuous emission and absorption of both light and infrared radiation, take place concurrently, as happens in the sun's *photosphere. The H⁻⁻ ion has recently been discovered.

Neutral hydrogen, usually denoted HI, occurs throughout interstellar space as filaments and clouds (*HI regions) of varying density. It is detected by means of the 21 cm hydrogen line emission. At low temperatures and when the hydrogen density is sufficiently high, pairs of hydrogen atoms can combine to form molecular hydrogen, which exists in discrete interstellar clouds. Molecular hydrogen is also a major constituent of the atmosphere of the *giant planets, with liquid hydrogen forming the bulk of the interiors of the giants.

See also hydrogen spectrum.

hydrogen alpha line (Hα line). *See* hydrogen spectrum.

hydrogen line (21 cm line). *See* HI region.

hydrogen region. *See* HI region; HII region; interstellar medium.

hydrogen spectrum. Emission and absorption spectra of hydrogen are relatively simple compared with spectra of heavier elements. Atomic hydrogen (HI) has a *line spectrum in which several series of lines can be distinguished. The Swiss physicist Johann Balmer showed, in 1885, that lines in the visible region of the hydrogen spectrum formed a series represented by the equation

$$1/\lambda = R(1/4 - 1/m^2)$$

λ is the wavelength of the line and m an integer greater than two. The constant R has a value of 1.096 78 × 10^{-7} m^{-1}, and is called the *Rydberg constant*.

This series of spectral lines is called the *Balmer series*. Its existence was first explained by Niels Bohr in his theory of the atom (1913). Bohr postulated that only a discrete number of orbits is allowed to an electron in an atom and that when the electron jumps from one orbit to another, a photon of radiation is emitted or absorbed (*see* energy level). The Balmer series is produced by transitions of the electron between the second permitted orbit, in order of distance from the nucleus, and higher orbits. Thus when $m = 3$, the spectral line is produced by a transition from the second to the third orbit (or from the third to the second orbit). This line is referred to as Hα (the *hydrogen alpha line*) and has a wavelength of 656.3 nm, i.e. it falls in the red region of the visible spectrum. The next line, Hβ ($m = 4$), is at 486.1 nm in the blue. Hγ and Hδ occur at 434.2 nm and 410.2 nm, respectively. The lines in such a series get closer together at shorter wavelengths and the Balmer series converges to a limit at 364.6 nm in the ultraviolet region of the spectrum. Emission lines are produced by transitions from higher levels to the second orbit; absorption lines result from transitions from the second orbit to higher orbits.

In general, lines in the hydrogen spectrum can be represented by the equation

$$1/\lambda = R(1/n^2 - 1/m^2)$$

where n and m are integers; n is called the *principal quantum number* and can have values from one to infinity. The Balmer series is formed for $n = 2$, i.e. the second orbit of the atom. Transitions from or to the first orbit, i.e. the ground state ($n = 1$), produce the *Lyman series* of spectral lines. This is in the far ultraviolet region. Lyman alpha, denoted Ly α ($n = 1$, $m = 2$), occurs at 121.6 nm; the series Ly α, Ly β, Ly γ . . . , converges to a value of 91.2 nm. The *Paschen series* is produced by transitions to or from the third permitted orbit ($n = 3$) and occurs in the near infrared region of the spectrum. The *Bracket series* occurs at longer infrared wavelengths involving the fourth orbit ($n = 4$).

hydrostatic equilibrium. A state of equilibrium in which the inwardly directed gravitational force in a star just balances the outwardly directed gas and radiation pressure. The star is thus held together but supported against collapse. The equation of hydrostatic equilibrium is given at *stellar structure.

hydroxyl radical. The chemical radical OH, consisting of an oxygen and hydrogen atom bound together. It was the first interstellar molecule to be detected (1963) in the radio region of the spectrum and is also associated with certain red and infrared stars (*see* maser source). The spectral lines—OH lines—occur at 18, 6.3, 5.0, 3.7, and 2.2 cm.

Hydrus (Water Snake). A small constellation in the southern hemisphere lying between the Large and Small Magellanic Clouds. Its brightest star, Beta Hydri, is of 3rd magnitude and is the nearest conspicuous star to the south pole. Abbrev.: Hyi; genitive form: Hydri; approx. position: RA 2h, dec -70°; area: 243 sq deg.

hyperbola. A type of *conic section that has an eccentricity greater than one. *See also* orbit.

hyperboloid reflector. A reflector whose surface is hyperboloid in shape, i.e. its curvature is that obtained by rotating a hyperbola about its axis. The secondary mirror in a *Cassegrain telescope is hyperboloid.

Hyperion. A satellite of Saturn, discovered in 1848. *See* Table 2, backmatter.

Hz. *Symbol for* hertz.

I

Iapetus. A satellite of Saturn, discovered in 1671. It appears brighter when west of the planet, presumably because it presents the same face to Saturn as it orbits and has hemispheres with different *albedos. Recent research indicates that it has a water ice surface. *See* Table 2, backmatter.

IAT. *Abbrev. for* international atomic time. *See* atomic time.

IAU. *Abbrev. for* International Astronomical Union.

IC. *Abbrev. for* Index Catalogue. *See* NGC.

Icarus. *Minor planet number 1566 that was discovered in 1949 and passed only 0.04 astronomical units (AU) from the earth in 1968. It belongs to the *Apollo group and has the smallest *perihelion distance (0.19 AU) of any minor planet, well within the orbit of Mercury. *See* Table 3, backmatter.

IF amplifier. *Short for* intermediate-frequency amplifier. *See* receiver.

IGY. *Abbrev. for* International Geophysical Year.

IIIaJ emulsion. *See* photography.

Ikeya-Seki comet (1965 VIII). A *comet that was discovered by two amateur Japanese comet searchers. At one time it was actually thought that it would hit the sun but it turned out to be a sungrazer with a perihelion distance of 0.0078 astronomical units, missing the sun's surface by 0.68 solar radii. The orbit is retrograde with an eccentricity of 0.99915.

image converter. *See* image tube.

image intensifier. *See* image tube.

image orthicon. *See* television camera.

image photon-counting system. *See* photon-counting system.

image tube (image intensifier). An evacuated electronic device that is used to intensify a faint optical image. The beam of light (or near-ultraviolet radiation) falls on a *photocathode so that electrons are liberated by the photoelectric effect. The electrons are accelerated by an electric field and may be detected and recorded by a variety of methods. They can be focused on a positively charged phosphor screen so that an optical image is produced, many times brighter than that of the original beam. This image can be made to fall on the photocathode of a second image tube and the intensification process repeated. Several image tubes can be linked in this way to form a *multistage image tube* but the coupling between phosphor screen and subsequent photocathode must be very efficient. The image on the final phosphor screen may then be photographed. The *electronographic camera and the *photon-counting system use slightly different techniques.

In *image converters* a beam of infrared radiation or x-rays is incident on the photosensitive surface. The liberated electrons can then produce a pictorial representation of a 'nonvisible' image.

Imbrium Basin. The second youngest lunar *basin, of which much of the second and third rings survive as mountain arcs (*see table at* mountains, lunar). The *ejecta blanket of the basin covers much of the *nearside and provides the major reference horizon for lunar stratigraphy. The basin was formed 3900 million years ago and was flooded with basalt lavas during at least the following 600 million years. This produced *Mare Imbrium* (*Sea of Rains*), which was visited by Lunokhod 1, and also *Palus Putredinis,* which was sampled by Apollo 15. *See also* Fra Mauro formation.

immersion. Entry of a celestial body into a state of invisibility during an *eclipse or *occultation.

inclination. 1. Symbol: *i*. The angle between the orbital plane of a planet, comet, or satellite and the plane of the *ecliptic. It is one of the *orbital elements and varies between 0 and 180°, being less than 90° for a body with *direct motion. *See also* Tables 1-3, backmatter.
2. The angle between any two planes, such as that between the orbital plane of a satellite and the primary's equatorial plane.

Index Catalogues (IC). *See* NGC.

Indus (Indian). A constellation in the southern hemisphere near Grus, the brightest stars being of 3rd magnitude. Abbrev.: Ind; genitive form: Indi; approx. position: RA 21h, dec -60°; area: 294 sq deg.

inequality. An irregularity in the orbital motion of a celestial body. The inequalities of the moon's motion are periodic terms whose sum gives the variation of either the spherical coordinates or the osculating elements of the lunar orbit. The principal inequalities of the moon's motion are *evection and *variation.

inertia. The property of a body by which it resists change in its velocity. It is inertia that causes a body to continue in a state of rest or of uniform motion in a straight line (*see* Newton's laws of motion). The force required to give a specific acceleration to a body depends directly on its inertia. It is through the property of inertia that the concept of the *mass of a body (its *inertial mass*) arises. *See also* Mach's principle.

inertial frame. A *frame of reference in which an isolated particle placed at rest will remain at rest and one in uniform motion will maintain a constant velocity. It is thus a frame of reference that is not subject to acceleration, as from a gravitational field, and in which Newton's first law of motion applies. According to Einstein's special theory of *relativity, which is concerned only with inertial frames, all laws of physics are the same in every inertial frame.

inertial mass. *See* inertia; mass.

inferior conjunction. *See* conjunction.

inferior culmination (inferior transit). *Another name for* lower culmination. *See* culmination.

inferior planets. The planets Mercury and Venus, which lie closer to the sun than the earth.

infrared astronomy. The study of radiation from space with wavelengths beyond the red end of the visible spectrum, i.e. from about 0.8 micrometre to 1 mm. Observations are possible from the ground through several *atmospheric windows up to about 17 micrometres but longer-wavelength observations require balloon-, rocket-, or satellite-platforms. The equipment used includes large modified reflecting telescopes (*see* infrared telescope) and solid-state infrared photon detectors, such as *photovoltaic devices. A major difficulty is the need to cool at least the infrared detectors, and possibly the telescope too, to reduce intrinsic thermal *noise. For this reason early observations in the 1960s and '70s were carried out from the ground or in short rocket or balloon flights.

The first satellite-borne infrared telescope is due for launch in 1981, as a joint Dutch-US-UK enterprise. The satellite, *IRAS,* is expected to provide the first deep sky survey over the waveband of approximately 5 to 200 micrometres and is expected to yield 10^6 to 10^7 sources.

Cosmic infrared radiation arises in two main ways: thermally, from sources with temperatures (of gas or dust) in the range of about 100 to 3000 kelvin, or nonthermally, as the radiation of high-energy electrons, interacting for example with a magnetic field to produce *synchrotron emission. Early ground-based surveys have shown many bright *infrared sources.

infrared radiation (IR radiation). *Electromagnetic radiation lying between the radio and the visible bands of the electromagnetic spectrum. The wavelengths range from about 0.8 micrometres to about 1 millimetre. The shorter wavelengths, i.e. those close to the red end of the visible region, are often called *near infrared,* with *far infrared,* being applied to the longer wavelengths.

infrared sources. Any of a variety of objects emitting a considerable amount of infrared radiation. Most lie within our Galaxy. Since the mid-1960s observations in the near infrared (2-20 micrometres) have revealed many thousands of stars that radiate much more strongly at these wavelengths than in the visible region (0·4-0·75 micrometres). The great majority emit *thermal radiation, are cool (temperature range about 500-2500 kelvin), and are necessarily large to be seen (note: black-body *luminosity is proportional to T^4). In many cases the infrared radiation arises from a circumstellar dust cloud heated by absorbed light from a central star. Other infrared stars, such as *Becklin's star, appear to be cool giant *protostars. Certain *active galaxies, particularly type II *Seyfert galaxies, also exhibit strong infrared emission from their nuclei. It is unclear in particular cases whether this radiation arises from dust heated by bright stars in the galactic nucleus or from an extension of the radio *synchrotron-emission spectrum produced by the interaction of relativistic electrons with the local magnetic field.

infrared telescope (flux collector). A reflecting telescope used to detect and study infrared radiation from space; refractors cannot be employed because glass is opaque at these wavelengths. The mirror surface does not, however, need to be made to the same accuracy as one used for optical work due to the longer wavelengths of infrared radiation. Since neither the eye nor the photographic plate is sensitive to long-wavelength infrared radiation, special detectors, usually *photocells, are required, which must be cooled to liquid-nitrogen or liquid-helium temperatures. These telescopes can detect sources partially or totally obscured at optical wavelengths by interstellar dust, and can identify those that are true infrared sources rather than highly reddened ordinary stars. Because water vapour in the earth's atmosphere absorbs infrared radiation, the most advanced work is carried out with specially designed telescopes, such as the *UK Infrared telescope, at high-altitude observatories. Infrared telescopes can operate day and night.

injection (insertion). The process of boosting a spacecraft into a particular orbit or trajectory. It is also the time of such action or the entry itself.

inner Lagrangian point. The Lagrangian point, L_1, through which *mass transfer occurs. *See also* equipotential surfaces.

inner planets. The terrestrial planets Mercury, Venus, Earth, and Mars, which lie closer to the sun than the main belt of minor planets.

insertion. *See* injection.

insolation. The exposure of any surface or body to solar radiation. It is usually quoted as the radiant flux received from the sun per unit area per unit time; it can therefore be considered as the heating effectiveness of solar radiation. Outside the earth's atmosphere it is equal to the *solar constant, varying by 1-2% with the sunspot cycle and by ±3.5% around the earth's orbit.

The area of the earth's surface over which a given amount of solar radiation is spread depends primarily on the sun's altitude: the greater the altitude the smaller the area; it also depends on the topography of the area and on atmospheric conditions. The sun's altitude varies both with time and latitude. The insolation will therefore be greatest in the middle of the day, will be greater during summer than in winter, and will be greater in equatorial regions than in higher latitudes.

instability strip. *Short for* Cepheid instability strip. *See* pulsating variables.

INT. *Abbrev. for* Isaac Newton telescope.

integrated magnitude. *See* total magnitude.

integration time. The period for which a noisy signal is averaged in order to improve the signal to noise ratio in an electronic system. *See* sensitivity.

intensity. 1. (radiant intensity). The *radiant flux emitted per unit solid angle by a point source in a given direction. It is measured in watts per steradian.

2. *Obsolete name for* field strength of a magnetic or electric field.

intensity interferometer. *See* interferometer.

interacting galaxies. Two, three, or (more rarely) four or more galaxies that show signs of mutual disturbance, such as tidal distortions or extruded filaments of luminous material (which can sometimes link to form bridges between the galaxies). Approximately 20% of *peculiar galaxies appear to be members of interacting groups. Some apparently isolated galaxies also show signs of disturbance but it has not yet been established whether these are being disrupted by invisible companions, such as massive black holes, or by unusually violent activity in their nuclei.

interference. 1. Unwanted signals picked up by a *radio telescope or other electronic equipment; the signals usually arise from terrestrial sources but occasionally come from the sun or Jupiter.

2. The interaction of two sets of waves to produce patterns of high and low intensity. For example, under certain conditions light from two different sources can interfere, producing bright and dark bands (*interference fringes*). For this to happen the two beams must be coherent, i.e. they must be in *phase and have the same *wavelength. They must intersect at a fairly small angle and not differ too much in intensity. In addition, they must not be polarized in mutually perpendicular planes.

The interference pattern of fringes formed at a particular position is the sum of the intensities of the two interacting waves at that position. The fringes occur because of differences in pathlength between interacting waves, i.e. because of unequal distances from source to interaction point. If the difference is a whole number of wavelengths, then wave peaks (or troughs) of the two interacting waves coincide and the waves reinforce one another, producing a bright fringe; this is termed *constructive interference*. If the path difference is an integral number of half wavelengths a peak coincides with a wave trough and a dark fringe results; this is *destructive interference*.

Interference is responsible for *diffraction patterns and is used in dispersing radiation in *diffraction gratings. Both constructive and destructive interference of light can be produced by means of thin films of uniform thickness. Waves of selected wavelength are reflected from the front and back surfaces of the film and by a suitable choice of film composition and thickness the waves will either be reinforced or will cancel each other.

Interference can occur for all forms of electromagnetic

radiation. Using pairs of *radio telescopes the waves from a radio source can be made to interfere. *See also* interferometer.

interference filter. An optical, ultraviolet, or infrared *filter in which unwanted wavelengths are removed by destructive *interference rather than by absorption. It produces a very narrow band of wavelengths, typically between one and ten nanometres. One form, the *multilayer filter*, consists of a stack of transparently thin films with different refractive indices. The films allow only a waveband of 1 nm or less to be transmitted right through. The remaining wavelengths are reflected and caused to interfere destructively.

interference fringes. *See* interference; radio telescope.

interferometer. An instrument in which a beam of light or other radiation is split into two or more parts and subsequently reunited after traversing a different pathlength so that an *interference pattern results; with visible wavelengths this consists of light and dark bands or rings.

In radio astronomy the *radio interferometer* is one of the two basic forms of *radio telescope. In optical astronomy *stellar interferometers* are used to measure small angles, such as the angular separation of a double star or more often the apparent diameter of a single star. In A.A. Michelson's original stellar interferometer the telescope objective was covered by a screen. The screen was pierced with two parallel slits whose distance apart was almost equal to the objective diameter and was also adjustable. The two components of a double star will each produce a set of interference fringes as a result of the double slit. By varying the distance, d, between the slits, the bright fringes of one pattern can be made to coincide with the dark fringes of the second so that a continuous line of light is seen. This occurs when the slit separation is equal to $\lambda/2\alpha$, where λ is the wavelength of the light and α is the angular separation of the stars. For a nearby giant or supergiant star, such as Betelgeuse or Antares, the two halves of the stellar disc produce the two interference patterns, which coincide when the apparent diameter equals $1.22\lambda/d$. With a few modifications Michelson used this technique in the late 1920s to measure the diameters of six stars, down to about 0.01 arc seconds.

The *intensity interferometer*, recently developed at Narrabri Observatory, Australia, is a much more sensitive instrument. It is capable of measuring the extremely small angles subtended by main-sequence stars with a precision of about ±5%. The faintest star that can be measured is one of magnitude +2.5. The minimum diameter is about 2×10^{-4} seconds of arc. Two very large optical reflectors, 670 cm in diameter, are mounted on a circular track and guided by computer so that they can track the star to be measured as it travels round the sky. The starlight is focused by each reflector on to a separate *photocell, having passed through a narrow-band filter, and is converted into an electric current. The two currents are brought together and their fluctuations studied. A small component of the fluctuations corresponds to fluctuations in the intensity of the starlight. When the two reflectors are close together there is a high correlation between the intensity fluctuations—they are almost identical. As the distance between the reflectors is increased, the correlation decreases. The apparent diameter of the star can be determined from measures of correlation as a function of reflector spacing.

See also speckle interferometry.

intergalactic medium. The matter contained in the space between the *galaxies, about which very little is known. It has only been positively identified in *clusters of galaxies where it is completely ionized and constitutes a significant proportion of the total observed mass of the cluster.

intermediate frequency (IF). *See* receiver.

intermediate population II stars. *See* populations I and II.

International Astronomical Union (IAU). An international association of astronomers that is the controlling body of world astronomy. It was founded in Brussels in 1919. A general assembly is held every three years, the first one occurring in 1922. In addition the IAU organizes periodic symposia for the discussion of current astronomical problems, produces several publications, and runs a telegram service for the rapid dispersal of information relating to transient phenomena such as nova and comets.

international atomic time (TAI; IAT). *See* atomic time.

international colour index. *See* colour index.

International Colour System. *See* magnitude.

International Geophysical Year (IGY). The period from July 1 1957 to December 31 1958 that was timed to coincide with an epoch of maximum *solar activity. It saw a concerted international effort to study the geophysical effects of enhanced solar activity, particularly regarding the earth's upper atmosphere. Sputnik 1, the first artificial satellite, was launched during this period.

International Solar Polar Mission (ISPM). A proposed joint

ESA/NASA mission in which two 300 kg spacecraft, launched simultaneously, will travel similar paths to Jupiter. They will then be propelled, by means of the slingshot action of Jupiter's gravitational field, at right angles to the plane of the *ecliptic and towards the sun. One probe will move over the northern hemisphere and the other over the southern hemisphere. (Energy requirements are too high to launch the craft directly out of the ecliptic plane.) The launch, by Space Shuttle, should take place sometime in 1983 and the mission is expected to last five years. ESA and NASA will each provide one spacecraft.

The higher latitudes of the heliosphere, the sun's environment, will thus be explored for the first time. The experiments aboard each probe will include instruments to measure the interplanetary magnetic field (*see* solar wind), and hence to map the sun's magnetic field, to study the movement of interstellar gas and dust in the sun's vicinity, and to measure features of the solar wind and other particle fluxes emitted by the sun.

International Sun-Earth Explorer (ISEE). A joint project of NASA and ESA for studying the structure and interactions within the earth's *magnetosphere and the solar phenomena producing such effects. The two satellites ISEE-1 and ISEE-2 were launched in tandem, in Oct. 1977, into a highly eccentric earth orbit and are at a controllable distance apart. ISEE-3 was launched in Aug. 1978 into an orbit lying outside the magnetosphere and measures the *solar wind, solar *flares, and *sunspots, unperturbed by the earth's influence. It is thus a reference point for simultaneous observations made by ISEE-1 and ISEE-2.

International Colour System. *See* magnitude.

International Ultraviolet Explorer (IUE). A satellite that is a joint project of NASA, ESA, and the SRC (Scientific Research Council) and was launched by NASA into an elliptical *geosynchronous orbit on Jan. 26 1978. Its 45 cm Cassegrain telescope focuses ultraviolet radiation on either of two grating spectrographs, one operating in the range 115-190 nanometres, the other 180-320 nm. The satellite is controlled from ground stations in Madrid and Maryland by real-time data link, which permits operation in a manner similar to that of a ground-based telescope. Over 6000 UV spectra were obtained in the first year, providing a considerable amount of detailed information. It has been found, for instance, that highly ionized species are present in the interstellar medium and also that stellar winds are more widespread in stars than originally thought.

International Years of the Quiet Sun (IQSY). The period 1964-65, when *solar activity was at a minimum, during which an intensive international programme was mounted by observatories all over the world to study solar and geophysical phenomena.

interplanetary medium. The matter contained in the solar system in the space between the planets. This space is filled with a tenuous ionized gas (the *solar wind) that streams outwards from the sun at supersonic speeds and in which is embedded the sun's magnetic field wound into a spiral by the sun's rotation. (This is often termed the *interplanetary magnetic field.*) At the earth the medium has a density of about eight atoms per cm^3 moving at a velocity of 300 to 400 km s^{-1}, and a magnetic flux density of about 5×10^{-5} gauss (5×10^{-9} tesla). The earth moves at supersonic speed relative to the interplanetary medium, creating a shock front called the *bow shock,* rather like that set up by an ocean liner on the surface of the sea. Irregularities in the density of the solar wind cause interplanetary scintillations (*see* scattering) on small *radio sources viewed through the medium.

interplanetary scintillations (IPS). *See* scattering; interplanetary medium.

interpulse. The weaker pulse of radio emission received after the main pulse and before the next main pulse in the signals from some *pulsars.

interstellar absorption. *See* interstellar extinction.

interstellar dust. *Cosmic dust occurring in the *interstellar medium.

interstellar extinction (interstellar absorption). The reduction in brightness, i.e. the *extinction, of light (and other radiation) from stars as a result of absorption and scattering of the radiation by interstellar dust. Although dust makes up only a small proportion of the *interstellar medium, its effect on starlight is considerable. The amount of extinction, usually measured in *magnitudes, depends on the direction of observation. It is a maximum (about one magnitude per kiloparsec distance for visible wavelengths) towards the centre of the Galaxy, where the density and extent of the dust is greatest. The extinction varies inversely with wavelength. Red light is thus less affected than blue light so that starlight appears reddened when observed through a dust cloud. Similarly, radio wavelengths can pass through the interstellar medium with ease compared with light wavelengths. The presence of obscuring matter between the stars was first conclusively demonstrated by Robert Trumpler in 1930.

interstellar hydrogen. *See* HI region; HII region.

interstellar medium (ISM). The matter contained in the region between the stars of the *Galaxy, constituting about 10% of the galactic mass. It is largely confined to a thin layer in the galactic plane and tends to be concentrated in the spiral arms. Several different constituents have been observed in the medium, including relatively cool (100 kelvin) clouds of neutral hydrogen—*HI regions—surrounded by more tenuous regions of hot gas (1000 - 10 000 kelvin). The presence of some very much hotter regions has been inferred from a diffuse background of soft *x-ray emission. There are also dense clouds of molecular hydrogen in which a number of other molecules and radicals have been observed (*see* molecular-line radio astronomy).

In addition to the gas, *cosmic dust* is to be found throughout the medium. The dust comprises small solid particles that are about 0.1 micrometre in size and are thought to be composed of carbon or silicates, with icy mantles in some cases. The dust causes dimming and reddening of starlight (*see* interstellar extinction) and also partial *polarization of starlight. The material of the interstellar medium is thought to come from the cast-off outer layers of old stars, for example, from *supernova explosions, while new stars are created in the dense dusty clouds (*see* stellar birth).

The medium is permeated by a flux of *cosmic rays (origin unknown) that spiral along the field lines of the *galactic magnetic field of a few microgauss and cause the *synchrotron emission that is the *galactic radio background radiation*. *Radio maps show that this radiation is largely confined to the galactic plane but has several *spurs,* in particular the *north galactic spur,* radiating away from it. The origin of the north galactic spur is still uncertain although it could be an old *supernova remnant or be due to a local distortion in the galactic magnetic field.

interstellar molecules. *See* molecular-line radio astronomy.

interstellar polarization. The partial *polarization of starlight caused by dust in the interstellar medium. Since only nonspherical particles with a nonrandom orientation can polarize light, the dust grains must be elongated and are thought to be partially aligned by a weak *galactic magnetic field.

interstellar reddening. *See* interstellar extinction; extinction.

interstellar scintillations (ISS). *See* scattering.

interstellar wind. *See* solar wind.

intramercurial planet. *See* Vulcan.

intrinsic colour index. *See* colour index.

intrinsic variable. *See* variable star.

invariable plane. The unvarying plane in space defined by the total *angular momentum, from orbital motion and axial rotation, of all bodies of the solar system. At present, the *ecliptic is inclined by 1.65° to the invariable plane.

inverse Compton emission. *See* Compton scattering.

Io. The innermost of the four large *Galilean satellites of Jupiter. It keeps one face permanently towards the planet (*see* synchronous rotation) as it orbits at a distance of 422 000 km, within Jupiter's *magnetosphere. Io is yellow-brown in colour, has a high *albedo (0.63), and has been shown to have extensive surface deposits resembling those of the Utah salt flats. A sodium cloud, detected spectroscopically in the vicinity of Io in its orbit, probably consists of atoms stripped from its surface by energetic electrons and protons from Jupiter's radiation belts. The cloud also contains sulphur and potassium, while a torus of hydrogen, oxygen, and sulphur ions encompasses the whole of Io's orbit. Early photographs from the *Voyager 1 probe showed the surface of Io to lack impact craters. An explanation for this was found when intense volcanic action was observed on the satellite: at least eight active volcanoes were seen to eject material over a very wide area. The surface is thus being continuously renewed. The volcanic eruptions are a possible source of the ionized sulphur in the torus of ions associated with Io.

An interaction between the Jovian magnetosphere and Io, which itself has an ionosphere, occasionally triggers electrical discharges that emit bursts of radio waves detectable at the earth. *See* Jupiter's satellites; Table 2, backmatter.

ion. An atom or molecule that has lost or gained one or more electrons, thus having a positive or negative charge.

ionization. Any of several processes by which normally neutral atoms or molecules are converted into *ions or by which one ion of an atom or molecule is changed to another ionic form, as by losing a second electron. Positive ions are commonly formed by impact of high-energy photons or particles with neutral atoms or molecules. The minimum energy required to remove an electron from an atom or molecule is the *ionization potential, I,* of that atom or molecule, and for ionization to occur the energy of the impacting photon or particle must exceed the ionization potential. It requires a higher energy (the *second ionization potential*) to remove a second more tightly bound electron, and an even greater energy to remove a third electron. At high temperatures ionization takes place by collisions between atoms with high kinetic energies. For equilibrium to exist in a system, such as an *emission nebula, ionization processes must be balanced by recombination of positive ions and electrons.

Because of the high temperatures in stars, much of the matter present is in an ionized state; often more than one electron has been lost so that the atom or molecule is doubly or triply ionized, etc. In astronomy the neutral state of an element is conventionally denoted by I (HI, for example, denotes neutral hydrogen) whereas II denotes singly charged ions (HII is ionized hydrogen) and III indicates a doubly charged ion, and so on. *See also* Saha ionization equation; plasma.

ionization potential. *See* ionization.

ionization zone. *See* pulsating variables.

ionosphere. A region in the earth's atmosphere that extends from about 60 km to over 500 km altitude and contains sufficient free electrons and ions—produced by the ionizing action of solar radiation—to disturb the propagation of *radio waves through it. The degree of ionization varies with time of day, season, latitude, and state of *solar activity. Several distinct layers may be identified in the ionosphere where the density of free electrons is higher than average. The three most important are the *D-layer* at a height of between 60 and 90 km, the *E-layer* at about 120 km, and the *F-layer* at about 250 km. During the day the F-layer splits into two further layers labelled F_1 (200 km) and F_2 (300-400 km) but these combine again at night. Radio waves of sufficiently low frequency are reflected by the E- and F-layers. The D-layer, being lower in the atmosphere where collisions between molecules and ions are more frequent, is more an absorber than a reflector.

The ionosphere is important in radio communications, allowing long-distance propagation at frequencies up to about 30 megahertz by successive reflections between it and the ground. In *radio astronomy, however, it makes ground-based observations almost impossible below 10 megahertz, causes *phase and *amplitude scintillations (*see* scattering) on *radio sources viewed through it, and alters the observed *polarization of radio waves by *Faraday rotation. Other bodies, including Mars and Venus and the Jovian satellite Io, are now known to possess ionospheres.

ionospheric scintillations. *See* scattering.

ion tail. *See* comet tails.

IPS. *Abbrev. for* interplanetary scintillations. *See* scattering.

IQSY. *Abbrev. for* International Years of the Quiet Sun.

IR. *Abbrev. for* infrared.

IRAS (Infrared Astronomy Satellite). *See* infrared astronomy.

iron. *Short for* iron meteorite.

iron meteorite (siderite). One of the main classifications of *meteorites, containing nickel-iron with small amounts of accessory minerals. The composition averages 90% iron, 10% nickel. The three main subdivisions are *octahedrites*, which have more than 6% nickel, the less common *hexahedrites*, with less than 6% nickel, and the nickel-rich *ataxites*. *See also* Widmanstätten patterns.

iron-peak elements. *See* nucleosynthesis.

irregular galaxy. *See* galaxies; Hubble classification.

irregular maria. *See* maria, lunar.

irregular variables. Stars, such as the majority of red giants, that are *pulsating variables showing slow luminosity changes either with no periodicity or with a very slight periodicity. The latter can be grouped into slow irregular variables of *spectral types F, G, K, M, C, and S that are usually giants, and slow irregular supergiants of spectral types M, C, and S.

Isaac Newton telescope (INT). The 2.5 metre (98 inch) reflecting telescope that was made from a spare blank for the 200 inch Hale telescope. It began work in 1967 at the *Royal Greenwich Observatory, Herstmonceux, but is due to be moved in 1979 to the new *Northern Hemisphere Observatory in the Canary Islands.

ISEE. *Abbrev. for* International Sun-Earth Explorer.

island universe. *Obsolete term for* galaxy.

ISM. *Abbrev. for* interstellar medium.

isophote. A line on a diagram joining points of equal flux density or intensity.

isotopes. Forms of an element in which the nuclei contain the same number of protons but different numbers of neutrons. For example, there are three isotopes of hydrogen: 'normal' hydrogen has a single proton, deuterium has one proton and one neutron, and tritium, which is a radioactive isotope, has one proton and two neutrons.

isotropy. The property by which all directions appear indistinguishable to an observer expanding with the universe. Isotropy implies *homogeneity but the reverse is not necessarily true. *See* cosmological principle.

ISS. *Abbrev. for* interstellar scintillations. *See* scattering.

IUE. *Short for* International Ultraviolet Explorer.

J

jansky. Symbol: Jy. The unit of *flux density adopted by the IAU in 1973 and used in radio astronomy. One jansky is equal to 10^{-26} watts per metre squared per hertz.

Jansky noise. *Another name for* cosmic noise. *See* antenna temperature.

Janus. The tenth and innermost satellite of Saturn, discovered when *Saturn's rings were edge-on to the earth in 1966. It is a faint object orbiting the planet just beyond the edge of the ring system. *See* Table 2, backmatter.

JD. *Abbrev. for* Julian date.

jet. A long thin linear feature of bright emission extending from a compact object, such as a galaxy. An example is that found in the giant elliptical galaxy M87 (*see* Virgo A).

Jet Propulsion Laboratory (JPL). An observatory at Goldstone, California, run by the California Institute of Technology. It is the site of NASA's Deep Space Communications Complex. Its chief instrument is a 64 metre (210 foot) radio dish.

Jewel Box (Kappa Crucis; NGC 4755). A brilliant *open cluster in the southern constellation Crux that contains over 100 stars of various colours. It appears to be a young cluster of an estimated age of 10 million years.

Jodrell Bank. The site at Macclesfield in Cheshire of the Nuffield Radio Astronomy Laboratory of the University of Manchester. The principal instrument is the 76.2 metre (250 foot) fully steerable radio dish, known as Mark I, which has been in operation since 1957. It has a computer-controlled *altazimuth mounting. There is also a smaller 48 metre elliptical reflector (Mark II), completed in 1964. Both will be part of the radio network *MTRLI, whose seven component telescopes will be linked to a computer at Jodrell Bank.

joule. Symbol: J. The SI unit of energy, equal to the work done when the point of application of a force of one newton is moved one metre in the direction of the force.

Jovian planets. A term derived from the Latin name for Jupiter and applied collectively to the *giant planets Jupiter, Saturn, Uranus, and Neptune.

Julian day at noon, UT, on March 1st

year	Julian day	year	Julian day
1800	2378556	1940	2429690
1820	2385861	1950	2433342
1840	2393166	1960	2436995
1860	2400471	1970	2440647
1880	2407776	1980	2444300
1900	2415080	1990	2447952
1920	2422385	2000	2451605

Jovian satellites. *See* Jupiter's satellites.

JPL. *Abbrev. for* Jet Propulsion Laboratory.

Julian calendar. The calendar that was established in 46 BC in the Roman Empire by Julius Caesar, with Sosigenes of Alexandria as his chief advisor. It reached its final form in about 8 AD under Augustus and was in general use in the West up to 1582, when theM *Gregorian calendar was instituted. Each year contained 12 months and there was an average of 365.25 days per year: three years of 365 days were followed by a leap year of 366 days. (Leap years were not correctly inserted until 8 AD.) Since the average length of the year was about 11 minutes 14 seconds longer than the 365.2422 days of the *tropical year, a discrepancy arose between the calendar year and the seasons, with an extra day 'appearing' about every 128 years.

Julian date (JD). The number of days that have elapsed since noon GMT on Jan 1 4713 BC (the *Julian day number*) plus the decimal fraction of the day that has elapsed since the preceding noon. The consecutive numbering of days makes the system independent of the length of month or year and the JD is thus used to calculate the frequency of occurrence or the periodicity of phenomena over long periods. The system was devised in 1582 by the French scholar Joseph Scalinger. A specific Julian date can be determined from astronomical tables of Julian day numbers: thus since March 1 1980, noon GMT, is tabulated as day number 244 4300 then March 17 1980, 6 p.m. GMT, is JD 244 4316.25 (see table). The *modified Julian date* (*MJD*) is found by subtracting 240 0000.5 from the Julian date.

June Lyrids. A minor *meteor shower, radiant: RA 278°, dec +35°, maximizing on June 16.

Juno. The third *minor planet to be discovered, found in 1804. With a diameter of 247 km it is the smallest of the first four minor planets to be found and numbered. It has a spectrum similar to that of the *stony-iron (siderolite) meteorites. *See* Table 3, backmatter.

Jupiter. The largest planet, orbiting the sun at a mean distance of 5.2 astronomical units in 11.9 years. It has an equatorial diameter of 142 800 km (11 times that of the earth) and a polar diameter of 135 500 km. This *oblateness results from a rotation period of less than 10 hours, shorter than that of any other planet. Jupiter has a mass more than twice that of all the other planets combined but its average density, only 1.3 times that of water, suggests that it contains a high proportion of the lightest elements, hydrogen and helium. Orbital and physical characteristics are given in Table 1, backmatter.

*Oppositions recur at 13 month intervals; the planet has an apparent diameter of 47 arc seconds at mean oppositions and shines at these times at about magnitude -2.4, brighter than every other night-time object other than the moon, Venus, and (rarely) Mars. Jupiter has 14 satellites, including the four *Galilean satellites visible with a minimum of optical aid, and also a ring system (*see* Jupiter's satellites). Telescopes show the disc to be crossed by bands of light and dark clouds, running parallel to the equator. These are called *zones* and *belts* respectively. Irregular spots and streaks are seen, their motion across the disc indicating planetary rotation periods varying between about 9 hours 51 minutes in equatorial regions and 9 hours 56 minutes at high latitudes. Except for the *Great Red Spot and the White Ovals, most of the markings are temporary, lasting for days or months.

Models for Jupiter's atmospheric and internal structure have been refined following the fly-bys of the spaceprobes *Pioneer 10 and 11 in 1973 and 1974 and the *Voyager probes in 1979; earlier spectroscopic work had, however, detected the presence of hydrogen, ammonia, and methane. It appears that the white or yellowish zones are areas of higher clouds supported by upward *convection of warm gases; the reddish-brown belts have descending glas flows and lower clouds. Eddies may give rise to the spots. At latitudes greater than 50°, the belt/zone system gives way to a mottled surface appearance corresponding to ascending and descending convection cells (gas columns). The higher cloud zones appear to comprise ammonia crystals while the lower cloud belts may contain sulphur compounds, hydrogen and ammonia polysulphides, or complex organic compounds, the latter possibly formed by lightning discharges in the atmosphere. The highest cloud features, such as the Great Red Spot, may be coloured red by traces of phosphorus brought up by convection from the lower atmosphere.

Jupiter radiates, as heat, about twice as much energy as it receives from the sun, indicating that there is an internal reservoir of thermal energy left over from its creation 4.6 thousand million years ago. In addition to the rapid rotation of Jupiter, the outflow of energy from the planet and a *greenhouse effect ensure that there is little variation in temperature between the equator and poles, or between the day and night hemispheres. Atmospheric temperatures increase from about -130°C at the cloud tops to about 30°C at the base of a lower cloud layer of water droplets and ice-crystals believed to lie about 70 km below the ammonia clouds. Over this same range the pressure rises from 0.5 to 4.5 (earth) atmospheres. All Jupiter's weather occurs in this upper skin of its 1000 km deep gaseous atmosphere, 99% of which consists of hydrogen and helium in solar proportions, i.e. in the ratio 88 to 12.

The hydrogen becomes liquid at a depth of 1000 km; this marks the interface between atmosphere and planetary interior. At a depth of 25 000 km, under an estimated three million atmospheres pressure, the hydrogen becomes metallic: the molecules dissociate into protons and a sea of unattached electrons. Temperatures may increase from 2000°C at 1000 km, to 11 000°C at 25 000 km, and to 30 000°C at the planet's centre where there may be a rocky core 10 to 20 times more massive than the earth. The bulk of Jupiter's interior is thus composed of liquid hydrogen, most of which is in metallic form.

A magnetic field, probably arising from a dynamo action within Jupiter's conductive metallic hydrogen, has a total strength about 17 000 times the earth's magnetic field. Jupiter's field is tilted by 11° relative to its rotation axis and is reversed in polarity relative to that of the earth. It supports a *magnetosphere, about 20 million km across, in which the five inner satellites orbit. The magnetosphere contains radiation belts possibly 10 000 times more intense than the earth's *Van Allen radiation belts.

Jupiter emits radio waves by three mechanisms: high-frequency thermal radio noise comes from the atmosphere along with infrared radiation; high-frequency nonthermal *synchrotron emission is generated by electrons in Jupiter's magnetic field; intense bursts of decametric radio waves are believed to arise from electrical discharges along Jovian magnetic field lines when *Io, with its conductive ionosphere, moves across them.

Jupiter probes. *See* Pioneer probes; Voyager probes.

Jupiter's comet family. A *comet family whose distribution of aphelion distances (the comets' furthest points from the sun) correlates with the mean distance of the planet Jupiter. It contains about 70 members with aphelion distances between four and eight astronomical units. Jupiter is capable of substantially changing the orbits of comets that happen to pass close by. *Brooks 2* passed within two Jovian radii of Jupiter's surface in July 1886. Its period and aphelion distance were changed from 31 years and 14 AU to 7 years and 5.4 AU. The formation of comet families is caused by such gravitational perturbations. Jupiter is also capable of throwing comets out of its family into long-period orbits (using a process similar to the one used to deflect spacecraft out of the solar system). Over 90% of the Jovian family move in direct orbits. Comets with periods less than 10 years have a mean inclination to the ecliptic of 12°. After a comet has been captured by Jupiter it will take another 200 to 400 orbits (depending on perihelion distance) before it decays completely. No comet can remain in the Jupiter family for more than about 4000 years so the family is being replenished continually.

Jupiter's satellites (Jovian satellites). A system of at least 14 satellites, although the last, found in 1975, has still to be confirmed and named. They can be divided conveniently into three groups: an outer group of five small bodies (Ananke, Carme, Pasiphae, Sinope, and the 14th satellite) that move in loosely bound retrograde orbits (*see* direct motion) with periods of about 700 days; an intermediate group of four small satellites (Leda, Himalia, Lysithea, and Elara) in approximately 250 day orbits with *inclinations close to 28°; and an inner group consisting of the satellite *Amalthea and the four large *Galilean satellites (Io, Europa, Ganymede, and Callisto) that lie close to the plane of Jupiter's equator. The members of the outer and intermediate groups may be captured *minor planets, although the capturing process that would place objects into one group rather than the other has not been defined. Possibly some of them were former members of the *Trojan group of minor planets.

All five inner satellites lie within Jupiter's *magnetosphere and are effective in sweeping up the charged particles found there, becoming intensely radioactive in the process. In particular, the orbits of Amalthea and *Io pass through the denser radiation belts and a cloud containing sodium, sulphur, and potassium in the vicinity of Io is thought to comprise atoms knocked from fields of salt deposits on its surface by energetic electrons and protons from the belts—a mechanism called *sputtering*. Few details of the irregularly shaped Amalthea were known until the *Voyager probes visited Jupiter in 1979. The Galilean satellites however have been studied throughout the telescopic era and more recently by the *Pioneer 10 and 11 spaceprobes, which returned hazy photographs in 1973 and 1974. More information came from the Voyager probes.

Io and Europa are comparable in size to the moon, with Ganymede and Callisto rather larger. Each appears to be in *synchronous rotation, keeping one face turned towards the planet. Their maximum surface temperatures vary between the 140 kelvin of Io to 167 kelvin for Callisto, probably as a result of differences in *albedo; the albedo of Io is 0.63, of Europa 0.60, of Ganymede 0.43, and of Callisto as low as 0.17. Io has the greatest density, 3.5 times that of water, with Europa 3.28 times, Ganymede 1.95 times, and Callisto only 1.62 times that of water. These density differences may reflect a loss of water by the innermost satellites as a result of heating during the formation of Jupiter itself, leaving Io with the extensive salt deposits mentioned above.

Europa probably contains a larger rocky core than Ganymede, but spectroscopic studies of both worlds suggest that their surfaces are covered in water ice or a rock/ice mixture. This is supported by radar studies of Ganymede. With the exception of the tenuous vapour cloud associated with Io, Ganymede is the only Jovian satellite known to possess an atmosphere, possibly of methane and ammonia. Callisto may resemble Ganymede in many respects and both may have extensive oceans of liquid water below their frozen exteriors. Information and photographs from the Voyager 1 spacecraft showed that there is considerable volcanic activity on Io. Europa, which was not closely observed by Voyager 1, might also have active volcanoes. There are indications of internal activity on Ganymede. Ganymede and Callisto, unlike Io, are both heavily cratered bodies.

As Voyager 1 moved inside the orbit of Amalthea, a *ring system* of rocky boulders was discovered around Jupiter, lying within the Roche limit of the planet at about 55 000 km above the cloud tops. *See* Table 2, backmatter.

K

Kappa Crucis. *See* Jewel Box.

Kappa Pavonis. *See* Pavo.

Kapteyn's selected areas. *See* selected areas.

Kapteyn's star (HD 33793; CD -45°1841). A red *subdwarf in the constellation Pictor that is a high-velocity star with a radial velocity of 242 km/s. It is one of the nearest stars to the sun and is the nearest subdwarf. m_v: 8.8; M_v: 10.8; spectral type: sdM0; distance: 3.91 pc.

K-corona. The inner part of

the solar *corona, responsible for the greater part of its intensity out to a distance of about two solar radii. It consists of rapidly moving free electrons and exhibits a linearly polarized *continuous spectrum, due to Thomson scattering, by the electrons, of light from the *photosphere. The *Fraunhofer lines are not present in the spectrum: they have been sufficiently broadened by the large *Doppler shifts, resulting from the rapid random motion of the electrons, so as to overlap and be indiscernible. The K-corona attains a temperature of around 2 000 000 kelvin at a height of about 75 000 km (see chromosphere). It may exhibit considerable structure, with *streamers* (regions of higher than average density) overlying *active regions and quiescent *prominences, and with *coronal holes (regions of exceptionally low density) overlying areas characterized by relatively weak divergent (and therefore primarily unipolar) magnetic fields.

Keel. Common name for *Carina.

Kellner eyepiece. See eyepiece.

kelvin. Symbol: K. The SI unit of temperature, equal to 1/273.16 of the *thermodynamic temperature of the triple point of water. A temperature in kelvin may be converted to one in degrees Celsius by the subtraction of 273.15.

Kelvin contraction. See Kelvin timescale.

Kelvin timescale. The time taken for a mass of gas to collapse under its own gravitation. It is given by

$$\tau_{KH} = GM^2/RL$$

where G is the gravitational constant and M, R, and L are the mass, radius, and luminosity of the object. Such a collapse—known as the *Helmholtz-Kelvin contraction*—was proposed in the late 19th century as the source of the sun's energy output. This theory was abandoned when the contraction timescale of only 10^7 years was shown to be too short. The energy source was subsequently shown to originate in nuclear-fusion reactions.

Keplerian telescope. The first major improvement of the *Galilean telescope, developed by the German astronomer and mathematician Johannes Kepler, in which a positive (convex) lens was used as the eyepiece in place of the negative lens that Galileo used (see illustration). This gave a larger though inverted field of view and higher magnifications.

Ray path in Keplerian refractor

Kepler's equation. *See* anomaly.

Kepler's laws. The three fundamental laws of planetary motion that were formulated by Johannes Kepler and were based on the detailed observations of the planets made by Tycho Brahe, with whom Kepler had worked. The laws state that

1. The orbit of each planet is an ellipse with the sun at one focus of the ellipse.

2. Each planet revolves around the sun so that the line connecting planet and sun (the radius vector) sweeps out equal areas in equal times (see illustration). Thus a planet's velocity decreases as it moves further from the sun.

Kepler's second law

3. The squares of the sidereal periods of any two planets are proportional to the cubes of their mean distances from the sun. Thus if the period, P, is measured in years and the mean distance, a, in *astronomical units, then $P^2 = a^3$ for any planet.

The first two laws were published in 1609 in *Astronomia Nova* and the third law in 1619 in *Harmonices Mundi*. The third law, sometimes called the *harmonic law*, allowed the relative distances of the planets from the sun to be calculated from measurements of the planetary orbital periods. Kepler's laws gave a correct description of planetary motion. The physical nature of the motion was not explained until Newton proposed his laws of motion and gravitation. From these laws can be obtained Newton's form of Kepler's third law:

$$P^2 = 4\pi^2 a^3 / G(m_1 + m_2)$$

where G is the gravitational constant, m_1 and m_2 are the masses of the sun and a planet, a is the semimajor axis of the planet, and P is its sidereal period.

Kepler's star (SN Oph 1604). A type I *supernova that was observed in Oct 1604 in the constellation Ophiuchus and could be seen with the naked eye for over a year. It was studied by astronomers in Europe, China, and Korea and its position was so accurately determined by Kepler and Fabricius that a small patch of nebulosity above the galactic plane could be identified as the remnant of the original 1604 supernova (Walter Baade, 1943). The light curve plotted from the original observations shows that the magnitude increased to a maximum of over -2.5, dropping to +4 in about 300 days, and that it was a type I supernova.

Keyhole nebula. *See* Eta Carinae.

kilo-. Symbol: k. A prefix to a unit, indicating a multiple of 1000 or 10^3 of that unit. For example, one kilogram equals 1000 grams.

kilogram. Symbol: kg. The SI unit of mass. It is defined as the mass of a prototype platinum-iridium cylinder kept at the International Bureau of Weights and Measures at Sèvres, France. The spelling *kilogramme* is also used.

kinetic energy. Energy possessed by a body by virtue of its motion, equal to the work that the body could do in coming to rest. In classical mechanics a body with mass *m* and velocity *v* has kinetic energy $\frac{1}{2}mv^2$. A body with moment of inertia *I* rotating with angular velocity ω has kinetic energy $\frac{1}{2}I\omega^2$.

Kirkwood gaps. Gaps in the distribution of orbits within the main belt of *minor planets, corresponding to the absence of orbits with periods that are simple fractions (1/4, 1/3, 2/5, 1/2, etc.) of the orbital period of Jupiter (see illustration). Minor planets with such orbital periods would be perturbed into other orbits by the regularly recurring gravitational pull of Jupiter. They are named after the American astronomer Daniel

Kirkwood gaps, indicated by fractions

Kirkwood who first explained them in 1857.

Kitt Peak National Observatory (KPNO). An observatory near Tucson, Arizona, at an altitude of 2064 metres, that has a very large assembly of telescopes. Like the *Cerro Tololo Interamerican Observatory, it is run by the Association of Universities for Research in Astronomy and began operations in 1960. The chief instruments are a 4 metre (158 inch) reflecting telescope—the *Mayall reflector*—which has a very low *focal ratio of f/2.6 and went into use in 1973, a 2.1 metre (84 inch) reflector operational in 1964, and a huge 1.5 metre (60 inch) *solar telescope, the *McMath solar telescope*, completed in 1962. An 11 metre (36 ft) radio telescope is operated by the *National Radio Astronomy Observatory.

K line. *See* H and K lines.

Kochab. *See* Ursa Minor.

Kohoutek (1973 XII). A comet discovered in March 1973 by Lubos Kohoutek when he was looking for Biela's comet. At discovery comet Kohoutek was 4.6 AU away from the sun and early predictions indicated that near perihelion, at its brightest, Kohoutek would reach a magnitude of about -12, comparable with the full moon. This would make it visible in daylight. Comets however are among the most unpredictable astronomical objects and Kohoutek only just attained naked-eye visibility. It was extensively investigated from Skylab, astronaut Gibson being impressed by its overall orange colour and its sunward spike.

KREEP. *See* moonrocks.

Kruger 60. *See* Table 7 (nearest stars), backmatter.

K stars. Stars of *spectral type K, orange-red in colour and with a surface temperature (T_{eff}) of about 3500 to 5000 kelvin. The spectral lines of calcium (Ca II and CA I) are strong and neutral metal lines are very prominent; molecular bands strengthen. Arcturus and Aldebaran are K stars.

L

Lacerta (Lizard). A small inconspicuous constellation in the northern hemisphere near Cygnus, lying partly in the Milky Way, the brightest stars being of 4th magnitude. It contains the prototype of the quasar-like *BL Lacertae objects. Abbrev.: Lac; genitive form: Lacertae; approx. position: RA 22.5h, dec +45°; area: 201 sq deg.

Lagoon nebula (M8; NGC 6523). An *emission nebula in the form of an HII region that lies about one kiloparsec away in the direction of Sagittarius. Its total apparent magnitude

(5.8) makes it easily visible with binoculars.

Lagrangian points. Five locations in space where a small body can maintain a stable orbit despite the gravitational influence of two much more massive bodies, orbiting about a common centre of mass. They are named after the French mathematician J.L. Lagrange who first suggested their existence in 1772. A Lagrangian point 60° ahead of Jupiter in its orbit around the sun, and another 60° behind Jupiter, are the average locations of members of the *Trojan group of minor planets. The three other Lagrangian points in the sun-Jupiter gravitational field do not permit stable minor planet orbits owing to the perturbing influence of the other planets. In any system these three points lie on the line joining the centres of mass of the two massive bodies and are in unstable equilibrium (*see* equipotential surfaces). Lagrangian points 60° ahead of and 60° behind the moon in its orbit of the earth are proposed as the site of existing gas and dust clouds and of future earth-orbiting space colonies and solar power stations.

Lalande 21185. *See* Table 7 (nearest stars), backmatter.

Laplace's nebular hypothesis. *See* nebular hypothesis.

Large Magellanic Cloud (LMC; Nubecular Major). *See* Magellanic Clouds.

Large Space Telescope. *See* Space Telescope.

Las Campanas Observatory. An observatory near La Serena, Chile, sited on Cerro Las Campanas, at an altitude of 2300 metres; the *seeing at this site is exceptional. The Observatory is owned by the Carnegie Institution of Washington and is operated as part of the *Hale Observatories. The principal instruments are the 2.5 metre (100 inch) du Pont Telescope, which went into operation in 1977, and the 1 metre (40 inch) Swope Telescope, which became operational in 1972.

last quarter (third quarter). *See* phases, lunar.

late-type stars. Relatively cool stars of *spectral types K, M, C (carbon), and S. They were originally thought, wrongly, to be at a much later stage of stellar evolution than *early-type stars.

latitude. 1. *Short for* celestial latitude. *See* ecliptic coordinate system.
 2. *Short for* galactic latitude. *See* galactic coordinate system.

launch vehicle. Any system by which the necessary energy is given to a satellite, spaceprobe, space station, etc., in order to insert it into the desired orbit or trajectory. Expendable multistage rockets have been used to date. Reusable vehicles, such as the *Space Shuttle, are being developed so that partial or complete recovery of the launcher should soon be possible.

Rocket propulsion is a form of jet propulsion: all the propellant is carried in the vehicle at take-off and the hot combustion gases, resulting from the mixture of fuel with reactant, are ejected at high speed through a nozzle to produce the necessary force—termed *thrust*—to lift the vehicle off the ground. The thrust exerted by the rocket engine is equal (in vacuum conditions) to the rate of change of *momentum of the expelled propellant gas. An engine is rated in terms of the *specific impulse,* i.e. the impulse (thrust times time) delivered per unit weight of propellant expended; it is measured in seconds. The specific impulse is (ideally) equal to the jet velocity divided by the sea-level acceleration due to gravity, g. Although the thrust is held constant, the rocket velocity will increase as its fuel is expended and it becomes lighter.

Single rockets, carrying only a very small payload, would only just achieve earth orbit. *Multistage rockets* have therefore been developed: in a three-stage rocket, one rocket plus propellant plus spacecraft is used as the payload of a larger rocket, which in turn is the payload of a still larger rocket. The empty tanks, engine, and structure of first the largest and then the second largest rocket are discarded at relatively low altitudes and velocities, after each completes its firing. The final stage then carries the spacecraft into the desired orbit, the satellite separating from the rocket stage when orbital velocity is reached. During the flight the rocket path is tilted from the vertical so that when it reaches the orbit altitude it is parallel to the earth's surface.

Modern launchers generally consist of two, three, or four stages; there are design variations in the type of fuel used, the means of carrying the fuel, and the process by which the tanks, etc., are discarded. The largest launcher is the giant 111 metre Saturn V that propelled the manned *Apollo craft into orbit.

launch window. An interval of time during which a planetary probe, etc., must be launched in order to attain a desired position at a desired time.

leading spot. *Another name for* preceding (p-) spot. *See* sunspots.

leap month. *See* lunar year.

leap second. *See* universal time.

leap year. A year that contains one more day than the usual *calendar year so that the average length of the year is brought closer to the *tropical year of 365.2422 days or to the *lunar year. In the Julian calendar a leap year of 366 days occurred once every four years when the year was divisible by four. In the *Gregorian calender this rule was modified so that century years are leap years only when they are divisible by 400. The additional day is now added at the end of February (Feb. 29).

Leda. A small satellite of Jupiter, discovered in 1974. *See* Jupiter's satellites; Table 2, backmatter.

Lemaître model. *See* cosmological models.

lens. A specially shaped piece of transparent material, such as glass, quartz, or plastic, bounded by two surfaces of a regular (usually spherical) but not necessarily identical curvature. A ray of light passing through a lens is bent as a result of *refraction. Spherical lenses may be either converging or diverging in their action on a light beam, depending on their shape (see illustration). *See also* focal length; aberration; objective; eyepiece; coating of lenses.

lenticular galaxy. *See* galaxies.

Leo (Lion). A large conspicuous zodiac constellation in the northern hemisphere near Ursa Major. The brightest stars are the 1st-magnitude *Regulus (α), three of 2nd magnitude including *Denebola (β) and the binary *Algeiba (γ), and several of 3rd magnitude. Regulus lies at the base of the *Sickle of Leo*, the other

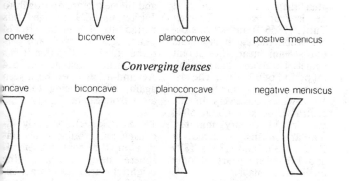

stars in the Sickle being Eta, Gamma, Zeta, Mu, and Epsilon Leonis. The area also contains the Mira star R Leonis, the dwarf nova X Leonis, and several galaxies including the spirals M65, M66, M95, and M96. Abbrev.: Leo; genitive form: Leonis; approx. position: RA 10.5h, dec +15°; area: 947 sq deg.

Leo Minor (Little Lion). A small inconspicuous constellation in the northern hemisphere near Leo, the brightest stars being of 4th magnitude. Abbrev.: LMi; genitive form: Leonis Minoris; approx. position: RA 10h, dec +35°; area: 232 sq deg.

Leonids (November swarm). A periodic *meteor shower that tends to be insignificant most years (ZHR about 10) but at 33 year intervals produces a spectacular meteor storm, radiant: RA 152°, dec +22°, maximizing on the two days around Nov. 17 each year. The associated *meteor stream is the dust debris from comet *Tempel-Tuttle (1866 I) and storms occur when the earth intersects the stream near the parent comet. This comet has an orbital period of 33 years. The last storm occurred on Nov. 17, 1966 when observers in the central USA saw about 5000 meteors in twenty minutes. Similar storms occurred in 1799, 1833, and 1866; 1899 and 1933 were years of only mediocre displays.

Leo systems. Two members of the *Local Group of galaxies.

lepton. *See* elementary particles.

Lepus (Hare). A constellation in the southern hemisphere near Orion. The brightest star is the very luminous remote Arneb (α), which is of 2nd magnitude. Abbrev.: Lep; genitive form: Leporis; approx. position: RA 5.5h, dec -20°; area: 290 sq deg.

Leviathan. The 183 cm (72 inch) reflecting telescope built 1842-45 by William Parsons, 3rd earl of Rosse. It was dismantled in 1908. *See* reflecting telescope.

Lexell's comet. A comet that was discovered by Messier in June 1770 but was named after the St. Petersburg mathematician who calculated its orbit. Prior to 1767 Lexell's comet had a period of 11.4 years; a close approach to Jupiter in 1770 changed this to 5.6 years and the next close approach to Jupiter in 1779 changed this period to a calculated 174 years. This last approach perturbed the comet so that it has a perihelion distance of 5.4 AU and it has never been seen again, always being too far away from earth.

Libra (Balance). A fairly inconspicuous zodiac constellation in the southern hemisphere near Scorpius, the brightest star, Beta Librae, being of 2nd magnitude and apparently green in colour. Delta

Librae is an *Algol variable (magnitude 4.8-6.1) with many other variables occurring in the globular cluster M5. Abbrev.: Lib; genitive form: Librae; approx. position: RA 15h, dec -15°; area: 538 sq deg. *See also* first point of Aries.

libration. The means by which 59% of the moon's surface is made visible, over a 30 year period, to the terrestrial observer despite the lunar day being of the same length as the *sidereal month (*see* synchronous rotation). *Physical libration is a real irregularity in the moon's rotation, arising from minor distortions in shape. *Geometrical librations,* including *diurnal librations, are apparent oscillations due to the moon being observed from slightly different directions at different times. A geometric *libration in longitude* of 7°45′ results from the inconstancy of the moon's orbital velocity, which follows from the *eccentricity of the moon's orbit and *Kepler's second law. This libration thus reveals part of the eastern or western farside limbs at different times during one month. In addition the 5°9′ inclination of the moon's orbital plane to the ecliptic produces a monthly geometrical *libration in latitude* and makes the moon's polar regions more readily observable.

Lick Observatory. An observatory on Mount Hamilton, California, at an altitude of 1280 metres. Endowed by James Lick, it was given to the University of California on the completion in 1888 of its (then) powerful 91 cm (36 inch) refracting telescope. The chief instrument today is the 3 metre (120 inch) reflecting telescope, with a comparatively large focal ratio of f/5, which became fully operational in 1959. Its mirror was one of the spare bases for the 200 inch Hale telescope.

life in the universe. *See* exobiology.

light. *Electromagnetic radiation to which the human eye is sensitive; light is thus also called *visible* or *optical radiation*. The wavelength range varies from person to person but usually lies within the limits of 380 to 750 nanometres. Light therefore forms a very narrow band of the electromagnetic spectrum between the infrared and ultraviolet bands. As the wavelength decreases, the colour of the light changes through the spectral hues of red, orange, yellow, green, blue, to violet. These colours, which may be seen in a rainbow or produced by a prism, form the *visible spectrum*. The eye is most sensitive to greenish-yellow light of wavelength about 500 to 550 nm. The earth's atmosphere is transparent to light so that astronomy at optical wavelengths has not had the problems with detection and observation as studies at other wavelengths. *See* atmospheric windows.

light cone. The representation of a light flash in *space-time. In three-dimensional space the wavefront of a light flash would be a sphere centred on the emitting point and growing in radius at the speed of light. In space-time (with one spatial coordinate suppressed) the vertex of a cone represents the time and place at which a flash is emitted and the cone itself describes the history of the propagating flash.

light curve. A graph of the brightness of a *variable star or other variable object plotted against time, the brightness usually being expressed in terms of apparent or absolute *magnitude. Different types of variable star can be distinguished by the shape of the curve; the period of the variable is one complete oscillation in brightness. The light curve of an *eclipsing binary has two minima: a shallower one (the *secondary minimum*) and a deeper one (the *primary minimum*), which occurs when the brighter star is eclipsed. The minimum will have a flat base if the eclipse is total or annular.

light-gathering power (light grasp). A measure of the ability of an optical telescope to collect light and thus discern fainter objects. It is proportional to the area of the telescope *aperture, i.e. to the square of the diameter of the primary mirror or objective.

light mantle. An avalanche of *highland material, possibly triggered by the formation of *secondary craters from *Tycho, that covers part of the *Taurus-Littrow region and was sampled by the crew of Apollo 17.

light pollution. One of the factors setting a limit on the faintness of stars that may be seen, photographed, or observed by telescope. Man-made light pollution includes street and domestic lighting and can be avoided. On a moonless night there are still three natural sources, of roughly equal contribution; these are *airglow and other atmospheric effects, the *zodiacal light arising from the scattering of sunlight by solar-system dust, and the background light of our Galaxy. *See also* seeing.

light-time. The time taken for light, radio waves, or other electromagnetic radiation to travel at the speed of light (3×10^5 km/s) between two celestial bodies. The light-time along the mean distance between sun and earth is 499 seconds. Since the light-time between celestial bodies and earth varies as the earth revolves in its orbit, a correction (the *equation of light*) must be made to calculations involving measurements of time.

light-year (l.y.). A unit of distance equal to the distance travelled through space in one tropical year by light, radio waves, or any other form of electromagnetic radiation. Since all electromagnetic radiation travels in a vacuum at the speed of light, 299 792 km/s, one light-year equals 9.4605×10^{12} km. Distances expressed in light-years give the time that radiation would take to cross that distance. One light-year equals 0.3066 parsecs, 63 240 astronomical units, or a parallax of 3.259 seconds of arc.

limb. The apparent edge of the sun, moon, or a planet, or any other celestial body with a detectable disc.

limb brightening. The increase in intensity at x-ray and radio wavelengths from the centre to the limb of the sun's disc. It arises for the same reason as the *limb darkening at visible wavelengths but has the opposite effect, since the radiation is emitted from the high-temperature upper *chromosphere/inner *corona rather than from the *photosphere.

limb darkening. The decrease in intensity at visible wavelengths from the centre to the limb of the sun's disc. Light from the central portion of the disc is radiated radially towards us whereas light from closer to the limb has to pass more obliquely through a proportionately greater thickness of the solar atmosphere. For a given limit of penetration we can therefore see deeper into the *photosphere at the centre of the disc than at the limb, where our line of sight passes through cooler gases. The limb consequently appears less bright.

Limb darkening has been detected in other stars, principally by the use of three-colour *photometry of the *light curves of eclipsing binaries (the effect being more pronounced in blue light than in red) and by *speckle interferometry.

A similar effect is observed in planets with deep atmospheres, notably Jupiter and Saturn, and arises from the scattering of sunlight by the atmospheric particles.

limiting magnitude. The faintest apparent *magnitude that may be observed through a telescope and/or recorded on a photographic plate or other device. It depends on the telescope aperture, on the sensitivity of the recording device, on atmospheric conditions, etc. Star catalogues usually list stars to a specific limiting magnitude.

linearly polarized wave (plane-polarized wave). See polarization.

linear momentum. See momentum.

line broadening. The production of broadened spectral lines by various effects. The natural width of a spectral line is determined by quantum mechanical uncertainty. Other factors can, however, produce extra line width, including *Doppler broadening, *pressure broadening, and the *Zeeman effect. Considerations of the *line profile can give information about the physical conditions of celestial objects.

line of apsides. *See* apsides.

line of cusps. The plane passing through the *cusps of a crescent moon or planet. *See also* equator of illumination.

line of inversion. *See* sunspots.

line of nodes. *See* nodes.

line-of-sight velocity. *See* radial velocity.

line profile. The plot of radiation intensity against wavelength or frequency for a spectral line. *See also* line broadening.

line receiver. A radio-astronomy *receiver that is either tuned to receive signals only in one narrow band of frequencies, the centre frequency being adjustable, or that has a broad *bandwidth divided up into many narrow channels of fixed frequency, the outputs of the channels being kept separate from one another. Line receivers are used in *molecular-line radio astronomy where the radio emissions are of narrow bandwidth. They are also used to study the distribution and relative velocity of neutral hydrogen (HI) in our own and in nearby spiral galaxies (*see* HI region).

line spectrum. A *spectrum consisting of discrete lines (*spectral lines*) resulting from radiation emitted or absorbed at definite wavelengths. Line spectra are produced by atoms or ionized atoms when transitions occur between their *energy levels as a result of emission or absorption of photons. The *Fraunhofer lines of the sun are an example of an absorption line spectrum.

Lion. Common name for *Leo.

lithosphere. *See* earth.

Little Dipper. A US name for *Ursa Minor.

Little Horse. Common name for *Equuleus.

Little Lion. Common name for *Leo Minor.

Lizard. Common name for *Lacerta.

LMC. *Abbrev. for* Large Magellanic Cloud. *See* Magellanic Clouds.

lobate ridge. A ramp-shaped feature occurring in the lunar *maria. Lobate ridges are the surviving remnants of the margins of thick lava flows. *Compare* wrinkle ridges.

lobate scarp. *See* Mercury.

lobe. *See* antenna.

Local Group. The *cluster of

MAIN MEMBERS OF LOCAL GROUP OF GALAXIES

galaxy	position (1950) RA (h,m)	dec(₀)	type	approx. distance (kpc)	absolute magnitude	approx. mass (M☉)
Galaxy	17 42	-29.0	Sb or Sc		-20(?)	2×10^{11}
LMC	05 24	-69.8	Irr I	53	-18.7	10^{10}
SMC	00 51	-73.2	Irr I	69	-16.7	2×10^9
M31 (And. gal)	00 40	+41.0	Sb	670	-21.1	3×10^{11}
M33 (Tri. gal)	01 31	+30.4	Sc	700	-18.8	10^{10}
M32 (NGC 221)	00 40	+40.6	E2	660	-16.3	3×10^9
NGC 205	00 38	+41.4	E5	690	-16.3	8×10^9
NGC 147	00 30	+48.2	Ep	690	-14.8	10^9
NGC 185	00 36	+48.1	Ep	690	-15.2	10^9
IC 1613	01 02	+01.8	Irr I	740	-14.8	2×10^8
NGC 6822	19 42	-14.9	Irr I	460	-15.6	3×10^9
Sculptor system	00 57	-34.0	E	85	-12	3×10^6
Fornax system	02 37	-34.7	E	190	-13	2×10^7
Leo I system	10 06	+12.6	E4	220	-11	4×10^6
Leo II system	11 11	+22.4	E1	220	-9.5	10^6
Draco system	17 19	+58.0	E	67	-8.5	10^5
UMi system	15 08	+67.3	E	67	-9	10^5
Maffei I	02 33	+59.4	E or S0	1200	-20	2×10^{11}

galaxies of which our *Galaxy is a member. It is a comparatively small cluster with 25 known members (1978) and some other possible members, most of which are dwarf ellipticals or irregular galaxies (see table). The most massive members are the Galaxy, the *Andromeda galaxy (M31), and (possibly) the recently discovered Maffei 1 (see Maffei galaxies). The other major members are the *Triangulum spiral (M33) and the Large *Magellanic Cloud. The group is irregular in shape with two notable subclusters: one around our Galaxy and the Magellanic Clouds, the other around M31 and M33.

local mean time. See local time.

local oscillator. See receiver.

local sidereal time (LST). The *sidereal time at a particular location. It can be determined by making a correction to *Greenwich Sidereal Time. The correction amounts to the addition of four minutes for each degree of longitude that the location is east of Greenwich, four minutes being subtracted for each degree west. A celestial object crosses the local meridian when the LST is equal to the object's *right ascension.

local standard of rest (LSR). The frame of reference, centred on the sun, in which the space velocities (see radial velocity) of all the stars average to zero. The *peculiar motions of these stars with respect to the LSR would also average to zero if the sun were at rest. The sun however is moving, relative to the LSR, at about 20 km s^{-1} towards the solar *apex. The stars reflect this solar motion in addition to their peculiar motions in their *proper motions. The LSR can also be considered as the reference frame, centred on the sun, that moves in a circular orbit around the galactic centre so that all neighbouring stars (i.e. those in circular orbits) are essentially at rest. The rotation velocity of the LSR is about 250 km s^{-1}.

Local Supercluster. A huge flattened cloud of *galaxies and *clusters of galaxies that is apparently centred on or near the *Virgo cluster and contains the *Local Group (including our Galaxy) as an outlying member. Its radius is believed to be about 30 megaparsecs. Whether the Local Supercluster represents a true physical system is still open to argument. The main evidence for its existence is the observed concentration of nearby bright galaxies towards a particular great circle in the sky.

Local System. *See* Gould Belt.

local time (local mean time). The *mean solar time at a particular location. It can be determined by adding four minutes to the *Greenwich Mean Time, i.e. *universal time, for each degree of longitude that the location is east of the Greenwich meridian or by subtracting four minutes for each degree west. *See also* local sidereal time; standard time.

long-focus photographic astrometry. A branch of *astrometry in which (usually) refracting telescopes of long focal lengths of eight metres or more and focal ratios of f/15 to f/20 are used to produce sharp photographs of both the star under consideration plus three or more distant reference stars. Accuracies of about 0.01 arc seconds can be reached. Used originally to measure stellar parallax, the technique is now also employed in the determination of orbital motion of binary stars.

longitude. 1. *Short for* celestial longitude. *See* ecliptic coordinate system; *See also* sun's longitude.

2. *Short for* galactic longitude. *See* galactic coordinate system.

longitude of the ascending node. *See* orbital elements.

long-period variables. A group of variable stars that are principally *Mira stars but sometimes the red *semiregular stars, type SRa, are included in the group.

Looped (or Loop) nebula. *See* 30 Doradûs.

loop prominence. *See* prominences.

Lorentz contraction (Lorentz-FitzGerald contraction). The contraction of a moving body in the direction of motion, put forward by H.A. Lorentz (1895) and independently by G.F. FitzGerald (1893) as an explanation for the result of the Michelson-Morley experiment (*see* ether). The contraction was later shown to be an effect of the special theory of *relativity (1905).

Lorentz transformation. A set of equations used in the special theory of *relativity to transform the coordinates of an event (x, y, z, t) measured in one inertial frame of reference to the coordinates of the same event (x', y', z', t') measured in another frame moving relative to the first at constant velocity v:

$$x = (x' + vt')/\beta$$
$$y = y'$$
$$z = z'$$
$$t = (t' + x'v/c^2)/\beta$$

β is the factor $\sqrt{(1 - v^2/c^2)}$ and c is the speed of light. When v is very much less than c, these equations reduce to those used in classical mechanics.

Lowell Observatory. An observatory at Flagstaff, Arizona, at an altitude of 2210 metres, set up by Percival Lowell in 1895.

lower culmination (lower transit; inferior culmination). *See* culmination.

LRV. *Abbrev. for* Lunar Roving Vehicle.

LSR. *Abbrev. for* local standard of rest.

LST. *Abbrev. for* local sidereal time.

luminosity. Symbol: L. The intrinsic or absolute brightness of a star, equal to the total energy radiated per second from the star, i.e. the total outflow of power. Luminosity is related to surface area and *effective temperature, T_{eff}, by a form of Stefan's law:

$$L = 4\pi R^2 \sigma T_{eff}^4$$

where σ is Stefan's constant and R is the stellar radius. Thus two stars with similar T_{eff} (i.e. of the same *spectral type) but greatly different luminosities must differ in size: they belong to different luminosity classes within that spectral type, as determined from their spectra. In the luminosity class of main-sequence stars, luminosity decreases as temperature decreases; the luminosity of *giant stars however increases with decreasing temperature: *red giants are much brighter than yellow giants; the luminosity of *supergiants drops and then rises with decreasing temperature. The luminosity of stars is in theory dependent on mass if their chemical composition is similar. It has been found that with the notable exception of highly evolved stars the *mass-luminosity relation is obeyed approximately.

Stellar luminosity is usually expressed in terms of absolute bolometric *magnitude:

$$M = -2.5 \log_{10} L + \text{constant}$$

The luminosity of a star can thus be compared with that of the sun ($L\odot$):

$$\log_{10}(L/L\odot) = 0.4(M\odot - M)$$

The sun's absolute bolometric magnitude ($M\odot$) is 4.75. There is a very great range in stellar luminosity from about a million times to less than one ten thousandth that of the sun's luminosity.

luminosity classes. *See* spectral types.

luminosity function. Symbol: $\Phi(M)$. The number of stars per cubic parsec or the number of galaxies per cubic megaparsec with absolute magnitudes $M \pm \frac{1}{2}$ or luminosities $L \pm \frac{1}{2}$. The luminosity function for stars within 10 parsecs of the sun shows a peak at an absolute magnitude of about +15: there is thus a predominance of intrinsically faint stars in the solar neighbourhood. The luminosity function for any given magnitude interval is not the same everywhere in the Galaxy: for instance, the luminosity function of a *globular cluster is different from that of an *open cluster. Stars of all *spectral types contribute to the value $\Phi(M)$. The specific luminosity function $\Phi(M,S)$ considers one spectral type at a time, each showing a peak distribution at a different mean magnitude.

The luminosity function of *radio sources is the number of sources per cubic megaparsec with a given radio emission. This radio emission or flux density is measured at a standard frequency in watts per unit solid angle.

luminosity-volume test (V/V_m test). A method of testing for evolutionary effects in a sample of extragalactic sources of known *redshift. All sources should be known over a particular region of the sky down to a certain flux density level. The volume, V, within the redshift of each object, together with the volume V_m within the maximum redshift out to which that object could have been and yet still remain in the sample, are computed from a cosmological model. If no evolution is present and the correct cosmological model is employed, then V ought to be uniformly distributed between 0 and V_m. The mean of V/V_m is then $\frac{1}{2}$.

Luna probes (Lunik probes). A series of Soviet lunar probes (see table) that included the first craft to reach the vicinity of the moon (1), the first to photograph the farside (3), the first crash lander (2), soft lander (9), orbiter (10), unmanned sample-return mission (16), and unmanned roving vehicle (17). *See also* Lunokhod; Zond probes; Lunar Orbiter probes; Surveyor; Ranger.

lunar calendar. *See* lunar year; calendar.

lunar cycle. *See* Metonic cycle.

lunar day. The rotation period (27.322 earth days) of the

Luna probes

Spacecraft	Launch date	Comments
Lunik 1	1959, Jan. 2	Missed moon by 6000 km
Lunik 2	1959, Sept. 12	Struck moon in Palus Putredinis
Lunik 3	1959, Oct. 4	First farside photographs
Luna 4	1963, Apr. 2	Missed moon by 8500 km
Luna 5	1965, May 9	Struck moon
Luna 6	1965, June 8	Missed moon by 160 000 km
Luna 7	1965, Oct. 4	Struck moon
Luna 8	1965, Dec. 3	Struck moon
Luna 9	1966, Jan. 31	Soft landed in Oc. Procellarum
Luna 10	1966, Mar. 31	Performed experiments in lunar orbit
Luna 11	1966, Aug. 24	Performed experiments in lunar orbit
Luna 12	1966, Oct. 22	Returned photographs from lunar orbit
Luna 13	1966, Dec. 21	Soft landed in Oc. Procellarum
Luna 14	1968, Apr. 7	Performed experiments in lunar orbit
Luna 15	1969, July 13	Struck moon in Mare Crisium
Luna 16	1970, Sept. 12	Returned 101 g soil from Mare Fecunditatis
Luna 17	1970, Nov. 17	Landed Lunokhod 1 in Mare Imbrium
Luna 18	1971, Sept. 2	Struck moon
Luna 19	1971, Sept. 28	Performed experiments in lunar orbit
Luna 20	1972, Feb. 14	Returned 30 g soil, Crisium highlands
Luna 21	1973, Jan. 8	Landed Lunokhod 2 in Mare Serenitatis
Luna 22	1974, May 29	Performed experiments in lunar orbit
Luna 23	1974, Oct. 28	Soft-landed but drill failure
Luna 24	1976, Aug. 15	Returned 170 g soil from Mare Crisium

moon, equal in length to the *sidereal month. The moon is thus in *synchronous rotation with the earth. *See also* libration.

lunar eclipse. *See* eclipse.

lunar grid. A postulated but questionable arrangement of preferentially aligned lineations that, if real, would imply global tectonism on the moon. The apparent pattern, however, may be the result of lighting artefacts and of *crater chains and *rilles associated with the major impact *basins, particularly *Imbrium.

lunar module (LM). *See* Apollo.

lunar month. *See* synodic month.

Lunar Orbiter probes. A series of five US space probes used for photographic reconnaissance of the Apollo landing sites (Lunar Orbiters 1, 2, and 3) and global mapping (4 and 5) of the moon from equatorial and polar orbits respectively. The probes were launched between Aug. 1966 and Aug. 1967. Each spacecraft was equipped with medium- and high-resolution cameras, the films from which were scanned in strips for transmission back to earth. Satellite tracking resulted in the discovery of *mascons. The spacecraft were crashed onto the moon on mission completion. *See also* Ranger; Surveyor; Luna probes; Zond probes.

lunar orbit rendezvous. *See* Apollo.

lunar parallax. *See* diurnal parallax.

lunar rocks. *See* moonrocks.

Lunar Roving Vehicle (Lunar Rover; Rover; LRV). The battery-driven car used to extend the sampling capabilities of *Apollos 15, 16, and 17. Transported to the moon in a compact form in the Lunar Module descent stage, the 213 kg Rover was equipped with a remote-controlled television camera and high-gain antenna for direct radio communication with earth. It was left behind on the moon after use. *See also* Lunokhod; Modular Equipment Transporter.

lunar year. A year of 12 *synodic months, each of 29.5306 days, i.e. a year of 354.3672 days. A *lunar calendar* is based solely on the moon's motion; it has a year of 354 days with a *leap year of 355 days and is composed of 12 months of 29 or 30 days. A *luni-solar calendar* is a lunar calendar that is brought into step with the solar or seasonal calendar by the intercalation (addition) of a 13th *leap month*.

lunation. *See* synodic month.

Lunik probes. *See* Luna probes.

luni-solar calendar. *See* lunar year.

luni-solar precession. *See* precession.

Lunokhod. An eight-wheeled Soviet unmanned lunar roving vehicle, soft-landed on the moon by *Lunas 17 and 21. Lunokhod 1 was equipped with a laser reflector for lunar-ranging experiments; it was active in Mare Imbrium for 10 months during which time it completed a 10 km traverse, performing photographic tasks, magnetic field measurements, and chemical (x-ray) and cosmic-ray analyses. Lunokhod 2 travelled 37 km in four months in the vicinity of Le Monnier crater on the borders of Mare Serenitatis. *See also* Lunar Roving Vehicle.

Lupus (Wolf). A constellation in the southern hemisphere near Centaurus, lying partly in the Milky Way, with several stars of 3rd magnitude. There are many naked-eye double stars. Abbrev.: Lup; genitive form: Lupi; approx. position: RA 15.3h, dec $-45°$; area: 334 sq deg.

Luyten 726-8, 789-6. *See* Table 7 (nearest stars), backmatter.

l.y. *Abbrev. for* light-year.

Lyman series. *See* hydrogen spectrum.

Lynx. A constellation in the northern hemisphere near Ursa Major, the brightest stars being one of 3rd magnitude and several of 4th magnitude. The area contains many faint double stars. Abbrev.: Lyn; genitive form: Lyncis; approx.

position: RA 8h, dec +47°; area: 545 sq deg.

Lyot filter (birefringent filter). A type of *interference filter devised by the French astronomer Bernard Lyot in 1933 for studying the solar corona (*see* coronagraph). Its action depends on the interference of the polarized light occurring in birefringent (double refracting) crystalline laminae: a stack of such laminae, of suitable thickness, transmits well-separated and very narrow wavebands of about 0.07 nanometres.

Lyra (Lyre). A constellation in the northern hemisphere near Cygnus, lying partly in the Milky Way, the brightest star being the blue zero-magnitude Vega. It contains the prototypes of the *Beta Lyrae (eclipsing binary) stars and RR Lyrae stars and the naked-eye pair Epsilon Lyrae, both of 4th magnitude and both double. It also contains the planetary *Ring nebula and a small globular cluster M56. Abbrev.: Lyr; genitive form: Lyrae; approx. position: RA 18.5h, dec +40°; area: 286 sq deg.

Lyrids (April Lyrids). A minor meteor shower, radiant: RA 272°, dec +32°, that maximizes on April 21. In the past it was much more active, the last great Lyrid shower occurring in 1803. Observations of Lyrids have been traced back 2500 years, Chinese observers describing a remarkable display in 15 BC. The associated *meteor stream has the same orbit as comet 1861 I. *See also* June Lyrids.

Lysithea. A small satellite of Jupiter, discovered in 1938. *See* Jupiter's satellites; Table 2, backmatter.

M

M87. *See* Virgo A.

M. *See* Messier catalogue.

Mach's principle. The idea put forward by Ernst Mach in the 1870s that the inertial properties of a particular piece of matter were in some way attributable to the influence of all the other matter in the universe. The *inertia of a body is its reluctance to change its state of rest or uniform motion in a straight line. According to Mach, inertia can be considered to arise from some (unspecified) interaction between the body and the background presence of distant stars and galaxies; a body in total isolation would not therefore possess inertia.

Maffei galaxies. Two galaxies that lie in Cassiopeia close to the galactic equator and are heavily obscured by interstellar dust. They were discovered as infrared objects by P. Maffei in 1968. *Maffei 1*, the brighter of the two, is a large elliptical or S0 galaxy with a total mass of about 2 ×

10^{11} solar masses (roughly that of our own Galaxy). It appears to lie about 1.2 megaparsecs away and is possibly an outlying member of the *Local Group. *Maffei 2* appears to be a medium-sized spiral galaxy. Its distance is uncertain but it is probably too distant to be a member of the Local Group.

Magellanic Clouds. Two comparatively small galaxies that are close neighbours (satellites) of our own Galaxy. Both are naked-eye objects but, being close to the south celestial pole, they are only visible from the southern hemisphere. They were first recorded in 1519 by Ferdinand Magellan. The *Large Magellanic Cloud (LMC)* has a total mass of 10^{10} solar masses and a diameter of about 12 000 parsecs; it lies in the constellation Dorado at a distance of about 53 000 parsecs. The *Small Magellanic Cloud (SMC)* has a total mass of approximately 2×10^9 solar masses and a diameter of about 9000 parsecs; it lies in the constellation Toucan and is some 69 000 parsecs away.

Although both galaxies are generally classified as irregular (Irr I) systems, there is some evidence of barred spiral structure (*see* Hubble classification), particularly in the LMC. Both galaxies are rich in young *population I objects and contain a much greater proportion of gas than our own Galaxy. Like the Galaxy they both possess population II haloes and have recently been found to be connected by a common envelope of neutral hydrogen. *See also* Local Group; dwarf galaxy.

magnetic field. The region surrounding a magnet, a conductor carrying an electric current, a stream of charged particles, etc., in which such a body or system exerts a detectable force. This force will be experienced by another magnetic substance or by a moving charged particle, such as an electron. The *magnetic flux density is a measure of the strength of the field, usually quoted in teslas or sometimes in gauss. *See also* geomagnetism; sunspot cycle; galactic magnetic field.

magnetic flux density (magnetic induction). Symbol: B. A vector quantity that is a measure of the strength of a magnetic field in a particular direction at a particular point. It may be given in terms of the force, F, on a charge, q, moving in the magnetic field:
$$B = F/qv \sin\theta$$
where v is the velocity of the charge, which is moving at an angle θ to the direction of the field. Magnetic flux density is usually measured in teslas. The *magnetic flux*, symbol: Φ, is the surface integral of B, i.e.
$$\Phi = \int B.dA = \int B\cos\alpha \, dA$$
where α is the angle between B and a line perpendicular to the surface. The magnetic flux density is thus the magnetic flux passing perpendicularly through unit area. Magnetic flux is measured in webers.

magnetic induction. See magnetic flux density.

magnetic stars. Stars, usually peculiar A stars (*Ap stars), that have detectable and often very large magnetic fields. Many of these stars show either periodic or irregular variations of the field together with a reversal of the polarity; others show irregular field variations but constant polarity. Magnetic variability is usually accompanied by very small changes in brightness. Magnetic stars are studied by measurements of the Zeeman splitting (*see* Zeeman effect) of spectral lines caused by a magnetic field; these give the average magnetic field over the entire disc. Periodic variations in the magnetic field could result from the magnetic axis of the star being tilted with respect to the rotational axis so that areas of different magnetic field strength are presented to the observer as the star rotates: this is the *oblique rotator theory*. *See also* spectrum variables.

magnetic storms. *See* geomagnetic storms.

magnetobremsstrahlung. *See* synchrotron emission.

magnetograph. An instrument used to map the distribution, strength, and polarity (direction) of magnetic fields over the sun's disc. It is usually an automatic scanning system employing instruments to detect and measure the Zeeman splitting of a selected spectral line (*see* Zeeman effect). The resulting diagram is called a *magnetogram*.

magnetohydrodynamics. The study of the behaviour of electrically conducting fluids, i.e. an ionized gas, a plasma, or some other collection of charged particles, in a magnetic field. The collective motion of the particles gives rise to an electric field that interacts with the magnetic field and causes the fluid motion to alter. This coupling between hydrodynamic forces and magnetic forces tends to make the fluid and the magnetic field lines move together.

magnetometer. Any of a variety of instruments used to measure the strength and direction of a magnetic field.

magnetopause. *See* magnetosphere.

magnetosheath. *See* magnetosphere.

magnetosphere. A region surrounding a planet, and bounded by the *magnetopause*, in which ionized particles are controlled by the magnetic field of the planet rather than by the sun's magnetic field, which is carried by the *solar wind. The region includes any radiation belts of the planet, such as the *Van Allen belts of the earth. The earth's magnetosphere extends to 60 000 km on the sunward side of the planet but is drawn out to

Earth's magnetosphere, shaped by the solar wind

many times this distance on the side away from the sun by the solar wind (see illustration). Above the magnetopause is a layer of turbulent magnetic field, the *magnetosheath,* enclosed by a shock wave where the smooth flow of solar-wind particles past the planet is first interrupted. Mercury and Jupiter are also known to possess magnetospheres.

magnifying power (magnification). The angle subtended by the image of an object seen through a telescope, divided by the angle subtended by the same object seen without telescopic aid. Where the object is too small or too distant to be resolved, the magnifying power may be calculated from the ratio of the focal lengths of the *objective (or *primary mirror) and *eyepiece, or from the *aperture divided by the diameter of the *exit pupil.

magnitude. A measure of the relative brightness of stars and other celestial objects, now extended to include measures of the relative radiation intensities of objects such as infrared sources. The brighter the object the lower its assigned magnitude. The *apparent magnitude,* symbol: *m,* is a measure of the brightness of a star as observed from earth. Its value depends on the star's *luminosity (i.e. its intrinsic bright-

POGSON SCALE OF MAGNITUDES

m_1-m_2	I_1/I_2	m_1-m_2	I_1/I_2
0.2	1.202	3.5	25.12
0.4	1.445	4.0	39.81
0.6	1.738	4.5	63.1
0.8	2.089	5.0	100
1.0	2.512	6.0	251
1.5	3.981	7.0	631
2.0	6.310	8.0	1585
2.5	10.0	9.0	3981
3.0	15.85	10.0	10 000

ness), its distance, and the amount of light absorption by interstellar matter between the star and earth.

In ancient times the visible stars were ranked in six classes of apparent magnitude: the brightest stars were of first magnitude and those just visible to the naked eye were of sixth magnitude. This system became inadequate as fainter stars were discovered with the telescope and instruments became available for measuring apparent brightness. In the 1850s it was proposed that the physiological response of the eye to a physical stimulus was proportional to the logarithm of that stimulus (Weber-Fechner law). A difference in apparent magnitude of two stars is thus proportional to the difference in the logarithms of their brightness, i.e. to the logarithm of the ratio of their brightness. In order to make the magnitude scale precise the English astronomer N.R. Pogson proposed, in 1856, that a difference of five magnitudes should correspond exactly to a brightness ratio of 100 to 1. (W. Herschel had shown this to be approximately true.) Hence two stars that differ by one magnitude have a brightness ratio of $^5(100):1$, i.e. a ratio, known as the *Pogson ratio*, of 2.512. A star two magnitudes less than another is $(2.512)^2$, i.e. 6.3 times brighter, and so on. In general the apparent magnitudes m_1 and m_2 of two stars with apparent brightness I_1 and I_2 are related by:

$m_1 - m_2 = 2.5 \log_{10}(I_2/I_1)$

The Pogson scale, based on the Pogson ratio, is now the universally adopted scale of magnitude (see table). The zero of the scale was established by assigning magnitudes to a group of standard stars near the north celestial pole, known as the *North Polar Sequence,* or more recently by photoelectric measurements. The class of magnitude-one stars was found to contain too great a range of brightness and zero and negative magnitudes were consequently introduced: the higher the negative number, the greater the brightness. The scale also had to be extended in the positive direction as fainter stars were discovered: the greatest apparent magnitude that can currently be measured is about +25.

Originally apparent magnitude was measured by eye— *visual magnitude,* m_{vis}— usually in conjunction with an instrument by which brightnesses could be compared. It is now measured much more accurately by photometric techniques but before that photographic methods were

Relative transmission of filters used in the UBV system

used. *Photographic magnitudes*, m_{pg}, are determined from the optical density of stellar images on ordinary film, i.e. film that has a maximum response to blue light. *Photovisual magnitudes*, m_{pv}, are measured using film that has been sensitized to light—yellowish green in colour—to which the human eye is most sensitive. In the early *International Colour System* these two magnitudes were measured with films having a maximum response to a wavelength of about 425 nanometres (m_{pg}) and 570 nm (m_{pv}), the magnitudes being equal for A0 stars.

A suitable combination of *photocell and *filter can select light or other radiation of a desired wavelength band and measure its intensity. These *photoelectric magnitudes* can be either *monochromatic* if a very narrow band is selected or *heterochromatic* for a broader band. The *UBV system* of stellar magnitudes is based on photoelectric photometry and has been widely adopted as the successor of the International System. The photoelectric magnitudes, denoted *U, B,* and *V*, are measured at three broad bands: *U* (ultraviolet radiation) centred on a wavelength of 360 nm, *B* (blue) centred on 420 nm, and *V* (visual, i.e. yellowish green) centred on 540 nm (see illustration). The zero point of the UBV system is defined in terms of standard stars having a carefully studied and agreed magnitude. The UBV system has been extended by the use of magnitudes at red and infrared wavelengths, denoted R and I respectively. R is measured at a wavelength of 680 nm and I at 825 nm. Another magnitude system used sometimes in preference to the UBV system is the *uvby system.

Apparent magnitude is a measure of the radiation in a particular wavelength band, say of blue light, received from the star. Apparent *bolometric magnitude*, m_{bol}, is a measure of the total radiation (ultraviolet, visible, and infrared) received from the star. The *bolometric correction (BC)*, is the difference between the bolometric and visual (V) magnitudes, apparent or absolute, and is always negative:

$$BC = m_{bol} - V = M_{bol} - M_v$$

It is equal to zero for the sun.

Apparent magnitude gives no indication of a star's *luminosity: a very distant very luminous star may have a similar apparent magnitude as a closer but fainter star. Luminosity is defined in terms of *absolute magnitude*, M, which is the apparent magnitude of a star if it were at a standard distance of 10 parsecs. It can be shown that the two magnitudes of a star are related to its distance d (in parsecs) or its *annual parallax, π (in arc seconds):

$$M = m + 5 - 5 \log d - A$$
$$M = m + 5 + 5 \log \pi - A$$

A is the interstellar *extinction. As with apparent magnitude there are values of absolute photoelectric magnitudes: M_U, M_B, and M_V; of bolometric magnitude: M_{bol}; and of photographic magnitudes: M_{pg} and M_{pv}. Knowledge of a star's bolometric magnitude enables its *luminosity to be found.

magnitude at opposition. A measure of the brightness of a planet or minor planet when at *opposition; it corresponds, normally, to the maximum brightness attained by the object during the *apparition since it is then viewed at its fullest phase and is near its closest point to the earth.

main lobe (main beam). *See* antenna.

main sequence. The principal sequence of stars on the *Hertzsprung-Russell diagram, running diagonally from upper left (high temperature, high luminosity) to lower right (low temperature, low luminosity) and containing about 90% of all known stars. A star spends most of its life on the main sequence. A newly formed star appears on the main sequence when it first achieves a stable state whereby its core temperature is sufficient for nuclear reactions to begin. It is then at age zero. The positions of the age-zero stars on the H-R diagram are specified by reference to the *zero-age main sequence*. As hydrogen is converted to helium, changes in chemical composition and stellar structure cause the star's position to shift slightly to the right from its zero-age position.

A star's position on the main sequence depends primarily on its mass, the more massive and thus more luminous stars occurring higher up the sequence. The star's lifetime there also depends on its mass, the more massive stars having much shorter lifetimes. After the star has consumed about 12% of the hydrogen in its core (*see* stellar evolution) it evolves away from the main sequence: its radius and luminosity increase and it eventually becomes a *red giant.

main-sequence fitting. A technique for measuring the distance modulus and hence distance for a star *cluster, especially an *open cluster: the main-sequence portion of a graph of colour index versus apparent magnitude for the cluster is superimposed on the main sequence of a standard *Hertzsprung-Russell diagram, the spectral types of the two curves being made to overlap. The difference between cluster magnitudes and the superimposed absolute magnitudes of the H-R diagram is approximately constant and equal to the distance modulus.

major axis. *See* semimajor axis.

major planets. The planets Mercury, Venus, Earth, Mars, Jupiter, Saturn, Uranus, Neptune, and Pluto, which have diameters significantly larger than the 1003 km diameter of Ceres, the largest *minor planet. *See* Table 1, backmatter.

Maksutov telescope (meniscus telescope). A *catadioptric telescope named after its Soviet inventor, who published its design in 1944. It differs from the *Schmidt telescope in that the *correcting plate is a deeply curved meniscus lens. Since the primary mirror is also spherical all three optical surfaces are simple to make. Its commonest form (see illustration) is a Cassegrain adaptation in which the image is reflected by an aluminized spot

Ray path in Maksutov telescope

on the meniscus lens back through a hole in the primary mirror. It performs well as a prime-focus camera. It can also be made in Newtonian and Gregorian configurations. Although having exceptional performance and compactness, the thick correcting plate limits it to relatively small sizes both because of its weight and the very thick blanks needed to make larger diameters.

manganese stars. *See* Ap stars.

mantle. *See* earth.

mare. (Latin: sea) *Singular of* maria.

maria, lunar (ring plains). Expanses of iron-rich basaltic lavas of low *albedo (i.e. comparatively dark) that were erupted onto the lunar surface between 3900 and 3000 million years ago. The maria are largely confined to the *nearside of the moon where the crust is thinnest. The youngest *basins were flooded to produce the *circular maria* with their associated *mascons. The *irregular maria* are shallower and fill the older less well-defined basins. Chemical variations between the maria are revealed in *reflectance spectra and by *colour difference photography. Age differences are apparent from their crater densities, which are invariably lower than in the older *highlands. Maria are mainly named after mental or meteorological states; the smaller ones are described as lakes, bays, and marshes (see table). *See also* moonrocks.

Lunar maria in approximate order of size:

Oceanus Procellarum
   Ocean of Storms
Mare Imbrium
   Sea of Rains
Mare Serenitatis
   Sea of Serenity
Mare Fecunditatis
   Sea of Fertility
Mare Tranquillitatis
   Sea of Tranquillity
Mare Crisium
   Sea of Crises
Mare Humorum
   Sea of Moisture
Mare Nectaris
   Sea of Nectar
Mare Frigoris
   Sea of Cold
Mare Orientale
   Eastern Sea

Mare Australe
   Southern Sea
Mare Cognitum
   Sea of Knowledge
Mare Nubium
   Sea of Clouds
Mare Marginis
   Border Sea
Mare Smythii
   Smyth's Sea
Mare Spumans
   Foaming Sea
Mare Undarum
   Sea of Waves
Mare Humboldtianum
   Humboldt's Sea
Mare Moscoviense
   Sea of Moscow
Mare Ingenii
   Sea of Ingenuity

maria, Martian

Mare Anguis
  Serpent Sea
Mare Vaporum
  Sea of Vapours
Sinus Medii
  Central Bay
Sinus Aestuum
  Bay of Heats
Sinus Roris
  Bay of Dews
Sinus Amoris
  Bay of Love
Sinus Iridum
  Bay of Rainbows
Palus Putredinis
  Marsh of Decay

Palus Somnii
  Marsh of Sleep
Palus Epidemiarum
  Marsh of Epidemics
Palus Nebularum
  Marsh of Mists

Lacus Mortis
  Lake of Death
Lacus Somniorum
  Lake of Dreams
Lacus Veris
  Lake of Spring
Lacus Autumni
  Lake of Autumn

maria, Martian. *See* Mars.

Mariner 2. A 202 kg US probe that made the first successful planetary fly-by, 34 830 km from Venus, on Dec. 14 1962. Its flight confirmed the existence of the *solar wind, measured Venus' temperature, but did not detect a Venusian magnetic field.

Mariner 4. A 261 kg US spacecraft that was the first successful probe to Mars. It passed the planet at a distance of 9844 km on July 14 1965, returning 22 photographs revealing an arid cratered surface.

Mariner 5. A 245 kg US spacecraft that passed Venus at a distance of 3990 km on Oct. 19 1967. It measured Venus' temperature and atmospheric profile, accurately determined its mass and diameter, but failed to detect any magnetic field.

Mariner 6, 7. Two identical 413 kg US probes that passed Mars at distances of 3412 km and 3424 km on July 31 and Aug. 5, 1969, respectively. Mariner 6 photographed part of Mars' equatorial region while Mariner 7 examined the southern hemisphere and southern polar cap. A total of 201 photographs were returned, while other experiments measured surface temperatures, atmospheric pressure and composition, and Mars' diameter.

Mariner 9. A 1031 kg US probe that became the first artificial satellite of another planet when it entered a 12-hour orbit of Mars on Nov. 13 1971. This orbit, ranging in altitude from 1650 km to 17 100 km approximately, allowed Mariner's two television cameras to return 7329 photographs before contact was terminated on Oct. 27 1972. The entire Martian surface was mapped and close-up views were obtained of Mars' two small natural satellites

*Phobos and *Deimos. Other equipment included infrared and ultraviolet spectrometers, used for atmospheric studies, and an infrared radiometer to measure surface temperatures.

The photographic reconnaissance was jeopardized by a violent planet-wide dust storm that developed in late September 1971. With the exception of four hazy spots, later found to be the summits of the principal Martian volcanoes of the *Tharsis Ridge, all surface features were obscured by dust when Mariner 9 arrived. As the dust settled during the following weeks, photographs revealed for the first time the extent of Martian volcanism and the existence of canyons, such as the *Valles Marineris, and channels, which suggest that water once flowed on Mars. The changing appearance of the *polar caps with the Martian seasons was also followed. See Mars; Mars, surface features.

Mariner 10. A US spacecraft that executed the first dual-planet mission by making close approaches to Venus and, three times, to Mercury. Weighing 503 kg, it carried two television cameras that returned the first close-up views of Venus and Mercury. Other instruments included an infrared radiometer to measure Mercury's surface temperature; an ultraviolet spectrometer to detect atmospheric airglow and another to measure atmospheric absorption of solar ultraviolet radiation; and two magnetometers to monitor magnetic field variations throughout the flight.

During the Venus fly-by at a distance of 5760 km on Feb. 5 1974, Mariner 10 returned 3500 photographs; those taken in ultraviolet radiation showing the planet's cloud cover and atmospheric circulation in unprecedented detail. Using the gravitational pull of Venus, the probe was placed on course for a Mercury fly-by on Mar. 29 1974. After course corrections two more visits to Mercury took place on Sept. 21 1974 and Mar. 19 1975, each after two solar orbits by Mercury and one by Mariner 10. The three Mercury encounters, at minimum distances of 740 km, 48 000 km, and 330 km respectively, provided more than 10 000 photographs of Mercury and led to the discovery of the planet's magnetic field. See also Mercury; Venus.

Mariner probes. A series of US planetary probes. Mariners 2 and 5 made close-up observations of Venus while *Mariner 10 became the first two-planet probe, visiting Venus and Mercury. *Mariners 4, 6, 7, and 9 were successful Mars probes, the latter becoming the first man-made orbiter of another planet. Mariners 11 and 12 were renamed *Voyagers 1 and 2. Mariners 1 (to Venus) and 8 (to Mars) suffered launch failure while Mariner 3 failed to achieve its intended trajectory to Mars.

Mariner's Compass. Common name for *Pyxis.

Markab (α Peg). A white giant that is the brightest star in the constellation Pegasus and, with Algenib, lies on the S side of the Great Square of Pegasus. m_v: 2.49; M_v: 0.0; spectral type: B9.5 III; distance: 32 pc.

Markarian galaxies. A number of galaxies that were discovered by the Soviet astronomer B.E. Markarian in the 1970s to have a strong ultraviolet continuum. The ultraviolet sources are either concentrated in a bright nucleus or are distributed throughout a more diffuse object. *Seyfert galaxies form a large component of the former group.

Mars. The fourth *planet from the sun and the nearest superior planet to the earth. It orbits the sun every 687 days at a distance that varies between 1.38 and 1.67 astronomical units. Its reddish 6794 km diameter disc is most favourably placed for observation during oppositions that occur with Mars near *perihelion (*see* Mars, oppositions). Mars rotates in 24h 37m 23s about an axis tilted by 24° to its orbital plane; these values help to make the Red Planet the most earthlike of all the other planets, even though it has only 28% of the earth's surface area. Mars has two small natural satellites: *Phobos and *Deimos. Orbital and physical characteristics are given in Table 1, backmatter.

Telescopes reveal an orangered surface with indistinct darker markings, once thought to be seas and named *maria*. White polar caps expand and contract with the Martian seasons. When it became apparent that the maria also varied in intensity and shape with the seasons, it was suggested that they were areas of lichen-like vegetation that flourished when water became available from the poles during each Martian spring. *Canals, charted by some observers, were postulated as an artificial planet-wide water distribution network. Observations by *Mariner spacecraft showed that the canals were a myth and reveal the maria as areas of darker bedrock on which the Martian winds deposit varying amounts of lightercoloured dust, so changing their appearance. For details of Martian topography *see* Mars, surface features; Mars, volcanoes; Mars, polar caps.

The Martian atmosphere has a surface pressure of about seven millibars, or 0.007 times the average pressure at the earth's surface, and extends to include an ionosphere at altitudes between 100 and 300 km. Daytime surface temperatures rarely climb above 0°C except during summer in the southern hemisphere. Most areas experience minimum temperatures near -100°C before sunrise each morning.

Tests by the *Viking spacecraft, which landed in 1976,

show an atmosphere comprising mainly carbon dioxide (95%), with nitrogen (2.7%), argon (1.6%), oxygen (0.15%), a variable trace of water vapour, and traces of carbon monoxide, krypton, and xenon. The water vapour sometimes freezes to form clouds of ice crystals above high topographic features, such as the volcanoes, or condenses as fog in low-lying areas, such as *Hellas Planitia and the *Valles Marineris.

More extensive surface obscurations occur during dust storms, which can engulf the whole planet—as happened when *Mariner 9 arrived to survey Mars from orbit in 1971. Storms appear to erupt each Martian year when summer arrives in the northern hemisphere and seasonal winds of hundreds of kilometres per hour whip powdery dust from the Martian surface to heights in excess of 25 km.

Mars has no radiation belts and only a weak magnetic field, at most 0.2% as strong as the earth's; this suggests that it lacks a molten nickle-iron core and may have no iron core at all. The surface rocks, however, appear to be rich in iron, resembling the minerals haematite and magnetite and lending the planet its red coloration. *See also* Viking probes.

Mars, oppositions. *Oppositions recur at an average interval of 2 years 50 days, although this varies by 30 days either way because of the *eccentricity of Mars' orbit. Favourable oppositions for observation of surface features recur every 15-17 years when Mars is near *perihelion and may approach within 56 million km of the earth, attaining an apparent diameter of 25 arc seconds. The next favourable opposition will occur on Sept. 28 1988. At *aphelion oppositions, Mars may be as much as 101 million km distant with a diameter of 14 arc seconds.

Mars, polar caps. Two variable white areas at the Martian poles, visible to earth-based observers but studied in detail by the *Mariner and *Viking Orbiter spacecraft. Each cap grows under a haze of cloud during winter for its hemisphere, reaching a latitude of 50° at its greatest extent. During summer it retreats to leave a small irregular patch about 500 km across. This residual cap probably consists of water ice, although the variable cap appears to be a frost of frozen carbon dioxide. At latitudes greater than 80°, the polar terrain exhibits a crater-free layered structure up to six km thick that possibly represents successive deposition of windborne dust.

Mars probes. A series of ill-fated Soviet probes to Mars. Contact was lost with Mars 1, launched on Nov. 1 1962, after it had travelled 106 million km. Mars 2 and 3 entered orbit around Mars on Nov. 27 and Dec. 2 1971 after detaching lander capsules: that from Mars 2 crashed and the other stopped transmitting within two minutes of touchdown. The orbiters made surface and atmospheric studies. Mars 4, an intended orbiter, passed the planet on Feb. 10 1974 after retrorocket failure. Mars 5 entered orbit two days later but ceased functioning after a few days of photographic reconnaissance. Mars 6 flew past Mars on Mar. 12 1974, ejecting a lander capsule that crashed on the surface. Mars 7 passed Mars three days before Mars 6 but its intended lander missed the planet completely.

Mars, surface features. Impact *craters dominate Mars' southern hemisphere—and may be as old as 3500 million years—but are thinly scattered over the younger and mainly volcanic surface of most of the northern hemisphere. In addition there are white polar caps, volcanic features (see Mars, polar caps; Mars, volcanoes), and areas, particularly in equatorial regions, that show evidence of running water in the past.

Many of the Martian craters show the effects of erosion by dust, which is generally of a higher *albedo than uneroded surface rocks and forms bright deposits in low-lying terrain such as crater floors and the southern hemisphere impact basins of *Argyre Planitia and *Hellas Planitia. In places the dust is swept into ridges or dunes by the wind. Bright or dark streaks occur in the lee of some craters where the prevailing wind has piled up the dust or has scoured the surface to reveal the underlying darker bedrock. Variations in dust deposition are responsible for albedo and outline changes of the darker markings, or maria, once attributed to life processes.

Southeast of the volcanoes of the *Tharsis Ridge lies the complex equatorial canyon system of *Valles Marineris, which measures 4500 km from east to west and 150 to 700 km from north to south. Individual canyons are up to 200 km wide and 7 km deep, dwarfing the maximum 28 km width and 2 km depth of earth's Grand Canyon. They appear to result from faulting and collapse of the Martian surface—a process that may be continuing today. Other fault systems are found elsewhere, all of them apparently associated with areas of volcanic activity.

Smaller valleys, or channels, meander for up to 1000 km or more across equatorial areas of Mars. They appear to have been formed by running water derived, at least in part, from rain falling on the planet's surface. Some channels begin in areas of collapsed terrain as though the water they carried was originally frozen as a subsurface layer of permafrost that was subsequently thawed and released by volcanic heating. The enigma is that neither rain nor surface water can survive on Mars given present atmospheric constraints. Possibly the channels indicate that

Mars has experienced spells of milder climatic conditions when the atmosphere was thicker than it is now. These spells, during which polar and subsurface ice may have been released, may be related to real fluctuations in the sun's output, to variations in the planet's axial tilt, or to periods of enhanced volcanic activity. Further evidence for the existence of a permafrost layer comes from the shape of the ejecta blankets surrounding some large craters.

Mars, volcanoes. Much of the northern hemisphere of Mars shows signs of volcanism with extensive lava plains sparsely pock-marked with impact craters: an indication of an age considerably less than the 3500 million years assigned to the heavily cratered southern hemisphere. This disparity may arise from the fact that the mean Martian surface lies at a lower level in the northern hemisphere.

The two main areas of volcanic activity, both situated on bulges in the Martian crust, are the *Tharsis Ridge—with the four principal volcanoes including the largest, *Olympus Mons—and the *Elysium Planitia. The volcanoes are of the shield type, having long gently sloping flanks. Counts of the numbers of impact craters on the slopes suggest ages between 200 million and 700 million years for the Tharsis Ridge volcanoes, and about 2500 million years for those of Elysium Planitia. The extraordinary size of the volcanoes may be related to the absence of plate tectonics on Mars, so that the crust remains locked in relation to volcano-inducing 'hot spots' in the mantle below. See also Mars, surface features.

Martian probes. See Mariner 4; Mariner 6, 7; Mariner 9; Mariner probes; Mars probes; Viking probes; Zond probes.

mascons. (*Shortened form of* mass concentrations.) Positive gravitational anomalies associated with the young circular lunar *maria, where relatively dense disc-shaped masses are now known to be located close to the lunar surface: because the gravitational attraction is somewhat higher over these regions, mascons perturb the orbits of lunar satellites. They were produced either by the mare basalt itself as it flooded the basins or by the uplifting of higher-density mantle material during *basin excavation. Mascons imply the existence of a rigid lunar crust for at least 3000 million years. The removal of crust during crater formation may also produce negative gravitational anomalies; these are observed in the younger large *craters and around the *Orientale Basin, which is only partially lava-filled. Since the circular maria are topographic lows, mascons may be partially isostatically compensated, i.e. they may have sunk under their excess density until some sort of equilibrium depth was reached

maser amplifier. A type of negative-resistance amplifier in which amplification is achieved by stimulating excited molecules to release photons in phase with the passing radiation. Amplifiers of this type have extremely low *noise figures, limited in practice by losses in the input and output circuits, and they therefore find application in *radio telescopes above one gigahertz.

maser source. A small celestial source of electromagnetic radiation, displaying narrow spectral lines and having an anomalously high *brightness temperature, in which the emission is thought to be by maser action: radiation at the maser frequency is amplified by stimulating excited molecules in the source to emit *photons in phase with the passing radiation. Several different types have been identified including *OH masers* radiating at 1.665 gigahertz (GHz), H_2O *masers* at 22.235 GHz, and *SiO masers* at 86.243 GHz.

Maser sources have been detected both in the atmospheres of old variable stars and in molecular clouds associated with newly formed or forming stars. The first maser source—an OH maser—was discovered in the Orion nebula in 1965. About 300 sources containing hydroxyl masers have since been found, about 20% of which also have water masers. Maser sources are usually very *variable and produce highly polarized radio emission from regions that are typically about one parsec in diameter.

mass. A measure of the quantity of matter in a body. The SI unit of mass is the kilogram. Mass is a property of matter that determines both the *inertia of an object, i.e. its resistance to any change in its motion or state of rest, and the gravitational field that it can produce. The former is called *inertial mass,* defined by *Newton's laws of motion, and the latter is *gravitational mass.* Inertial and gravitational mass have been found to be equivalent. This led to Einstein's principle of *equivalence between inertial and gravitational forces. See also weight; stellar mass; relativity, special theory.

mass discrepancy. See missing mass.

mass loss. The loss of mass by a star during its evolution by one or more processes. Mass is known to flow slowly into space from many stars in the form of a *stellar wind. Mass can also be ejected as the shell of a *planetary nebula or in a violent *nova or *supernova explosion. *Mass transfer can occur in close binary systems. Mass loss could possibly prevent an originally massive star from evolving to a *neutron star or *black hole.

mass-luminosity relation (M-L relation). An approximate relation between the mass and luminosity of main-sequence stars, predicted by Eddington in 1924. Although having some basis in theory it is obtained empirically from a graph of absolute bolometric magnitude against the logarithm of mass (in solar units), i.e. $m/m\odot$, for a large number of binary stars. Most points lie on an approximately straight line. Since a star's absolute bolometric magnitude is a function of the logarithm of its luminosity (in solar units), i.e. $L/L\odot$, this line is represented by the M-L relation:

$$\log(L/L\odot) = n \log(m/m\odot)$$
$$L/L\odot = (m/m\odot)^n$$

n averages about 3 for bright massive stars, about 3.5 for sun-type stars, and about 2 for dim red dwarfs of low mass. The relation holds for main-sequence stars, which all have similar chemical composition; it is not obeyed by *white dwarfs (degenerate matter) or red giants (extended atmospheres).

mass number. Symbol: A. The number of protons and neutrons (i.e. nucleons) in the nucleus of a particular atom. It is the number closest to the mass of a nuclide in atomic mass units, i.e. in units equal to 1/12 of the mass of a carbon-12 atom. *See also* atomic number.

mass transfer. A process occurring in close binary stars when one of the stars has expanded to such an extent that mass can be transferred to its companion. This mass may strike the more compact companion directly or may orbit the star and eventually form a ring of matter—an *accretion disc*. The mass transfer takes place through the inner Lagrangian point, lying between the stars, and occurs when the larger star fills its Roche lobe (*see* equipotential surfaces).

matching (impedance matching). Arranging electrical impedances so that maximum power is transferred from one device to another. This occurs when the input impedance of one device equals the output impedance of the other to which it is connected. *See also* feeder.

matter era. *See* big bang theory.

Maunder minimum. *See* sunspot cycle.

Max Planck Institute for Astronomy (MPIA). A West German institute set up in 1969 to oversee the design and installation of two similar reflecting telescopes, 2.2 metres (87 inches) in diameter. One will be sited in the northern hemisphere on the Calar Alto mountain in Almeria, Spain, at an altitude of 2168 metres; the other will be at a southern hemisphere site, probably on Cerro La Silla, Chile.

Max Planck Institute for Radio Astronomy. An observatory at Effelsberg, near Bonn. It has the world's largest fully steerable radio dish, a 100 metre (328 foot) paraboloid, which has been in operation since 1971. It has a computer-controlled *altazimuth mounting.

Mayall reflector. See Kitt Peak National Observatory.

McDonald Observatory. An observatory of the University of Texas at Fort Davis, Texas, at an altitude of 2081 metres. It has a 2.7 metre (107 inch) and a 2.1 metre (82 inch) reflecting telescope.

McMath telescope. See Kitt Peak National Observatory; solar telescope.

mean anomaly. See anomaly.

mean daily motion. The angle through which a celestial body would move in one day if its orbital motion were uniform.

mean density of matter. Symbol: ϱ_0. The factor that determines the dynamical behaviour of the *universe, i.e. whether it is open or closed. The *density parameter*, Ω_0, is a dimensionless quantity given by
$$\Omega_0 = 8\pi G\varrho_0/(3H_0^2)$$
It is related to the *deceleration parameter, q_0, by
$$\Omega_0 = 2q_0$$
If the universe is a continuously expanding open system, Ω_0 must be ≤ 1; if it exceeds unity the universe is closed and must eventually collapse. The *critical density* for which Ω_0 is unity is 5×10^{-27} kg m^{-3} for a *Hubble constant H_0 equal to 55 km s^{-1} megaparsec^{-1}. Various estimates involving the mass to luminosity ratios of galaxies lead to values for Ω of about 0.01 due to galaxies. The *missing-mass problem for *clusters of galaxies indicates the presence of hidden matter and suggests that the above value may be an underestimate. Various tests have shown that values of Ω_0 equal to one can be accommodated within observational measurements.

mean equinox. The position of the vernal *equinox obtained when a correction is made to the *true equinox for the small periodic displacements resulting from *nutation. Coordinates in star charts and catalogues are measured from the mean equinox at the time when the sun's mean longitude is 280° (occurring about Jan. 1) for the fundamental epoch which in current use is 1950.0.

mean life. See half-life.

mean noon. See noon.

mean parallax. The *parallax obtained by statistical methods for a large group of stars with different apparent magnitudes and proper motions.

mean position. See true position.

mean sidereal time. See sidereal time.

mean solar day. *See* mean solar time.

mean solar time (mean time). Time as indicated on an accurate clock and measured with reference to the uniform motion of the *mean sun. The *mean solar day* is the interval of 24 hours between two successive passages of the mean sun across the meridian. The mean solar second is 1/86 400 of the mean solar day. The difference between mean solar time and *apparent solar time on any particular day is the *equation of time. Since the mean sun is an abstract and hence unobservable point, mean solar time is defined in terms of *sidereal time. It can be determined at any instant and on any meridian from the observed *local sidereal time by means of the constant ratio of 1.002 737 909 between an interval of mean solar time and one of sidereal time. This ratio has been calculated from the day lengths: one mean solar day is equal to 24h 3m 56.555s of mean sidereal time, one mean sidereal day to 23h 56m 4.091s of mean solar time.

Mean solar time was originally devised in order to provide a uniform measure of time based on the earth's rate of rotation, which was assumed constant. Since the discovery of variations in the rotation rate, mean solar time could not be assumed to have a constant rate and the system of *ephemeris time was developed. *See also* Greenwich Mean Time; universal time; atomic time.

mean sun. An abstract reference point that has a constant rate of motion and is used in timekeeping in preference to the real sun whose observed motion is nonuniform. This nonuniformity results primarily from the earth's elliptical orbit around the sun, which causes the earth's orbital speed, and hence the sun's apparent speed, to vary (*see* Kepler's laws). Secondly the direction of the sun's apparent motion is along the ecliptic and is thus inclined to the direction parallel to the celestial equator along which solar time is measured. An additional seasonal variation is therefore introduced when the sun's motion is measured relative to the equator.

The mean sun follows a circular orbit around the celestial equator so that it moves eastwards at a constant speed and completes one circuit in the same time—one *tropical year—as the apparent sun takes to orbit the ecliptic, if the slight secular acceleration of the sun is ignored. The mean sun is defined by an expression for its right ascension, which fixes its position among the stars at every instant and is a means of defining a measure of *mean solar time.

mean time. *Short for* mean solar time.

mega-. Symbol: M. A prefix to

a unit, indicating a multiple of 10^6 (i.e. one million) of that unit. For example, one megaparsec is one million parsecs.

meniscus lens. A lens in which the two surfaces have different or sometimes equal curvatures but, unlike a concave or convex lens, curve in the same general direction. A *positive meniscus* is thicker at the centre than at the edges; a *negative meniscus* is thinner at the centre. *See illustration at* lens.

meniscus telescope. *See* Maksutov telescope.

Menkalinan. *See* Auriga.

Mensa (Table Mountain). A small inconspicuous constellation in the southern hemisphere, close to the south pole, the brightest stars being of 5th magnitude. It contains part of the Large Magellanic Cloud. Abbrev.: Men; genitive form: Mensae; approx. position: RA 5.5h, dec -78°; area: 153 sq deg.

Merak (β UMa). One of the seven brightest stars in the *Plough and one of the two *Pointers.

Mercury. The innermost and, with a diameter of 4880 km, the second smallest planet, orbiting the sun in 88 days at an average distance of 0.39 astronomical units (AU). It has no natural satellite. The distance from the sun varies between 0.31 and 0.47 AU because of Mercury's high orbital *eccentricity of 0.21. Maximum *elongation from the sun varies between 28° and 18°, so that Mercury can be observed only when low in the twilight sky. Telescopes reveal a disc, between 5 and 15 arc seconds across, showing vague markings and *phases varying from full at superior conjunction to new at inferior conjunction. *Transits across the sun's disc recur at intervals of between 7 and 13 years, the next being on Nov. 13 1986. Orbital and physical characteristics are given in Table 1, backmatter.

Mercury's axial rotation period of 58.6 days, determined by radar in 1965, corresponds exactly to 2/3 of its orbital period, probably because of tidal coupling between the extra gravitational force of the sun at *perihelion and irregularities in Mercury's almost perfectly spherical form. This period is close to half the planet's *synodic period; this explains why early observers, who saw the same face of the planet at favourable oppositions, concluded erroneously that it had a *synchronous rotation period of 88 days.

*Mariner 10, which made three close approaches to Mercury in 1974 and 1975, measured a maximum daytime temperature of 190°C; this may reach 450°C near perihelion. At night, temperatures plunge to -180°C beneath the most tenuous of atmospheres, consisting mainly of minute traces of helium and argon. Mariner found that Mercury has a *magnetosphere supported by a magnetic field about 1% as strong as the earth's; it is not clear whether this is generated within the planet's core or arises from permanent magnetization within the crust. Mercury's density, 5.4 times that of water, implies that it has a large nickle-iron core, perhaps 3600 km across.

Photography of half of Mercury's surface by Mariner 10 shows it to resemble the moon in being heavily cratered with intervening areas of lava-flooded plains, called *maria*. The largest impact feature detected is the 1300 km diameter *Caloris Basin,* similar to the moon's Mare *Imbrium. Antipodal to Caloris is an area of chaotic terrain created, it is believed, by shock-waves from the same impact. Some recent craters possess radiating systems of light-coloured surface streaks, similar to the lunar *rays. The *ejecta blankets surrounding many craters are half the size of those around lunar craters of similar diameter, probably because Mercury has twice the moon's surface gravity. Of particular interest, and not seen on the moon or Mars, are lines of cliffs, called *lobate scarps,* up to 3 km high and 500 km long. These may have resulted from wrinkling of the silicate-rich crust as Mercury's core cooled and contracted billions of years ago.

Mercury, perihelion. *See* advance of the perihelion.

Mercury project. The pioneering US space project to orbit a manned craft, to investigate man's reactions to and ability to adapt to the weightlessness of space, and to recover both men and spacecraft. Mercury MR3, launched May 5 1961 by a Redstone rocket, carried the first American, Alan Shepard, into the upper atmosphere in a suborbital hop. The first true orbital flight was made by John Glenn in Mercury MA6, launched Feb. 20 1962 by an Atlas rocket. The longest flight, 34.3 hours, was made by Gordon Cooper in the last capsule in the series, Mercury MA9, launched May 15 1963. *See also* Gemini project; Apollo.

Mercury, transits. *See* transit.

meridian. 1. A great circle on the earth's surface at right angles to the equator and passing through the north and south poles.
2. *Short for* celestial meridian. The projection of the observer's terrestrial meridian on to the *celestial sphere. It is thus the great circle passing through the north and south *celestial poles and the observer's *zenith and intersecting the observer's horizon at the north and south points (*see* cardinal points).

meridian angle. *See* hour angle.

meridian circle (transit circle). An instrument by which the altitude of a star may be measured with very great accuracy. It is essentially a telescope mounted so as to rotate in a vertical north-south plane about a horizontal axis fixed in an east-west direction. It thus points to the observer's *meridian and is used to observe stars as they cross the meridian (*see* culmination). The star's altitude as it crosses the meridian is measured by means of a very accurately graduated circular scale, set in the vertical plane. When a correction is made for atmospheric refraction, the true altitude is found. The star's *declination can then be determined if the latitude of the observatory is accurately known.

If the exact *sidereal time of culmination is also measured, the *right ascension of the star can be found. Alternatively if the right ascension of the star is known precisely its time of culmination will allow any correction necessary to be made to a sidereal clock. Instruments in which only the time of culmination can be measured, i.e. those with no graduated circle, are known as *transit instruments*. Modern instruments use electronic means to determine both time of culmination and altitude.

Meridian circles and transit instruments are subject to errors, which affect the measuring accuracy. *Inclination error* and *azimuthal error* arise when the rotational axis deviates from the horizontal or from the east-west direction, respectively; these errors can be corrected by reversing the instrument. *Collimation error* occurs when the optical axis of the telescope does not intersect the rotational axis at right angles. Observational data must be corrected for these errors. The errors can be reduced by means of the *broken-back transit:* a prism is placed at the intersection of the rotational and optical axes of the telescope so that the eyepiece is at one end of the horizontal axis.

meridian passage. The passage of a star across the meridian (*see* culmination) or of an inferior planet across the sun (*see* transit).

Merope. See Pleiades.

mesosphere, mesopause. *See* atmospheric layers.

Messier catalogue. A catalogue of the brightest 'nebulae' prepared by the French astronomer Charles Messier and printed in final form in 1784. Messier compiled the list in order to avoid confusion of these cloud-like objects with comets, for which he was a keen searcher. 103 celestial objects, which appeared fuzzy and extended in telescopes of the time, were listed and given a number preceded by the letter M. This is the object's *Messier number,* an example being M42: the Orion nebula. Six more objects were listed in 1786. It was later found that most of the objects were not true *nebulae but galaxies and star clusters. See Table 8, backmatter; NGC.

Messier number. See Messier catalogue.

Me stars. Cool red stars whose spectra show emission lines of hydrogen in addition to the molecular bands of titanium oxide characteristic of *M stars. Many are long-period variables, such as *Mira stars.

MET. Abbrev. for Modular Equipment Transporter.

metagalaxy. The total recognized assemblage of galaxies and intergalactic space, i.e. the entire contents of the universe, both within and beyond the limits of observation, together with the region of space it occupies.

meteor. The streak of light seen in the clear night sky when a small particle of interplanetary dust (a *meteoroid) burns itself out in the earth's upper atmosphere. On a clear moonless night at a time away from *meteor showers the dark-adapted eye can see about 10 per hour, the eye detecting meteors down to about fifth visual *magnitude. The most probable magnitude seen is +2·5, 75% of the meteors observed having magnitudes in the range 3·75 to 0·75. The visual rate maximizes at about 04·00 local time when the observer is on the leading side of the earth, 'ploughing' into the cosmic dust cloud. The meteoroid enters the atmosphere at velocities between 11 and 74 km s^{-1} depending on whether the earth just catches up the particle or they have a head-on collision. The ablating meteoroid not only leaves behind it a *train of excited atoms that de-excite to produce the brief blaze of light (a *visual meteor*) but it also produces a column of ionized atoms and molecules (both atmospheric and meteoric) that can act as reflectors of radar pulses transmitted from ground-based telescopes—a *radio meteor*. The meteoroid dissipates its energy, distributes its disaggregated atoms and molecules, and produces the visual and radio meteor in the region between 70 km and 115 km above the earth's surface. The mean altitude of maximum luminosity and electron density is 97 km.

A large percentage of the mass influx to the earth's atmosphere (some 16 000 tonnes per year) is made up of particles in the size range that produce visual and radio meteors, i.e. one millionth to one million grams. The meteoroid has a kinetic energy considerably in excess of the energy required to vaporize itself. Air molecules striking the meteoroid have a kinetic energy of about 400 electronvolts. As the binding energy of the atoms in the meteoroid is a few electronvolts, a hundred or so meteoroid atoms boil off for each absorbed air molecule. The meteoroid starts to ablate at about 115 km. These ablated atoms then collide with the surrounding air molecules and by losing energy in these collisions produce *excitation and *ionization. After ten or so collisions the energy of the ablated atom has dropped below the excitation potential of the air molecules and ionization and excitation no longer occur. The ablated atom has moved about 50 cm from the meteoroid path during this process. The small meteoroid is completely evaporated after intercepting an air mass about 1% of its own mass. Owing to these processes the meteoroid *train has a length of between 7 and 20 km and a width of about 100 cm. A 4th-magnitude meteor contains around 10^{18} electrons and 10^{18} positive ions. The atoms quickly de-excite and electrons and positive ions recombine and become attached to other atmospheric molecules. The time duration is related to the magnitude but is usually less than a second.

The meteor spectrum is a low-excitation one, ionization potentials lying in the range 1·9 to 13·9 volts (equivalent to temperatures roughly 1600 to 4800 kelvin). The Ca II lines, H and K, are the main features of fast meteors whereas Na I, Mg I, and Fe I are dominant in slow ones.

meteorite. Interplanetary debris that falls to the earth's surface. The first documented one was the stone that fell near Einsisheim, Alsace in 1492 but it wasn't until 1803 that meteorites were accepted by the scientific community as being extraterrestrial. One that is seen to hit the ground is known as a *fall;* one that is discovered accidentally some time after having fallen is termed a *find.* Over 2000 meteorites have been recovered and roughly six falls and 10 finds are added to the list annually. About 500 hit the earth each year; the majority go unrecorded, falling in oceans, deserts, and other uninhabited regions.

Meteorites can be crudely divided into three types: *stony meteorites, which subdivide into *chondrites and *achondrites, *iron meteorites, which subdivide into hexahedrites, octahedrites, and ataxites, and *stony-iron meteorites. Iron meteorites have densities of about $7·8$ g cm^{-3} and contain on average 91% iron, 8% nickel, and 0·6% cobalt. Stony meteorites have densities about $3·4$ g cm^{-3} and on average contain 42% oxygen, 20·6% silicon, 15·8% magnesium, and 15·6% iron, no other element exceeding 2%. Stony-irons are intermediate in composition. The relative percentages of each type are shown in the table. The fall percentages give a reasonable approximation to the actual meteorite population of the solar system. However the fragile *carbonaceous chondrites mostly disintegrate on entry.

The largest iron meteorite that has been found is the 60 tonne *Hoba meteorite, the largest stony meteorite being

meteorite types		finds	falls
chondrites } stones		38.8	84.8
achondrites		0.9	7.9
stony-irons		5.8	1.7
irons		54.5	5.6

one tonne, part of the Norton County, Kansas achondrite. The found meteorite might only represent a small percentage of the in-space mass of the body. Meteorite entry is accompanied by a brilliant *bolide, the meteorite usually being retarded to *free-fall velocity at a height of about 20 km. It hits the ground at about 300 km per hour. The surface of the meteorite can easily be heated up to several thousand degrees kelvin during entry. This usually produces a black glassy fusion crust, which covers the meteorite. Most meteorites fragment on entry, scattering pieces over an elliptical area: the carbonaceous chondrite meteorite *Allende,* which fell in 1969 in Mexico, scattered five tonnes of material over an area 48 km long by 7 km wide.

Meteorites are named after the nearest topographical feature to the fall point. Recognizing meteorite finds relies mainly on analysing chemically for the presence of nickel, which is present in all iron meteorites. Etched and polished irons also show *Widmanstätten patterns. Stony meteorites contain nickel-iron particles that can be extracted magnetically from a crushed sample.

Meteorites bring to our notice a type of rock existing in the solar system but different from anything that occurs in the outer shell of the planet earth. Most meteorites are believed to be fragments of *minor planets. Two problems arise: firstly, meteorites only seem to have had a short exposure to cosmic rays and secondly, the dynamics by which a meteorite gets from the asteroid belt to the earth's orbit is nontrivial. The carbonaceous chondrites are thought to originate as parts of cometary *nuclei. The ages of meteorites have been obtained by radioactive-dating methods and have been estimated at up to $4 \cdot 7 \times 10^9$ years. This indicates that meteorites were formed at about the same time as the solar system.

meteorite craters. Circular impact structures produced by meteorite bombardment, usually known as *astroblemes if they occur in the earth's crust. *See* craters.

meteoroid. The collective terms applied to meteoritic material in the solar system, usually replaced by the terms *micrometeorite for particles with mass less than 10^{-6} gram and *meteorite for bodies with mass greater than about 10^5 grams. The majority of the mass of the meteoroid cloud around the sun is made up of particles with individual masses between 10^{-7} and 10^{-3} gram. Meteoroids are usually produced by the collisional fragmentation of *meteor stream particles. In the main they move around the sun in low-inclination direct orbits. The space density of meteoroids varies as about $1/r^2$, where r is the distance from the sun. Individual meteoroids in the mass range 10^{-6} to 10^4 grams are fragile crumbly rocky dust particles with a composition similar to carbonaceous chondrite meteorites. *See also* meteor.

meteor shower. The increase in observed rate of appearance of *meteors (*shower meteors*) when the earth crosses through a *meteor stream; a particular shower usually occurs at the same time each year. The increase in rate varies from stream to stream: it is a function of mass of the originating comet, age of stream, position of intersection point with respect to stream axis, and orbital perihelion. With many showers the rate varies from year to year as the meteoroids are not evenly distributed around the stream. The ratio between shower rate and sporadic-meteor rate is a function of the mass of the meteoroids being observed. This ratio can be as high as 10 for visual meteors, is about one to two for radio meteors, and is negligible for satellite-observed meteors. Shower meteors appear to emanate from a *radiant, the shower being named after the constellation that contains the radiant. *See also* Table 4, backmatter.

meteor storm (meteor swarm). A *meteor shower with an enormous rate of appearance that occurs when the earth intersects a new *meteor stream close to the originating comet. The most famous was the Leonid storm of Nov. 12-13,1833 when the rate exceeded 10 000 visual meteors an hour, single observers often seeing 20 a second.

meteor stream. The annulus of *meteoroids around the orbit of a decaying *comet. Dust particles are pushed away from the cometary *nucleus by gas pressure at the time when the nucleus is near *perihelion. The particles then have slightly different orbital parameters to the comet. Stream-formation time depends on perihelion distance and meteoroid size, being anything from tens to hundreds of years. New streams only have meteoroids close to the comet and only produce *meteor showers periodically. Streams are about 10 times thicker at aphelion than at perihelion. They contain anywhere between 10^{12} grams and 10^{16} grams of dust and have volumes in the range 10^{21} to 10^{25} km^3.

Metonic cycle (lunar cycle). A

period of 19 years (tropical) after which the phases of the moon will recur on the same days of the year: the period contains 6939.60 days, which is very nearly equal to 235 *synodic months, i.e. 6939.69 days. Since it is also almost equal to 20 *eclipse years, i.e. 6932.4 days, it is possible for a series of four or five eclipses to occur on the same dates at intervals of 19 years. The cycle was discovered by the Greek astronomer Meton in the fifth century BC and was used in determining how intercalary months could be inserted into a lunar calendar so that the calendar year and tropical (seasonal) year were kept in step.

metre. Symbol: m. The usual scientific unit of length and distance. It is the length equal to 1 650 763.73 wavelengths in vacuum of the radiation corresponding to the transition between the levels $2p_{10}$ and $5d_5$ of the krypton-86 atom. This definition came into use in 1960. Before that, the metre was the length of a prototype metre bar.

Michelson interferometer. *See* interferometer.

Michelson-Morley experiment. *See* ether.

micro-. Symbol: μ. A prefix to a unit, indicating a fraction of 10^{-6} (i.e. one millionth) of that unit. For example, one micrometre is 10^{-6} metre.

microdensitometer. An optical instrument that measures and records small changes in the transmission density on a photographic plate, etc. If the density is linearly proportional to the amount of light that originally fell on the sample, as it is in an electronograph, this gives a direct recording of light intensity.

micrometeorite. A cosmic dust particle of mass less than about 10^{-6} gram and diameter less than 0·1 mm. On impact with the earth's atmosphere the heat absorbed by the particle from atmospheric friction is insufficient to raise it to boiling point. The ratio of heat radiated to heat absorbed is proportional to the inverse of the radius of a particle: those larger than 10^{-6} gram ablate and form *meteors; smaller ones don't. The micrometeorite will be decelerated to a normal free-fall velocity and then drift to the surface of the earth. On the moon, however, no deceleration occurs and they impact the surface with the normal geocentric velocity.

Microscopium (Microscope). A small inconspicuous constellation in the southern hemisphere near Sagittarius, the brightest star being of 4th magnitude. Abbrev.: Mic; genitive form: Microscopii; approx. position: RA 21h, dec -35°; area: 210 sq deg.

microwave background radiation. The remnant of the radia-

Microwave background radiation spectrum

tion content of the *primordial fireball associated with the hot *big bang. Observed as *black-body radiation characteristic of a temperature of three kelvin, the microwave background is nearly equal in intensity from all directions of the sky. Small variations in this intensity are detected due to the motion of the Galaxy and to *Compton scattering in hot intracluster gas. The earth's atmosphere complicates observations over the black-body peak, which occurs at a wavelength of about 2.5 mm (see illustration), necessitating balloon- or satellite-borne instruments. A microwave background was first considered by George Gamow in the 1940s but the concept was generally neglected until the radiation was discovered in 1965 by Arno Penzias and Robert Wilson.

*Cosmological redshift implies that these photons were more energetic in the past and at redshifts greater than about 1000 they were energetic enough to photoionize hydrogen (*see* big bang theory). The mean free path of the photons was then drastically reduced by scattering on the free electrons present at that time. This essentially prevents our learning anything about earlier times from the microwave background, although it is hoped that small-scale variations in the microwave background may relate to fluctuations in the distribution of matter. Such fluctuations might later have become galaxies. The microwave background is the dominant radiation in deep intergalactic space and has an energy density of 4×10^{-14} joules per metre cubed.

microwaves. Very high frequency radio waves. Their frequencies usually lie in the range 3×10^{11} hertz (just beyond the infrared band of the electromagnetic spectrum) to about 3×10^8 hertz. The wavelength range is therefore about 1 mm to 1 m.

Milky Way. A dense band of faint stars that extends right round the *celestial sphere, dividing it into roughly equal parts. Its central line marks the central plane of our *Galaxy, inclined at 62° to the celestial equator. Although the vast majority of the stars are too faint to be seen individually, they are collectively visible on a clear moonless night as a diffuse band of light. The Milky Way is seen because the sun lies close to the central plane of the galactic disc. Since this region of the Galaxy is highly flattened, a much greater depth of stars is visible in directions along the plane (i.e. towards the Milky Way) than in other directions. The Milky Way has a distinctly patchy appearance; it also varies considerably in width and brightness and is noticeably brighter towards Sagittarius (the direction of the galactic centre). Many of the apparent gaps are due to *dark nebulae, such as the Coalsack, along our line of sight, which prevent stars behind them from being seen.

Milky Way System. *See* Galaxy.

milli-. Symbol: m. A prefix to a unit, indicating a fraction of 10^{-3} (i.e. 1/000) of that unit. For example, one millisecond is 10^{-3} second.

Mills cross antenna. An interferometer (*see* radio telescope) using two long narrow *antennae mounted in the form of a cross and producing a single pencil beam; it is named after B.Y. Mills. One arm of the cross may be shortened to give a *T-antenna*, which has an antenna pattern very similar to that of the full cross but whose *sensitivity is less.

Mimas. A satellite of Saturn, discovered in 1789. *See* Table 2, backmatter.

Mimosa. *See* Beta Crucis.

mini black hole. *See* black hole.

minimum-energy orbit. *Another name for* Hohmann transfer. *See* transfer orbit.

minor axis. The least distance across an ellipse, crossing the major axis perpendicularly at the centre of the ellipse. The length is usually given as $2b$.

minor-planet families. *See* Hirayama families.

minor planets (asteroids; planetoids). Small solar-system bodies, most (perhaps 95%) of which orbit the sun in a main belt (the asteroid belt) between the orbits of Mars and Jupiter. They orbit at distances of between 2.17 and 3.3 astronomical units (AU) and with periods between three and six years. They also include members of the *Amor, *Apollo, and *Trojan groups, as well as the unique object *Chiron, which have orbits very different from those in the main belt. The distribution of orbits within the belt is modified by *Kirkwood gaps. Like the planets most minor planets follow nearly circular orbits close to the plane of the *ecliptic, though the average minor planet orbit does have a larger *eccentricity (0.15) and a greater *inclination (9.7°) than those of most planets. A few objects, such as *Hidalgo, follow highly eccentric and inclined orbits more akin to those of some periodic *comets.

Minor planets shine by reflected sunlight but only one—Vesta—is bright enough to be seen with the naked eye. More than 100 000 minor planets may be bright enough for photographic observation and discovery but only 2100 had received official minor planet numbers by 1978. A permanent number is given only when the orbit has been calculated accurately; the object's position can then be determined for any desired time. Many more, with less well-determined orbits, have been assigned temporary designations. Most numbered minor planets have received names also, usually the prerogative of the discoverer.

Until recently the diameters of even the largest minor planets could only be roughly estimated. Since 1970 studies of the polarization of reflected sunlight and of the brightness at visible and infrared wavelengths for a large number of minor planets has led to improved values for their *albedos; from these values the diameters have been recalculated. Minor planets vary in size from the largest and first to be discovered (the 1003 km diameter *Ceres) to objects smaller than 1 km: there probably exists a continuous distribution down to the size of *meteoroids. Most are thought to be irregular in shape, causing them to vary in brightness as they rotate every few hours. Smaller objects predominate: there are only some 200 objects larger than 100 km diameter and about 10 larger than 300 km.

The total mass of the minor planets is estimated at 2.4×10^{21} kg (approximately 0.0004 of the earth's mass). This indicates that the hypothesis that they are the remains of a single large planet which exploded is untenable. Probably they represent the debris from collisions between a handful of small bodies that condensed between Mars and Jupiter during the formation of the solar system.

As well as providing improved values for their diameters, studies of minor-planet spectra and polarizations show that the surface rocks are diverse and are similar to those of *meteorites. It is now believed therefore that most meteorites are fragments of minor planets. The majority of minor planets resemble *carbonaceous chondrites while others appear to have a metallic or a silicaceous surface. The latter are more common near the inner edge of the main asteroid belt while carbonaceous types predominate near the outer edge. This is an important clue to the composition of the pre-solar nebula from which the proto-minor planets condensed. The minor planets are being considered as targets for *in situ* studies by future space missions. *See* Table 3, backmatter; Hirayama families.

Mintaka. *See* Orion.

minute of arc (arc minute; arc min). *See* arc second.

Mira or **Mira Ceti (the Wonderful;** *o* **Cet).** A red giant in the constellation Cetus that is a long-period *variable star with an average period of 331 days. It was known to early astronomers for its regular appearances and disappearances and was established in the 17th century as a variable star. It is the prototype of the *Mira stars. Its radius varies by 20% during its cycle and at maximum size and brightness is over 330 times the sun's radius. The surface temperature varies from about 2600 kelvin (K) at maximum luminosity down to about 1900 K. The visible light emitted varies considerably, by about 8 magnitudes, over the cycle (see illustration). The average maximum apparent magnitude, m_v, lies between 3 and 4, although it may reach 2nd magnitude (2.1 in 1969) and reached 1.2 in 1779. At minimum it is about 8th to 10th magnitude. Most of the star's energy is emitted at infrared rather than visible wavelengths.

Mira is both a visual binary with a faint peculiar companion that is also variable and an optical double. M_V: -1.0; spectral type: M6e-M9e III; distance: 40 pc.

Mirach (β And). A red giant that is the second-brightest star in the constellation An-

Light curve of Mira

dromeda. It forms an optical double with a 10th-magnitude star at 91″ distance. m_v: 2.07; M_V: 0.2; spectral type: M0 III; distance: 24 pc.

Miranda. The innermost and smallest of the five satellites of Uranus, discovered in 1948. *See* Table 2, backmatter.

Mira stars (Mira Ceti variables). Long-period *pulsating variables, either red giants or red supergiants, that have periods ranging from about 80 to over 600 days and a range in brightness beginning at about 2.5 magnitudes and sometimes exceeding 10 magnitudes. They are numerous, *Mira Ceti being the prototype. The shape of the *light curve is not constant, the maximum brightness varying quite considerably between periods (*see* Mira). Because of the large amplitude they are easily recognizable, their high luminosity permitting detection at great distances.

Although their visual range is large, the range in bolometric and infrared *magnitudes is very much less, many of them being infrared objects. In terms of their spectra, 90% can be classified as *Me stars, the rest being either *carbon stars (Ce) or zirconium (Se) stars, i.e. there are bright emission lines present in the spectra in addition to molecular bands. The pulsations of these huge stars are not very stable. There is evidence of shock waves developing within the tenuous atmosphere and travelling outwards, thus heating the gas and causing the production of emission lines.

Mirfak (α Per). A remote creamy-yellow supergiant that is the brightest star in the constellation Perseus. m_v: 1.80; M_V: -4.1; sprectral type: F5 Ib; distance: 150 pc.

mirror. An optical element consisting of a glass or glass-like substrate on which is deposited a thin but uniform coating of highly reflective material. Aluminium is now usually used in telescope mirrors (*see* aluminizing). A very high quality reflecting surface can also be made from a multi-layered stack of very thin films. The films have alternately high and low *refractive indices and are of such a thickness and spacing that almost all the incident light is reflected.

Mirror shapes may be concave (converging) or convex (diverging) and are generally spherical or paraboloid in form. The mirror base must be accurately ground and the surface polished to within one quarter of the wavelength used, i.e. to about 0.1 micrometres, if good quality images are to be produced. Paraboloid mirrors do not suffer from *spherical aberration but for wide fields have severe *coma. Spherical mirrors are free of coma and astigmatism if a suitable stop is used; they do however suffer from spherical aberration. *See also* primary mirror.

Mirzam. *See* Canis Major.

missing mass. A discrepancy between the mass of a self-gravitating system inferred by dynamical studies and the mass inferred by summing the contributions of luminous objects in the system. The dynamical mass of a rich *cluster of galaxies exceeds that of the component galaxies by a factor of about 10. The dynamical mass is determined by measurement of the size and *velocity dispersion of the cluster. The luminous mass is found from the integrated effect of the observed galaxies, employing galaxy masses for a given morphological type as determined by, for example, studies of binary galaxies.

The missing mass cannot exist in the form of a smoothly distributed gas at any temperature without contradicting observations. It may consist of stars of very low mass, and consequently low luminosity, or perhaps of compact cold objects.

mixer (frequency changer). *See* receiver.

Mizar (ζ UMa). A white star that is one of the brighter stars of the *Plough. It was the first *spectroscopic binary to be identified (E.C. Pickering, 1889) and has a period of 20.5 days; the B component is also a spectroscopic binary, period 361 days. It forms an optical double with the 4th-magnitude *Alcor, separation 14″. m_V: 2.12; M_V: 0.0; spectral type: A2 V; distance: 26 pc.

MK system (MKK system; Yerkes system). *See* spectral types.

M-L relation. *See* mass-luminosity relation.

MMT. *Abbrev. for* Multiple Mirror Telescope.

modified Julian date (MJD). *See* Julian date.

Modular Equipment Transporter (MET). A hand-cart used to extend the sampling capability of *Apollo 14. *See also* Lunar Roving Vehicle.

modulus of distance. *See* distance modulus.

moldavite. *See* tektites.

molecular-line radio astronomy. The study of celestial regions containing molecules by means of the radio emissions from the molecules themselves; unlike optical emissions, radio emissions are not obscured by material in the interstellar medium. A *radio telescope using a *line receiver is best suited to the purpose. Many different molecules have now been detected in the Galaxy in relatively high density regions of dust and cool gas. The first interstellar molecules—CH, CH⁺, and CN— were detected optically in the

OBSERVED INTERSTELLAR MOLECULES

di—, triatomic	4—,5—,6— atomic	7—,8—,9—,11— atomic
H_2	NH_3	CH_3NH_2
OH	H_2CO	CH_3CHO
CH	H_2CS	CH_3C_2H
CH^+	H_2C_2	CH_2CHCN
CN	$HNCO$	HC_5N
CO	C_3N	
CS		
NS	CH_4	$HCOOCH_3$
SO	CS_2NH	CH_3C_2CN
SiO	NH_2CN	
SiS	$HCOOH$	CH_3OCH_3
	HC_3N	CH_3CH_2OH
H_2O	HC_4	HC_7N
H_2S		
HCO	CH_3OH	HC_9N
HCO^+	CH_3CN	
HCN	NH_2CHO	
HNC		
HN_2^+		
HC_2		
SO_2		
OCS		

1940s. This list has been greatly extended by radio-astronomy measurements (see table). Most of the molecules are organic, i.e. based on carbon, but some inorganic species, such as silicon oxide (SiO), have also been detected.

The radio emissions usually come from transitions between different rotational states of the molecule and lie in the gigahertz region of the radio spectrum. Some clouds of molecules appear to emit by maser action (*see* maser source); for example, the *OH line* at 1.667 gigahertz in some radio sources is observed to be very narrow and highly *polarized: characteristic features of a maser. Molecular-line radio astronomy is especially useful in determining cosmic isotope ratios.

Molonglo Radio Observatory. An observatory at Hoskinstown, New South Wales, run by the University of Sydney. The main instrument is a *Mills cross antenna, operational since 1960. Each arm is about 1.6 km long and has a series of cylindrical parabolic reflectors about 12 metres wide.

moment of inertia. Symbol: *I.* A property of any rotating body by which it resists any attempt to make it stop or speed up. It is given by the sum of the products of the mass, *m*, of each particle in the body and the square of the particle's perpendicular distance, *r*, from the axis of rotation, i.e.

$$I = \Sigma mr^2$$

The summation is effected by integration for a solid body. If the body is rotating with angular velocity ω, the kinetic energy of the body is $\frac{1}{2}I\omega^2$ and its angular momentum is $I\omega$.

moment of momentum. *See* angular momentum.

momentum. 1. (linear momentum). Symbol: *p*. A property of a body moving in a straight line. It is equal in magnitude to the product of the mass and velocity (*mv*) of the body. It is a vector quantity, having the same direction as the velocity. *See also* conservation of momentum; Newton's laws of motion.

2. *See* angular momentum.

Monoceros (Unicorn). An inconspicuous equatorial constellation near Orion, lying in the Milky Way, the brightest stars being of 4th magnitude. It contains the triple star Beta Monocerotis and the double Epsilon and several variables including R Monocerotis, which lights up the *Hubble nebula. There are many clusters, including the open cluster NGC 2244, and there is the gaseous *Rosette nebula. Abbrev.: Mon; genitive form: Monocerotis; approx. position: RA 7h, dec 0°; area: 482 sq deg.

monochromatic. Of or producing a single wavelength or very narrow band of wavelengths.

monochromatic magnitude. *See* magnitude.

monochromator. An instrument in which one narrow band of wavelengths is isolated from a beam of light or other radiation. This is usually achieved by means of a diffraction grating or prism, which spreads the beam out into its component wavelengths, and an exit slit through which the desired waveband may pass. Changes in the intensity of the monochromatic beam can then be investigated.

Monster. Less common name for *Hydra.

month. The period of the moon's revolution around the earth with reference to some specified point in the sky (see table). The differences in the monthly periods result from the complicated motion of the moon.

moon (or Moon). The only natural satellite of the earth, visible by virtue of reflected sunlight. It has a diameter of 3476 km, lies at a mean distance from the earth of 384 400 km, and completes one orbit around the earth in 27.322 days (*see* sidereal month; month). Its mass in relation to that of the earth is 1:81.3, i.e. 0.0123. It is in

moon

month	reference point	length (days)
tropical	equinox to equinox	27.321 58
sidereal	fixed star to fixed star	27.321 66
anomalistic	apse to apse	27.554 55
draconic	node to node	27.212 22
synodic	new moon to new moon	29.530 59

*synchronous rotation, i.e. it keeps the same face, the *nearside, towards the earth, although more than 50% of the moon's surface can be seen as a result of *libration. The same face is not however always turned towards the sun, the length of the solar day on the moon being equal to the moon's *synodic period, i.e. 29.530 59 days. During this period—the *synodic month—the moon exhibits a cycle of *phases, reaching an apparent magnitude of -12.7 at full moon. It can also undergo *eclipses and produce *occultations.

The moon's mean orbital velocity is 10 km s^{-1}. Its orbital motion is however subject to *secular acceleration and periodic *inequalities due to the gravitational attraction of the sun, earth, and other planets (*see also* evection; variation; annual equation; parallactic inequality). In addition its *perigee and *nodes move eastward and westward with periods of 8.85 years and 18.61 years, respectively: this results in different lengths for the sidereal, *anomalistic, and *draconic months. Physical and orbital characteristics are given in the table.

Although earth-based observations have revealed much lunar information, the moon has now been carefully photographed, measured, and sampled by the *Apollo spacecraft and by *Ranger, *Lunar Orbiter, *Surveyor, *Luna, and *Zond spacecraft. It has an extremely tenuous atmosphere: a collisionless gas in which helium, neon, argon, and radon were detected by Apollo instruments. This near lack of atmosphere, together with the absence of an appreciable global magnetic field, exposes the *regolith to extremes of temperature (-180°C to +110°C) and to the *solar wind, *cosmic rays, *meteorites, and *micrometeorites. The crustal rocks were magnetized early in lunar history; this randomly orientated remnant magnetism was detected by small satellites launched from Apollo command modules.

The moon's interior structure is investigated from seismograms of *moonquakes and meteoritic and artificial impacts. The temperature of the central core may reach 1500°C but a metallic iron composition is precluded by the moon's moment of inertia.

Moon's Physical and Orbital Characteristics

mass	7.35×10^{25} g
diameter	3476 km
mean density	3.34 g cm^{-3}
surface gravity	0.016 53 g
mean albedo	0.07
mean visual opposition magnitude	-12.7
interior structure: crust	60 km
mantle	1200 km
core radius	500 km
approximate age	4600 million years
mean horizontal parallax	3422".46
moment of inertia	0.395±0.005
mean distance	384 400 km
maximum	406 730 km
minimum	364 400 km
mean orbital velocity	10.1 km s^{-1}
escape velocity	2.38 km s^{-1}
mean eccentricity	0.0549
inclination of orbital plane to ecliptic	5°8′43″
inclination of equator to ecliptic	1°32′1″
sidereal month	27.321 66 days
anomalistic month	27.554 55 days
draconic month	27.212 22 days
synodic month	29.530 59 days

The moon's *centre of mass is displaced towards earth by 2.5 km because of a thick farside crust.

The moon was formed out of refractory materials about 4600 million years ago. Its outer layers were melted and differentiated to produce mafic cumulates (iron magnesium silicates) and a feldspathic crust (calcium aluminium silicates), below which radioactive elements were concentrated. The first 700 million years of lunar history constituted a period of intense bombardment, culminating in a cataclysm during which the *basins and larger *highland craters were formed. The eruption of radioactive basalts (KREEP), or their excavation during formation of the Imbrium Basin, resulted in a concentration of natural radioactivity in the western hemisphere, as measured by gamma-ray spectrometers. A zone of melting then moved inwards towards the moon's centre to generate a sequence of basaltic lavas that produced the *maria. Since then the moon has been relatively quiescent.

The origin of the moon is still debatable. In the *fission hypothesis,* the moon formed from one or more fragments that broke off from the earth; this is now considered unlikely. *Capture theories* propose formation elsewhere and subsequent capture by the earth's gravitational field; these are improbable but could explain the absence of a metallic core and why the earth has such a massive companion. It seems more likely that the moon may have formed as part of a double planet or may have coalesced from a swarm of minor earth satellites.

See also moon, surface features; moonrocks.

moonquakes. Localized disturbances inside the moon, detected by the *Apollo seismic network. Rare but strong moonquakes occur in the crust, although they are much less intense than earthquakes by a factor of about 1000. Weaker but more numerous moonquakes occur at depths of 900-1000 km and are triggered at *perigee by lunar *tides. Seismic signals on the moon are extended by scattering in the dry and porous lunar crust. Deep moonquakes provide information, through their arrival times, of the structure of the lunar interior: a 60 km crust, a 1200 km mantle, and a central partially molten core 500 km in radius.

moonrocks (lunar rocks). Samples of the moon returned by *Apollo and *Luna spacecraft. Lunar soils, or fines, consist of glasses, aggregates, and meteorite debris in addition to recognizable fragments of larger rock types. Hand specimens include mare basalts (low titanium content from the western *maria, high titanium content from the eastern maria, and aluminous basalts from Mare Fecunditatis), non-mare basalt (also known as *KREEP* because of its high content of potassium, rare earth elements, and phosphorus), and anorthositic rocks from the *highlands. Rare lunar-rock types include granite, ultramafic green glass, pyroxenite, norite, troctolite, and dunite. Most highland rocks have been impact metamorphosed (to produce cataclastic, recrystallized, and impact melt textures), but cumulates are also observed and these date back to the formation of the first lunar crust.

Lunar rocks are depleted in volatile and siderophile elements. Strong correlations exist between elements that are geochemically similar to one another (e.g. iron and manganese) or that are both excluded by major minerals (e.g. potassium and lanthanum). The major lunar minerals are calcic plagioclase, clino- and ortho-pyroxene, olivine, and ilmenite. Minor phases include spinels, K-feldspar, troilite, metallic iron, quartz, tridymite, cristobalite, apatite, and whitlockite. Uniquely lunar minerals include armalcolite, tranquillityite, and pyroxferroite. No hydrated phases, highly oxidized minerals, diamonds, or mineralization have yet been found.

moon, surface features. The surface of the moon is characterized by light-coloured *highlands interspersed with much lower *albedo (i.e. darker) and less rugged *maria. The oldest and largest features on the moon are the vast impact *basins. These grade into large *craters, with which highland areas are saturated. There is no evidence for continental drift on the moon (*but see* lunar grid) but *faults, *rilles, *wrinkle and *lobate ridges, *domes, *ghost craters, and volcanic plateaux are the expressions of past tectonic and igneous activity. Impact features include *rays, *dark halo craters, *secondary craters (including *crater clusters and *crater chains), and textured *ejecta blankets.

Moreton wave. *See* flares.

Morgan's classification. A classification scheme for galaxies, devised by W.W. Morgan, in which galaxies are grouped according to the composite spectra of their component stars and also by their form, i.e. spiral, barred spiral, elliptical, irregular, etc., by their degree of concentration, and by their orientation, which varies from face-on to edge-on.

morning star. A term normally applied to *Venus when it shines brightly in the eastern sky before sunrise. Such periods last while Venus moves between inferior *conjunction and superior conjunction and is located west of the sun. The term occasionally refers to spells of morning visibility of Mercury or other planets.

mountains, lunar. Raised regions on the moon, most of which consist of crustal blocks uplifted to define the multiple ring structures of impact *basins; these survive either as *ring mountains,* such as the Apennines, or as isolated peaks, such as La Hire (see table). In comparison the lunar *highlands are topographically higher than the *maria but are characterized more by the elevated rims of craters and the rolling hills of basin *ejecta blankets. Discrete mountains not associated with impact basins include the central peaks of *craters, such as Tsiolkovsky, and volcanic domes and plateaux. Despite their jagged appearance, few mountains have slopes that exceed 12°.

mounting. The structure that supports an astronomical telescope, designed so that the telescope may be pointed at almost every part of the heavens, preferably without forcing an observer into awkward positions. All telescope mountings must therefore be constructed so that the tube can turn about two axes at right angles. There are two main types: the *equatorial mounting, with one axis parallel to the earth's axis, and the *altazimuth mounting, with one axis vertical. Since high magnifications exaggerate any movement, the support must

lunar mountains	location
Alps	N Imbrium, part of second ring
Altai Scarp	SW Nectaris, part of third ring
Apennines	SE Imbrium, part of third ring
Apollonius	S Crisium, part of third ring
Aristarchus Plateau	Procellarum, Volcanic province
Carpathians	S Imbrium, part of third ring
Caucasus	W Imbrium, part of third ring
Cordillera	Orientale basin, third ring
Haemus	S Serenitatis, part of second ring
Harbinger	SW Imbrium, part of third ring
La Hire	Imbrium, isolated peak, part of first ring
Marius Hills	Procellarum, Volcanic province
Piton	Imbrium, isolated peak
Pyrenees	W Nectaris, part of first ring
Riphaeus	W Cognitum
Rook	Orientale basin, second ring
Rümker Hills	Sinus Roris, volcanic province
Spitzbergen	N Imbrium, part of first ring
Straight	N Imbrium, part of first ring
Taurus	E Serenitatis, part of second ring
Tenerife	N Imbrium, part of first ring

be rigid. In addition, because the earth is turning through 0.25° per minute, means must be provided to move the telescope smoothly in the opposite direction; this counterbalances the diurnal motion of a star and keeps it in the field of view.

Mount Stromlo Observatory. An observatory on Mount Stromlo, near Canberra, at an altitude of 768 metres. It is run by the Australian National University. Its principal instrument is a 1.9 metre (74 inch) reflecting telescope.

Mount Wilson Observatory. *See* Hale Observatories.

moving cluster. An *open cluster, such as the *Hyades or the *Ursa Major cluster, for which a distance can be derived from the individual radial velocities and proper motions of the member stars. The method is often termed *moving-cluster parallax*. The stars in a cluster share a common space motion and, if the system is sufficiently close to earth, their paths will appear to diverge from or converge towards a point in space (point P in the illustration). The direction of P is parallel to the space motion of the cluster. Considering each star in turn (illustration B):

$$v_r = v\cos \alpha$$
$$v_t = v\sin \alpha = 4.74\mu/p$$

where v v_r, v_t, are respectively the space, *radial, and tangential velocities of the star in km/s, μ is the proper motion in seconds of arc per year, and p is the *parallax in seconds of arc. Determination of α, μ, and v_r for each star gives a val-

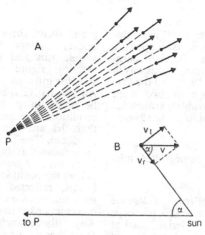

Moving cluster parallax

ue for the parallax of the system.

moving-cluster parallax. *See* moving cluster.

MS stars. *See* S stars.

M stars. Stars of *spectral type M, cool red stars with surface temperatures (T_{eff}) less than about 3500 kelvin. Molecular bands are prominent in the spectra, with titanium oxide (TiO) bands dominant by M5. Lines of neutral metals are also present. Antares and Betelgeuse are M stars. *See also* carbon stars; S stars.

MTRLI. *Abbrev. for* Multi Telescope Radio Linked Interferometer. A network of seven radio-telescope dishes that will be linked to a computer at *Jodrell Bank and can be used, in different com-binations, to build up high-resolution maps of radio sources. Two 25 metre dishes have yet to be completed at sites a few kilometres from Jodrell Bank. The remaining five, already operational, are the two big dishes, 76 and 38 metres in size, at Jodrell Bank itself, the elliptical mesh dish 24 km away at Wardle near Nantwich, the 25 metre dish at the Royal Radar Establishment at Defford, near Malvern, and the 25 metre dish at Knockin, Oswestry. The latter, operational since 1977, has a design identical to that of the two dishes in the process of construction.

Mullard Radio Astronomy Observatory. The radio observatory of the Cavendish Laboratory, Cambridge University, near Cambridge. The chief instruments are two interferometers: the Five Kilometre telescope (*see* aperture synthesis) with its eight paraboloid reflectors, which was in operation in 1971, and the One Mile telescope of two fixed and one movable paraboloid, which came into use in 1964.

multilayers. *See* coating of lenses; interference filter, mirror.

Multiple Mirror Telescope (MMT). A reflecting telescope of totally new design, sited at the Smithsonian Astrophysical Observatory, on Mount Hopkins, near Tucson, Arizona at an altitude of 2382 metres. Its construction has been a joint project of the University of Arizona and the Harvard-Smithsonian Center for Astrophysics. The telescope saw first light in 1978. It comprises six identical mirrors, 183 cm (72 inches) in diameter, that are arranged in a hexagonal array on an *altazimuth mounting. The MMT has the light-gathering power of a 4.5 metre (176 inch) single mirror telescope and is thus the third largest telescope in the world; however it cost considerably less than a single 4.5 metre mirror to construct. It is designed for use at both visible and infrared wavelengths.

The mirrors are mounted symmetrically about a central axis. Their images are sent to a six-sided hyperbolic secondary mirror system on the central axis, which produces a single image at a common 'Cassegrain' focus; there is one secondary system for visible wavelengths and another, of identical figure but smaller size, for infrared wavelengths. An alternative focus can also be used. In perfect *seeing conditions star images of less than 0.7 arc seconds can be produced. The six mirrors remain focused at the same point by means of a feedback mechanism: the position of six laser beams, reflected by the six main mirrors are monitored by detectors linked to a computer. Any discrepancies arising from thermal or gravitational stresses are corrected by computer-controlled repositioning of the secondary mirror systems. The whole telescope can be smoothly guided to very great accuracies by computer-controlled drive mechanisms.

If the multiple mirror concept, with its complicated alignment and guidance system, proves reliable it might well become the prototype for future giant telescopes.

multiple star. A group of three or more stars, such as *Castor or *Alpha Centauri, that are bound together by their mutual gravitational attraction. (*Binary stars are sometimes considered as multiple stars.) Orbital motions within multiple systems are complex. For instance, systems containing two or more close binaries in orbit about each other appear to be quite common.

multiprobe. *See* Pioneer Venus probes.

multistage rocket. *See* launch vehicle.

muon. Symbol: μ. An *elementary particle (a lepton) with the same *charge and *spin as an *electron but with a mass that is 207 times greater. It decays into electrons and neutrinos with a lifetime of two microseconds. *Cosmic-ray showers detected on earth are comprised mainly of muons produced by *pion decay. Muons do not interact strongly with matter and hence a large proportion of them reach the earth's surface. The *antiparticle of the muon is the antimuon.

mural quadrant. *See* quadrant.

Musca (Fly). A small constellation in the southern hemisphere near Crux, lying in the Milky Way, the brightest stars being of 3rd magnitude. Abbrev.: Mus; genitive form: Muscae; approx. position: RA 13h, dec -70°; area: 138 sq deg.

N

nadir. The point on the *celestial sphere that lies directly beneath an observer and is therefore unobservable. It is diametrically opposite the observer's *zenith and with the zenith lies on the observer's *celestial meridian.

naked singularity. *See* singularity.

Nancy. French site of the Radio Observatory of Nancy, which is part of the Observatory of Paris. The chief instrument is a 305 metre (1000 foot) partially steerable (meridian transit) array, which has been in operation since 1964.

nano-. Symbol: n. A prefix to a unit, indicating a fraction of 10^{-9} of that unit. For example, one nanometre is 10^{-9} metres.

Naos. *See* Puppis.

NASA. *Abbrev. for* National Aeronautics and Space Administration. The US civilian agency that was formed in Oct. 1958 and is responsible for all nonmilitary aspects of the US space programme. Its major achievements so far have been the manned *Apollo programme, which followed the *Mercury and *Gemini projects, the *Skylab space station, and a large number of artificial *satellites and *planetary probes. It is now developing a reusable space transportation system—the *Space Shuttle—to replace the expendable and hence costly launch rockets it has used to date. It is collaborating on several projects with the *European Space Agency (ESA). NASA offers a satellite launcher service to other countries.

National Geographic Society—Palomar Observatory Sky Survey. *See* Palomar Sky Survey.

National Radio Astronomy

Observatory. A radio observatory with most of its telescopes at Green Bank, West Virginia. The principal instruments include a partially steerable (meridian transit) 91 metre (300 foot) paraboloid dish, completed in 1962, a fully steerable 43 metre (140 foot) dish, completed in 1965, and a 1.6 km baseline interferometer. The Observatory also operates the Very Large Array in New Mexico (*see* aperture synthesis) and a radio telescope at *Kitt Peak, Arizona.

Nautical Almanac. *See* Astronomical Ephemeris.

nautical twilight. *See* astronomical twilight.

n-body problem. Any problem in celestial mechanics that involves the determination of the trajectories of *n* point masses whose only interaction is gravitational attraction. The bodies in the solar system are an example if it is assumed that the masses of the planets, etc., are concentrated at their *centres of mass. A general solution exists for the *two-body problem and in special cases a solution can be found for the *three-body problem. The complete solution for a larger number is normally considered insoluble.

neap tide. *See* tides.

nearest stars. The stars nearest to the sun. They are either comparable in luminosity to or fainter than the sun. Most are less massive than the sun and a large proportion belong to binary or triple star systems. *See* Table 7, backmatter.

nearside. The hemisphere of the moon that is permanently turned towards the earth because of the tidal synchronization of the *lunar day and the *sidereal month. The lunar *librations allow the terrestrial observer also to see 9% of the *farside,* the hemisphere that is permanently turned away from the earth. The farside was first photographed by the Soviet *Luna 3 in 1959; it has no major *maria.

nebula. A cloud of interstellar gas and dust that can be observed either as a luminous patch of light—a bright nebula—or as a dark hole or band against a brighter background—a dark nebula. Interstellar clouds cannot usually be detected optically but various processes can cause them to become visible.

With *emission nebulae, ultraviolet radiation, generally coming from nearby hot stars, ionizes the interstellar gas atoms and light is emitted by the atoms as they interact with the free electrons in the nebula. Emission nebulae can be in the form of *HII (ionized hydrogen) regions, *planetary nebulae, or *supernova remnants. With *reflection nebulae light from a nearby star or stellar group is scattered (irregularly reflected) by the dust grains in the cloud. Reflection and emission nebula are bright nebulae. In contrast *dark nebulae are detected by what they obscure: the light from stars and other objects lying behind the cloud along our line of sight is significantly decreased or totally obscured by interstellar *extinction. Dark nebulae contain approximately the same mixture of gas and dust as bright nebulae but there are no nearby stars to illuminate them.

The term 'nebula' was originally applied to any object that appeared fuzzy and extended in a telescope: over 100 were listed in the 18th-century *Messier catalogue. The majority of these objects were later identified as *galaxies and star *clusters. Galaxies are still occasionally and misleadingly referred to as extragalactic nebulae.

nebular hypothesis. A theory for the origin of the solar system put forward by the French mathematician Pierre-Simon de Laplace in 1796, and similar to a suggestion made by the philosopher Immanuel Kant in 1755. It was proposed that the solar system formed from a cloud, or nebula, of material collapsing under its own gravitational attraction. Rotation of the nebula caused it to form a disc that, because of the law of conservation of *angular momentum, rotated more rapidly as the contraction progressed. Laplace suggested that rings of material became detached from the spinning disc when the velocity at its edge exceeded a critical value, and that the material in these rings later coalesced to form the planets. The central product of the contraction is the sun, while the planetary satellites may have formed from further rings shed by the condensing planets.

The hypothesis, popular throughout the 19th century, went out of favour when it could not explain why the sun has almost 99.9% of the mass of the solar system but only 3% of the total angular momentum, and because calculations showed that the rings would not condense to form planets. However, in a modified form, it is the basis of most modern ideas for the formation of the sun and planets. *See* solar system, origin. *Compare* encounter theories.

negative lens. A lens or lens system with a diverging action on a light beam.

Neptune. The eighth major planet, orbiting the sun every 165 years at an almost constant distance of 30.1 astronomical units. It is the most distant *giant planet and resembles *Uranus in many respects; it does however have a slightly smaller diameter (49 500 km: determined by stellar *occultation observations), a greater mass (17.2 earth-masses), and consequently a higher density (1.7 times that of water). Its two satellites, *Triton and *Nereid, move in unusual orbits. Orbital and physical characteristics are given in Table 1, backmatter.

Neptune was discovered in 1846 by J.G. Galle at the Berlin Observatory on the basis of predictions supplied by the Frenchman Urbain Leverrier, who had analysed observed perturbations in the motion of Uranus. Similar predictions had been made by the English mathematician John Couch Adams.

The planet returns to *opposition two days later each year, appearing as a magnitude 7.7 object visible in binoculars. No markings are seen on its greenish disc, only 2.3 arc seconds across. The presence of methane and hydrogen has been revealed by spectroscopic study of sunlight reflected from what is probably a visible surface of methane clouds. Recent spectroscopic measurements suggest that the rotation period, about an axis tilted by 29° to the vertical, may be a few hours longer than the 15.8 hours normally credited.

The atmospheric and internal structure may be almost identical to those of Uranus, with a rocky core 16 000 km in diameter encased in an 8000 km thick ice jacket and another 8000 km of, mainly, molecular hydrogen. To account for the greater overall mass, the core may be more massive than that of Uranus.

Neptune has an effective temperature of -220°C, a few degrees warmer than is to be expected at its solar distance and indicating that it, like Jupiter and Saturn, radiates more energy than it receives from the sun. Possibly this heat is generated by resistance to tides raised by its nearby massive satellite, Triton.

Nereid. A small satellite of Neptune with a highly eccentric orbit that takes it between 2 and 10 million km from the planet. It was discovered in 1949. *See* Triton; Table 2, backmatter.

Net. Common name for *Reticulum.

neutrino. Symbol: ν. A stable *elementary particle with zero *charge and *spin ½. Neutrinos are hypothesized to have zero *rest mass and according to the special theory of relativity travel at the speed of light. Two types of neutrinos are known: one is produced in association with an *electron (ν_e) and the other in association with a *muon (ν_μ); a third neutrino is assumed to be associated with the recently discovered massive tau lepton. The *antiparticle of the neutrino is the *antineutrino* ($\bar{\nu}$).

Wolfgang Pauli first postulated the existence of the neutrino in the 1930s to explain the violation of the laws of conservation of energy and momentum in *beta decay. Neutrinos are leptons and thus only have a weak interaction with matter. Neutrinos produced in nuclear reactions in the centres of stars can thus escape without further collision. *See also* neutrino detection experiments.

neutrino detection experiments. Experiments to detect *neutrinos released during energy-producing nuclear reactions occurring in the centres of stars. Neutrino interaction with other particles is highly improbable so that the great majority can escape from the star and travel unhindered through space. They thus provide direct evidence of conditions in the stellar core.

An experiment to detect *solar neutrinos* was first set up by Raymond Davies, Jr., in the 1960s and is still in progress. It is being conducted deep underground in the Homestake Gold Mine near Lead, South Dakota. A large quantity of the dry-cleaning fluid perchloroethylene, C_2Cl_4, is exposed for several months to the presumed flux of solar neutrinos. A very small (but calculable) number of the high-energy particles should react with the chlorine atoms:

$$^{37}Cl + \nu_e \rightarrow {}^{37}Ar + e^-$$

The amount of radioactive argon, ^{37}Ar, produced by the reaction gives a measure of the neutrino flux. The flux is measured in *solar neutrino units (SNU)*, where one SNU is the flux giving rise to one neutrino-induced event per second per 10^{36} atoms of chlorine. The average flux detected is 1.75 ± 0.4 SNU. This is much lower than the predicted value of 4.47 ± 1.5 SNU.

The expected neutrinos arise from one of the rarer pathways in the *proton-proton chain reaction whereby hydrogen is converted into helium. (The proton-proton chain is the most important energy-producing reaction in the sun.) The neutrinos are emitted with high energy during the decay of boron-8, which in turn results from a reaction between nuclei of helium-3 and helium-4. The rate of neutrino production has been calculated from the expected conditions in the solar core and is very temperature-dependent.

The negative results of the above experiment are very difficult to explain. Assuming that the experiment is being carefully conducted, is sufficiently sensitive, and is based on correct premises, the near-lack of neutrinos implies that the condition in the sun's interior and/or the nuclear reactions taking place there are not fully appreciated. Experiments are now planned in the US and the USSR to detect the less energetic but more numerous neutrinos that are released in the earliest step of the proton-proton chain. These neutrinos can be absorbed by nuclei of gallium-71 to produce germanium-71; the amount of ^{71}Ge formed will then be a measure of the rate of a more fundamental step in the energy-producing reaction.

neutron. Symbol: n. An *elementary particle (a baryon) that is present in the *nucleus of all atoms except hydrogen.

It has zero *charge, *spin ½, and a *rest mass of 1.6749×10^{-24} grams—slightly greater than that of the *proton. The absence of charge enables the neutron to penetrate atoms easily since it has a negligible electromagnetic interaction with the constituents of the atom. Unlike free protons, free neutrons are unstable: they *beta decay into protons, electrons, plus antineutrinos with a mean life of approximately 930 seconds. The small difference between the neutron and proton mass provides the energy for this decay. The neutron is stable, however, when bound in a nucleus. The *antiparticle of the neutron is the *antineutron*. *See also* isotopes; nucleosynthesis; s-process; r-process; neutron star.

neutron star. A star that has undergone *gravitational collapse to such a degree that most of its material has been compressed into *neutrons. Neutron stars were postulated in the 1930s by a number of astronomers including Landau and Zwicky. They are thought to form when the mass of the stellar core remaining after a *supernova explosion exceeds the *Chandrasekhar limit, i.e. 1.44 solar masses. Such a core is not stable as a *white dwarf star since even the pressure of *degenerate electrons cannot withstand the strong gravity. Collapse continues until the mean density reaches 10^{17} kg m^{-3} and the protons and electrons coalesce into neutrons, supporting the star against further contraction.

With diameters of only 20-30 km, intense magnetic fields (10^8 tesla), and extremely rapid spin, young neutron stars are believed to be responsible for the *pulsar phenomenon. Their strong gravity is also thought to give rise to rapid heating of material observed in some *x-ray binary systems. Models of the structure of neutron stars have been derived from sudden changes in pulsar spin rates—so called glitches. A typical neutron star may have an atmosphere only a few centimetres thick, under which is a crystalline crust one kilometre in depth. Beneath the crust the material is thought to act like a superfluid of neutrons (having zero viscosity) all the way through to a solid crystalline core.

New General Catalogue. *See* NGC.

new moon. *See* phases, lunar.

New Style (NS). The current method of measuring dates using the *Gregorian calendar as distinct from the former method—*Old Style* (OS)—that was based on the *Julian calendar. Britain and the American colonies switched calendars on Sept 14 1752, the 11 days from Sept 3 to Sept 13 having been eliminated from that year.

newton. Symbol: N. The SI unit of force, equal to the force that gives a mass of one kilogram an acceleration of

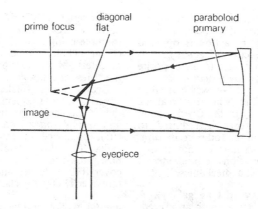

Ray path in Newtonian telescope

one metre per second per second.

Newtonian telescope. The first *reflecting telescope to be built, developed in about 1670 by Issac Newton from the ideas of Zucchi and Gregory: Newton thought that there was no way in which the *chromatic aberration of refracting telescopes could be corrected. The Newtonian has a paraboloid *primary mirror and a flat *diagonal mirror* that reflects the light into an *eyepiece mounted on the side of the tube (see illustration). A 45° prism can be used as an alternative to the diagonal mirror. The Newtonian is still the form used by many amateurs although somewhat inconvenient to use in larger form. Newton's original telescope had a mirror only 2.5 cm in diameter and made of *speculum metal; it had a focal length of about 15 cm and magnified some 30 times.

Newton's law of gravitation. *See* gravitation.

Newton's laws of motion. The three fundamental laws concerning the motion of bodies, that were formulated by Isaac Newton and published together with the law of *gravitation in *Principia,* 1687. The laws are

1. Every body continues in a state of rest or of uniform motion in a straight line until that state is changed by the action of a force on the body.

2. The rate of change of linear *momentum is proportional to the applied force, F, and occurs in the same direction as that of the force, i.e.
$F = d(mv)/dt = m(dv/dt) = ma$
where m is the mass, v the ve-

locity, and *a* the resulting acceleration of the body.

3. Every action is opposed by a reaction of equal magnitude that acts in an opposite direction to the action.

The first law was conceived by Galileo who first realized the falsity of the Greek notion that a force is required to maintain a body in motion. Newton's laws of motion and of gravitation are fundamental to *celestial mechanics.

N galaxy (N-type galaxy). A distant galaxy with an exceedingly bright (and sometimes variable) nucleus; they were originally identified by W.W. Morgan in 1958 as one of the optical counterparts of powerful radio sources. Many radio-quiet N galaxies have since been discovered. X-ray emission has also been detected from several N-types. N galaxies are believed to be less energetic examples of the *quasar phenomenon. *See also* Seyfert galaxy.

NGC. Short for *New General Catalogue of Nebulae and Clusters of Stars,* prepared by the Danish astronomer J.L.E. Dreyer at Armagh Observatory, Ireland, and published in 1888. It was based on an earlier catalogue compiled by the Herschel family. The original work listed 7840 currently known nebulae, clusters, and galaxies (not yet identified as such). Each object is numbered by *right ascension and its position and description given. The number is preceded by the letters NGC; this is the object's *NGC number,* an example being NGC 1976: the Orion nebula. Some of the brightest objects also appear in the *Messier catalogue.

Dreyer listed further discoveries in two supplements to the NGC. These were the first and second *Index Catalogues* (*IC*), published in 1895 and 1908. The three catalogues, now published as one volume, cover the whole sky and contain over 13 000 objects.

nightglow. *See* airglow.

Nix Olympica. *Earlier name for* Olympus Mons, assigned by telescopic observers to the bright-albedo feature they saw in its position.

nm. *Short for* nanometer, i.e. 10^{-9} metre.

nodes. The two points at which the orbit of a celestial body intersects a reference plane, usually the plane of the *ecliptic or *celestial equator. When the body, such as the moon or a planet, crosses the reference plane from south to north it passes through the *ascending node;* crossing from north to south the body passes through its *descending node.* The line joining these two nodes is the *line of nodes.*

A celestial body may not cross the reference plane at the same points every time: the line of nodes moves slowly forwards or backwards around the reference plane, i.e. in the same or in the opposite direc-

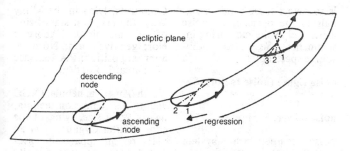

Regression of moon's nodes

tion to the body. This is termed *progression* or *regression of the nodes*. The moon's nodes regress along the ecliptic, making one complete rotation in 18.61 years (see illustration). This period is equal to the period of nutation.

nodical month. *See* draconic month.

noise. 1. Unwanted fluctuations that occur in the output of any electronic or electrical device and, in a measuring instrument, ultimately limit its sensitivity. Noise can be caused by many different factors, for example by insufficient regulation of the voltage of a power supply, but the most important source is usually that produced by random agitation of electrons (Brownian motion) in a resistance, i.e. *thermal noise*.

In electrical circuits a resistance at temperature T produces an exchangeable *noise power* in *bandwidth B of $P = kTB$, where k is the Boltzmann constant. If this noise power is generated in the first stages of the measuring instrument it is amplified along with the signal, which results in uncertainty in the measurement. The contribution to the noise power of any part of an electric circuit can be specified by its *noise figure* or *noise factor*, F, which is defined by

$$F = (P_1 + P_0)/P_0$$

where P_1 is the extra noise power generated by the circuit and P_0 is the noise power present at the input terminals of the circuit due to the source resistance, i.e. the thermal noise. Thus if the circuit adds no extra noise, it has a noise figure of unity. F is usually expressed in decibels. *See also* sensitivity.

2. Fluctuations of a random nature in which the amplitude of the deflection from a given reference level cannot be predicted from one moment to the next but whose statistical properties may be well-defined. Signals from *radio sources are of this kind, and in a *radio telescope the problem

noise figure

is to detect these random signals in the presence of noise generated in the measuring instruments. *See also* radiosource spectrum.

noise figure (noise factor). *See* noise.

noise power. *See* noise.

noise temperature (system temperature). *See* sensitivity.

nonrelativistic. Describing any phenomenon, object, etc., for which the effects of general or special *relativity can be disregarded or do not apply.

nonthermal emission (nonthermal radiation). Electromagnetic radiation, such as *synchrotron emission, that is produced, like *thermal emission, by the acceleration of electrons or other charged particles but is nonthermal in origin, i.e. its spectrum is not that of a perfect *black-body radiator.

noon. The time of day at which the sun crosses the local meridian and is at its highest point above the horizon. This time differs from 12.00 *local time by up to about 15 minutes before or after midday, the amount—the *equation of time—depending on the time of year. *Mean noon* is the time at which the *mean sun crosses the meridian and always occurs at 12.00 local time.

Norma (Rule; Level). A small inconspicuous constellation in the southern hemisphere near Centaurus, lying in the Milky Way, the brightest stars being of 4th magnitude. Abbrev.: Nor; genitive form: Normae; approx. position: RA 16h, dec -50°; area: 165 sq deg.

North America nebula (NGC 7000). An *emission nebula, nearly 2° in size, lying in the constellation Cygnus. It is an *HII region apparently associated with the exceptionally luminous and remote A2 supergiant *Deneb.

north celestial pole. *See* celestial poles.

Northern Cross. *See* Cygnus.

Northern Crown. Common name for *Corona Borealis.

Northern Hemisphere Observatory. A new international observatory on La Palma in the Canary Islands to which the 2.5 metre *Isaac Newton telescope is due to be moved in 1979. There will be two additional telescopes, one of one metre (39 inches) which is in the process of construction, and the other of 4.2 metres (165 inches).

north galactic pole. *See* galactic pole.

north galactic spur. *See* interstellar medium.

north point. *See* cardinal points.

north polar distance (NPD). The angular distance of a

celestial object from the north *celestial pole, measured along the object's hour circle. It is the complement of the object's *declination (90° - δ) and is sometimes used instead of this coordinate in the *equatorial coordinate system.

North Polar Sequence. *See* magnitude.

North Star. *See* Polaris.

nova. A highly evolved star in which there is a sudden and unpredictable increase in brightness by maybe 10 magnitudes or more (see table). Maybe 10 to 15 occur each year in our Galaxy. The brightness increases to a maximum within days or sometimes weeks and then declines to a value probably close to its faint pre-nova magnitude, indicating that the eruption did not disrupt the bulk of the star. The decline can range from months for *fast novae* to many years for the lower *amplitude *slow novae*. Hydrogen-rich gas is ejected from the star resulting in the tremendous outflow of heat and light. The ejected matter forms a rapidly expanding envelope of gas that can become visible as the nova fades. At later stages of the eruption the spectra of most novae show the bright forbidden lines characteristic of very low density *emission nebulae.

It is now thought that novae are members of close binary systems: the presence of a companion might even be a necessary condition for a star to be a nova. One component is usually a *white dwarf and is the source of the eruption; the other is a very close cooler redder star from which matter is slowly streaming into a disc-shaped orbit around the degenerate star (*see* mass transfer). This matter slowly accumulating on the star's hydrogen-rich upper surface layer somehow provides the energy to trigger the nova explosion, possibly by the runaway nuclear burning of hydrogen.

Nova are now named by constellation and year of observation; before 1925 they

nova	type	before	m_v max	after	max M_v
Nova Persei 1901 (GK Per 2)	fast	13.5	+0.2	13.2	-8.3
Nova Aquilae 1918 (V603 Aql 3)	fast	10.6	-1.1	10.9	-8.4
Nova Pictoris 1925 (RR Pic)	slow	12.7	+1.2	9	-6.4
Nova Herculis 1934 (DQ Her)	slow	14.3	+1.4	13.8	-6.2
Nova Cygni 1975 (V1500 Cyg)	fast	>21	+1.5		-9.5

were numbered in order of observation. They may also be given a variable star designation, as for example with DQ Herculis, i.e. Herculis 1934. *See also* recurrent nova; dwarf novae; supernova; x-ray transients.

Nova Mon 1975 (A 06020-00). *See* x-ray transients.

November swarm. *See* Leonids.

NPD. *Abbrev. for* north polar distance.

NS. *Abbrev. for* New Style.

N stars. *See* carbon stars.

N-type galaxy. *See* N galaxy.

Nubecular Major, Minor. *Alternative names for* Large and Small Magellanic Clouds, respectively. *See* Magellanic Clouds.

nuclear fission. A process in which an atomic nucleus splits into two smaller nuclei, usually with the emission of neutrons or gamma rays. Fission may be spontaneous or induced by neutron or photon bombardment. It occurs in heavy elements such as uranium, thorium, and plutonium. The process is accompanied by the release of large amounts of energy.

nuclear fusion. A process in which two light nuclei join to yield a heavier nucleus. An example is the fusion of two hydrogen nuclei to give a deuterium nucleus plus a positron plus a neutrino; this reaction occurs in the sun. Such processes take place at very high temperatures (millions of kelvin) and are consequently called *thermonuclear reactions*. With light elements, fusion releases immense amounts of energy: fusion is the energy-producing process in stars.

Iron and lighter elements evolve energy in fusion reactions whereas heavier elements require an input of energy to maintain the reaction. Thus although most elements up to iron can be formed by fusion reactions in stars (*see* nucleosynthesis), heavier elements must be synthesized by other nuclear reactions. *See also* proton-proton chain reaction; carbon cycle.

nuclear reactions. Reactions involving changes in atomic nuclei. Nuclear reactions include *nuclear fusion and fission and transformations of nuclei by bombardment with particles or photons. The energy released by nuclear reactions is equivalent to a decrease in mass according to Einstein's equation ($E = mc^2$) and appears in the form of kinetic energy of the particles or electromagnetic radiation.

nuclear timescale. The total nuclear-energy resources of a star divided by the rate of energy loss. For typical stars, the nuclear timescale is of the order of 10^{10} years, although the rate of energy loss may change during the star's lifetime if the reaction processes alter.

nucleon. One of the constituents of the atomic nucleus, i.e. a proton or a neutron.

nucleosynthesis. The creation of the *elements by nuclear reactions. To unravel the complex situation, the measured *cosmic abundances of the elements are interpreted in terms of their nuclear properties and the set of environments (temperature, density, etc.) in which they can be synthesized. Nucleosynthesis began when the temperature of the primitive universe had dropped to about 10^9 kelvin (K). This occurred approximately 100 seconds after the big bang (see big bang theory). Protons fused with neutrons to form deuterium nuclei and deuterium nuclei could then fuse to form helium. Very few heavier elements are thought to have been formed at this time.

It is now known that helium and the heavier elements are synthesized in stars and that this nucleosynthesis has occurred continuously in the Galaxy for many thousands of millions of years as a by-product of *stellar evolution. While a star remains on the *main sequence, hydrogen in its central core will be converted to helium by the *proton-proton chain reaction or the *carbon cycle; the core temperature is then about 10^7 K.

When the central hydrogen supplies are exhausted, the star will begin to evolve off the main sequence. Its core, now composed of helium, will contract until a temperature of 10^8 K is reached; carbon-12 can then be formed by the *triple alpha process, i.e. by helium burning. In stars more than twice the sun's mass a sequence of reactions, involving further nuclear fusion, produces elements with a mass number, A, that is a multiple of four: ^{16}O, ^{20}Ne, ^{24}Mg, and then, at temperatures increasing up to about 3.5×10^9 K, ^{28}Si to ^{56}Fe. Even higher temperatures will trigger reactions by which almost all elements up to a mass number of 56 can be synthesized. The *iron-peak elements,* i.e. ^{56}Fe, ^{56}Ni, ^{56}Co, etc., represent the end of the nucleosynthesis sequence by nuclear fusion: further fusion would require rather than liberate energy since nuclei with this mass number have the maximum binding energy per nucleon.

The formation of nuclei with $A \geq 56$ requires nuclear reactions involving neutron capture: neutrons can be captured at comparatively low energies owing to their lack of charge. If there is a supply of free neutrons in a star, produced as by-products of nuclear-fusion reactions, the *s-process can slowly synthesize nuclei up to ^{209}Bi. An intense source of neutrons allows the *r-process to generate nuclei up to ^{254}Cf, or higher, in a very short period. Such intense neutron fluxes arise in *supernovae and it is by means of these explosions, and by novae and other mass-loss phenomena, that the elements are precipitated into and enrich the interstellar medium.

nucleus. 1. The kilometre-sized body that is the permanent portion of a *comet and is thought by most researchers to be the fount of all cometary activity. The density is about 1.1 g cm^{-3}. It contains about 75% by mass ice (mainly water ice but with liberal amounts of CO_2, CO, CH_4, and NH_3 ices) and 25% by mass dust, thought to have a composition similar to *carbonaceous chondritic *meteorites. It is often described as a *dirty snowball*. A comet of mass 10^{18} grams would have a nucleus of diameter about 6 km. No one has seen a cometary nucleus. Even comet Halley, which has a calculated mass of 3×10^{19} grams and a nucleus diameter of 35 km, did not exhibit a nucleus when it *transited the sun. The nucleus of dirty ice is surrounded by a thin (a few cm) layer of insulating dust from which the ice has sublimated. This enables it to survive about 200 perihelion passages. At each passage the nucleus looses on average a layer of material one metre thick. This material forms the *coma and *comet tails and also any associated *meteor stream.

A less likely model for the nucleus, which was prevalent up to the 1950s, regarded it as a vast collection (a swarm) of tiny particles. In this model a 10^{18} gram comet was thought to contain about 10^{25} particles, each of about 0.1 mm diameter, separated by tens of metres and occupying a volume of space about 100 000 km across. *See also* head.

2. The small core of an *atom, consisting of *protons and *neutrons bound together by strong nuclear forces. The nucleus has a positive charge equal to Ze, where e is the magnitude of the electron charge and Z the number of protons present—the atomic number. The total number of protons plus neutrons is the mass number, A. A given element is characterized by its atomic number but may, within limits, have different numbers of neutrons in its nuclei, giving rise to different *isotopes of the element. The total mass of the protons and neutrons bound together in a nucleus is less than when the particles are unbound. This mass difference is equivalent to the energy required to bind the particles together. Nuclei are represented by their chemical symbols to which numbers are attached, as with $^4_2$He where 2 is the atomic number of helium (2 protons) and 4 the mass number of the particular isotope.

3. (galactic nucleus). *See* Galaxy.

Nuffield Radio Astronomy Laboratory. *See* Jodrell Bank.

number count. *See* source count.

Mirzam. *See* Canis Major.

missing mass. A discrepancy between the mass of a self-gravitating system inferred by dynamical studies and the mass inferred by summing the contributions of luminous objects in the system. The dynamical mass of a rich *cluster of galaxies exceeds that of the component galaxies by a factor of about 10. The dynamical mass is determined by measurement of the size and *velocity dispersion of the cluster. The luminous mass is found from the integrated effect of the observed galaxies, employing galaxy masses for a given morphological type as determined by, for example, studies of binary galaxies.

The missing mass cannot exist in the form of a smoothly distributed gas at any temperature without contradicting observations. It may consist of stars of very low mass, and consequently low luminosity, or perhaps of compact cold objects.

mixer (frequency changer). *See* receiver.

Mizar (ζ UMa). A white star that is one of the brighter stars of the *Plough. It was the first *spectroscopic binary to be identified (E.C. Pickering, 1889) and has a period of 20.5 days; the B component is also a spectroscopic binary, period 361 days. It forms an optical double with the 4th-magnitude *Alcor, separation 14″. m_V: 2.12; M_V: 0.0; spectral type: A2 V; distance: 26 pc.

MK system (MKK system; Yerkes system). *See* spectral types.

M-L relation. *See* mass-luminosity relation.

MMT. *Abbrev. for* Multiple Mirror Telescope.

modified Julian date (MJD). *See* Julian date.

Modular Equipment Transporter (MET). A handcart used to extend the sampling capability of *Apollo 14. *See also* Lunar Roving Vehicle.

modulus of distance. *See* distance modulus.

moldavite. *See* tektites.

molecular-line radio astronomy. The study of celestial regions containing molecules by means of the radio emissions from the molecules themselves; unlike optical emissions, radio emissions are not obscured by material in the interstellar medium. A *radio telescope using a *line receiver is best suited to the purpose. Many different molecules have now been detected in the Galaxy in relatively high density regions of dust and cool gas. The first interstellar molecules—CH, CH⁺, and CN— were detected optically in the

OBSERVED INTERSTELLAR MOLECULES

di—, triatomic	4—,5—,6— atomic	7—,8—,9—,11— atomic
H_2	NH_3	CH_3NH_2
OH	H_2CO	CH_3CHO
CH	H_2CS	CH_3C_2H
CH^+	H_2C_2	CH_2CHCN
CN	HNCO	HC_5N
CO	C_3N	
CS		
NS	CH_4	$HCOOCH_3$
SO	$C S_2NH$	CH_3C_2CN
SiO	NH_2CN	
SiS	HCOOH	CH_3OCH_3
	HC_3N	CH_3CH_2OH
H_2O	HC_4	HC_7N
H_2S		
HCO	CH_3OH	HC_9N
HCO^+	CH_3CN	
HCN	NH_2CHO	
HNC		
HN_2^+		
HC_2		
SO_2		
OCS		

1940s. This list has been greatly extended by radio-astronomy measurements (see table). Most of the molecules are organic, i.e. based on carbon, but some inorganic species, such as silicon oxide (SiO), have also been detected.

The radio emissions usually come from transitions between different rotational states of the molecule and lie in the gigahertz region of the radio spectrum. Some clouds of molecules appear to emit by maser action (*see* maser source); for example, the *OH line* at 1.667 gigahertz in some radio sources is observed to be very narrow and highly *polarized: characteristic features of a maser. Molecular-line radio astronomy is especially useful in determining cosmic isotope ratios.

Molonglo Radio Observatory. An observatory at Hoskinstown, New South Wales, run by the University of Sydney. The main instrument is a *Mills cross antenna, operational since 1960. Each arm is about 1.6 km long and has a series of cylindrical parabolic reflectors about 12 metres wide.

moment of inertia. Symbol: *I*. A property of any rotating body by which it resists any attempt to make it stop or speed

up. It is given by the sum of the products of the mass, m, of each particle in the body and the square of the particle's perpendicular distance, r, from the axis of rotation, i.e.
$$I = \Sigma mr^2$$
The summation is effected by integration for a solid body. If the body is rotating with angular velocity ω, the kinetic energy of the body is $\frac{1}{2}I\omega^2$ and its angular momentum is $I\omega$.

moment of momentum. *See* angular momentum.

momentum. 1. (linear momentum). Symbol: p. A property of a body moving in a straight line. It is equal in magnitude to the product of the mass and velocity (mv) of the body. It is a vector quantity, having the same direction as the velocity. *See also* conservation of momentum; Newton's laws of motion.
2. *See* angular momentum.

Monoceros (Unicorn). An inconspicuous equatorial constellation near Orion, lying in the Milky Way, the brightest stars being of 4th magnitude. It contains the triple star Beta Monocerotis and the double Epsilon and several variables including R Monocerotis, which lights up the *Hubble nebula. There are many clusters, including the open cluster NGC 2244, and there is the gaseous *Rosette nebula. Abbrev.: Mon; genitive form: Monocerotis; approx. position: RA 7h, dec 0°; area: 482 sq deg.

monochromatic. Of or producing a single wavelength or very narrow band of wavelengths.

monochromatic magnitude. *See* magnitude.

monochromator. An instrument in which one narrow band of wavelengths is isolated from a beam of light or other radiation. This is usually achieved by means of a diffraction grating or prism, which spreads the beam out into its component wavelengths, and an exit slit through which the desired waveband may pass. Changes in the intensity of the monochromatic beam can then be investigated.

Monster. Less common name for *Hydra.

month. The period of the moon's revolution around the earth with reference to some specified point in the sky (see table). The differences in the monthly periods result from the complicated motion of the moon.

moon (or Moon). The only natural satellite of the earth, visible by virtue of reflected sunlight. It has a diameter of 3476 km, lies at a mean distance from the earth of 384 400 km, and completes one orbit around the earth in 27.322 days (*see* sidereal month; month). Its mass in relation to that of the earth is 1:81.3, i.e. 0.0123. It is in

month	reference point	length (days)
tropical	equinox to equinox	27.321 58
sidereal	fixed star to fixed star	27.321 66
anomalistic	apse to apse	27.554 55
draconic	node to node	27.212 22
synodic	new moon to new moon	29.530 59

*synchronous rotation, i.e. it keeps the same face, the *nearside, towards the earth, although more than 50% of the moon's surface can be seen as a result of *libration. The same face is not however always turned towards the sun, the length of the solar day on the moon being equal to the moon's *synodic period, i.e. 29.530 59 days. During this period—the *synodic month—the moon exhibits a cycle of *phases, reaching an apparent magnitude of -12.7 at full moon. It can also undergo *eclipses and produce *occultations.

The moon's mean orbital velocity is 10 km s^{-1}. Its orbital motion is however subject to *secular acceleration and periodic *inequalities due to the gravitational attraction of the sun, earth, and other planets (*see also* evection; variation; annual equation; parallactic inequality). In addition its *perigee and *nodes move eastward and westward with periods of 8.85 years and 18.61 years, respectively: this results in different lengths for the sidereal, *anomalistic, and *draconic months. Physical and orbital characteristics are given in the table.

Although earth-based observations have revealed much lunar information, the moon has now been carefully photographed, measured, and sampled by the *Apollo spacecraft and by *Ranger, *Lunar Orbiter, *Surveyor, *Luna, and *Zond spacecraft. It has an extremely tenuous atmosphere: a collisionless gas in which helium, neon, argon, and radon were detected by Apollo instruments. This near lack of atmosphere, together with the absence of an appreciable global magnetic field, exposes the *regolith to extremes of temperature (-180°C to +110°C) and to the *solar wind, *cosmic rays, *meteorites, and *micrometeorites. The crustal rocks were magnetized early in lunar history; this randomly orientated remnant magnetism was detected by small satellites launched from Apollo command modules.

The moon's interior structure is investigated from seismograms of *moonquakes and meteoritic and artificial impacts. The temperature of the central core may reach 1500°C but a metallic iron composition is precluded by the moon's moment of inertia.

Moon's Physical and Orbital Characteristics

mass	7.35×10^{25} g
diameter	3476 km
mean density	3.34 g cm^{-3}
surface gravity	0.016 53 g
mean albedo	0.07
mean visual opposition magnitude	-12.7
interior structure: crust	60 km
mantle	1200 km
core radius	500 km
approximate age	4600 million years
mean horizontal parallax	3422″.46
moment of inertia	0.395±0.005
mean distance	384 400 km
maximum	406 730 km
minimum	364 400 km
mean orbital velocity	10.1 km s^{-1}
escape velocity	2.38 km s^{-1}
mean eccentricity	0.0549
inclination of orbital plane to ecliptic	5°8′43″
inclination of equator to ecliptic	1°32′1″
sidereal month	27.321 66 days
anomalistic month	27.554 55 days
draconic month	27.212 22 days
synodic month	29.530 59 days

The moon's *centre of mass is displaced towards earth by 2.5 km because of a thick farside crust.

The moon was formed out of refractory materials about 4600 million years ago. Its outer layers were melted and differentiated to produce mafic cumulates (iron magnesium silicates) and a feldspathic crust (calcium aluminium silicates), below which radioactive elements were concentrated. The first 700 million years of lunar history constituted a period of intense bombardment, culminating in a cataclysm during which the *basins and larger *highland craters were formed. The eruption of radioactive basalts (KREEP), or their excavation during formation of the Imbrium Basin, resulted in a concentration of natural radioactivity in the western hemisphere, as measured by gamma-ray spectrometers. A zone of melting then moved inwards towards the moon's centre to generate a sequence of basaltic lavas that produced the *maria. Since then the moon has been relatively quiescent.

The origin of the moon is still debatable. In the *fission hypothesis,* the moon formed from one or more fragments that broke off from the earth; this is now considered unlikely. *Capture theories* propose formation elsewhere and subsequent capture by the earth's gravitational field; these are improbable but could explain the absence of a metallic core and why the earth has such a massive companion. It seems more likely that the moon may have formed as part of a double planet or may have coalesced from a swarm of minor earth satellites.

See also moon, surface features; moonrocks.

moonquakes. Localized disturbances inside the moon, detected by the *Apollo seismic network. Rare but strong moonquakes occur in the crust, although they are much less intense than earthquakes by a factor of about 1000. Weaker but more numerous moonquakes occur at depths of 900-1000 km and are triggered at *perigee by lunar *tides. Seismic signals on the moon are extended by scattering in the dry and porous lunar crust. Deep moonquakes provide information, through their arrival times, of the structure of the lunar interior: a 60 km crust, a 1200 km mantle, and a central partially molten core 500 km in radius.

moonrocks (lunar rocks). Samples of the moon returned by *Apollo and *Luna spacecraft. Lunar soils, or fines, consist of glasses, aggregates, and meteorite debris in addition to recognizable fragments of larger rock types. Hand specimens include mare basalts (low titanium content from the western *maria, high titanium content from the eastern maria, and aluminous basalts from Mare Fecunditatis), non-mare basalt (also known as *KREEP* because of its high content of potassium, rare earth elements, and phosphorus), and anorthositic rocks from the *highlands. Rare lunar-rock types include granite, ultramafic green glass, pyroxenite, norite, troctolite, and dunite. Most highland rocks have been impact metamorphosed (to produce cataclastic, recrystallized, and impact melt textures), but cumulates are also observed and these date back to the formation of the first lunar crust.

Lunar rocks are depleted in volatile and siderophile elements. Strong correlations exist between elements that are geochemically similar to one another (e.g. iron and manganese) or that are both excluded by major minerals (e.g. potassium and lanthanum). The major lunar minerals are calcic plagioclase, clino- and ortho-pyroxene, olivine, and ilmenite. Minor phases include spinels, K-feldspar, troilite, metallic iron, quartz, tridymite, cristobalite, apatite, and whitlockite. Uniquely lunar minerals include armalcolite, tranquillityite, and pyroxferroite. No hydrated phases, highly oxidized minerals, diamonds, or mineralization have yet been found.

moon, surface features. The surface of the moon is characterized by light-coloured *highlands interspersed with much lower *albedo (i.e. darker) and less rugged *maria. The oldest and largest features on the moon are the vast impact *basins. These grade into large *craters, with which highland areas are saturated. There is no evidence for continental drift on the moon (*but see* lunar grid) but *faults, *rilles, *wrinkle and *lobate ridges, *domes, *ghost craters, and volcanic plateaux are the expressions of past tectonic and igneous activity. Impact features include *rays, *dark halo craters, *secondary craters (including *crater clusters and *crater chains), and textured *ejecta blankets.

Moreton wave. *See* flares.

Morgan's classification. A classification scheme for galaxies, devised by W.W. Morgan, in which galaxies are grouped according to the composite spectra of their component stars and also by their form, i.e. spiral, barred spiral, elliptical, irregular, etc., by their degree of concentration, and by their orientation, which varies from face-on to edge-on.

morning star. A term normally applied to *Venus when it shines brightly in the eastern sky before sunrise. Such periods last while Venus moves between inferior *conjunction and superior conjunction and is located west of the sun. The term occasionally refers to spells of morning visibility of Mercury or other planets.

mountains, lunar. Raised regions on the moon, most of which consist of crustal blocks uplifted to define the multiple ring structures of impact *basins; these survive either as *ring mountains,* such as the Apennines, or as isolated peaks, such as La Hire (see table). In comparison the lunar *highlands are topographically higher than the *maria but are characterized more by the elevated rims of craters and the rolling hills of basin *ejecta blankets. Discrete mountains not associated with impact basins include the central peaks of *craters, such as Tsiolkovsky, and volcanic domes and plateaux. Despite their jagged appearance, few mountains have slopes that exceed 12°.

mounting. The structure that supports an astronomical telescope, designed so that the telescope may be pointed at almost every part of the heavens, preferably without forcing an observer into awkward positions. All telescope mountings must therefore be constructed so that the tube can turn about two axes at right angles. There are two main types: the *equatorial mounting, with one axis parallel to the earth's axis, and the *altazimuth mounting, with one axis vertical. Since high magnifications exaggerate any movement, the support must

lunar mountains	location
Alps	N Imbrium, part of second ring
Altai Scarp	SW Nectaris, part of third ring
Apennines	SE Imbrium, part of third ring
Apollonius	S Crisium, part of third ring
Aristarchus Plateau	Procellarum, Volcanic province
Carpathians	S Imbrium, part of third ring
Caucasus	W Imbrium, part of third ring
Cordillera	Orientale basin, third ring
Haemus	S Serenitatis, part of second ring
Harbinger	SW Imbrium, part of third ring
La Hire	Imbrium, isolated peak, part of first ring
Marius Hills	Procellarum, Volcanic province
Piton	Imbrium, isolated peak
Pyrenees	W Nectaris, part of first ring
Riphaeus	W Cognitum
Rook	Orientale basin, second ring
Rümker Hills	Sinus Roris, volcanic province
Spitzbergen	N Imbrium, part of first ring
Straight	N Imbrium, part of first ring
Taurus	E Serenitatis, part of second ring
Tenerife	N Imbrium, part of first ring

be rigid. In addition, because the earth is turning through 0.25° per minute, means must be provided to move the telescope smoothly in the opposite direction; this counterbalances the diurnal motion of a star and keeps it in the field of view.

Mount Stromlo Observatory. An observatory on Mount Stromlo, near Canberra, at an altitude of 768 metres. It is run by the Australian National University. Its principal instrument is a 1.9 metre (74 inch) reflecting telescope.

Mount Wilson Observatory. *See* Hale Observatories.

moving cluster. An *open cluster, such as the *Hyades or the *Ursa Major cluster, for which a distance can be derived from the individual radial velocities and proper motions of the member stars. The method is often termed *moving-cluster parallax*. The stars in a cluster share a common space motion and, if the system is sufficiently close to earth, their paths will appear to diverge from or converge towards a point in space (point P in the illustration). The direction of P is parallel to the space motion of the cluster. Considering each star in turn (illustration B):

$$v_r = v\cos \alpha$$
$$v_t = v\sin \alpha = 4.74\mu/p$$

where v, v_r, v_t, are respectively the space, *radial, and tangential velocities of the star in km/s, μ is the proper motion in seconds of arc per year, and p is the *parallax in seconds of arc. Determination of α, μ, and v_r for each star gives a val-

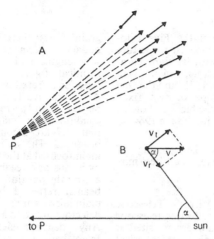

Moving cluster parallax

ue for the parallax of the system.

moving-cluster parallax. *See* moving cluster.

MS stars. *See* S stars.

M stars. Stars of *spectral type M, cool red stars with surface temperatures (T_{eff}) less than about 3500 kelvin. Molecular bands are prominent in the spectra, with titanium oxide (TiO) bands dominant by M5. Lines of neutral metals are also present. Antares and Betelgeuse are M stars. *See also* carbon stars; S stars.

MTRLI. *Abbrev. for* Multi Telescope Radio Linked Interferometer. A network of seven radio-telescope dishes that will be linked to a computer at *Jodrell Bank and can be used, in different combinations, to build up high-resolution maps of radio sources. Two 25 metre dishes have yet to be completed at sites a few kilometres from Jodrell Bank. The remaining five, already operational, are the two big dishes, 76 and 38 metres in size, at Jodrell Bank itself, the elliptical mesh dish 24 km away at Wardle near Nantwich, the 25 metre dish at the Royal Radar Establishment at Defford, near Malvern, and the 25 metre dish at Knockin, Oswestry. The latter, operational since 1977, has a design identical to that of the two dishes in the process of construction.

Mullard Radio Astronomy Observatory. The radio observatory of the Cavendish Laboratory, Cambridge University, near Cambridge. The chief instruments are two interferometers: the Five Kilometre telescope (*see* aperture synthesis) with its eight paraboloid reflectors, which was in operation in 1971, and the One Mile telescope of two fixed and one movable paraboloid, which came into use in 1964.

multilayers. *See* coating of lenses; interference filter, mirror.

Multiple Mirror Telescope (MMT). A reflecting telescope of totally new design, sited at the Smithsonian Astrophysical Observatory, on Mount Hopkins, near Tucson, Arizona at an altitude of 2382 metres. Its construction has been a joint project of the University of Arizona and the Harvard-Smithsonian Center for Astrophysics. The telescope saw first light in 1978. It comprises six identical mirrors, 183 cm (72 inches) in diameter, that are arranged in a hexagonal array on an *altazimuth mounting. The MMT has the light-gathering power of a 4.5 metre (176 inch) single mirror telescope and is thus the third largest telescope in the world; however it cost considerably less than a single 4.5 metre mirror to construct. It is designed for use at both visible and infrared wavelengths.

The mirrors are mounted symmetrically about a central axis. Their images are sent to a six-sided hyperbolic secondary mirror system on the central axis, which produces a single image at a common 'Cassegrain' focus; there is one secondary system for visible wavelengths and another, of identical figure but smaller size, for infrared wavelengths. An alternative focus can also be used. In perfect *seeing conditions star images of less than 0.7 arc seconds can be produced. The six mirrors remain focused at the same point by means of a feedback mechanism: the position of six laser beams, reflected by the six main mirrors are monitored by detectors linked to a computer. Any discrepancies arising from thermal or gravitational stresses are corrected by computer-controlled repositioning of the secondary mirror systems. The whole telescope can be smoothly guided to very great accuracies by computer-controlled drive mechanisms.

If the multiple mirror concept, with its complicated alignment and guidance system, proves reliable it might well become the prototype for future giant telescopes.

multiple star. A group of three or more stars, such as *Castor or *Alpha Centauri, that are bound together by their mutual gravitational attraction. (*Binary stars are sometimes considered as multiple stars.) Orbital motions within multiple systems are complex. For instance, systems containing two or more close binaries in orbit about each other appear to be quite common.

multiprobe. *See* Pioneer Venus probes.

multistage rocket. *See* launch vehicle.

muon. Symbol: μ. An *elementary particle (a lepton) with the same *charge and *spin as an *electron but with a mass that is 207 times greater. It decays into electrons and neutrinos with a lifetime of two microseconds. *Cosmic-ray showers detected on earth are comprised mainly of muons produced by *pion decay. Muons do not interact strongly with matter and hence a large proportion of them reach the earth's surface. The *antiparticle of the muon is the antimuon.

mural quadrant. *See* quadrant.

Musca (Fly). A small constellation in the southern hemisphere near Crux, lying in the Milky Way, the brightest stars being of 3rd magnitude. Abbrev.: Mus; genitive form: Muscae; approx. position: RA 13h, dec -70°; area: 138 sq deg.

N

nadir. The point on the *celestial sphere that lies directly beneath an observer and is therefore unobservable. It is diametrically opposite the observer's *zenith and with the zenith lies on the observer's *celestial meridian.

naked singularity. *See* singularity.

Nancy. French site of the Radio Observatory of Nançy, which is part of the Observatory of Paris. The chief instrument is a 305 metre (1000 foot) partially steerable (meridian transit) array, which has been in operation since 1964.

nano-. Symbol: n. A prefix to a unit, indicating a fraction of 10^{-9} of that unit. For example, one nanometre is 10^{-9} metres.

Naos. *See* Puppis.

NASA. *Abbrev. for* National Aeronautics and Space Administration. The US civilian agency that was formed in Oct. 1958 and is responsible for all nonmilitary aspects of the US space programme. Its major achievements so far have been the manned *Apollo programme, which followed the *Mercury and *Gemini projects, the *Skylab space station, and a large number of artificial *satellites and *planetary probes. It is now developing a reusable space transportation system—the *Space Shuttle—to replace the expendable and hence costly launch rockets it has used to date. It is collaborating on several projects with the *European Space Agency (ESA). NASA offers a satellite launcher service to other countries.

National Geographic Society—Palomar Observatory Sky Survey. *See* Palomar Sky Survey.

National Radio Astronomy

Observatory. A radio observatory with most of its telescopes at Green Bank, West Virginia. The principal instruments include a partially steerable (meridian transit) 91 metre (300 foot) paraboloid dish, completed in 1962, a fully steerable 43 metre (140 foot) dish, completed in 1965, and a 1.6 km baseline interferometer. The Observatory also operates the Very Large Array in New Mexico (*see* aperture synthesis) and a radio telescope at *Kitt Peak, Arizona.

Nautical Almanac. *See* Astronomical Ephemeris.

nautical twilight. *See* astronomical twilight.

n-body problem. Any problem in celestial mechanics that involves the determination of the trajectories of *n* point masses whose only interaction is gravitational attraction. The bodies in the solar system are an example if it is assumed that the masses of the planets, etc., are concentrated at their *centres of mass. A general solution exists for the *two-body problem and in special cases a solution can be found for the *three-body problem. The complete solution for a larger number is normally considered insoluble.

neap tide. *See* tides.

nearest stars. The stars nearest to the sun. They are either comparable in luminosity to or fainter than the sun. Most are less massive than the sun and a large proportion belong to binary or triple star systems. *See* Table 7, backmatter.

nearside. The hemisphere of the moon that is permanently turned towards the earth because of the tidal synchronization of the *lunar day and the *sidereal month. The lunar *librations allow the terrestrial observer also to see 9% of the *farside,* the hemisphere that is permanently turned away from the earth. The farside was first photographed by the Soviet *Luna 3 in 1959; it has no major *maria.

nebula. A cloud of interstellar gas and dust that can be observed either as a luminous patch of light—a bright nebula—or as a dark hole or band against a brighter background—a dark nebula. Interstellar clouds cannot usually be detected optically but various processes can cause them to become visible.

With *emission nebulae, ultraviolet radiation, generally coming from nearby hot stars, ionizes the interstellar gas atoms and light is emitted by the atoms as they interact with the free electrons in the nebula. Emission nebulae can be in the form of *HII (ionized hydrogen) regions, *planetary nebulae, or *supernova remnants. With *reflection nebulae light from a nearby star or stellar group is scattered (irregularly reflected) by the dust grains in the cloud. Reflection and emission nebula are bright nebulae. In contrast *dark nebulae are detected by what they obscure: the light from stars and other objects lying behind the cloud along our line of sight is significantly decreased or totally obscured by interstellar *extinction. Dark nebulae contain approximately the same mixture of gas and dust as bright nebulae but there are no nearby stars to illuminate them.

The term 'nebula' was originally applied to any object that appeared fuzzy and extended in a telescope: over 100 were listed in the 18th-century *Messier catalogue. The majority of these objects were later identified as *galaxies and star *clusters. Galaxies are still occasionally and misleadingly referred to as extragalactic nebulae.

nebular hypothesis. A theory for the origin of the solar system put forward by the French mathematician Pierre-Simon de Laplace in 1796, and similar to a suggestion made by the philosopher Immanuel Kant in 1755. It was proposed that the solar system formed from a cloud, or nebula, of material collapsing under its own gravitational attraction. Rotation of the nebula caused it to form a disc that, because of the law of conservation of *angular momentum, rotated more rapidly as the contraction progressed. Laplace suggested that rings of material became detached from the spinning disc when the velocity at its edge exceeded a critical value, and that the material in these rings later coalesced to form the planets. The central product of the contraction is the sun, while the planetary satellites may have formed from further rings shed by the condensing planets.

The hypothesis, popular throughout the 19th century, went out of favour when it could not explain why the sun has almost 99.9% of the mass of the solar system but only 3% of the total angular momentum, and because calculations showed that the rings would not condense to form planets. However, in a modified form, it is the basis of most modern ideas for the formation of the sun and planets. *See* solar system, origin. *Compare* encounter theories.

negative lens. A lens or lens system with a diverging action on a light beam.

Neptune. The eighth major planet, orbiting the sun every 165 years at an almost constant distance of 30.1 astronomical units. It is the most distant *giant planet and resembles *Uranus in many respects; it does however have a slightly smaller diameter (49 500 km: determined by stellar *occultation observations), a greater mass (17.2 earth-masses), and consequently a higher density (1.7 times that of water). Its two satellites, *Triton and *Nereid, move in unusual orbits. Orbital and physical characteristics are given in Table 1, backmatter.

Neptune was discovered in 1846 by J.G. Galle at the Berlin Observatory on the basis of predictions supplied by the Frenchman Urbain Leverrier, who had analysed observed perturbations in the motion of Uranus. Similar predictions had been made by the English mathematician John Couch Adams.

The planet returns to *opposition two days later each year, appearing as a magnitude 7.7 object visible in binoculars. No markings are seen on its greenish disc, only 2.3 arc seconds across. The presence of methane and hydrogen has been revealed by spectroscopic study of sunlight reflected from what is probably a visible surface of methane clouds. Recent spectroscopic measurements suggest that the rotation period, about an axis tilted by 29° to the vertical, may be a few hours longer than the 15.8 hours normally credited.

The atmospheric and internal structure may be almost identical to those of Uranus, with a rocky core 16 000 km in diameter encased in an 8000 km thick ice jacket and another 8000 km of, mainly, molecular hydrogen. To account for the greater overall mass, the core may be more massive than that of Uranus.

Neptune has an effective temperature of -220°C, a few degrees warmer than is to be expected at its solar distance and indicating that it, like Jupiter and Saturn, radiates more energy than it receives from the sun. Possibly this heat is generated by resistance to tides raised by its nearby massive satellite, Triton.

Nereid. A small satellite of Neptune with a highly eccentric orbit that takes it between 2 and 10 million km from the planet. It was discovered in 1949. *See* Triton; Table 2, backmatter.

Net. Common name for *Reticulum.

neutrino. Symbol: ν. A stable *elementary particle with zero *charge and *spin ½. Neutrinos are hypothesized to have zero *rest mass and according to the special theory of relativity travel at the speed of light. Two types of neutrinos are known: one is produced in association with an *electron (ν_e) and the other in association with a *muon (ν_μ); a third neutrino is assumed to be associated with the recently discovered massive tau lepton. The *antiparticle of the neutrino is the *antineutrino* ($\bar{\nu}$).

Wolfgang Pauli first postulated the existence of the neutrino in the 1930s to explain the violation of the laws of conservation of energy and momentum in *beta decay. Neutrinos are leptons and thus only have a weak interaction with matter. Neutrinos produced in nuclear reactions in the centres of stars can thus escape without further collision. *See also* neutrino detection experiments.

neutrino detection experiments. Experiments to detect *neutrinos released during energy-producing nuclear reactions occurring in the centres of stars. Neutrino interaction with other particles is highly improbable so that the great majority can escape from the star and travel unhindered through space. They thus provide direct evidence of conditions in the stellar core.

An experiment to detect *solar neutrinos* was first set up by Raymond Davies, Jr., in the 1960s and is still in progress. It is being conducted deep underground in the Homestake Gold Mine near Lead, South Dakota. A large quantity of the dry-cleaning fluid perchloroethylene, C_2Cl_4, is exposed for several months to the presumed flux of solar neutrinos. A very small (but calculable) number of the high-energy particles should react with the chlorine atoms:

$$^{37}Cl + \nu_e \rightarrow {}^{37}Ar + e^-$$

The amount of radioactive argon, ^{37}Ar, produced by the reaction gives a measure of the neutrino flux. The flux is measured in *solar neutrino units (SNU),* where one SNU is the flux giving rise to one neutrino-induced event per second per 10^{36} atoms of chlorine. The average flux detected is 1.75 ± 0.4 SNU. This is much lower than the predicted value of 4.47 ± 1.5 SNU.

The expected neutrinos arise from one of the rarer pathways in the *proton-proton chain reaction whereby hydrogen is converted into helium. (The proton-proton chain is the most important energy-producing reaction in the sun.) The neutrinos are emitted with high energy during the decay of boron-8, which in turn results from a reaction between nuclei of helium-3 and helium-4. The rate of neutrino production has been calculated from the expected conditions in the solar core and is very temperature-dependent.

The negative results of the above experiment are very difficult to explain. Assuming that the experiment is being carefully conducted, is sufficiently sensitive, and is based on correct premises, the near-lack of neutrinos implies that the condition in the sun's interior and/or the nuclear reactions taking place there are not fully appreciated. Experiments are now planned in the US and the USSR to detect the less energetic but more numerous neutrinos that are released in the earliest step of the proton-proton chain. These neutrinos can be absorbed by nuclei of gallium-71 to produce germanium-71; the amount of ^{71}Ge formed will then be a measure of the rate of a more fundamental step in the energy-producing reaction.

neutron. Symbol: n. An *elementary particle (a baryon) that is present in the *nucleus of all atoms except hydrogen.

It has zero *charge, *spin ½, and a *rest mass of 1.6749×10^{-24} grams—slightly greater than that of the *proton. The absence of charge enables the neutron to penetrate atoms easily since it has a negligible electromagnetic interaction with the constituents of the atom. Unlike free protons, free neutrons are unstable: they *beta decay into protons, electrons, plus antineutrinos with a mean life of approximately 930 seconds. The small difference between the neutron and proton mass provides the energy for this decay. The neutron is stable, however, when bound in a nucleus. The *antiparticle of the neutron is the *antineutron*. See also* isotopes; nucleosynthesis; s-process; r-process; neutron star.

neutron star. A star that has undergone *gravitational collapse to such a degree that most of its material has been compressed into *neutrons. Neutron stars were postulated in the 1930s by a number of astronomers including Landau and Zwicky. They are thought to form when the mass of the stellar core remaining after a *supernova explosion exceeds the *Chandrasekhar limit, i.e. 1.44 solar masses. Such a core is not stable as a *white dwarf star since even the pressure of *degenerate electrons cannot withstand the strong gravity. Collapse continues until the mean density reaches 10^{17} kg m^{-3} and the protons and electrons coalesce into neutrons, supporting the star against further contraction.

With diameters of only 20-30 km, intense magnetic fields (10^8 tesla), and extremely rapid spin, young neutron stars are believed to be responsible for the *pulsar phenomenon. Their strong gravity is also thought to give rise to rapid heating of material observed in some *x-ray binary systems. Models of the structure of neutron stars have been derived from sudden changes in pulsar spin rates—so called glitches. A typical neutron star may have an atmosphere only a few centimetres thick, under which is a crystalline crust one kilometre in depth. Beneath the crust the material is thought to act like a superfluid of neutrons (having zero viscosity) all the way through to a solid crystalline core.

New General Catalogue. *See* NGC.

new moon. *See* phases, lunar.

New Style (NS). The current method of measuring dates using the *Gregorian calendar as distinct from the former method—*Old Style* (OS)—that was based on the *Julian calendar. Britain and the American colonies switched calendars on Sept 14 1752, the 11 days from Sept 3 to Sept 13 having been eliminated from that year.

newton. Symbol: N. The SI unit of force, equal to the force that gives a mass of one kilogram an acceleration of

Ray path in Newtonian telescope

one metre per second per second.

Newtonian telescope. The first *reflecting telescope to be built, developed in about 1670 by Issac Newton from the ideas of Zucchi and Gregory: Newton thought that there was no way in which the *chromatic aberration of refracting telescopes could be corrected. The Newtonian has a paraboloid *primary mirror and a flat *diagonal mirror* that reflects the light into an *eyepiece mounted on the side of the tube (see illustration). A 45° prism can be used as an alternative to the diagonal mirror. The Newtonian is still the form used by many amateurs although somewhat inconvenient to use in larger form. Newton's original telescope had a mirror only 2.5 cm in diameter and made of *speculum metal; it had a focal length of about 15 cm and magnified some 30 times.

Newton's law of gravitation. *See* gravitation.

Newton's laws of motion. The three fundamental laws concerning the motion of bodies, that were formulated by Isaac Newton and published together with the law of *gravitation in *Principia*, 1687. The laws are

1. Every body continues in a state of rest or of uniform motion in a straight line until that state is changed by the action of a force on the body.

2. The rate of change of linear *momentum is proportional to the applied force, F, and occurs in the same direction as that of the force, i.e.
$F = d(mv)/dt = m(dv/dt) = ma$
where m is the mass, v the ve-

locity, and a the resulting acceleration of the body.

3. Every action is opposed by a reaction of equal magnitude that acts in an opposite direction to the action.

The first law was conceived by Galileo who first realized the falsity of the Greek notion that a force is required to maintain a body in motion. Newton's laws of motion and of gravitation are fundamental to *celestial mechanics.

N galaxy (N-type galaxy). A distant galaxy with an exceedingly bright (and sometimes variable) nucleus; they were originally identified by W.W. Morgan in 1958 as one of the optical counterparts of powerful radio sources. Many radio-quiet N galaxies have since been discovered. X-ray emission has also been detected from several N-types. N galaxies are believed to be less energetic examples of the *quasar phenomenon. *See also* Seyfert galaxy.

NGC. Short for *New General Catalogue of Nebulae and Clusters of Stars*, prepared by the Danish astronomer J.L.E. Dreyer at Armagh Observatory, Ireland, and published in 1888. It was based on an earlier catalogue compiled by the Herschel family. The original work listed 7840 currently known nebulae, clusters, and galaxies (not yet identified as such). Each object is numbered by *right ascension and its position and description given. The number is preceded by the letters NGC; this is the object's *NGC number,* an example being NGC 1976: the Orion nebula. Some of the brightest objects also appear in the *Messier catalogue.

Dreyer listed further discoveries in two supplements to the NGC. These were the first and second *Index Catalogues* (*IC*), published in 1895 and 1908. The three catalogues, now published as one volume, cover the whole sky and contain over 13 000 objects.

nightglow. *See* airglow.

Nix Olympica. *Earlier name for* Olympus Mons, assigned by telescopic observers to the bright-albedo feature they saw in its position.

nm. *Short for* nanometer, i.e. 10^{-9} metre.

nodes. The two points at which the orbit of a celestial body intersects a reference plane, usually the plane of the *ecliptic or *celestial equator. When the body, such as the moon or a planet, crosses the reference plane from south to north it passes through the *ascending node;* crossing from north to south the body passes through its *descending node.* The line joining these two nodes is the *line of nodes.*

A celestial body may not cross the reference plane at the same points every time: the line of nodes moves slowly forwards or backwards around the reference plane, i.e. in the same or in the opposite direc-

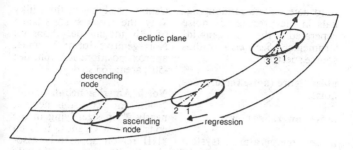

Regression of moon's nodes

tion to the body. This is termed *progression* or *regression of the nodes*. The moon's nodes regress along the ecliptic, making one complete rotation in 18.61 years (see illustration). This period is equal to the period of nutation.

nodical month. See draconic month.

noise. 1. Unwanted fluctuations that occur in the output of any electronic or electrical device and, in a measuring instrument, ultimately limit its sensitivity. Noise can be caused by many different factors, for example by insufficient regulation of the voltage of a power supply, but the most important source is usually that produced by random agitation of electrons (Brownian motion) in a resistance, i.e. *thermal noise*.

In electrical circuits a resistance at temperature T produces an exchangeable *noise power* in *bandwidth B of $P = kTB$, where k is the Boltzmann constant. If this noise power is generated in the first stages of the measuring instrument it is amplified along with the signal, which results in uncertainty in the measurement. The contribution to the noise power of any part of an electric circuit can be specified by its *noise figure* or *noise factor*, F, which is defined by

$F = (P_1 + P_0)/P_0$

where P_1 is the extra noise power generated by the circuit and P_0 is the noise power present at the input terminals of the circuit due to the source resistance, i.e. the thermal noise. Thus if the circuit adds no extra noise, it has a noise figure of unity. F is usually expressed in decibels. *See also* sensitivity.

2. Fluctuations of a random nature in which the amplitude of the deflection from a given reference level cannot be predicted from one moment to the next but whose statistical properties may be well-defined. Signals from *radio sources are of this kind, and in a *radio telescope the problem is to detect these random signals in the presence of noise generated in the measuring instruments. *See also* radio-source spectrum.

noise figure (noise factor). *See* noise.

noise power. *See* noise.

noise temperature (system temperature). *See* sensitivity.

nonrelativistic. Describing any phenomenon, object, etc., for which the effects of general or special *relativity can be disregarded or do not apply.

nonthermal emission (nonthermal radiation). Electromagnetic radiation, such as *synchrotron emission, that is produced, like *thermal emission, by the acceleration of electrons or other charged particles but is nonthermal in origin, i.e. its spectrum is not that of a perfect *black-body radiator.

noon. The time of day at which the sun crosses the local meridian and is at its highest point above the horizon. This time differs from 12.00 *local time by up to about 15 minutes before or after midday, the amount—the *equation of time—depending on the time of year. *Mean noon* is the time at which the *mean sun crosses the meridian and always occurs at 12.00 local time.

Norma (Rule; Level). A small inconspicuous constellation in the southern hemisphere near Centaurus, lying in the Milky Way, the brightest stars being of 4th magnitude. Abbrev.: Nor; genitive form: Normae; approx. position: RA 16h, dec -50°; area: 165 sq deg.

North America nebula (NGC 7000). An *emission nebula, nearly 2° in size, lying in the constellation Cygnus. It is an *HII region apparently associated with the exceptionally luminous and remote A2 supergiant *Deneb.

north celestial pole. *See* celestial poles.

Northern Cross. *See* Cygnus.

Northern Crown. Common name for *Corona Borealis.

Northern Hemisphere Observatory. A new international observatory on La Palma in the Canary Islands to which the 2.5 metre *Isaac Newton telescope is due to be moved in 1979. There will be two additional telescopes, one of one metre (39 inches) which is in the process of construction, and the other of 4.2 metres (165 inches).

north galactic pole. *See* galactic pole.

north galactic spur. *See* interstellar medium.

north point. *See* cardinal points.

north polar distance (NPD). The angular distance of a

celestial object from the north *celestial pole, measured along the object's hour circle. It is the complement of the object's *declination (90° - δ) and is sometimes used instead of this coordinate in the *equatorial coordinate system.

North Polar Sequence. *See* magnitude.

North Star. *See* Polaris.

nova. A highly evolved star in which there is a sudden and unpredictable increase in brightness by maybe 10 magnitudes or more (see table). Maybe 10 to 15 occur each year in our Galaxy. The brightness increases to a maximum within days or sometimes weeks and then declines to a value probably close to its faint pre-nova magnitude, indicating that the eruption did not disrupt the bulk of the star. The decline can range from months for *fast novae* to many years for the lower *amplitude *slow novae*. Hydrogen-rich gas is ejected from the star resulting in the tremendous outflow of heat and light. The ejected matter forms a rapidly expanding envelope of gas that can become visible as the nova fades. At later stages of the eruption the spectra of most novae show the bright forbidden lines characteristic of very low density *emission nebulae.

It is now thought that novae are members of close binary systems: the presence of a companion might even be a necessary condition for a star to be a nova. One component is usually a *white dwarf and is the source of the eruption; the other is a very close cooler redder star from which matter is slowly streaming into a disc-shaped orbit around the degenerate star (*see* mass transfer). This matter slowly accumulating on the star's hydrogen-rich upper surface layer somehow provides the energy to trigger the nova explosion, possibly by the runaway nuclear burning of hydrogen.

Nova are now named by constellation and year of observation; before 1925 they

nova	type	before	m_v max	after	max M_v
Nova Persei 1901 (GK Per 2)	fast	13.5	+0.2	13.2	-8.3
Nova Aquilae 1918 (V603 Aql 3)	fast	10.6	-1.1	10.9	-8.4
Nova Pictoris 1925 (RR Pic)	slow	12.7	+1.2	9	-6.4
Nova Herculis 1934 (DQ Her)	slow	14.3	+1.4	13.8	-6.2
Nova Cygni 1975 (V1500 Cyg)	fast	>21	+1.5		-9.5

were numbered in order of observation. They may also be given a variable star designation, as for example with DQ Herculis, i.e. Herculis 1934. *See also* recurrent nova; dwarf novae; supernova; x-ray transients.

Nova Mon 1975 (A 06020-00). *See* x-ray transients.

November swarm. *See* Leonids.

NPD. *Abbrev. for* north polar distance.

NS. *Abbrev. for* New Style.

N stars. *See* carbon stars.

N-type galaxy. *See* N galaxy.

Nubecular Major, Minor. *Alternative names for* Large and Small Magellanic Clouds, respectively. *See* Magellanic Clouds.

nuclear fission. A process in which an atomic nucleus splits into two smaller nuclei, usually with the emission of neutrons or gamma rays. Fission may be spontaneous or induced by neutron or photon bombardment. It occurs in heavy elements such as uranium, thorium, and plutonium. The process is accompanied by the release of large amounts of energy.

nuclear fusion. A process in which two light nuclei join to yield a heavier nucleus. An example is the fusion of two hydrogen nuclei to give a deuterium nucleus plus a positron plus a neutrino; this reaction occurs in the sun. Such processes take place at very high temperatures (millions of kelvin) and are consequently called *thermonuclear reactions*. With light elements, fusion releases immense amounts of energy: fusion is the energy-producing process in stars.

Iron and lighter elements evolve energy in fusion reactions whereas heavier elements require an input of energy to maintain the reaction. Thus although most elements up to iron can be formed by fusion reactions in stars (*see* nucleosynthesis), heavier elements must be synthesized by other nuclear reactions. *See also* proton-proton chain reaction; carbon cycle.

nuclear reactions. Reactions involving changes in atomic nuclei. Nuclear reactions include *nuclear fusion and fission and transformations of nuclei by bombardment with particles or photons. The energy released by nuclear reactions is equivalent to a decrease in mass according to Einstein's equation ($E = mc^2$) and appears in the form of kinetic energy of the particles or electromagnetic radiation.

nuclear timescale. The total nuclear-energy resources of a star divided by the rate of energy loss. For typical stars, the nuclear timescale is of the order of 10^{10} years, although the rate of energy loss may change during the star's lifetime if the reaction processes alter.

nucleon. One of the constituents of the atomic nucleus, i.e. a proton or a neutron.

nucleosynthesis. The creation of the *elements by nuclear reactions. To unravel the complex situation, the measured *cosmic abundances of the elements are interpreted in terms of their nuclear properties and the set of environments (temperature, density, etc.) in which they can be synthesized. Nucleosynthesis began when the temperature of the primitive universe had dropped to about 10^9 kelvin (K). This occurred approximately 100 seconds after the big bang (see big bang theory). Protons fused with neutrons to form deuterium nuclei and deuterium nuclei could then fuse to form helium. Very few heavier elements are thought to have been formed at this time.

It is now known that helium and the heavier elements are synthesized in stars and that this nucleosynthesis has occurred continuously in the Galaxy for many thousands of millions of years as a by-product of *stellar evolution. While a star remains on the *main sequence, hydrogen in its central core will be converted to helium by the *proton-proton chain reaction or the *carbon cycle; the core temperature is then about 10^7 K.

When the central hydrogen supplies are exhausted, the star will begin to evolve off the main sequence. Its core, now composed of helium, will contract until a temperature of 10^8 K is reached; carbon-12 can then be formed by the *triple alpha process, i.e. by helium burning. In stars more than twice the sun's mass a sequence of reactions, involving further nuclear fusion, produces elements with a mass number, A, that is a multiple of four: ^{16}O, ^{20}Ne, ^{24}Mg, and then, at temperatures increasing up to about 3.5×10^9 K, ^{28}Si to ^{56}Fe. Even higher temperatures will trigger reactions by which almost all elements up to a mass number of 56 can be synthesized. The *iron-peak elements,* i.e. ^{56}Fe, ^{56}Ni, ^{56}Co, etc., represent the end of the nucleosynthesis sequence by nuclear fusion: further fusion would require rather than liberate energy since nuclei with this mass number have the maximum binding energy per nucleon.

The formation of nuclei with $A \geq 56$ requires nuclear reactions involving neutron capture: neutrons can be captured at comparatively low energies owing to their lack of charge. If there is a supply of free neutrons in a star, produced as by-products of nuclear-fusion reactions, the *s-process can slowly synthesize nuclei up to ^{209}Bi. An intense source of neutrons allows the *r-process to generate nuclei up to ^{254}Cf, or higher, in a very short period. Such intense neutron fluxes arise in *supernovae and it is by means of

nucleus these explosions, and by novae and other mass-loss phenomena, that the elements are precipitated into and enrich the interstellar medium.

nucleus. 1. The kilometre-sized body that is the permanent portion of a *comet and is thought by most researchers to be the fount of all cometary activity. The density is about 1.1 g cm^{-3}. It contains about 75% by mass ice (mainly water ice but with liberal amounts of CO_2, CO, CH_4, and NH_3 ices) and 25% by mass dust, thought to have a composition similar to *carbonaceous chondritic *meteorites. It is often described as a *dirty snowball*. A comet of mass 10^{18} grams would have a nucleus of diameter about 6 km. No one has seen a cometary nucleus. Even comet Halley, which has a calculated mass of 3×10^{19} grams and a nucleus diameter of 35 km, did not exhibit a nucleus when it *transited the sun. The nucleus of dirty ice is surrounded by a thin (a few cm) layer of insulating dust from which the ice has sublimated. This enables it to survive about 200 perihelion passages. At each passage the nucleus looses on average a layer of material one metre thick. This material forms the *coma and *comet tails and also any associated *meteor stream.

A less likely model for the nucleus, which was prevalent up to the 1950s, regarded it as a vast collection (a swarm) of tiny particles. In this model a 10^{18} gram comet was thought to contain about 10^{25} particles, each of about 0.1 mm diameter, separated by tens of metres and occupying a volume of space about 100 000 km across. *See also* head.

2. The small core of an *atom, consisting of *protons and *neutrons bound together by strong nuclear forces. The nucleus has a positive charge equal to Ze, where e is the magnitude of the electron charge and Z the number of protons present—the atomic number. The total number of protons plus neutrons is the mass number, A. A given element is characterized by its atomic number but may, within limits, have different numbers of neutrons in its nuclei, giving rise to different *isotopes of the element. The total mass of the protons and neutrons bound together in a nucleus is less than when the particles are unbound. This mass difference is equivalent to the energy required to bind the particles together. Nuclei are represented by their chemical symbols to which numbers are attached, as with $^4_2$He where 2 is the atomic number of helium (2 protons) and 4 the mass number of the particular isotope.

3. (galactic nucleus). *See* Galaxy.

Nuffield Radio Astronomy Laboratory. *See* Jodrell Bank.

number count. *See* source count.

nutation. A slight periodic but irregular movement superimposed on the precessional circle (*see* precession) traced out by the celestial poles. It results from the varying distances and relative directions of the moon and sun, which continuously alter the strength and direction of their gravitational attraction on the earth. The primary component is lunar nutation, which causes the celestial pole to wander by ±9 seconds of arc from its mean position in a period of 18.6 years.

Changes in *right ascension and *declination arising from precession are perceptibly modified by nutation.

nutation in right ascension. *Former name for* equation of the equinoxes. *See* sidereal time.

O

OAO. *Abbrev. for* Orbiting Astronomical Observatory.

O-association (OB-association). *See* association.

Oberon. The outermost satellite of Uranus, discovered in 1787. *See* Table 2, backmatter.

objective (object lens). The lens or lens system in a *refracting telescope that faces the observed object. The focal plane in which the image forms is termed the *prime focus*. *See also* aperture.

objective prism. A narrow-angle prism placed in front of the primary mirror or objective lens of a telescope. The prism disperses (*see* dispersion) the incident light very slightly so that each star image is spread out into a small *spectrum. The spectra of a field of stars can thus be photographed in a single exposure and the chief spectral features of the stars can be quickly assessed (*see* spectral types).

oblateness (ellipticity). The degree of flattening of a celestial object, such as a planet, from a true spherical form, largely as a result of rotation. It is numerically equal to the difference between the equatorial and polar diameters of the object divided by the equatorial diameter. Saturn, with an oblateness of 0.098, is the most oblate planet.

oblique rotator theory. *See* magnetic stars.

obliquity of the ecliptic (obliquity). Symbol: ε. The angle at which the *ecliptic is inclined to the *celestial equator. This angle, equal to the tilt of the earth's axis, is about 23°27′. *Precession and *nutation cause this angle to change between the extremes of about 21°55′ and 24°18′, altering (at present diminishing) by about 0″.47 per year. Its value on 1950 Jan 1 was 23°26′45″.

observatory. A structure built primarily for astronomical observation, equipped nowadays

occultation

with optical, radio, and/or infrared telescopes and, in the larger observatories, the associated equipment with which photographic, spectrographic, photometric, and other such measurements are made. Observatories date back to the earliest civilizations: it is thought, for example, that Stonehenge and the Great Pyramid, Cheops, were both observatories. The sites of modern observatories are selected very carefully to give the best *seeing conditions with the minimum of atmospheric and light pollution. Most are sited in dry and/or mountainous areas where the atmosphere is very thin and the absorption effects of water vapour are reduced. Radio observatories, less hampered by seeing conditions, clouds, and light pollution, must still be isolated from electrical interference. Observatories are now being built in both northern and southern hemispheres and installed with highly refined equipment. *See* Table 11, backmatter.

Instruments can now be carried into space, away from the disturbing and absorbing effects of the earth's atmosphere. The greatest hopes for future astronomical studies lie in manned laboratories, such as *Skylab and *Salyut, and in unmanned remote-controlled equipment, such as *x-ray astronomy satellites and the proposed *Space Telescope.

occultation. Complete or partial obscuration of an astronomical object by the moon or a planet. A solar eclipse is strictly an occultation. The precise timings of occultations provide information about planetary atmospheres, the dimensions of extended visible, radio, and x-ray objects, and the positions of objects, such as distant radio sources. *See also* grazing occultation; eclipse.

O-cluster. An *open cluster that contains main-sequence O and B stars of high luminosity. These bright and presumably very massive stars evolve more rapidly than less massive stars; their presence in a cluster suggests that it is comparatively young (10 million years or less) and that it may contain stars still in the process of formation. Examples are NGC 2264 and *h and Chi Persei. *T Tauri stars are often observed in these clusters. *See also* association.

octahedrite. *See* iron meteorite.

Octans (Octant). An inconspicuous constellation containing and surrounding the south celestial pole, the brightest stars being of 4th magnitude. The 5th-magnitude Sigma Octantis is the nearest naked-eye star to the south pole. Abbrev.: Oct; genitive form: Octantis; approx. position: RA 0 - 24h, dec -80°; area: 291 sq deg.

ocular. *Obsolete name for* eyepiece.

Of stars. Young peculiar O stars belonging to earlier subdivisions than O5. In addition to a well-developed absorption spectrum they show selectively enhanced lines of ionized helium (He II) and nitrogen (N III) that can vary in intensity in an irregular manner. The brightest Of star known is Zeta Puppis, which is embedded in the *Gum nebula.

OGO. *Abbrev. for* Orbiting Geophysical Observatory.

OH line. *See* molecular-line radio astronomy; hydroxyl radical.

OH maser. *See* maser source.

Olbers' paradox. Why is the sky dark at night? Heinrich Olbers in 1826, and earlier J.P.L. Chesaux in 1744, pointed out that an infinite and uniform universe, both unchanging and static, would produce a night sky of the same surface brightness as the sun: every line of sight would eventually strike a star, a typical example of which is the sun. This theoretical argument is obviously in disagreement with observation. The observable universe is, however, neither uniform, unchanging, nor static and does not extend infinitely back into the past. The paradox is then resolved since the *redshift of extragalactic radiation (i.e. the diminution in its energy) and in particular the youth of our universe make the background radiation field at optical wavelengths very low indeed—less than about 10^{-21} times the surface brightness of the sun.

Old Style (OS). *See* New Style.

Olympus Mons. The largest volcano on Mars and possibly the largest in the solar system. With an estimated age of 200 million years, it may be the youngest of the volcanoes on the *Tharsis Ridge. It has a base 550 km across and rises to an 80 km diameter caldera at an elevation of 25 km above the surrounding plains. Around much of the base is an escarpment up to 6 km high. For comparison, the largest volcano on earth, Hawaii's Mauna Kea, rises only 9.3 km from its 120 km diameter base on the floor of the Pacific Ocean. *See also* Mars, volcanoes.

Omega nebula (Swan nebula; M17; NGC 6618). An *emission nebula with a conspicuous bar that lies in the direction of Sagittarius, very close to Scutum. Its apparent magnitude is 7. It is a double radio source.

Oort cloud. A cloud of *comets that move in near-circular orbits round the sun in a zone with heliocentric distances between 30 000 and 100 000 AU, i.e. in a zone far beyond Pluto's orbit. Since the nearest star to the sun—Proxima Centauri—is 270 000 AU away, these comets can be perturbed by passing stars; this changes their orbits and sends them in towards our sun (or out towards the perturbing star). These perturbed comets have periods of hundreds of thousands of years. The Oort cloud is so distant that it acts as a refrigerated cometary reservoir.

Its existence was proposed by J.H. Oort who thought that comets originated from the disintegration of a planet some time in the early history of the solar system. After this the major planets, Jupiter in particular, drove the majority of the comets either into the inner solar system, where they decayed, or into the outer regions. Of this latter group 95% would have been lost from the solar system, only 5% remaining in the Oort cloud. The number of comets in the cloud is uncertain but it could be somewhere in the range 10^7 to 10^{11} with a total mass in the region 10^{25} to 10^{28} grams. The orbits are orientated at random. A cometary cloud of radius 100 000 AU has a half life of at least 1.1×10^9 years. The existence of the cloud shows up very clearly when the number distribution of comets is considered as a function of the reciprocal of the semimajor axes of the orbits.

Oort's constants. *See* Oort's formulae.

Oort's formulae. Two mathematical expressions derived by Jan H. Oort that describe the effects of differential *galactic rotation on the radial velocities (v_r) and tangential velocities (v_t) of stars at an average distance r from the sun. If each star is moving in a circular orbit about the galactic centre, and if r is small in comparison with the sun's distance from the centre, then

$$v_r = Ar \sin 2l''$$
$$v_t = Br + Ar \cos 2l''$$

where l'' is the galactic longitude (*see* galactic coordinate system) and A and B are *Oort's constants*. Since v_t is not directly measurable the second equation is usually replaced by the expression for *proper motion, μ:

$$\mu = 0 \cdot 211 \, (B - A \cos 2l'')$$

The equations are valid for μ measured in arc seconds per year, velocities in km s^{-1}, and distances in parsecs. The most widely accepted values for Oort's constants are

A: 15 (km s^{-1})/kiloparsec
B: -10 (km s^{-1})/kpc

Most stars, including the sun, do not move in precisely circular orbits but once a correction has been made for the sun's velocity with respect to the average velocity of stars in the solar neighbourhood (assumed to be circular), the measured variation of v_r and μ with l'' is sufficiently close to that predicted by the equations to support the theory that the general rotation pattern is circular and galactocentric.

opacity. A measure of the ability of a gaseous, solid, or liquid body to absorb radiation. It is the ratio of the total radiant energy received by the body to the total energy transmitted through it.

The opacity of a star depends on temperature, density, and chemical composition. Various processes contribute to the opacity: energy absorption by electrons bound to an atom, allowing them to jump to a higher level or to escape from the atom as free electrons; absorption by free electrons; scattering of photons of radiation by free electrons or atoms. The opacity of the sun is seen to drop rapidly to zero at the sun's limb.

open cluster (galactic cluster). A diffuse cluster of stars that contains at most a few hundred stars and sometimes less than twenty. Examples visible to the naked eye are the *Hyades and the *Pleiades. About 1000 open clusters are known. They are *population I systems and occur in or close to the plane of the Galaxy. The brightest stars in an open cluster can be either red or blue giants, depending upon its age. Stars in the older clusters, such as M67 in Cancer, are similar in appearance to those in *globular clusters, though with some subtle differences due to the higher metal content of the material from which they were formed.

Open clusters are more loosely bound systems than globular clusters and they tend to be gradually dispersed by the combined effects of the differential rotation of the Galaxy and perturbations due to close encounters with interstellar clouds. Calculations suggest that many will not survive more than one or two circuits of the Galaxy. Hence most open clusters are comparatively young systems. Some, such as NGC 2264, are less than 10 million years old and in these clusters star-formation is probably still taking place (*see* O-cluster). *See also* moving cluster; Hertzsprung-Russell diagram; turn-off point.

open universe. *See* universe; deceleration parameter.

Ophiuchids. A minor *meteor shower, radiant: RA 260°, dec -20°, maximizing on June 20.

Ophiuchus (Serpent Bearer). A large equatorial constellation near Scorpius, lying partly in the Milky Way. The brightest star is the 2nd-magnitude white giant Ras Alhague (α), which lies close to *Ras Algethi (Alpha Herculis). The area contains the supernova remnant of *Kepler's star, the dark *Ophiuchus nebula, and many globular clusters. Abbrev.: Oph; genitive form: Ophiuchi; approx. position: RA 17h, dec -7°; area: 948 sq deg.

Ophiucus nebula. A *dark nebula in the constellation Ophiucus, about 250 parsecs away.

opposition. The moment at which a body in the solar system has a longitude differing from that of the sun by 180°, so that it lies opposite the sun in the sky and crosses the meridian at about midnight (see illustration at *elongation*). The term also applies to the alignment of the two bodies at this moment. Although the *inferior planets cannot come to opposition, it is the most favourable time for observation of the other planets since they are then observable throughout the night and are near their closest point for that *apparition. *See also* synodic period.

optical axis. The imaginary line passing through the midpoint of a lens, mirror, or system of such elements and on which lies the focal point of parallel paraxial rays.

optical depth. Symbol: τ. A measure of the absorption of radiation of a particular wavelength as it passes through a gaseous medium. If the initial radiation flux Φ_0 is reduced to Φ_x after passing a distance x through the medium then

$$\Phi_x/\Phi_0 = \exp(-\tau)$$

where τ is the optical depth for the radiation wavelength. If τ equals zero the medium is transparent; if τ is much greater or much less than one the medium is *optically thick* or *optically thin,* respectively.

optical double. 1. *See* double star.
   2. *See* double galaxy.

optical libration. *Another name for* geometrical libration. *See* libration.

optical pathlength. The distance, d, travelled by a light beam multiplied by the refractive index, n, of the medium through which the light has passed. If the light traverses more than one medium, then the optical pathlength is the sum

$$(n_1 d_1 + n_2 d_2 ...)$$

for each medium. It gives the effective pathlength in terms of wavelength of light.

optical pulsars. *Pulsars that flash at visible wavelengths. Two are known: the Crab pulsar (*see* Crab nebula) appears as a 16th magnitude star but is pulsing with a period of 0.033 seconds; the optical flashes of the *Vela pulsar, about 26th magnitude, were discovered by looking for faint pulses with exactly the radio period (0.089 seconds) of its radio emission.

optical temperature. *See* effective temperature.

optical window. *See* atmospheric windows.

orbit. The path followed by a celestial object or an artificial satellite or spaceprobe that is moving in a gravitational field. For a single object moving freely in the gravitational field of a massive body the orbit is a *conic section, i.e. it is either elliptical, parabolic, or hyperbolic. Most orbits are elliptical, most planetary orbits being almost circular. A parabolic or hyperbolic orbit results in the object escaping from the vicinity of a massive body with a velocity equal to or greater than *escape velocity. *See also* orbital elements; Kepler's laws.

orbital elements. The six parameters that uniquely specify the position and path of a celestial body in its orbit as a function of time and that can be established by observation (see illustration). The shape and size of the orbit are specified by the *eccentricity, e,* and for an elliptical orbit by the *semimajor axis, a*. The orientation of the orbit in space is specified firstly by the *inclination, i,* of the orbital plane to the reference plane, usually that of the ecliptic, and secondly by the *longitude of the ascending node,* Ω; the latter is the angular distance from the vernal *equinox, , to the ascending *node, N. The orientation of the orbit in the orbital plane is usually specified by the angular distance, ω, between the periapsis, P (*see* apsides), and the ascending node. For an orbit round the sun the periapsis is the *perihelion and the angular distance ω is the *argument of the perihelion*. It is measured along the direction of motion.

The eccentric *anomaly is used to determine the position of the body in its orbit. To calculate the position as a function of time the sixth orbital element is used. This is the *time of periapsis* (or for a solar orbit *perihelion*) passage, t_0, by means of which Kepler's equation (*see* anomaly) can be solved. To determine the orbit of a binary star system of unknown mass a seventh element, the *period, must be established.

orbital velocity. The velocity of a satellite or other orbiting body at any given point in its orbit. It is also the velocity required by a satellite to enter an orbit around a body. The

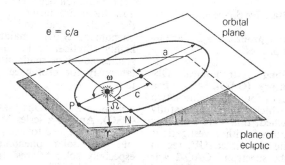

Orbital elements

orbital velocity, v, is given by the expression
$$v = \sqrt{[gR^2(2/r - 1/a)]}$$
where R is the radius of the orbited body, r is the distance from the *centre of mass of the system (i.e. from the approximate centre of the primary), a is the *semimajor axis of the orbit, and g is the standard acceleration of gravity. For a circular orbit, $r = a$ and the circular velocity is given by
$$v = \sqrt{(gR^2/r)}$$
To escape from an orbit a must tend to infinity and the escape velocity is then given by
$$v_e = \sqrt{(2gR^2/r)}$$
The orbital period for an elliptical orbit is given by
$$P = 2\pi a^{3/2}/\sqrt{gR^2}$$

orbit determination. The derivation of the orbit of a celestial object, such as a comet, from several observations of its position by using the laws of celestial mechanics. At least three observations are needed to determine the *orbital elements and many methods have been used.

Orbiter. *See* Space Shuttle.

Orbiter probes. *See* Lunar Orbiter probes.

Orbiting Astronomical Observatory (OAO). Any of a series of four US satellites, two of which were successfully launched, that were primarily concerned with investigations in the ultraviolet (UV) region of the spectrum. OAO-2 was launched on Dec. 7 1968 and continued operating for over four years. In one group of experiments a UV survey of about one-sixth of the sky was achieved by means of four telescopes plus UV-sensitive television camera tubes, the *Celescope experiment*, which produced pictures in the wavelength range 115-320 nanometres. A catalogue of bright UV sources was compiled from the data. In the other experimental package measurements were made of the UV luminosity and spectra of a large number of preselected targets.

OAO-3 was launched on Aug. 21 1972 and was named *Copernicus. It carried a 0.9 metre telescope used with a high-resolution photoelectric spectrometer for UV measurements and also had x-ray equipment on board.

Orbiting Geophysical Observatory (OGO). Any of a series of US geophysical satellites, first launched in Sept. 1964, for studying the earth's atmosphere, ionosphere, magnetic field, radiation belts, etc., and how these are influenced by the *solar wind and other phenomena. Moving mainly in highly elliptical orbits, many studies could be made simultaneously for prolonged periods.

Orbiting Solar Observatory (OSO). Any of a series of US satellites designed to study the sun and solar phenomena, especially solar *flares, from earth orbit. The first one (OSO-1) was launched in

Orion and bright neighbouring stars

March 1962, OSO-8 being orbited in 1975. The quality and range of the observations has improved as the series has progressed, with high spectral and spatial resolution in ultraviolet and x-ray spectral regions in later craft. Instruments have included a *coronagraph, *spectroheliograph, ultraviolet and x-ray *spectrographs, *photometers, gamma-ray detectors, and particle-flux sensors.

Orientale Basin. The youngest lunar *basin, only partially filled by mare lava. The concentric ring structure of *mountains and the *ejecta blanket (the Hevelius formation) are exceptionally well preserved. Orientale is situated on the western limb of the moon and is only visible from earth at times of favourable *libration.

Orion. A very conspicuous constellation on the celestial equator, close to the Milky Way, that is visible as a whole from most parts of the world and can be used to find the positions of many bright neighbouring stars (see illustration). All the leading stars, apart from *Betelgeuse, are very hot, white, and luminous. The 1st-magnitude *Rigel (β) and Betelguese (α) and 2nd-magnitude *Bellatrix (γ) and Saiph (κ) form the quadrilateral outline inside which are the three remote 2nd-magnitude stars of *Orion's Belt,* Alnilam (ε), Alnitak (ζ), and Mintaka (δ). South of the Belt is *Orion's Sword,* in which lies the naked-eye *Orion nebula and the adjacent nebula M43. The area contains many double stars including Rigel, several variable stars, and the dark *Horsehead nebula. An immense shell of ionized hydrogen, *Barnard's loop,* surrounds most of Orion. Abbrev.: Ori; genitive form: Orionis; approx. position: RA 5.5h, dec 0°, area: 594 sq deg.

Orion arm. *See* Galaxy.

Orion association (I Orionis). A large O-association (*see* association) that contains over 1000 stars and lies in the region of the Orion nebula. There are at least three clusters in the association: the Lambda Orionis cluster, the Belt group around Epsilon Orionis, and the *Trapezium cluster (often classified as a multiple star).

Orionids. An active *meteor shower, radiant: RA 96°, dec +15°, that maximizes on Oct. 21 with a *zenithal hourly rate of about 30. The activity has been observed to change by as much as a factor of four between one year and the next. The shower has been the centre of controversy firstly because it was thought that the *radiant was stationary during the three weeks of activity. This view has now been abandoned: all showers have radiants that move with respect to the stars as the earth moves along its orbit. Secondly the shower is possibly associated with Halley's comet and the Eta *Aquarids because of similarities in orbital elements.

Orion nebula (Great Nebula in Orion; M42; NGC 1976). The nearest and thus one of the brightest *emission nebulae in the sky, about 500 parsecs distant. It is just visible to the naked eye as a luminous patch, 35 arc minutes across, in the centre of Orion's Sword. It is a young HII (ionized hydrogen) region containing *O stars, *T Tauri stars, and other extreme *population I stars and is ionized by the four brightest O and B stars of the *Trapezium. Some of the light from the nebula is starlight scattered by dust grains in the cloud.

Behind the emission nebula is a cloud of molecular hydrogen that contains several sources of infrared radiation; one is *Becklin's star, which is almost certainly a collapsing *protostar. These two regions of hydrogen appear to be in physical contact, the ionized gas possibly expanding into the neutral region. The HII region is also an x-ray source.

Orion's Belt. *See* Orion.

Orion's Sword. *See* Orion.

orrery. An instrument that demonstrates the movements of some or all of the planets around the sun, in their correct relative periods, by means of an elaborate, usually clock-driven system of gears. The movements of satellites, such as the moon, may be incorporated in the mechanism.

orthoscopic eyepiece. *See* eyepiece.

OS. *Abbrev. for* Old Style. *See* New Style.

oscillating universe (pulsating universe). A closed *universe in which collapse is followed by a re-expansion.

osculating orbit. The truly elliptical orbit that a celestial object would follow if the perturbing forces of other bodies were to disappear so that it was subject only to the central gravitational field of a single massive body. In general elliptical motion is an approximation to the real motion of members of the solar system. If all forces except the central gravitational force disappeared at time t, then the *osculating elements* would be the *orbital elements of the ellipse followed by a body at a given instant after time t. For a real body in perturbed motion the instantaneous osculating elements are not constant but are functions of time. The osculating elements are often used to describe the perturbed motion of a body.

OSO. *Abbrev. for* Orbiting Solar Observatory.

O stars. Stars of *spectral type O that are very hot massive blue stars, emitting copious quantities of ultraviolet radiation, and with surface temperatures (T_{eff}) from about 28 000 to over 40 000 kelvin. Strong lines of singly ionized helium (HeII) dominate the spectra. Lines of doubly and triply ionized elements are also present with lines of neutral helium (HeI) and hydrogen strengthening in later subdivisions. No O0, O1, or O2 stars have been discovered and only a few O3 and O4 stars are known. Most O stars are found in the spiral arms of galaxies, often in O *associations, and are very fast rotators. They have short lifetimes of about three to six million years. *See also* Of stars; ultraviolet stars.

outburst, cometary. *See* Schwassmann-Wachmann 1.

outer planets. The planets Jupiter, Saturn, Uranus, Neptune, and Pluto, which lie further from the sun than the main belt of minor planets. With the exception of Pluto, all are *giant planets.

Out-of-Ecliptic (OOE) mission. Earlier name for *International Solar Polar Mission.

Owl nebula (M 97; NGC 3587). A relatively large *planetary nebula in the constellation Ursa Major, about 600 parsecs away.

Ozma project. A pioneering project in the search for extraterrestrial life. The 26 metre radio telescope at Green Bank, West Virginia, was used by Frank Drake in 1960 for about 150 hours in an unsuccessful attempt to detect signals—the 21 cm hydrogen line emission—from other civilizations. The targets were the two nearest and most sunlike stars, Tau Ceti and Epsilon Eridani. *See also* exobiology.

ozonosphere (ozone layer). *See* atmospheric layers.

P

Painter. Common name for *Pictor.

pair production. The conversion of a high-energy *photon, or gamma ray, in the field of an atomic nucleus into an *electron and positron, which are formed simultaneously. *See also* annihilation.

Pallas. The second *minor planet to be discovered, found in 1802. With a diameter of 608 km it is the second largest minor planet and has a spectrum similar to that of the *carbonaceous chondrite meteorites. *See* Table 3, backmatter.

Palomar Observatory. The observatory sited on Mount Palomar, California, that became one of the Hale Observatories in 1970. The principal instruments are the 200 inch (508 cm) reflecting telescope and the 48 inch (122 cm) Schmidt telescope. *See* Hale Observatories.

Palomar Sky Survey. A photographic star atlas of the northern sky and part of the southern sky to a *declination of -33°. It was prepared at the Mount Palomar Observatory, California, with finance from the National Geographic Society, using the 48″ Schmidt telescope. It was released 1954-58. It consists of 935 pairs of photographic prints: one of each pair was made from an exposure on a blue-sensitive plate, the other was made from an exposure taken through a red filter. It includes stars to a limiting magnitude of 21 (20 in the southernmost plate). *See also* Southern Sky Survey.

parabola. A type of *conic section with an eccentricity equal to one. *See also* paraboloid; orbit.

parabolic velocity. The velocity of an object following a *parabolic trajectory* around a massive body. Its velocity at a given distance from the massive body is equal to the *escape velocity at that distance.

paraboloid. A curved surface formed by the rotation of a parabola about its axis. Cross sections along the central axis are circular. A beam of radiation striking such a surface parallel to its axis is reflected to a single point on the axis (the focus), no matter how wide the aperture (see illustration). A paraboloid mirror is thus free of *spherical aberration; it does however suffer from *coma. Paraboloid surfaces are used in reflecting telescopes and radio telescopes. Over a small area a paraboloid differs only slightly from a sphere. A paraboloid mirror can therefore be made by deepening the centre of a spherical mirror.

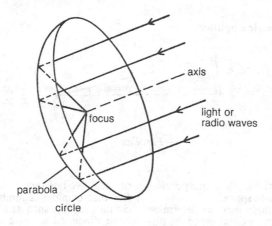

Paraboloid surface

parallactic angle. *See* astronomical triangle.

parallactic ellipse. *See* annual parallax.

parallactic inequality. A minor periodic term in the moon's motion that results from the gravitational attraction of the sun and causes the moon to be ahead at first quarter and to lag behind at last quarter (*see* phases, lunar). The inequality can be used to calculate *solar parallax. *See also* evection; variation.

parallactic motion. *See* parallax.

parallax. The angular displacement in the apparent position of a celestial body when observed from two widely separated points. It is thus the angle that the baseline connecting two viewpoints would subtend at the object (see illustration). It is very small in value and is usually expressed in arc seconds. If the baseline is of a fixed length then as the distance to the celestial object decreases, its parallax will increase accordingly. If the parallax can be measured then so can the distance. 'Parallax' is thus often used synonymously with 'distance'.

The observer's position on earth changes with the daily rotation of the earth, the annual revolution of the earth around the sun, and the long-term motion of the sun and solar system relative to the background stars. Each motion produces a different measure of parallax: *diurnal parallax, *annual parallax, and *secular parallax, respectively. The continual change in the apparent position of a celestial object, produced by the observer's changing position, is termed *parallactic motion* and must be distinguished

parametric amplifier

Parallax

from the star's own *peculiar motion in space.

Methods used to determine the parallax and hence the distance of celestial bodies require an accurate knowledge of the baseline length. The baseline for diurnal parallax—the earth's radius—can only be used for distance measurements within the solar system. The baselines used in annual and secular parallax are sufficiently great for stellar distances to be measured. Other methods based on the concept of parallax and used in distance determination include *dynamic parallax, *moving-cluster parallax, and *rotational parallax.

Distances determined indirectly from stellar brightness are also sometimes called parallaxes: in *spectroscopic parallax* the absolute magnitude of a main-sequence star is deduced from the spectral type of the star using the *Hertzsprung-Russell diagram and this together with the apparent magnitude gives the *distance modulus and hence the distance. *See also* distance determination.

parametric amplifier. A type of negative-resistance amplifier that employs a nonlinear circuit element, such as a varactor diode, to act as a time-varying capacitance. These amplifiers have low *noise figures at high frequencies and are useful in *radio telescopes in the range 1-30 gigahertz.

paraxial. Lying close to the optical axis of a lens, mirror, or system of such elements.

Parkes. Site of the *Australian National Radio Astronomy Observatory.

parking orbit. An orbit in which a satellite or spacecraft can be placed before being injected into a final orbit or desired trajectory.

parsec. *Short for* parallax second. Symbol: pc. A unit of length normally used for distances beyond the solar system. It is the distance at which the mean radius of the earth's orbit (one astronomical unit) subtends an angle of one arc second. It is thus the distance at which a star would have an *annual parallax of one arc second. A star with a parallax of p seconds is at a distance of

d parsecs, given by $d = 1/p$ (accurate up to distances of about 30 pc). One parsec equals 30.857×10^{12} km, 206 265 astronomical units, and 3.2616 light-years.

partial eclipse. *See* eclipse.

particle. *See* elementary particle; alpha particle; beta particle.

pascal. Symbol: Pa. The SI unit of pressure, equal to a pressure of one newton per square metre.

Paschen series. *See* hydrogen spectrum.

Pasiphae. A small satellite of Jupiter, discovered in 1908. *See* Jupiter's satellites; Table 2, backmatter.

pathlength. *See* optical pathlength.

path of totality (zone of totality). *See* eclipse.

Pavo (Peacock). A constellation in the southern hemisphere near Grus, the brightest stars being the 2nd-magnitude spectroscopic binary Alpha Pavonis and two 3rd-magnitude stars. It contains several variables, including the bright *Cepheid variable Kappa Pavonis (a W Virginis star), and the large bright globular cluster NGC 6752. Abbrev.: Pav; genitive form: Pavonis; approx. position: RA 19.5h, dec -65°; area: 378 sq deg.

Pavonis Mons. A 19 km high volcano on the *Tharsis Ridge of Mars. It has a base diameter of 400 km, the diameter of the summit caldera being 45 km. *See* Mars, volcanoes.

payload. The total mass (or in some cases the mass of the experimental and operational equipment) of a satellite, spacecraft, etc., that is carried into orbit by a launch vehicle. It is that part of the total launcher mass that is not necessary for the operation of the launcher. It is usually a small or very small fraction of the total launcher mass.

pc. *Symbol for* parsec.

P Cygni stars. A small group of hot variable stars whose prototype is the B1e star P Cygni. Their spectra show numerous strong emission lines, like those of *Be stars and *Wolf-Rayet stars, and sharp absorption lines that have a *blueshift. The blueshifted absorption components probably originate in an expanding shell of low-density matter continuously ejected from the hot star. Many, like P Cygni, suffer random outbursts. *See also* Hubble-Sandage variables.

Peacock. Common name for *Pavo.

peculiar galaxies. Galaxies that do not conveniently fit into the regular classification schemes of normal galaxies. These comparatively rare objects include *interacting galaxies, single irregular or disturbed galaxies, *compact galaxies, *dwarf galaxies, *ring galaxies, and spiral galaxies with peculiar spiral arms or even with one or three arms.

peculiar motion. 1. The motion of an individual star relative to a group of neighbouring stars.
2. That part of the *proper motion of a star that results from its actual movement in space; it is the star's proper motion after elimination of the effects of the sun's motion.

peculiar stars. Stars with spectral features that do not correspond exactly with the usual classification of *spectral types. They are designated by a 'p' after their spectral type, as with *Ap stars.

Pegasus. An extensive conspicuous constellation in the northern hemisphere near Cygnus, with three 2nd-magnitude stars (Alpha, Beta, and Epsilon Pegasi) and four of 3rd magnitude. The stars *Markab (α), *Scheat (β), and *Algenib (γ) together with Alpha Andromedae (*Alpheratz) form the distinctive *Great Square of Pegasus*. The area also contains the globular cluster M15. Abbrev.: Peg; genitive form: Pegasi; approx. position: RA 23h, dec +20°; area: 1121 sq deg.

pencil beam. *See* antenna.

penumbra. 1. *See* umbra.
2. *See* sunspots.

penumbral eclipse. *See* eclipse.

periapsis. *See* apsides.

periastron. *See* apastron.

pericynthion. The point in the lunar orbit of a satellite launched from earth that is nearest the moon. The equivalent point in the lunar orbit of a satellite launched from the moon is the *perilune*. *Compare* apocynthion.

perigalacticon. The point in a star's orbit around the *Galaxy that is nearest the galactic centre. The furthest point is the *apogalacticon*.

perigee. The point in the orbit of the moon or an artificial earth satellite that is nearest the earth and at which the body's velocity is maximal. Strictly, the distance to the perigee is taken from the earth's centre. At perigee the moon's distance is 364 400 km, about 42 000 km closer than at *apogee.

perihelion. The point in the orbit of a planet, comet, or artificial satellite in solar orbit that is nearest the sun. The earth is at perihelion around Jan. 3. The perihelion distance, q, and *aphelion distance Q, of major and minor planets are given in Tables 1 and 3, backmatter. *See also* advance of the perihelion.

perihelion distance. The distance from the focus of a parabolic trajectory around the sun to the vertex.

perihelion passage, time of. *See* orbital elements.

perijove. *See* apojove.

perilune. *See* pericynthion.

perimartian. *See* apojove.

period. The time interval between two successive and similar phases of a regularly occurring event. The period of *rotation or of *revolution of a planet, etc., is the time to complete one rotation on its axis or one revolution around its primary. The period of a *binary star is the time observed for the companion to orbit the primary. The periods range from less than a day to several thousand years, *visual binaries having very much longer periods than *spectroscopic or *photometric binaries: over half the known spectroscopic binaries have periods of less than 10 days. The period of a regular intrinsic *variable star or an *eclipsing binary is the time between two successive maxima or minima on the *light curve. It can vary from less than an hour to more than 500 days.

period-colour relation. *See* Cepheid variables.

period-density relation. *See* pulsating variables.

periodic comet. A *comet that has been seen more than once.

period-luminosity relation (P-L relation). A relation showing graphically how the period of the light variation of a *Cepheid variable depends on its mean luminosity: the

Period-luminosity relation

longer the period the greater the luminosity and hence mean absolute magnitude of the star. The relation was discovered by Henrietta Leavitt in 1912 and much subsequent work was done in calibrating the graph. It was found, in the 1950s, that there were two types of Cepheid (classical Cepheids and the less luminous W Virginis stars) whose P-L relations are approximately parallel, as shown on the graph. The relation results from the dependence of both luminosity and period of light variation, i.e. of pulsation, on stellar radius.

The relation is an invaluable means of determining the distances of Cepheids and hence of their surroundings: the absolute magnitude of a particular Cepheid is determined from the measured value of its period, and the distance can then be found from an independent measure of its apparent magnitude (*see* distance modulus). Cepheids occur in many star clusters within our own Galaxy and because of their great luminosity can be observed in nearby galaxies up to about three million parsecs away. They can therefore be used for measuring an immense range of distances. *See* distance determination.

period-spectrum relation. *See* Cepheid variables.

perisaturnian. *See* apojove.

Perseids. A major *meteor shower, radiant: RA 46°, dec +58°, maximizing on Aug. 12 with a *zenithal hourly rate (ZHR) of about 70. Perseid meteors can be seen throughout the two weeks each side of maximum and are bright and flaring, with fine *trains. The geocentric velocity of the meteoroids is about 60 km s^{-1}. The *meteor stream is closely associated with comet 1862 III. Although the display is regarded as being constant in hourly rate there are exceptions, a ZHR of 250 being observed in 1921 and a ZHR of less than 10 in 1911 and 1912. The shower has been regularly observed for over 100 years.

Perseus. A fairly conspicuous constellation in the northern hemisphere near Cassiopeia, lying in the Milky Way. It has two 2nd-magnitude stars, *Mirfak (α) and the prototype eclipsing binary *Algol (β), and several of 3rd magnitude. It also contains the naked-eye Double Cluster *h and Chi Persei, several open clusters including the just-visible M34, the *California nebula, and the galaxy NGC 1275, Perseus A, which is a powerful radio source. Abbrev.: Per; genitive form: Persei; approx. position: RA 3.5h, dec +45°; area: 615 sq deg.

Perseus A. *See* Perseus cluster.

Perseus arm. *See* Galaxy.

Perseus cluster. A rich *cluster of galaxies lying about 70 megaparsecs away in the direction of the constellation Perseus. It contains several strong radio sources, three of which have been identified with the galaxies IC 310, NGC 1265, and NGC 1275. NGC 1275 is the most intense radio source (3C 84A or *Perseus A*) and occurs at the centre of the cluster. It is a supergiant elliptical galaxy that is surrounded by a large x-ray emitting cloud of hot gas. This x-ray source (3U 0316 + 41 or *Perseus X-1*) extends over approximately the same region as the halo of faint stars surrounding NGC 1275. The *missing mass is thought to be about 20 times the observed mass of the component galaxies.

Perseus X-1. *See* Perseus cluster.

persistent train. *See* train.

perturbation. A small disturbance that causes a system to deviate from its equilibrium state. Periodic perturbations cancel out over the period involved and do not affect the system's stability. Secular perturbations have a progressive effect on the system that can cause it eventually to become unstable.

A single planet orbiting the sun would follow an elliptical orbit according to Kepler's laws; in reality the orbit of one planet is perturbed from its elliptical orbit by the gravitational effects of the other planets. Likewise the revolution of the moon around the earth is perturbed mainly by the sun and to a much lesser extent by other bodies. The assumptions that the sun is the only perturbing body and that the earth orbits the sun in a fixed elliptical orbit lead to a simplified theory of lunar motion. The orbits of comets and minor planets are strongly perturbed when the body passes close to a major planet, such as Jupiter. The influence of a perturbing body can be calculated from the orbital elements of an *osculating orbit, to which corrections are made.

phase. 1. The appearance of the illuminated surface of the moon or a planet at a particular time during its orbit. Only the moon and *inferior planets can exhibit those phases in which half or less of the illuminated hemisphere is visible from earth. *See also* phases, lunar.
2. The fraction of one complete cycle of a regularly recurring quantity that has elapsed with respect to a fixed datum point. The *phase difference*, ϕ, is the difference in phase between two electrical oscillations, wavetrains, etc., of the same frequency (i.e. coherent signals) and is usually expressed in terms of part of one complete cycle or wavelength. 'Phase difference' is often referred to simply as 'phase', as it is when considering the coherent signals in the two arms of a radio interferometer.

phase angle. The angle between the lines connecting the moon and sun and the moon and earth or connecting a plan-

Phase angles for superior and inferior planet

et and the sun and the planet and earth (see overleaf). At full or new moon the phase angle is 0° or 180°, respectively. At first and last quarters it is 90°. For the superior planets phase angle is greatest at quadrature. For the inferior planets it is greatest (180°) at inferior conjunction and equals 90° at greatest *elongation.

phase defect. The angular amount by which the illuminated area of the moon differs from a circular disc at full moon due to the inclination of the moon's orbital plane to the *ecliptic.

phase difference. *See* phase.

phase rotator. A device that adds extra pathlength to one arm of an interferometer (*see* radio telescope), especially one used for *aperture synthesis, in order to keep the *phase difference close to zero as a *radio source moves across the sky.

phase-sensitive detector. An electronic device that detects a weak periodic signal in the presence of *noise. The signal plus noise is multiplied by a reference waveform whose shape and frequency are similar to those of the required signal. Components near zero frequency are produced. When the output is smoothed by a low-pass *filter a steady signal results, whose amplitude is related to the amplitude of the required signal and to the *phase difference between the

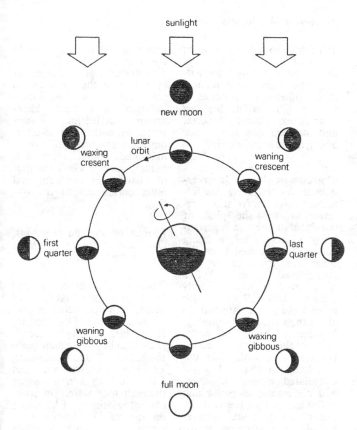

Moon's phases, outer circle showing phases as seen from earth

required signal and the reference waveform. The device is used, for example, in a *phase-switching interferometer to detect the component at the switching frequency in the output of the first detector.

phases, lunar. The *phases exhibited by the moon (see illustration). A *new moon* occurs when the moon is at *conjunction and the *nearside is totally unilluminated by the sun. As the moon moves eastwards in its orbit around the earth, the sunrise *terminator crosses the nearside from east to west to produce a *crescent moon*. The crescent waxes to *first quarter* when the moon, at *quadrature, is half illuminated, through a *gibbous moon*, and finally to *full moon,* when the moon is at *opposition and the nearside is fully illuminated. The sunset *terminator then follows to produce a waning gibbous moon, *last quarter,* a waning crescent, and the next new moon. *See also* synodic month; eclipse.

phase-switching interferometer. A correlating interferometer (*see* correlation receiver) in which the *phase of the signal in one arm is periodically reversed before being added to the signal from the other arm. The phase reversal is achieved by means of a *phase switch,* which simply adds an extra half wavelength of path every alternate half cycle of the switching frequency, between 10 and 1000 hertz. The switching has no effect on noise signals, which are uncorrelated in the two arms. Correlated signals, however, add differently in phase and antiphase so that the first detector output in the *receiver contains a component at the switching frequency with an amplitude dependent on the power, and relative phase in the two arms, of the correlated signal.

This component is detected by means of a *phase-sensitive detector whose smoothed output contains no steady deflection unless a correlated signal is present. As a *radio source moves through the antenna beam, interference fringes appear in the output of the phase-sensitive detector; the fringes are similar to those in the output of an ordinary radio interferometer, except that in this case their average is zero rather than being superimposed on a steady background deflection. Phase-switching makes the interferometer much less sensitive to man-made *interference and reduces the effect on its output of changes in the gain and other parameters of the system.

Phecda or **Phekda** (γ UMa). One of the seven brightest stars in the *Plough.

Phobos. The innermost and larger of the two satellites of Mars, both discovered in 1877 by Asaph Hall. Observations from Mars-orbiting spacecraft show it to be an irregular potato-shaped body, measuring $19 \times 21 \times 27$ km with a mass of 1.0×10^{16} kg and a density twice that of water. It orbits above the equator of Mars at a distance of 9350 km from the centre of the planet every 7.65 hours, keeping its long axis pointing towards Mars. The low density, together with a low *albedo of 0.06, suggest that it resembles the water-rich *carbonaceous chondrite meteorites, though it is difficult to see how such an object could have formed at the distance of Mars from the sun. Possibly Phobos and *Deimos, Mars' other small satellite, formed as more distant *minor planets and were later captured by Mars in some way. The orbit of Phobos is contracting because of tidal drag, and the satellite may decay from orbit within another 100 million years.

Phobos has a *regolith-covered cratered surface with the two largest craters, *Stickney* and *Hall,* having diameters of 10 and 6 km. A network of pitted grooves, typically 100-200 metres wide and 20-30 metres deep, radiate around the satellite from Stickney and are thought to have arisen from severe internal fracturing and heating caused by the impact event that formed the crater. Chains of small craters parallel to the orbit of Phobos possibly represent the secondary impacts of clumps of material swept up from orbit around Mars after being ejected from the satellite during earlier primary impacts. *See* Table 2, backmatter.

Phoebe. The outermost satellite of Saturn, discovered in 1898, and the only one to follow a retrograde orbit (*see* direct motion) around the planet. *See* Table 2, backmatter.

Phoenicids. A minor southern hemisphere *meteor shower, radiant: RA 15°, dec -55°, maximizing on Dec. 4.

Phoenix. A constellation in the southern hemisphere near Grus, the brightest star, Ankaa (α) being of 2nd magnitude. Beta Phoenicis is a binary, both components being of 4th magnitude. Abbrev.: Phe; genitive form: Phoenicis; approx. position: RA 1h, dec -50°; area: 469 sq deg.

Phosphorus. The Greek name for *Venus when it was seen as a *morning star. *Compare* Hesperus.

photocathode. An electrode in an electronic device, such as a photocell, photomultiplier, or image tube, that emits electrons when a beam of electromagnetic radiation strikes the surface. By a suitable choice of photocathode material, a reasonable response may be obtained from near-infrared wavelengths to low-energy x-ray wavelengths. The electrons result from the *photoelectric effect. As many as 30% of the incident photons can be converted into electrons although the percentage is usually lower when taken over a wide spectral region. The current of resulting electrons increases linearly with radiation intensity over a wide range of intensities.

photocell. An electronic device that converts a light signal, or a signal in the near-infrared or ultraviolet, into an equivalent electrical signal. The earliest devices were *photoelectric cells*. These are evacuated and work by the *photoelectric effect. Electrons are emitted from a *photocathode when a beam of radiation falls on the surface. The electrons are attracted to a positively charged anode so that a current flows in an external circuit.

photoconductive cell

Most photocells now consist of a layer of semiconductor sandwiched between two electrical contacts. When light, infrared, or ultraviolet radiation falls on the sandwich its electrical conductivity increases markedly as a result of photoconductivity, i.e. by the creation of electron-hole pairs by the incident photons; the device is often called a *photoconductive cell*. A current therefore flows in the external circuit of the device. The current is proportional to the amount of radiation falling on the cell and may be measured by means of a galvanometer.

Devices, such as the *solar cell, that operate by the photovoltaic effect are also known as photocells.

photoconductive cell. *See* photocell.

photoelectric cell. *See* photocell.

photoelectric effect. An effect whereby electrons are emitted by material exposed to electromagnetic radiation above a certain frequency. This frequency usually lies in the ultraviolet region of the spectrum for most solids, but can occur in the visible region. The number of ejected electrons depends on the intensity of the incident radiation. The electron velocity is proportional to the radiation frequency. Radiation energy is transferred in the form of *photons to the electrons within the atoms of the material. If the energy is above a certain level the most weakly bound electrons can escape. The photoelectric effect is utilized in certain forms of *photocell and in *photomultipliers.

photoelectric magnitude. *See* magnitude.

photographic emulsion. A light-sensitive layer of silver halide crystals suspended in gelatin and coated on a flexible or rigid base, from which a permanent record of a light or near-ultraviolet image, focused on the surface, may be produced. The layer can be made sensitive to near-infrared radiation. Photographic emulsions have high resolution, wide spectral response, good dimensional stability and permanence, and reasonably high signal/noise ratio images. The light from a star, well below the visual threshold, can be recorded during a long exposure: exposure times of over an hour may be used. Emulsions do however have a very low *quantum efficiency compared to electronic recording devices; they can only record a very small fraction of the incident photons of radiation. In addition a linear relation between intensity of incident light and resulting image density exists for only a narrow range of exposures ('exposure' here means intensity of illumination times exposure time); at shorter or longer exposures the relation is nonlinear and reciprocity failure results. *See also* photography.

photographic magnitude. *See* magnitude.

photographic zenith tube (PZT). A specially designed telescope that is used for the accurate determination of time by most of the world's major time services. The telescope is fixed in a vertical position so that the observed objects are stars at or near the *zenith of the observatory. A photographic plate is driven across the field of view of the telescope at the exact speed of the star image. As the carriage moves, the plate is marked at several positions at known clock times. The positions of the star image, relative to these marks, gives the correction to the clock. The plate carriage is reversed by 180° halfway through the observation to remove any instrumental errors that might have occurred in the recording of the star image.

photography. There are two broad categories of astronomical photography: wide-angle photography of large areas of the sky, and photography of comparatively near solar-system bodies. A camera or photographic plate, suitably mounted on a telescope, must be carefully driven at a constant (usually sidereal) rate in order to maintain a steady image and a constant field of view (*see also* guide telescope). Photography of the brighter solar-system objects is fairly straightforward. Wide-angle studies are somewhat more complicated: lenses of short *focal ratio (less than f/5) should be used together with a *photographic emulsion that must be able to respond to the very low light levels and the long exposure times of an hour or more that are involved. Long exposures can be badly affected by stray light so that *light pollution must be minimal: the sites of both major observatories and small amateur instruments must therefore be carefully selected both to reduce pollution and to have the best possible *seeing.

Plates or films for long-exposure use must have high-contrast fine-grained emulsions with low reciprocity failure: they must be able to respond uniformly to a wide range of light intensity. One such emulsion, highly suited for astronomical purposes, is Kodak type *IIIaJ emulsion* (blue-green sensitive) and its panchromatic version, *IIIaF*. The resulting photographic images are very uniform and have a very high resolution; they can compete very successfully with the two-dimensional images produced by electronic detectors. With the *Schmidt telescope a very great area of sky can be photographed at one time: Schmidt plates have been used in the *Southern and *Palomar Sky Surveys. Special emulsions have also been prepared for near-infrared photography.

photoionization. The *ionization of an atom or molecule by photons of electromagnetic radiation. A photon can only remove an electron if the photon energy exceeds the first *ionization potential of the atom or molecule. The excess energy is shared between the electron and the ion so that the electron can leave the atom with considerable velocity. If the radiation is of sufficiently high energy more strongly bound electrons will be removed, leaving the resulting ion in an excited state. *See also* recombination-line emission.

photometer. An instrument by which the intensity and various other properties of light, and also of infrared and ultraviolet radiation, may be measured. In a visual photometer the intensity of a light source is compared by eye to that of a standard light source. The eye can detect very slight differences in intensity (i.e. in brightness) so that fairly accurate measurements can be made. For greater accuracies or when the intensity is constantly changing, physical receptors such as the *photocell, *photomultiplier, or *bolometer, must be used. In these devices the incident radiation is converted into an electrical signal whose magnitude can be determined very precisely.

photometric binary. A *binary star that can be detected by virtue of periodic changes in its light curve. *See* eclipsing binary.

photometry. The measurement of light intensities or energy flux, usually by comparison with a reference source. In stellar photometry this amounts, primarily, to the determination of the apparent brightness of sources of radiation. At optical, infrared, and near-ultraviolet wavelengths, brightness is measured in terms of apparent *magnitude. At other wavelengths measurements of *flux density are made.

In some cases the eye can make comparisons, with a fair degree of accuracy (to 0.1 magnitude) between the known brightness of a reference star and the unknown brightness of the source under study. The greater accuracies now required in astronomy are achieved with electronic instruments or specially prepared photographic emulsions. In optical and infrared astronomy, magnitudes are determined over a selection of wavelength bands. Present systems, such as the internationally accepted UBV system (*see* magnitude), use a combination of a suitable *filter, a *photocell or *photomultiplier, etc., plus an electronic amplifier. The light intensity is thus measured for a particular wavelength band as a greatly increased (amplified) electrical signal. The U, B, and V magnitudes are determined over three relatively broad bands in the ultraviolet, blue, and yellow-green (visual) spectral regions. Magnitudes can also be measured for various infrared bands. Narrower wavelength bands are used in the *uvby system. Extremely narrow bands, now obtainable over the entire optical and infrared regions, can be studied by means of *interference filters: the distribution of, say, hydrogen in a source can be determined by isolating a particular spectral line of hydrogen.

Photoelectric photometry can be used to study an isolated star, region, etc. In photographic photometry, measurements can be made of objects on a photographic image of a wide angle of sky; the photographs are taken at wavelengths varying from near-infrared to near-ultraviolet (depending on emulsion). Highly resolved features can be registered, often of much greater quality than can be obtained electronically. The image, for a given photographic exposure time, varies in density (and size) by an amount that, within relatively wide limits, depends on the original brightness. Densities can be measured accurately by means of microdensitometers, etc.

photomultiplier. An evacuated electronic device used to convert a low-intensity light signal into an electrical signal and to amplify this signal very considerably. It contains a *photocathode from which electrons are liberated by the incident light photons. The electrons are accelerated down a tube by a series of positively charged electrodes to which an increasingly high potential is applied. At each electrode a number of additional electrons are released by each impacting electron (by secondary emission). The electrons are all collected at the final electrode in the chain, which is held at a very high potential, and a large current pulse is produced.

photon. Symbol: γ. A quantum of *electromagnetic radiation. The photon can be considered as an *elementary particle with zero *rest mass, *charge, and *spin, travelling at the speed of light. A beam of electromagnetic radiation must thus be thought of as a beam of photons, with the intensity of the radiation proportional to the number of photons present. The energy of the photons, and hence of the radiation, is equal to the product, $h\nu$, of the *Planck constant and the frequency of the radiation. Photons with sufficient energy can ionize atoms and disintegrate nuclei. Although photons have no rest mass they do have momentum and thus exert a pressure, usually referred to as *radiation pressure.

photon-counting system. A recently developed and highly efficient all-electronic system for detecting and recording light or ultraviolet radiation. Astronomical observations can be recorded in very much shorter times than those required for photographic exposures of the sky. It is essentially a high-gain multistage *image tube in which a *television camera records the two-dimensional image produced on the final phosphor screen. The gain of the device is such that each electron liberated from the photocathode can be detected as a burst of light by the television camera. The signal from the camera is fed into a computer for analysis.

photosphere. The visible surface of the *sun and source of the *absorption spectrum that is characteristic of most stars. The solar photosphere is a stratum several hundred kilometres thick from which almost all the energy emitted by the sun is radiated into space. Within the photosphere the temperature falls from 6000 kelvin just above the *convective zone to 4000 kelvin at the *temperature minimum*, where the photosphere merges with the *chromosphere.

The intensity of the photosphere decreases at visible wavelengths from the centre to the limb of the disc (*see* limb darkening). The intensity is due to the radiation emitted, principally by negative *hydrogen ions (H$^-$), at depths of up to a few hundred kilometres; a continuous range of wavelengths, from ultraviolet to near-infrared, is produced. At higher levels, where the density of H$^-$ ions is too low for appreciable *opacity, the lower temperature gives rise to the absorption of radiation at discrete wavelengths. The *Fraunhofer lines of the resulting absorption spectrum have provided the key to determining the chemical composition of the photosphere, since a direct comparison can be made with the laboratory spectra of known elements under various conditions.

Regions of the photosphere (and lower chromosphere) several thousand kilometres in diameter rhythmically rise and fall with a period of about five minutes, attaining a maximum velocity of about 0.5 km s^{-1}. These vertical oscillations are thought to be produced by the outward propagation of low-frequency sound waves, generated by turbulence in the *convective zone. A longer-period (2h 40m) pulsation has recently been announced but has yet to be confirmed.

See also granulation; supergranulation; sunspots; faculae.

photovisual magnitude. *See* magnitude.

photovoltaic cell. An electronic device consisting of a junction between a metal and a semiconductor or between two opposite polarity (p- and n-type) semiconductors. When the junction is exposed to radiation, usually in the range near-ultraviolet to infrared frequencies, electron-hole pairs are generated and can cross the junction as a result of the built-in electric field of the (unbiased) junction. A forward voltage thus appears across the illuminated junction and power can be delivered to an external circuit.

physical double. 1. *See* double star.
2. *See* double galaxy.

physical libration. Slight variations in the moon's rotation on its axis that are caused by minor irregularities in shape. They amount to less than two minutes of arc but allow the moon's principal moments of inertia, and hence the distribution of mass within the moon, to be calculated. The mean orientation of the moon with respect to its physical librations defines the moon's *prime meridian. See also libration.

Pic du Midi Observatory. A French observatory in the Pyrénées at an altitude of 2860 metres. It is an important site for solar observations.

pico-. Symbol: p. A prefix to a unit, indicating a fraction of 10^{-12} of that unit. For example, one picosecond is 10^{-12} seconds.

Pictor (Painter). A small inconspicuous constellation in the southern hemisphere near the Large Magellanic Cloud, the brightest stars being the 3rd-magnitude Alpha Pictoris and two of 4th magnitude. Abbrev.: Pic; genitive form: Pictoris; approx. position: RA 5.5h, dec -50°; area 247 sq deg.

Pioneer probes. A series of US solar-system probes. Pioneers 1-3 were intended lunar probes in 1958, although Pioneer 2 suffered launch failure and Pioneers 1 and 3 fell short of the moon but did measure the extent of the *Van Allen radiation belts. Pioneer 4 began to orbit the sun after passing the moon and was followed into solar orbit in 1960 by Pioneer 5, which returned data about the *solar wind and solar *flares.

Pioneer 6, launched in 1965, was joined in solar orbit by Pioneer 7 in 1966, Pioneer 8 in 1967, and Pioneer 9 in 1968. Together they formed a network of 'solar weather stations' between 0.75 and 1.12 astronomical units from the sun. Monitoring *solar activity and the solar wind, they provided early warnings of solar flares during the *Apollo programme.

Pioneer 10, launched on Mar. 3 1972, made the first fly-by of Jupiter at a distance of 130 300 km on Dec 4 1973, after becoming the first probe to traverse the asteroid belt. It returned photographs of three of the *Galilean satellites of Jupiter (*Europa, *Ganymede, and *Callisto), as well as of the Jovian atmosphere and the *Great Red Spot. Other experiments measured the strength and extent of Jupiter's radiation belts. Pioneer 11, launched on a similar mission on Apr. 5 1973, passed 42 940 km below Jupiter's south pole on Dec. 3 1974 and reached a top speed of 171 000 km per hour, faster than any previous man-made object. It is now on course for a fly-by of Saturn in Sept. 1979. *See*

Pioneer Venus probes. Two US probes, an orbiter and a *multiprobe,* launched in May and August 1978, respectively, towards Venus. The orbiter, Pioneer Venus 1, entered a highly inclined elliptical orbit around Venus on Dec. 4 1978, approaching to within 150 km of the surface. It is designed to make observations over 243 days (Venus' rotational period) and do radar mapping of the surface, cloud studies at ultraviolet and infrared frequencies, and magnetometer studies.

The multiprobe, Pioneer Venus 2, released one large probe on Nov. 15 of 316 kg and three small probes on Nov. 19, each of 93 kg. These probes penetrated the Venusian atmosphere on Dec. 9 at different locations and continuously gathered and relayed data as they approached and hit the surface. One small probe survived the hard landing and continued transmitting for a further hour. Pioneer Venus 2—the interplanetary 'bus'—entered the atmosphere on Dec. 9, shortly after its payload, as an upper-atmosphere probe.

Preliminary data has revealed a surprisingly large amount of primordial argon and neon in Venus' atmosphere, three main cloud layers between altitudes of 70-50 km, a strong bow shock wave that holds the solar wind off, an immense canyon near the equator, and violent lightning on the dark side of the planet.

pions. Symbol: π. A triplet of *elementary particles (mesons) that have *spin zero and exist in three states: neutral, positively charged, and negatively charged. The magnitude of the charge is equal to that of the electron. The charged pions have a mass of 2.4886×10^{-25} grams, the neutral pion mass being slightly less at 2.4066×10^{-25} grams. Charged pions decay into *muons and *electrons with a mean life of 2.6×10^{-8} seconds; the neutral pion converts, usually directly, into gamma-ray photons with an average lifetime of 8.4×10^{-15} seconds. *See also* cosmic rays.

Pisces (Fishes). A large but inconspicuous zodiac constellation mainly in the northern hemisphere near Pegasus that contains the *first point of Aries. It has one 3rd-magnitude star and several of 4th magnitude. Abbrev.: Psc; genitive form: Piscium; approx. position: RA 1h, dec +15°; area: 889 sq deg.

Pisces Australids. A minor southern hemisphere *meteor shower, radiant: RA 340°, dec -30°, maximizing on July 30.

Piscis Austrinus (Southern Fish). A small constellation in the southern hemisphere near Grus, the brightest stars being of 4th magnitude apart from the 1st-magnitude *Formalhaut. Abbrev.: PsA; genitive form: Piscis Austrini; approx. position: RA 22.5h, dec -30°; area: 245 sq deg.

Piscis Volans. *See* Volans.

plages (formerly sometimes termed **bright flocculi**). Bright patches in the solar *chromosphere, at a higher temperature than their surroundings, that occur in areas where there is an enhancement of the relatively weak vertical magnetic field. They are approximately coincident with photospheric *faculae (though unlike the faculae they are visible, in the monochromatic light of certain strong *Fraunhofer lines, anywhere on the sun's disc) and may be considered as defining the extent of *active regions.

Planck constant. Symbol: h. A fundamental constant with the value $6.626\ 196 \times 10^{-34}$ joule seconds. According to quantum theory, *electromagnetic radiation has a dual nature. Although many phenomena, including reflection and refraction, can be explained in terms of the wavelike nature of radiation, radiation may also be considered to be composed of discrete packets of energy called *photons, so that it acts as a stream of particles. The particle-like and wavelike properties are related by *Planck's law*, in which the Planck constant is the constant of proportionality:
$$E = h\nu = hc/\lambda$$
E is the energy of the photons and ν and λ the frequency and wavelength of the wave; c is the speed of light in a vacuum. The Planck constant appears in most equations of quantum theory and quantum mechanics, including Planck's radiation law for *black bodies.

Planck's law. *See* Planck constant.

Planck's radiation law. *See* black body.

plane-polarized wave (linearly polarized wave). *See* polarization.

planet. A body that orbits the sun or another star and shines only by the light that it reflects. In the solar system there are nine known *major planets and numerous *minor planets, all of which probably share a common origin with the sun about 4600 million years ago (*see* solar system, origin). Planets of other stars are too faint to be seen from earth but very massive planets may be detected around the nearest stars by their perturbing effect on stellar *proper motions, as is the case with *Barnard's star.

planetarium. An optical instrument by means of which an artificial night sky can be projected on the interior of a fixed dome so that the positions of the sun, moon, planets, stars, etc., may be shown and their motions, real or apparent, demonstrated. The instruments were first built by the Carl Zeiss firm from 1913. The term also refers to the building housing such an instrument.

planetary aberration. The difference between the true and apparent positions of a planet, etc., arising from *light-time.

planetary alignment. The rough alignment of the outer planets, Jupiter to Neptune, that takes place approximately every 178 years and will next be most in evidence in the period 1981-82.

This planetary configuration can be used to direct spaceprobes on a relatively short duration trajectory, swinging past two or more of the planets. A probe is given sufficient energy (i.e. velocity) to reach Jupiter. Jupiter's strong gravitational field swings the passing probe around the planet, away from its original direction of motion. With the correct initial velocity and flight path, the probe will be swung through an angle to take it towards Saturn or towards Uranus. Saturn, Uranus, and Neptune have a similar but smaller action on a passing probe. The energy and time required for such a mission is low, because of the planets' proximity, thus reducing the cost of the programme.

This *slingshot action* of the giant planets is being used in the *Voyager project. Voyager 1 flew by Jupiter in March 1979 and was shot towards Saturn, where it will make a close encounter with the satellite Titan in Nov. 1980. The probe cannot then proceed to Uranus, for dynamical reasons. Voyager 2 will pass Jupiter in July 1979 at twice the distance of Voyager 1; if Voyager 1's Titan encounter is successful, Voyager 2 will be made to fly by Saturn (Aug. 1981) on a path that will take it to Uranus (in early 1986) and then to Neptune (1990). The slingshot action has been used on a more modest scale in the flights of *Pioneers 10 and 11 and *Mariner 10.

planetary nebula. An expanding and usually symmetrical cloud of gas that has been ejected from a dying star, possibly in the form of a *red giant. The ejection process is not yet fully understood but the gas is thought to have formed all or part of the outer layer of the star. The gas cloud is ionized by the compact hot burnt-out stellar core that remains in the centre of the cloud; the cloud is detected by virtue of the resulting light emission. Planetary nebulae are therefore a class of *emission nebulae. They are usually ring-shaped or sometimes hourglass-shaped and are generally less than 50 000 years old. The name refers to their resemblance to planetary discs rather than point-like stars under low magnification. They have a large size range: the smallest objects have a starlike appearance on photographs— and are thus called *stellar planetaries*—but can be identified by the characteristic spectral emission lines. Planetary nebulae occur in isolation and usually lie close to the galactic plane, concentrated towards the galactic centre; they are classified as disc population objects (*see* populations I and II). They are believed to be a normal evolutionary stage of single stars of about one to four solar masses.

Although planetary nebulae are less massive and more symmetrical than *HII (ionized hydrogen) regions, their optical spectra are similar. There are bright emission lines of oxygen, hydrogen, nitrogen, and other components, the characteristic green of the inner region being due to doubly ionized oxygen and the red of the outer periphery resulting from singly ionized nitrogen and from hydrogen alpha emission. About 1000 planetary nebulae are known, the *Ring nebula in Lyra being a typical example. *See also* nebula.

planetary precession. *See* precession.

planetary probe. An unmanned spacecraft designed and equipped to study conditions on or in the vicinity of the planets, their satellites, and other members of the solar system, and in the interplanetary medium itself, and to transmit the information back to earth. At the time of writing, probes have made successful visits to all the major planets of the solar system outwards as far as Jupiter. One probe, Pioneer 11, is on course for Saturn while two Voyager spacecraft are due to visit Saturn, and possibly Uranus, after passing Jupiter in 1979. *See* Mariner probes; Mars probes; Pioneer probes; Pioneer Venus probes; Venera probes; Viking probes; Voyager probes; Zond probes.

Most planetary and interplanetary probes have used solar panels to derive power for communications and internal operation. However, spacecraft that travel further from the sun than the orbit of Mars receive comparatively little solar energy and must rely on other power systems, such as the radioisotope thermoelectric generators carried by Pioneers 10 and 11 and Voyagers 1 and 2. Most probes carry a main rocket propulsion system for course corrections as well as smaller thrusters for attitude adjustments. Orientation is often maintained by reference to a bright star, *Canopus being a frequent choice.

planetary system. A system of planets and other bodies, such as comets and meteoroids, that orbits a star. The sun and its planetary system together comprise the *solar system. Planetary systems may be common in the Galaxy, being formed along with stars from the gravitational contraction of interstellar clouds of gas and dust.

planetesimals. Bodies, ranging in size from less than a millimetre to many kilometres, that are thought to have formed the

planetesimal theory

planets of the *solar system by a process of *accretion.

planetesimal theory. See encounter theories.

planetoids. See minor planets.

planetology. The study of the planets of the solar system—their history, internal and surface structures, and atmospheres—with particular regard as to how and why these differ from planet to planet.

planets, origin. See solar system, origin.

planet X. Unofficial designation for a hypothetical planet orbiting the sun beyond the orbit of Pluto. Various suggestions for its mass and distance have been made, most of them being based on analyses of apparent perturbations by it on the orbits of planets and comets. It now appears that the planets are not detectably perturbed by an unknown body; planet X, if it exists, must therefore be small and/or very remote. However, analysis of the orbits of very long period comets suggests that there may be two earth-mass planets at distances of about 53 and 83 astronomical units.

planisphere. A two-dimensional map projection centred on the northern or southern pole of the celestial sphere that shows the principal stars of the constellations, the Milky Way, etc., and is equipped with a movable overlay to indicate the stars visible at a particular time on any day of the year for a particular zone of terrestrial latitude.

planoconcave, planoconvex. See lens.

plasma. A state of matter consisting of ions and electrons moving freely. A plasma is thus an ionized gas that can be formed at high temperatures (as in stars) or by *photoionization (as in interstellar gas). The properties of plasmas differ from those of neutral gases because of the effects of magnetic and electric fields on the charged species and because of interactions between the species. See also magnetohydrodynamics.

plate tectonics. See earth.

Platonic year. See precession.

Pleiades (Seven Sisters; M45; NGC 1432). A young *open cluster about 120 parsecs away in the constellation Taurus. It contains maybe 3000 stars of which six are easily visible to the naked eye: *Alcyone, Maia, Atlas, Electra, Merope, and Taygete; *Pleione could once have been brighter. The brightest stars are blue-white (B or Be) and highly luminous; the less bright ones being mainly A and F stars. Gas and dust surrounding the brightest stars reflect the starlight thus producing faint *reflection nebulae.

The Plough

Pleione (28 Tau). One of the brighter stars of the *Pleiades. It developed a surrounding envelope that was first observed in 1938, reached maximum intensity in 1945, and was just visible in 1954; another shell started to develop in 1972. It is thought to be unstable owing to its fast rotation rate. Spectral type: B8pe.

Plough (*US* **Big Dipper**). A group of stars in *Ursa Major that contains the seven brightest stars in that constellation, nearly all with similar magnitudes (m_v), and thus has a very distinctive shape (see illustration). The seven stars are *Alioth (ε), *Dubhe (α), *Alcaid (η), *Mizar (ζ), Merak (β, m_v: 2.36), Phecda (γ, m_v: 2.43), and Megrez (δ, m_v: 3.31), in order of brightness. Alcaid and Dubhe have different values and directions of *proper motion than the other five so that the shape of the Plough is slowly but continuously changing.

P-L relation. *See* periodluminosity relation.

Pluto. The ninth and smallest major planet, discovered in 1930 by Clyde Tombaugh at the Lowell Observatory in Arizona. Its 248 year orbit has a higher *inclination (17°) and a greater *eccentricity (0.25) than any other planet. Although it has the greatest mean planetary distance of about 39 astronomical units (AU) from the sun, it approaches within 30 AU at *perihelion; between 1979 and 1999 Pluto will therefore be closer to the sun than Neptune. Orbital and physical characteristics are given in Table 1, backmatter.

*Oppositions occur one or two days later each year, but Pluto's 14th magnitude disc is too small to be measured using earth-based telescopes. Until 1978 the best estimate of its size was about 5900 km, a value based on negative observations of a predicted occultation of a star by Pluto during an *appulse in 1965. However, examination of photographs of Pluto in 1978 revealed that it has a very close satellite, provisionally called *Charon*, orbiting at a distance of about 20 000 km every 6.4 days—a period previously identified in the small light variation of the planet and ascribed to its axial rotation. The orbital parameters of Charon enable *Kepler's third law to be used to estimate Pluto's mass as about 0.2% that of the earth. Apparently Pluto is a lowdensity body about 2500 km across, while Charon, about ten times fainter, may be 800 km in diameter; it may thus be the largest satellite in proportion to the size of its planet in the solar system, a title previously credited to our own moon. Because of their proximity, both Pluto and Charon are probably in *synchronous rotation about their common centre of mass.

At a temperature of -230°C Pluto is too cold and small to have an appreciable atmosphere. Infrared spectroscopy suggests that part of the surface, at least, has a covering of high-albedo methane ice; it is possible that the whole globe may be a low-density ice-ball comparable in size and composition to some of the satellites of the *giant planets. Indeed it is thought possible that Pluto was once a satellite of Neptune but escaped after a gravitational interaction with *Triton, forcing the latter into its present unusual retrograde orbit.

Pogson ratio. *See* magnitude.

Pointers. The two stars in the *Plough that point toward *Polaris and the north pole: Dubhe and Merak.

point source. A celestial object whose angular extent cannot be measured. *See also* radio source structure.

polar axis. 1. The axis in an *equatorial mounting that is parallel to the earth's axis and hence points towards the celestial poles.
2. *See* poles.

polar caps. White deposits of ices in the vicinity of the rotational poles of earth and Mars. Those of earth consist of water ice and water, while the caps of Mars comprise both water ice and frozen carbon dioxide. The caps of both planets show seasonal variations. *See also* Mars, polar caps.

polar diagram. *Another name for* antenna pattern. *See* antenna.

polar distance. *Short for* north polar distance.

Polaris (North Star; α UMi). A remote creamy-yellow supergiant that is the brightest star in the constellation Ursa Minor. It lies about one degree from the north celestial pole and is the present *pole star. Its position is found by means of the *Pointers in the Plough. Polaris is a *Cepheid variable (period 3.97 days). It is also the primary (A) of a multiple star system: it is a spectroscopic binary (period 29.6 years) and also a visual binary with a 9th-magnitude F3 V companion (B). There are two other fainter components (C and D) of the system. m_v: 2.02; M_V: -4.5; spectral type: F8 Ib; distance: 200 pc; position (1950): RA 1h 57m, dec 89°0′.

polarization. The degree to which the direction of the electric or magnetic vector in an *electromagnetic wave (or similar quantity in another transverse wave) changes in a regular fashion. The vector thus shows a less random orientation in the plane perpendicular to the direction of wave motion than occurs in an unpolarized (normal) wave. Waves in which the electric vectors are entirely vertical or horizontal with respect to the direction of motion are described as vertically or horizontally *plane* (or *linearly*) *polarized*. In general, both polarizations are present and the wave is then *elliptically polarized* in the right-handed or left-handed sense accordingly as the resultant vector rotates clockwise or anticlockwise when viewed along the direction of motion of the wave.

Radio emissions from celestial sources are usually partially polarized, i.e. the waves can be considered to be composed of a completely unpolarized component plus a small polarized component. *Synchrotron emission, however, is strongly polarized. The general situation can be described by the four *Stokes' parameters (I, Q, U, and V)*, which are defined in such a manner that specifying their four values uniquely describes the state of polarization. *See also* interstellar polarization.

polarization curve. A curve showing the relationship

Motion of north celestial pole

between the angle of reflection of light from a surface and the degree of polarization of the reflected light. It depends on the surface characteristics (texture, composition). It is used to deduce the *albedo of the surfaces of minor planets so that their diameters may be calculated.

polar orbit. A satellite orbit that passes over the north and south poles of the earth.

poles. The two points on a sphere lying 90° above or below all points on a particular great circle. The *polar axis* is the imaginary straight line between the poles. *See* celestial poles; galactic poles.

poles of the ecliptic. *See* ecliptic.

pole star. The star closest to the north or south celestial pole at any particular time. As a result of *precession the position of each pole traces out a circle in the sky in a period of about 25 800 years. There is thus a sequence of stars, including *Polaris, *Vega, and *Thuban, that have been and will become again the northern pole star (see illustration). The altitude of the pole star is approximately equal to the local latitude.

Pollux (β Gem). An orange giant that is the brightest star in the constellation Gemin and is the nearest giant to the earth. It has several optical companions none of which are physically related to it. m_V: 1.15; M_V: 1.0; spectral type: K0 III; distance: 10.7 pc.

Poop. Common name for *Puppis.

populations I and II. The two classes into which stars and other celestial objects can be divided according to age, space distribution, and metal content. The classification was first made by W. Baade in 1944. He proposed that population I stars are the young metal-rich highly luminous stars found in spiral arms of galaxies—i.e. strongly concentrated in the galactic plane—and associated with interstellar gas and dust. In contrast population II stars are old metal-deficient stars located in the galactic centre, in the surrounding galactic halo, and in a disc coextending with the galactic plane but much thicker. It is now known that there is a continuum of populations existing between Baade's two classes (see table). The origin of population types is connected with the dynamic and chemical evolution of the Galaxy. Put very simply the earliest stars, formed from the original hydrogen-helium gas cloud, were *halo population II stars* followed, after futher gas contraction produced a relatively thin extended disc, by *intermediate population II stars;* contraction of the remaining gas into an even thinner disc produced the *disc population,* which may be young population II stars or old population I stars. Successive generations of population I stars have subsequently formed from interstellar matter that has been slowly enriched with metals and other elements produced in *supernovae explosions. The youngest *extreme population I stars* contain 20 to 30 times the metal abundance of the oldest halo objects.

pores. *See* sunspots.

Porrima (γ Vir). A visual binary with cream components that lies in the constellation Virgo and has a highly eccentric orbit (period 171 years). The separation, about 5″ in the 1970s and 6″.6 at maximum, is decreasing and by 2010 the star will appear single except in the largest telescopes. Porrima is also an optical double. m_V: 3.6 (A and B), 2.75 (AB); spectral type: F0 V (A and B).

position. The location of an object on the celestial sphere, i.e. its observed direction in the sky. It is given in terms of two coordinates, usually *right ascension and *declination. *See also* true position.

positional astronomy. *See* astrometry.

position angle. 1. The direction in the sky of one celestial body with respect to another, measured from 0° to 360° in an easterly direction from north. It is used, for example, to give

CHARACTERISTICS OF POPULATIONS I AND II

population group	extreme population I	older population I	disc population	intermediate population II	halo population II
typical objects	interstellar gas, dust HII regions T Tauri stars O associations O, B stars supergiants young open clusters classical Cepheids	sun strong-line stars A stars Me dwarfs giants older open clusters	weak-line stars planetary nebulae galactic bulge novae RR Lyrae stars (P<0.4 days)	high-velocity stars long-period variables	globular clusters extreme metal-poor stars (subdwarfs) RR Lyrae stars (P>0.4 days) Type II Cepheids
distribution	spiral arms; extremely patchy	patchy	smooth	smooth	smooth
*z, Z, age			increasing →		
concentration to Galactic centre	none	little	considerable	strong	strong
Galactic orbits	circular	approx. circular	slightly eccentric	eccentric	very eccentric
brightness, metals/H ratio			increasing →		

Adapted from A. Blaauw
*z, Z are mean distance from Galactic plane, mean of velocity components perpendicular to Galactic plane

the position of the fainter component of a *visual binary with respect to the brighter component.

2. The angle at which the axis of a planet or star, or some other line on a celestial body, is inclined to the *hour circle passing through the centre of the body; it is measured eastwards from 0° to 360° from north.

position circle. *See* circle of position.

position micrometer. *See* filar micrometer.

positive lens. A lens or lens system with a converging action on a light beam.

positron. *See* electron.

potential energy. The energy possessed by a body or system by virtue of its position or configuration. It is equal to the work done by the system changing from its given state to some standard state. In a gravitational field, a mass m placed at a height h above a standard level (say the surface of the earth) has potential energy mgh, where g is the acceleration of gravity.

power. The rate at which *energy is expended by a source or work is done. It is measured in watts, i.e. in joules per second.

power, optical. The reciprocal of the focal length in metres of a lens or spherical mirror. It gives a measure of the relative focusing or defocusing powers of optical elements. A converging lens or mirror has a positive power; one that is diverging in action has a negative power. The power of two thin lenses in contact is equal to the sum of the separate powers. For a thick lens, account must be taken of refractive index.

power pattern. *See* antenna.

power spectrum. A plot of the power contained in a signal versus the frequency, or of the power in a spatial distribution versus the wavenumber (reciprocal of wavelength). It may be obtained by taking a *Fourier transform of the signal or distribution and squaring the amplitude of each component in the result.

Poynting-Robertson effect. An effect whereby small interplanetary dust particles (radius 10^{-6} to 10^{-2} metres) slowly spiral into the sun. The particles interact with solar radiation (by absorbing and re-radiating energy) so that they lose orbital momentum; this causes a decrease in orbital velocity so that they move into progressively smaller orbits around the sun. For a particle moving in a circular orbit that is d astronomical units from the sun, the time taken to fall into the sun is given by

$$t = 7 \times 10^6 \, \varrho r d^2 \text{ years}$$

where r and ϱ are particle radius (in cm) and density (in g cm^{-3}), respectively.

*Radiation pressure on the particles opposes the Poynting-Robertson effect and can overcome it for very small particles (radius less than 10^{-6} m). These particles are blown out of the solar system by the action of the radiation pressure aided by the *solar wind. Although these two processes remove dust from the solar system, the dust level is constantly replenished by disintegrating comets, etc.

p-p chain. *See* proton-proton chain reaction.

p-process. A process of *nucleosynthesis by which certain heavy proton-rich nuclei are thought to be produced.

Praesepe (Beehive; M44; NGC 2632). A large *open cluster in the constellation Cancer that contains over 100 stars. It can almost be resolved by the naked eye but is best viewed with a very low-power telescope or with binoculars.

preamplifier. An amplifier used in a *radio telescope to boost the signals from the *antenna before they are fed to the *receiver.

preceding spot (*p*-spot). *See* sunspots.

precession. The slow periodic change in the direction of the axis of rotation of a spinning body due to the application of an external force (a torque). The extremities of the axis trace out circles in one complete precessional period if the rotational speed and applied torque are both constant. The earth's axis precesses with a period of 25 800 years, the axial extremities, the celestial poles, tracing out circles about 23.5° in radius on the celestial sphere (see illustration). This radius is equal to the inclination of the earth's equatorial plane to the plane of the ecliptic. The precession results primarily from the gravitational attraction of the sun and moon on the *equatorial bulge of the nonspherical earth. The sun and moon and also the planets pull on the bulge in an attempt to bring the equatorial plane into near coincidence with the plane of the ecliptic. The result of this gravitational action and the opposing rotation of the earth is a precessional force that acts perpendicularly to both these actions.

The combined effect of the sun and moon leads to *lunisolar precession,* which together with the much smaller *planetary precession* gives the *general precession* of the earth's axis and of the equinoxes. The general precession is about 50.27 arc seconds per year. Thus one precessional circle is completed about every 25 800 years, this period being the *Platonic year.* The precession of the earth's axis leads not only to the *precession of the equinoxes and the consequential change in stellar coordinates but also to the changing position of the celes-

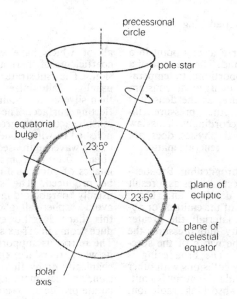

Precession of the earth

tial poles and hence to the slow change in the two *pole stars.

precession, constant of. The ratio of the luni-solar *precession to the cosine of the *obliquity of the ecliptic. It is about 54.94 seconds of arc per year and varies very slightly with time.

precession of the equinoxes. The slow continuous westward motion of the *equinoxes around the *ecliptic that results from the *precession of the earth's axis. It was first described by Hipparchus in the second century BC. As the axis precesses, the celestial equator, which lies in the plane perpendicular to the axis, moves relative to the ecliptic. The two points at which the planes intersect (the equinoxes) thus move round the ecliptic in the precessional period of about 25 800 years, the annual rate being about 50.27 seconds of arc. Since the *equatorial and *ecliptic coordinate systems use the vernal *equinox as their zero point, stellar coordinates in these systems change with time. Star catalogues and charts should therefore list stellar positions for a given *epoch (1900, 1950, 2000, etc.).

precision catalogue. *See* star catalogue.

pressure. The force per unit

pressure broadening

surface area at any point in a gas or liquid. The pressure of a gas is proportional to temperature and density: at constant temperature, as the density is increased the pressure increases accordingly. This law of classical physics does not apply to *degenerate matter.

pressure broadening. Broadening of spectral lines as a result of high density, and hence pressure, of the emitting or absorbing material: the greater the density and pressure, the greater the width of the spectral lines. The broadening is caused by collisions with other atoms while the atom is emitting or absorbing radiation. *See also* line broadening.

primary. The celestial body that is nearest to the centre of mass of a system of orbiting bodies. The other members, called *secondaries,* appear to orbit the primary, which is the most massive in the system. In fact all members move round the common centre of mass. The earth is the moon's primary. *See also* visual binary.

primary cosmic rays (primary radiation). *See* cosmic rays.

primary mirror. The optically worked mirror in a *reflecting telescope that faces the observed object. The focal plane in which its image forms is termed the *prime focus.* Modern mirror-substrate materials are glass/ceramic compounds or glass-like materials, such as fused quartz or Pyrex, all of which have very low coefficients of thermal expansion. The substrate is now usually *aluminized rather than silvered to obtain the reflecting surface. The surface must be accurately ground and polished to within one quarter of the wavelength used, i.e. to about 0.1 micrometres, if images of good quality are to be produced. The shape is usually *paraboloid in form to avoid *spherical aberration; this shape does however produce *coma in off-axis images. The mirror is supported over its whole rear face, either mechanically or by fluid suspension, and can maintain an accurate profile up to considerable diameters without sagging. *See also* cell; mounting.

prime focus. *See* primary mirror; objective.

prime meridian. The arbitrary circle of zero longitude on the earth or some other planet or mapped celestial body. The earth's prime meridian is the *Greenwich meridian. See also* Greenwich Mean Time.

primeval atom. The earliest phases of the universe. *See* big bang theory.

primeval nebula. The cloud of interstellar gas—mainly hydrogen and helium—and dust grains from which the solar system developed. *See* nebular hypothesis; solar system, origin.

prime vertical. The *vertical

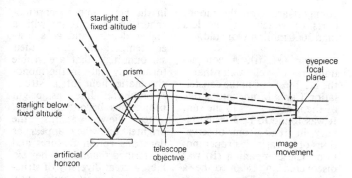

Optical path of prismatic astrolabe

circle that passes through an observer's zenith at right angles to the *celestial meridian and intersects the horizon at the east and west points (*see* cardinal points).

primordial fireball (ylem). The radiation-dominated phase of the universe. *See* big bang theory.

prismatic astrolabe. An astrometric instrument for determining the position of a celestial body by the accurate timing of its passage across a *vertical circle of the celestial sphere. It usually consists of an artificial horizon (such as a mercury surface) and a prism, which are placed in front of a telescope. The prism is fixed in position so that a fixed altitude, usually 45° or 60°, is used. Parallel rays of light falling on the artificial horizon and the prism are each reflected by the prism into the telescope (see illustration). As the celestial body moves towards the fixed altitude, the two images approach each other from top and bottom of the field of view. At the fixed altitude the two images lie close together on the same horizontal line through the centre of the field. The time is then recorded, usually electronically.

prism transit. *Another name for* broken-back transit. *See* meridian circle.

probe (spaceprobe). *See* planetary probe.

Procellarum, Oceanus (Ocean of Storms). A vast irregular *mare—possibly an impact depression, the *Gargantuan Basin*—that covers a major part of the moon's western *nearside. It was the landing site for *Lunas 9 and 13, *Surveyors 1 and 3, and *Apollo 12. It is one of the youngest areas on the moon, parts of which may be less than 3000 million years old.

Procyon (αCMi). A conspicuous creamy-yellow star that is the brightest one in the constellation Canis Minor and is one of the nearest stars to the sun. It was found to be a visual binary in 1840 from observations of its variable proper motion. The companion (B) was discovered in 1896 to be a *white dwarf. The period is 40.6 years, the separation 4".6. There is a third 12th-magnitude component (C) at 81". Procyon is also a spectroscopic binary, period 40.23 years. m_V: 0.34; M_V: 2.7; spectral type: F5 IV-V (A), wdF8 (B); mass: 1.74 (A), 0.63 (B) times solar mass; radius: 1.7 (A), 0.01 (B) times solar radius; distance: 3.5 pc.

Prognoz. A series of USSR unmanned observatories, first orbited in 1972. Prognoz-6 was launched in Sept. 1977 into a highly elliptical earth orbit. They have investigated the effect of solar activity on the interplanetary medium and the earth's magnetosphere by measuring solar radiation (both electromagnetic and corpuscular) and magnetic fields in near-earth space. They have also studied galactic ultraviolet radiation, x-rays, and gamma rays.

Progress craft. *See* Salyut.

prominences. Clouds of gas in the sun's upper *chromosphere/inner *corona, with a higher density and at a lower temperature than their surroundings, that are visible (ordinarily only in the monochromatic light of a strong *Fraunhofer line) as bright projections beyond the limb. When viewed against the brighter disc they appear as dark absorption features and are termed *filaments*. They exhibit a great diversity of structure and are most conveniently classified according to their behaviour, as either *quiescent* or *active*.

Quiescent prominences are particularly long-lived and are among the most stable of all solar features. They may persist for several months before breaking up or, less frequently, blowing up and have been known to reform at the same location with an almost identical configuration. Typically they are a couple of hundred thousand kilometres long, several tens of thousands of kilometres high, and several thousand kilometres thick. They occur in high *heliographic latitudes, where they are supported by the horizontal magnetic field separating the polar field from the adjacent fields of opposite polarity that were formerly associated with the *f*-spots of active regions (*see* sunspot cycle). They attain their greatest frequency a few years after the minimum of the sunspot cycle, when their average latitude is around ±50°. Thereafter they appear in increasingly high latitudes, reaching the polar regions shortly after sunspot maximum. Then, after a brief discontinuity, they reappear around latitude ±50° and remain there in small numbers until a few years after the next minimum, when they again progress poleward.

Active prominences are relatively short-lived and may alter their structure appreciably over a matter of minutes. There are many characteristic types, for example *surges and *sprays, in which chromospheric material is ejected into the inner corona, and *loop prominences* and *coronal rain*, in which the reverse occurs. Loop prominences are impulsive events that often accompany *flares, while coronal rain represents the return of flare-ejected material. In developing active regions *arch filaments* are usually present. These tend to connect regions of opposite polarity across the line of inversion and gradually ascend while material descends along both sides of the arch.

Intermediate between quiescent and active prominences are the so-called *active-region filaments*. These invariably lie along the line of inversion between the vertical magnetic fields of opposite polarity in well-developed active regions. Though long-lived, they may be distinguished from quiescent filaments by an almost continuous flow of material along their axis.

With the exception of surges and sprays, prominences may be regarded as an efficient (though limited) heat sink for the highly ionized corona, their material condensing out of it and then descending along the predominantly vertical magnetic field lines.

proper motion. Symbol: μ. The apparent angular motion per year of a star on the celestial sphere, i.e. in a direction perpendicular to the line of sight. It results both from the actual movement of the star in space—its *peculiar motion—and from the star's motion relative to the solar system. It was first detected in 1718 by Halley. The considerable distances of stars reduce their apparent motion to a very small amount, which for most stars is considered negligible. It is only after thousands of years that differences in the directions of proper motion cause groups of stars to change shape appreciably.

The proper motion of a star, usually a nearby star, can be determined if its cumulative effect over many decades produces a measurable change in the star's position. It is quoted in seconds of arc per year, often in terms of two components: proper motion in right ascension (μ_a) and in declination (μ_d). *Barnard's star has the greatest proper motion (10″.3 per year). If the distance, d (in parsecs), of the star is known then the velocity along the direction of proper

motion (the *tangential velocity*, v_t) can be found:
$v_t = \mu d$ AU/year
$= 4.74\mu d$ km s^{-1}

proportional counter. A gas-filled device for the detection of x-rays and other ionizing radiation and the counting of charged particles. It is similar to but operates at a lower voltage than the *Geiger counter. Radiation entering the tube ionizes the gas along its path. The size of the output pulse of the counter is proportional to the number of ion pairs produced by the ionizing event and thus to the energy of the radiation. It is possible to differentiate between various forms of ionizing radiation.

proton. Symbol: p. A stable *elementary particle (a hadron) that has a positive *charge, equal in magnitude to that of the *electron. It forms the nucleus of the hydrogen atom and is present in differing numbers in all nuclei. The rest mass of the proton is 1.6726×10^{-24} gram, approximately 1836 times that of the electron. It has spin ½. The antiparticle of the proton is the *antiproton*. *See also* hydrogen; proton-proton chain reaction.

proton-proton chain reaction (p-p chain). A series of *nuclear fusion reactions by which energy can be generated in stars. The overall effect of the chain reaction is the conversion of hydrogen nuclei to helium nuclei with the release of an immense amount of energy. The reactions require a temperature of 10^7 kelvin and are therefore believed to be the major source of energy in the sun and in all main-sequence stars that are cooler than the sun. The *carbon cycle predominates in hot stars.

The sequence of steps of the

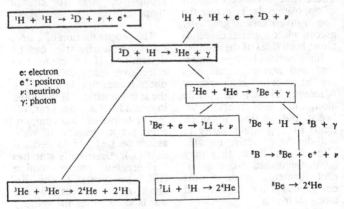

Proton-proton chain reaction

p-p chain is shown in the illustration, with the principal sequence outlined in black. All three possible chains occur simultaneously but in what is thought to be 99.75% of the reactions, the sequence begins with two hydrogen nuclei (i.e. protons) combining to emit a positron and a neutrino and forming a nucleus of deuterium. The deuterium then combines with a proton to yield a nucleus of helium-3 and a photon. About 86% of the helium-3 nuclei will combine with another helium-3 to give helium-4, liberating two protons. *See also* neutrino detection experiments.

protoplanet. An evolving planet in the process of *accretion, together with any of its satellites forming by accretion at the same time in the vicinity of the planet.

protostar. A stage in the evolution of a young star after it has fragmented from a gas cloud but before it has collapsed sufficiently for nuclear reactions to begin. This phase may take from 10^5 to 10^7 years, depending on the mass of the star. *See also* Bok globules; Hayashi line; Herbig-Haro objects; stellar birth; T Tauri stars.

Proxima Centauri. The nearest known star to the sun. It is a *flare star that lies in the constellation Centaurus and is a component of the triple star *Alpha Centauri. It is an intense source of low-energy x-ays and high-energy ultraviolet, i.e. XUV wavelengths. m_v: 10.7; M_v: 15.1; spectral type: M5eV; mass 0.1 times solar mass; radius: 0·093 times solar radius; distance: 1.31 pc (4.26 ly); position (1975): RA 14h 28m, dec -62° 34'.

p-spot. *Short for* preceding spot. *See* sunspots.

Ptolemaic system. A model that attempted to explain the observed motions of the sun, moon, and planets and predict their future positions. Originally proposed by Apollonius of Perga in the third century BC and developed by Hipparchus, it was completed by Claudius Ptolemaeus of Alexandria in the second century AD. Ptolemy assumed a *geocentric system, possibly for ease of calculation, in which the moon, Mercury, Venus, the sun, Mars, Jupiter, and Saturn moved around the earth in circular orbits of increasing radius. The position of the earth, E, was somewhat offset from the centre of the orbit, C (see illustration). The orbiting body moved at a uniform rate about a point A rather than about C. A, C, and E were colinear and AC = CE. Although this improved the accuracy of earlier theories, Ptolemy also had to allow the moon or planet to move in a small circle (an *epicycle*) the centre of which moved around the circumference of a larger circle (the *deferent*), which was the body's orbit. By a suitable choice of the relative size of epicycle to

Ptolemaic system

deferent, the relative rates of motion in these circles, and the distance CE, Ptolemy was able to predict planetary positions to within about 1°. Comparison between prediction and observation were much simpler with a geocentric rather than a heliocentric system. It is not known however whether Ptolemy believed in an actual geocentric solar system, as did his later followers. The Ptolemaic system, with minor modifications, was in use until the *Copernican system was finally accepted.

Pulkovo Astronomical Observatory. An astronomical observatory of the Academy of Sciences of the USSR, built near Leningrad and opened in 1839. It was completely destroyed in World War II, after which it was rebuilt. It reopened in 1954 with the surviving instruments repaired and modernized and some new instruments.

pulsar. A very regularly 'pulsing' source of radiation, which almost certainly originates from a rotating *neutron star. Pulsars were originally discovered at radio wavelengths, and two of these have now been detected at optical and gamma-ray wavelengths; x-ray astronomy has turned up the related class of x-ray pulsars.

The first radio pulsar, PSR 1919+21, was detected by Antony Hewish and Jocelyn Bell (now Burnell) during a pioneering study of interplanetary scintillation (*see* scattering). Over 300 radio pulsars are now known, of which 155 were discovered in 1977 in a pulsar survey made at Molonglo Radio Observatory. The periods range from 0.033 seconds (the Crab pulsar) to 4 seconds and can be measured to accuracies of, typically, one part in 10^{10}. The width of the pulse is usually a few per cent of the pulsar's period but can reach 50%. Almost every known pulsar is a single star: even the one *binary pulsar owes none of its emission characteristics to its companion.

The total number of pulsars in the Galaxy, based on the Australian pulsar survey, is now estimated at over one million. Their rate of formation is now calculated as maybe 25 per century. Pulsars, together with a smaller number of black holes, are currently thought to originate in *supernova explosions. Astronomers are finding difficulty however in reconciling the rate of formation of pulsars with the estimated supernova production rate in the

Galaxy of only two to three per century; it is possible that theories on pulsar origin may have to be revised.

The received pulses occur when a beam of radio waves, emitted by a rotating neutron star, sweeps past the earth in an identical manner to the flashes produced by a lighthouse lamp. This beam of radiation is comprised of *synchrotron emission. It arises from electrons moving in the neutron star's strong magnetic field (about 10^8 tesla), whose axis lies along a different direction to the pulsar's rotation axis. There is still dispute over the emission site of the beam: it may be near the star, at the magnetic poles, so that radio waves are beamed down the magnetic axis; alternatively it may be further out, near the speed of light cylinder, where the magnetic field is rotating at almost the speed of light.

Despite their impressive regularity, the periods of all radio pulsars are very gradually increasing as the neutron star loses rotational energy. The central star of the *Crab nebula—the fastest and youngest pulsar—is slowing at the rate of one part in a million per day. The slow-down rate of both this pulsar and the young *Vela pulsar are occasionally interrupted by *glitches* (or *spin-ups*): temporary changes in rotation rate. These are thought to be due to rearrangements of the crust or core of the neutron star.

X-ray pulsars occur in close binary systems when gas from a companion is channelled on to the neutron star at its magnetic poles: the x-rays appear to pulse on and off as these hot gas patches are exposed by the star's rotation. The gas flow affects the neutron star's spin and as a result all x-ray pulsars are gradually speeding up. Although some have typical pulsar periods (0.7 to 8 seconds) there are many 'slow' x-ray pulsars (with periods from 100 seconds to 14 minutes), whose rotation must somehow have been strongly 'braked'.

pulsating universe. *See* oscillating universe.

pulsating variables. *Variable stars that periodically brighten and fade as their outer layers go through a cycle of expansion and contraction, i.e. the star pulsates. The period of light variation is equal to the period of pulsation and is normally approximately constant. The spectrum of the star also changes periodically due to changes in surface temperature. The pulsation cycle is demonstrated by variations in *radial velocity. All these variations are shown in the diagram at *Cepheid variables. The pulsation period, P, was shown by Eddington to be related to mean stellar density, ϱ, by the *period-density relation*, and hence to the star's radius, R, and mass, M:

$$P = 1/\sqrt{\varrho} = \sqrt{(R^3/M)}$$

Since *luminosity is also proportional to radius, pulsa-

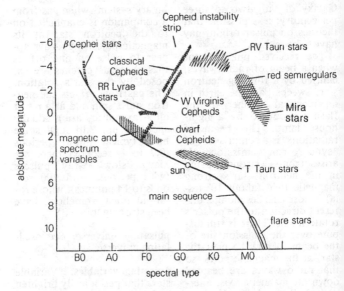

Hertzsprung-Russell diagram of pulsating and other variable stars

tion period should be related to luminosity (*see* period-luminosity relation). Types of pulsating variable include the *Cepheid variables (classical and W Virginis stars), *RR Lyrae stars, *dwarf Cepheids, *Delta Scuti stars, long-period *Mira stars, *semiregular variables (including *RV Tauri stars), *irregular variables, and *Beta Cephei stars.

A star pulsates because there is a small imbalance between gravitational force and outward directed pressure so that it is not in hydrostatic equilibrium. When it pulsates it expands past its equilibrium size until the expansion is slowed and reversed by gravity. It then overshoots its equilibrium size again until the contraction is slowed and reversed by the increased gas pressure within the star. The fact that the pulsations do not die away as energy is dissipated means that some process is continuously converting heat energy into mechanical energy to drive the pulsations. The centre of the star is not involved in the pulsation. For most pulsating variables (probably excluding Beta Cephei stars) the driving force is a valve mechanism in the form of a region of changing *opacity near the stellar surface where the pulsation amplitude is greatest. This region is an *ionization zone* in which atoms, mainly helium or hydrogen, are partially ionized. Normally atoms become transparent, letting heat escape, as they are compressed. In an ionization zone the opacity increases with compression and the zone effectively acts as a heat engine: as long as it lies at the required depth below the surface it can drive the pulsations. If the zones are too close to the surface, as in a very hot star, or if too deep, then pulsations cannot occur. The period of pulsation is complicated by the fact that the gas is thought to be able to undergo oscillations (as in an open musical pipe) either in a fundamental mode or as a harmonic of the fundamental.

The distribution of pulsating variables on the *Hertzsprung-Russell diagram differs from that of normal stars. Most lie on a nearly vertical band, the *Cepheid instability strip*, that extends upwards from the main sequence (and possibly below it) and merges into a broader instability region at top right (see illustration). Stars reach these instability regions by various evolutionary paths and may turn up there several times during their evolution. The existence of such instability regions indicates that with certain combinations of stellar luminosity and effective temperature a state of pulsation rather than of rest is favoured.

Pump. Common name for *Antlia.

Puppis (Poop; Stern). A constellation in the southern hemisphere near Canis Major, lying partly in the Milky Way. It was once part of the constellation *Argo. The brightest star is the extremely remote and luminous 2nd-magnitude Naos (Zeta). It also contains several open clusters including the just-visible M47, the bright planetary nebula NGC 2438, and the supernova remnant *Puppis A,* which is a strong radio and x-ray source. Abbrev.: Pup; genitive form: Puppis; approx. position: RA 7.5h, dec -30°; area: 673 sq deg.

Puppis A. *See* Puppis.

Pyxis (Mariner's Compass). A small inconspicuous constellation in the southern hemisphere near Vela, lying partly in the Milky Way. It once formed part of the Mast of Argo, an obsolete division of *Argo. It has one 3rd-magnitude and two 4th-magnitude stars and contains the peculiar *recurrent nova T Pyxidis. Abbrev.: Pyx; genitive form: Pyxidis; approx. position: RA 9h, dec -30°; area: 221 sq deg.

PZT. *Abbrev. for* photographic zenith tube.

Q

QSO. *Abbrev. for* quasi-stellar object. *See* quasar.

QSS. *Abbrev. for* quasi-stellar radio source. *See* quasar.

quadrant. An instrument dating back to antiquity and used for measuring altitudes. It remained the most important astronomical instrument until the telescope was invented. It consisted of a 90° graduated arc (a quarter circle) with a swivelling arm to which a sighting mechanism was attached. In the *mural quadrant* the graduated arc, often very large, was attached to a wall and was orientated along the observer's meridian. The mural quadrant was therefore the forerunner of the *meridian circle.

Quadrantids. A major *meteor shower, radiant: RA 232°, dec +50°, that maximizes on Jan. 3 when the *sun's longitude is 282·8°. The *zenithal hourly rate (ZHR) at maximum is about 110 and is constant from year to year. The *meteor stream is narrow, the ZHR being greater than half maximum for only 17 hours: the *meteor stream's cross-sectional diameter about 1·7 million km compared with 20 million and 13 million km for those of the *Perseids and *Geminids. The stream has its ascending *node close to the orbit of Jupiter and there is a possibility that Quadrantid meteors could be seen from the surface of that planet too. Owing to gravitational and radiational perturbations the large 'visual' *meteoroids are in a slightly different orbit to the smaller 'radio' meteoroids. The shower is named after the obsolete constellation quadrans Murali found in early 19th-century star atlases. The radiant is actually in Boötes.

quadrature. The position, or aspect, of a body in the solar system when its *elongation from the sun, as measured from the earth, is 90°, i.e. when the angle sun-earth-planet is a right angle.

quantum. The minimum amount by which certain properties of a system, such as its energy or angular momentum, can change. The value of the property cannot therefore vary continuously but must change in steps: these steps are equal to or are integral multiples of the relevant quantum. This idea is the basis of *quantum theory and quantum mechanics. The *photon, for example, is a quantum of *electromagnetic radiation and the graviton is a proposed quantum of the gravitational field (*see* quantum gravitation).

quantum efficiency. A measure of the efficiency of conversion or utilization of light or other radiation. It is given by the proportion of incident *photons, of a particular frequency, that are converted by a device into a useful form, such as a permanent record. Photographic emulsions, at best, record only a few photons out of every thousand falling on them. Devices using a *photocathode combined with an electronic recording medium have a much higher efficiency.

quantum gravitation. A theory of gravitation in which the gravitational interactions of matter are described not in terms of the geometry of space-time, as in the general theory of *relativity, but by the exchange of *elementary particles called *gravitons*. The graviton, which is as yet only hypothetical, may be considered the *quantum of the gravitational field. It has zero charge, zero rest mass, and a spin of 2. The graviton is analogous to the *photon, which is the quantum of the electromagnetic field.

quantum theory. The theory put forward by the German physicist Max Planck in 1900. Classical physics regarded all changes in the physical properties of a system to be continuous. By departing from this viewpoint and allowing physical quantities such as energy, angular momentum, and action to change only by discrete amounts, quantum theory is born. It grew out of Planck's attempts to explain the form of the curves of intensity against wavelength for a *black-body radiator. By assuming that energy could only be emitted and absorbed in discrete amounts, called *quanta, he was successful in describing the shape of the curves. Other early applications of quantum theory were Einstein's explanation of the *photoelectric effect and Bohr's theory of the atom (*see* hydrogen spectrum; energy level). The more precise mathematical theory that developed in the 1920s from quantum theory is called *quantum mechanics*. Relativistic quantum mechanics resulted from the extension of quantum mechanics to include the special theory of *relativity. *See also* quantum gravitation.

quark. One of (at present) four different hypothetical particles that might be the structural units from which *elementary particles (excluding leptons) are formed. In the usual scheme, quarks all have spin 1/2, charges that are multiples of one third, and other innate qualities (colour). Despite intensive searches, no quarks have yet been observed.

At present there are tentative suggestions that states of condensed matter, such as *neutron stars, may consist of a soup of free quarks.

quasar. A compact extragalactic object that looks like a point of light but emits more energy than a hundred supergiant galaxies. The name is a contraction of *quasi-stellar object* (*QSO*), and early papers on the subject sometimes refer to *quasi-stellar galaxies* (*QSG*). About 1% of quasars are also radio sources, referred to as *QSS* (*quasi-stellar radio sources*).

Quasars, of which over 600 are known, were discovered in 1963 as the optical counterparts to some powerful radio sources. The spectra of these 'stars' were peculiar, with bright emission lines, apparently of an unknown element, superimposed on a continuum. Maarten Schmidt finally recognized the pattern of lines in one of these objects (3C 273) as the Balmer series of hydrogen redshifted to $z = 0.158$ (*see* redshift), showing that it was impossible for this to be a star within the Galaxy. Other quasars were then discovered to have far greater redshifts. The record is currently held by OQ 172, whose redshift of 3.53 indicates that it is receding from us with over 90% of the speed of light. Astronomers today interpret the quasar redshifts as Doppler shifts (*see* Doppler effect) arising from the expansion of the universe, making them the most distant and hence the youngest extragalactic objects we observe.

Many quasar spectra also show absorption lines, which sometimes have a wide range of redshifts up to the value of the emission redshifts. Those close to the emission redshift presumably arise from clouds of matter ejected from the quasar, while lower-redshift systems may be due to galaxies lying in the line of sight.

To be visible at such great distances, quasars must be exceedingly luminous: many have absolute *magnitudes brighter than -27. Most of their energy is actually emitted in the infrared. There is however a great range in luminosity. In addition quasars themselves are often *variable by a factor of two or greater, on timescales of a few weeks to a year. This indicates that their light-producing regions cannot be more than a light-year across, which poses several problems in explaining their energy generation. The only process known to be efficient enough is some sort of accretion on to a supermassive *black hole (of the order of 10^9 solar masses) that is located at the nucleus of the quasar.

Some of the nearer quasars are surrounded by a 'fuzz', believed to be a surrounding galaxy too faint to be detected around the more distant objects. This gives support to the suggestion that quasars are the nuclei of *active galaxies, and are the most violent examples in a sequence that ranges through *BL Lacertae objects, *N galaxies, and *radio galaxies to *Seyfert galaxies and normal galaxies like our own.

quasi-stellar galaxy (QSG). *See* quasar.

quasi-stellar object (QSO). *See* quasar.

quasi-stellar radio source (QSS). *See* quasar.

quiescent prominence. *See* prominences.

quiet sun. The term applied to the sun around the minimum of the *sunspot cycle, when the absence of *active regions allows relatively 'quiet' conditions to prevail. The sun is not then free of all disturbance but there are few (if any) concentrations of intense magnetic flux to produce significant departures from a uniform distribution of radiative flux across the disc. *Compare* active sun.

R

RA. *Abbrev. for* right ascension.

radar astronomy. The study of celestial bodies (as yet) within the solar system by means of the faint reflections from them of powerful high-frequency radio transmissions aimed in their directions from the earth. The time interval between transmission of a signal and reception of the reflected signal is an accurate measure of intervening distance. The determination of planetary distances has led to a precise value for the *astronomical unit. The transmitted frequency may be chosen to penetrate the atmosphere of a planet in order to establish a profile of the surface, which is otherwise hidden from view. In addition, rotation periods can be determined from the Doppler shift produced in pulses bounced off different parts of an object. The rotation periods of Venus and Mercury were determined by radar measurements.

radar meteor. *Another name for* radio meteor. *See* meteor.

radial velocity (line-of-sight velocity). Symbol: v_r. The vel-

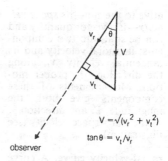

Space velocity and its components

ocity of a star along the line of sight of an observer. It is calculated directly from the Doppler shift (*see* Doppler effect) in the lines of the star's spectrum: if the star is receding there will be a *redshift in its spectral lines and the radial velocity will be positive; an approaching star will produce a *blueshift and the velocity will be negative. About 60% of stars have values of about ±20 km/s. Radial velocity is usually given in terms of the sun's position so that the earth's orbital motion has no effect.

The velocity and direction of motion of the sun relative to the *local standard of rest can be determined from a statistical analysis of the observed radial velocities of nearby stars: stars in the region of the solar *apex have predominantly positive velocities while those around the antapex have predominantly negative ones.

A star's velocity in space relative to the sun—its *space velocity*—is a vector quantity and can be split into two components: its radial velocity and its tangential velocity, v_t, along the direction of *proper motion. Measurements of these components give both the magnitude, V, and direction θ, of the star's space velocity (see illustration).

radial-velocity curve. A curve showing how the radial velocity of a *spectroscopic binary changes during one revolution.

radian. A unit of angle used in plane geometry that is the angle subtended at the centre of a circle by an arc of the circle equal in length to the circle's radius. Thus 2π radians equals 360°, 1 radian equals 57°.296, and 1° equals 0.017 radians. Solid angles, used in spherical geometry, are measured in steradians.

radiant. The point on the celestial sphere, usually given in the coordinates *right ascension and *declination, from which a *meteor appears to originate. Shower meteors all seem to radiate from this point and the *trains of a series of shower meteors, extrapolated backwards, intercept at the radiant. Its position is found by calculating the intercept with the celestial sphere of the vector sum of the meteor's velocity vector and the earth's velocity vector on collision. The radiant position changes slightly with time as the earth moves through the *meteor stream. Meteor showers are usually named after the constellation that contains the radiant.

radiant flux. The total power emitted or received by a body in the form of radiation. It is measured in watts.

radiation. *See* electromagnetic radiation; energy transport.

radiation belts. One or more doughnut-shaped regions in the *magnetosphere of a planet in which energetic electrons and the ionized nuclei of atoms, chiefly protons from hydrogen atoms, are trapped by the planet's magnetic field. The particles within the belt, whose axis is the magnetic axis of the planet, spiral along magnetic field lines, travelling backwards and forwards between reflections which occur as they approach the magnetic poles. This motion produces *synchrotron emission from the particles. The particles are either captured from the *solar wind or are formed by collisions between *cosmic rays and atoms or ions in the planet's outer atmosphere.

Only two planets, the earth and Jupiter, are known to possess radiation belts, although there is some evidence that Saturn also has them. The magnetic field of Mercury appears to be too weak to sustain belts. The radiation belts of Jupiter are many times more intense than the *Van Allen radiation belts of the earth and pose a threat to spacecraft systems in the vicinity of the planet: they would certainly be very hazardous to manned exploration of the *Galilean satellites of Jupiter.

radiation era. *See* big bang theory.

radiation pressure. The very small pressure exerted on a surface by light or other electromagnetic radiation due to the transfer of momentum from the *photons to the surface. Although radiation pressure has a negligible effect on large particles or on dense matter, the effect is considerable with very small particles, which are driven away from the radiation source. Thus solar radiation pressure, aided by the *solar wind, clears tiny particles of dust (<1 μm) from the solar system. *See also* Poynting-Robertson effect.

radiation resistance. A fictitious resistance in which the power radiated away from or collected by an *antenna is considered to be dissipated. For a perfect antenna made of superconducting material, the ohmic losses (power dissipation arising from resistance) in the elements themselves are zero; the resistive part of the impedance presented at the feed point is then equal to the radiation resistance. In any practical antenna the ohmic losses are kept as small as possible in comparison to the radiated power. Maximum power transfer between transmitter or *receiver and the antenna is effected when the *feeder has a characteristic impedance equal to the radiation resistance.

radiation temperature. *See* effective temperature.

radiative transport, zone. *See* energy transport.

radioactive dating. The dating of rocks (and also fossils and archaeological remains) by the accurate determination of the quantities of a long-lived radioactive isotope and its stable decay product in a sample. Assuming that the parent radioisotope was present at the time of formation of the rock, etc., then the number of daughter isotopes produced by *radioactive decay of the parent depends only on the *half-life of the parent and the age of the sample. Half-lives must therefore be known with great accuracy for precise dating and should range from about 10^5 to about 10^{10} years. In addition, there should be no loss or gain of parent or daughter isotope during the time the 'radioactive clock' is operating; if this condition is only partially satisfied, allowances must be made. Pairs of isotopes used in radioactive dating include potassium-40 which decays to argon-40 with a half-life of 1.25×10^9 years, and rubidium-87 which decays to strontium-87 with a half-life of 4.88×10^{10} years.

radioactive decay. The spontaneous transformation of one atomic nucleus into another with the emission of energy. The energy is released in the form of an energetic particle, usually an *alpha particle, *beta particle (i.e. an electron), or positron, sometimes accompanied by a gamma-ray photon. The unstable isotopes of an element that can undergo such transformations are called radioactive isotopes, or *radioisotopes*. The emission of a particle from the nucleus of a radioisotope results in the production of an isotope of a different element, as in the beta decay of carbon-14 to nitrogen-14 or the alpha decay of radium-226 to radon-222. The isotope produced is itself often radioactive.

The average time taken for half a given number of nuclei of a particular radioisotope to decay is the *half-life of that radioisotope; values range from a fraction of a second to thousands of millions of years.

radio astrometry. A branch of *astrometry in which the coordinates of celestial objects are measured with high precision using a *radio telescope. It has become more accurate than its optical counterpart in recent years with the advent of interferometers having very long and accurately determined *baselines.

radio astronomy. The study of celestial bodies by means of the *radio waves they emit. Extraterrestrial radio waves were first detected in 1931 by the American radio engineer Karl Jansky. The first *radio telescope was built and operated by Grote Reber in the late 1930s but radio astronomy only developed after the war. It has benefited greatly from advances in electronics and solid-state physics and is now a major branch of observational astronomy.

radio frequency. *See* radio waves.

radio galaxy. An extragalactic radio source identified with an optical galaxy whose radio-power output lies in the range 10^{35} to 10^{38} watts. These sources often show double structure (*see* radio-source structure) and are to be distinguished from normal galaxies, such as M31, whose radio emissions roughly follow the optical contours of the galaxy and whose radio powers are much lower, in the range 10^{30} to 10^{32} watts. Radio galaxies with diameters greater than one megaparsec are now known. An example of such a *giant radio galaxy* is 3C 236, which is a double radio source whose radio emission spans almost six megaparsecs of space, making it one of the biggest radio galaxies in the universe.

At the high-luminosity end of the range, the distinction between radio galaxy and *quasar becomes hazy. The energy mechanism of radio galaxies is not understood but supermassive *black holes have been postulated as the source

FREQUENCY ALLOCATIONS FOR RADIO ASTRONOMY
(1971 W.A.R.C.)

21.850-21.870 MHz	worldwide exclusive
37.75-38.25 MHz	secondary allocation (fixed and mobile, primary)
150.05-153 MHz	primary allocation (met. aids, secondary)
406.1-410 MHz	primary (shared with fixed and mobile, excl. aeronautical)
	[406-406.1 allocated to low-power emergency radio beacons]
1400-1427 MHz	worldwide exclusive
1660-1670 MHz	primary, shared with met. aids
2690-2700 MHz	worldwide exclusive [broadcasting satellites allocated band up to 2690 MHz]
4990-5000 MHz	primary, shared with fixed and mobile
10.60-10.68 GHz	primary, shared with fixed and mobile
10.68-10.70 GHz	worldwide exclusive (also radiolocation in U.K.)
15.35-15.40 GHz	worldwide exclusive
22.21-22.26 GHz	shared with fixed and mobile [includes H_2O at 22.235 GHz]
23.6-24.0 GHz	worldwide exclusive [includes NH_3 at 23.7 GHz]
31.3-31.5 GHz	worldwide exclusive
33.0-33.4 GHz	primary, shared with radio navigation
86-92 GHz	primary [includes HCN at 86.3 and 88.6 GHz]
130-140 GHz	primary
230-240 GHz	primary

of the prodigious power output.

radio heliograph. A *radio telescope designed for making maps of the *radio sun. The radio heliograph at *Culgoora, Australia, is a particular example.

radioisotope. *See* radioactive decay.

radioisotope thermoelectric generator. *See* thermoelectric generator.

radio map. *See* radio-source structure.

radio meteor (radar meteor). *See* meteor.

radiometer. A device that measures the total energy or power received from a body in the form of radiation, especially infrared radiation. In radio astronomy, a radiometer measures the total received radio *noise power.

radio receiver. *See* receiver.

radio source. A celestial object whose radio emissions may be detected with a *radio telescope. Within our solar system these sources include Jupiter and the sun, both of which emit powerful radio *bursts. Galactic radio sources include *pulsars, *supernova remnants, *HI and *HII regions, *maser sources, and the galactic centre itself. *Synchrotron emission from *cosmic rays in the Galaxy gives rise to a diffuse background radiation, centred broadly on the galactic plane but having several spurs away from it (*see* interstellar medium). This background radiation may limit the *sensitivity of a radio telescope, especially at frequencies below about 400 megahertz.

Extragalactic radio sources include spiral galaxies (whose HI emissions may be used to map the distribution of hydrogen within them), *radio galaxies, and *quasars. An extragalactic radio source is often classified according to the spatial distribution of the radio emission shown on a radio map of the source. *See* radio-source structure; radio-source spectrum.

radio-source catalogue. A numbered list of *radio sources detected in a survey at a given frequency down to a limiting *flux density. A well-known example is the *third Cambridge catalogue*, 3C. The catalogue usually lists the source positions and *flux densities together with any other information that may be available about angular sizes, etc. From one or more catalogues *source counts can be compiled, in which the total number of sources per unit solid angle (steradian) above a given flux density is plotted as a function of the flux density. These counts are important in *cosmology since different mathematical models of the universe predict different source counts: they provide one method of comparing theory with the real world.

radio-source spectrum. The variation of the *flux density of a *radio source with frequency. The emissions from most radio sources are *noise signals displaying continuous spectra that rise or fall with increasing frequency more or less smoothly. Sources of *thermal and *synchrotron emission both give spectra of this type, sometimes showing a low-frequency cut-off due to self-absorption (*see* synchrotron emission). The spectra of the radiation from *maser sources or from clouds of interstellar molecules, such as some *HI regions, have discrete lines of bright emission superimposed on a background continuum.

The continuous spectra are often classified according to the *spectral index*, α, which for a source whose flux density is S is defined by

$$S_\alpha \nu_\alpha \text{ or } S^\alpha \nu^{-\alpha}$$

where ν is the frequency. Both definitions are in use and it is important to be clear which is intended in each case.

radio-source structure. The spatial distribution of the radio emission from a *radio source. It may be partially inferred from a *radio map* of the source, which is the *convolution of the true source distribution projected onto a plane and the *antenna pattern of the *radio telescope, and which is often plotted in contours of constant *brightness temperature or *flux density; such a map is called a *contour map*. Where the angular size of the source is smaller than the antenna pattern, the radio map reveals no details of its structure; the source is then said to be *unresolved* and is de-

Double radio source

Bent double radio source

Radio-trail source

scribed as a *point source*. In the converse case, the source is said to be *extended*.

Maps of extragalactic radio sources often show two distinct bright features that may or may not be joined by a faint *bridge* of diffuse emission (see illustration). These are *double radio sources*. The bright features can sometimes be subdivided into distinct spots of bright emission, or *hot spots*, with trailing extensions. *Cygnus A is a fine example. Sometimes the trailing extensions and the bridge merge into one and do not lie on the straight line defined by the bright heads. The source is then described as a *bent double radio source*. The angle of bending may become so large that the trailing extensions merge into one another or become almost parallel, forming a *radio-trail source*. *See also* jet; beam.

radio sun. The sun mapped at radio frequencies. *See* corona.

radio telescope. An instrument for recording and measuring the *radio-frequency emissions from celestial *radio sources. All radio telescopes consist of an *antenna, or system of antennas, connected by *feeders to one or more *receivers. The antennas may be in the form of *dishes or simple linear *dipoles. The receiver outputs may be displayed on suitable devices such as pen-recorders or recorded on tape for further analysis by computer.

The antenna system may consist of two separated units whose electrical signals are conveyed by feeders to a common receiver, forming an *interferometer*. The antenna units are often mounted on an east-west *baseline and are arranged to point in the same direction. A radio source, in its passage through the antenna beam, then produces *interference fringes* at the output of the receiver. The fringes are sinusoidal variations whose maximum amplitude depends on the *flux density of the source and whose period depends on both the radio wavelength and the spatial separation of the antenna units. *See also* aperture synthesis; array; resolving power; sea interferometer.

radio-trail source. *See* radiosource structure.

radio waves. Travelling electromagnetic disturbances whose frequency and wavelength lie in the radio region of the electromagnetic spectrum, extending from about 20 kilohertz up to about 300 gigahertz. Frequencies within this range are called *radio frequencies*. *Radio astronomy may in principle be carried out at any radio frequency but in practice observations are confined to certain 'radio-astronomy bands' set aside for that purpose by international agreement to minimize *interference from other services (see table). These bands lie within the radio window.

radio window. *See* atmospheric windows.

radius of curvature. Symbol: r, R. The radius of the sphere whose surface contains the surface of a spherical mirror, lens, or wavefront. The *curvature* of the mirror, etc., is given by the reciprocal of the radius of curvature, i.e. by $1/r$. The radius of curvature of a mirror is twice its *focal length, f. That of a lens is given by

$1/f = (n - 1)(1/r_1 - 1/r_2)$

where n is the refractive index; it is assumed that, in the direction of light travel, the radii, r_1 and r_2, for the convex and concave lens surfaces are positive and negative, respectively.

radius vector. The line, of length r, joining a point on a curve to a reference point, such as the origin in spherical and cylindrical coordinate systems. In planetary motion the radius vector is the line between the position of a planet and the focus of its orbit, i.e. the sun's position. *See also* Kepler's laws.

radome. A protective covering over a radio *antenna that is transparent to radio waves at the frequency in use.

Ram. Common name for *Aries.

Ramsden disc (Ramsden circle). *See* exit pupil.

Ramsden eyepiece. *See* eyepiece.

Ranger. An American space programme, originally intended to hard-land instruments on the lunar surface but upgraded in 1963 into purely photographic missions. The first five Ranger craft were launched between August 1961 and October 1962. Ranger 6, launched Jan. 30 1964, reached its target but suffered camera malfunction. The last three (Rangers 7, 8, and 9) were launched on July 28 1964, Feb. 17 1965, and Mar. 21 1965. They obtained the first close-ups of the moon, revealing the existence of boulders and metre-sized *craters in the *regolith. Over 4300, 7100, and 5800 photographs were transmitted by Rangers 7, 8, and 9 respectively, leading to a great advance in knowledge of the moon. *See also* Lunar Orbiter probes; Surveyor; Luna probes; Zond probes.

Ras Algethi (α Her). A visual binary in the constellation Hercules. The primary (A) is a red supergiant that varies irregularly in magnitude from 3.0 to 4.0, the 6th-magnitude yellow giant companion (B) appearing green by contrast. B is a spectroscopic binary, period 51.6 days. Spectral type: M5 I(A), G5 III(?) + F2(?)(B).

Ras Alhague. *See* Ophiuchus.

RATAN 600. *See* Academy of Sciences Astrophysical Observatory.

rays, lunar. High *albedo (bright) streaks that radiate out from the youngest moon *craters and help to assign them to the Copernican System—the most recent era in lunar history. Rays consist of crater *ejecta that has not yet been darkened by radiation or by admixture with local material. The most prominent systems are associated with *Tycho and Giordano Bruno.

Razin effect. The reduction of the *synchrotron emission from a relativistic electron due to the *refractive index of the surrounding medium being less than one.

R Coronae Borealis stars. A small group of *variable stars in which the brightness decreases—typically by about four *magnitudes—at irregular intervals and often quite rapidly, returning to its normal maximum often slowly and with considerable fluctuation. The frequency of occurrence of the decrease varies greatly between the stars. The prototype of the group is R Coronae Borealis, which drops from 6th magnitude usually down to 14th. They are supergiants that have carbon-rich atmospheres and an abnormally low proportion of hydrogen. The decrease in brightness apparently arises from the strong absorption of light by layers of carbon particles occasionally formed in the outer layers of the stellar atmosphere and gradually blown away by radiation pressure. Some R Coronae Borealis stars, including the prototype, are also *pulsating variables.

receiver. An electronic device that, in radio astronomy, detects and measures the radio-frequency signals picked up by the *antenna of a radio telescope. A receiver that measures the total *noise power from the antenna and from itself is called a *total-power receiver*. This is in distinction to other types that measure, for example, the correlated power in two antennas (*see* phase-switching interferometer; Dicke-switched receiver; correlation receiver).

Radio-astronomy receivers usually use the *superheterodyne* technique: the radio-frequency signals at frequency f_1 are combined in a *mixer* with signals from a *local oscillator* at frequency f_0 to produce a combination signal whose amplitude faithfully follows that of the radio-frequency signal but whose frequency is $f_1 - f_0$. This is the *intermediate frequency, IF,* which is amplified in the *IF amplifier* and presented to the first *detector*. The output of this nonlinear detector is a voltage that depends on the input power.

In a total-power receiver the output from the detector becomes the receiver output, having first been smoothed by a low-pass *filter to reduce the fluctuations due to the system noise (*see* sensitivity). In a switching receiver the first detector output is passed to a *phase-sensitive detector before smoothing in order to pick out the component at the switching frequency that is buried in the noise. *See also* line receiver.

recession of the galaxies. An important discovery made by Edwin Hubble in 1929 and based on the fact that the spectral lines of all galaxies beyond the *Local Group are systematically shifted towards longer wavelengths. These observed *redshifts are interpreted as being due to the Doppler shift and imply that all distant galaxies are moving away from our own with speeds (*recession velocities*) roughly proportional to their distances from us. Thus the universe as a whole appears to be expanding. If it is assumed that matter is neither created nor destroyed in the universe it follows that the universe was denser in the past. *See* Hubble's law; expanding universe; big bang theory.

recombination-line emission. Electromagnetic radiation resulting from the *recombination* of a free electron (one not bound to a nucleus) and an ionized atom. The spectrum of the radiation consists of a continuum from the transition of free electrons to bound states and a number of distinct recombination lines, each one due to a transition between two *energy levels in the atom. The emission lines occur at radio, infrared, and optical frequencies.

The energy levels are numbered in sequence starting at one for the ground state. Recombination lines are then classified according to the number of the energy level to which the electron moves and the number of energy levels that it passes. Recombination-line emission occurs, for example, in *emission nebulae.

recurrent nova. A star that suffers a series of violent nova-like outbursts at observed average intervals from about 15 to 80 years. The change in brightness is smaller and the decline in brightness more pronounced than with *novae. Like novae they are thought to be close binary systems with a similar sort of energy outburst. Possibly all novae recur but the periods between outbursts of the most powerful sort are too long for more than one eruption to have been observed. T Pyxidis (1890, 1902, 1920, 1944, 1965) RS Ophiuchi (1901, 1933, 1958, 1967), and T Coronae Borealis (1866, 1946) are recurrent novae.

reddening. *See* extinction.

red giant. A *giant star with a surface temperature between 2000-3000 kelvin and a diameter 10-100 times greater than the sun. Red giants are believed to be one of the final phases in the evolution of a normal star, reached only when its central hydrogen has been used up. The star resorts to burning hydrogen in a shell around its dense inert helium core, which results in a rapid inflation of the outer layers of atmosphere. Since red giants are so distended, gravity has only a small effect on their surface layers and many lose considerable amounts of mass into space in the form of *stellar winds.

Similarly, red giants are often *variable stars because their surface layers slowly expand and contract. Because these stars are so large, their pulsations usually take about a year to complete; red giants therefore belong to the class of *Mira stars, or long-period variables. There are indications that some red giants end this phase of evolution as *planetary nebulae, gently puffing off their distended atmospheres at low velocities (a few km s^{-1}) and leaving their dense cores exposed as *white dwarf stars, radiating away their heat into space. *See also* stellar evolution.

Red Planet. *Another name for* Mars.

redshift. A displacement of spectral lines towards longer wavelength values; for an optical line, the shift would be towards the red end of the visible spectrum. The *redshift parameter*, z is given by the ratio $\Delta\lambda/\lambda$, where $\Delta\lambda$ is the observed increase in wavelength of the radiation and λ is the wavelength of the spectral line at the time of emission from a source, i.e. the wavelength in the 'normal' terrestrial spectrum.

The redshifts of astronomical objects within the Galaxy are interpreted as Doppler shifts (*see* Doppler effect) caused by movement of the source away from the observer. The value of z is then v/c, where v is the relative *radial velocity and c is the speed of light. The redshifts of extragalactic sources, including *quasars, are also usually interpreted in terms of the Doppler effect, i.e. in terms of the expansion of the universe. The redshift parameter of a distant galaxy thus gives its velocity of recession; since recessional velocities can be very great, the relativistic expression for redshift must be used (*see* Doppler effect). From measurements of galactic redshifts it has been possible to calculate the distances of galaxies, using *Hubble's law.

The redshift of spectral lines occurs in all regions of the electromagnetic spectrum: ultraviolet radiation can be redshifted to visible wavelengths and light to infrared wavelengths. The redshifts described above represent a loss of energy by the photons of radiation in overcoming the effects of recession or expansion. There is another mechanism, however, by which redshifts can be produced, i.e. by which photons can lose energy—the presence of a strong gravitational field. This *gravitational redshift was predicted by Einstein in his general theory of relativity. Although the redshifts of galaxies are often interpreted as due to the relativistic Doppler effect alone, both the expansion and the gravitational field of the universe are involved. *See also* cosmic scale factor.

Red Spot. *See* Great Red Spot.

red variables. The large group of variable stars, including *Mira stars and *irregular variables, that lie in the top right of the Hertzsprung-Russell diagram of *pulsating variables.

reflectance. The ratio of the total radiant power (in watts) reflected by a body to that incident on it.

reflectance spectra. *Spectra by means of which remote chemical analyses may be made, based on the *reflectance efficiency of a planetary or satellite surface at various wavelengths, particularly near-infrared wavelengths.

reflecting telescope (reflector). A telescope employing a mirror (the *primary mirror) to bring the light rays to a focus. The various configurations include the *Cassegrain, *Newtonian, *Gregorian, and *coudé telescopes; there are also the *Maksutov and *Schmidt telescopes, which are *catadioptric. These instruments differ in the additional optical system used to bring the image to a convenient point where it can be viewed through an *eyepiece, photographed, or otherwise analysed. The *aberrations in reflectors can be reduced to a very low level by suitable shaping of the primary mirror and with the possible use of a *correcting plate. In addition a mirror, unlike a lens, suffers no *chromatic aberration.

The first reflector was a Newtonian. In the period following its introduction in about 1670, reflecting telescopes with mirrors made from *speculum metal displaced the simple *Galilean and *Keplerian *refracting telescopes whose performance was limited by the chromatic aberration of their single-lens objectives. It was William Herschel who developed a special ability in making speculum-metal mirrors of really accurate figure. In 1783 he built a reflector with an aperture of 46 cm and in 1789 one of 122 cm. He used these telescopes for a wide range of observations, making many important advances in stellar astronomy.

The end of the era was marked by the construction in 1842-45 of an 183 cm diameter speculum-metal mirror by William Parsons, third Earl of Rosse, at his family seat of Birr Castle, at Parsonstown, Eire. This was the largest reflector built until the Mount Wilson 254 cm reflector was commissioned 75 years later; it was known as Leviathan. Parsons was the last of the gifted amateurs to lead the main stream of astronomical development, which in future would require the resources of large corporate bodies.

In the latter half of the 19th century, refracting telescopes with *achromatic objectives overtook reflectors in performance and held their lead until the turn of the century. By then the silvered glass mirror had been found to give a great improvement upon the reflectivity of speculum metal, which rapidly became obsolete. It was also realized that refractors had reached a real size limit at about 100 cm diameter: a lens can only be supported around its edge and larger lenses sag under their own weight to an unacceptable degree. A primary mirror can be supported over its whole rear face and thus escapes this limit.

The modern era opened with the construction of the giant reflecting telescopes at Mount Wilson and then Mount Palomar (*see* Hale Observatories). There are now 12 reflectors of *aperture exceeding 2.5 metres, led by the 6 metre Soviet telescope at Zelenchukskaya in the Caucasus mountains. For this telescope, advantage has been taken of the developments in drive technology pioneered by radio astronomers to return from the *equatorial to the simple *altazimuth mounting.

reflection. A phenomenon occurring when a beam of light or other wave motion strikes a surface separating two different media, such as air and glass or air and metal. Part of the wave has its direction changed, according to the laws of reflection, so that it does not enter the second medium. A rough surface will produce diffuse reflection. A smooth surface, as of polished metal, will reflect the radiation in a regular manner. The laws of reflection state firstly that the incident beam, the reflected beam, and the normal (the line perpendicular to the surface at the point of incidence) lie in the same plane, and secondly that the two beams are inclined at the same angle to but on either side of the normal.

reflection grating. *See* diffraction grating.

reflection nebula. A bright cloud of interstellar gas and dust that lies near and somewhat to one side of a star or stellar group, usually of *spectral type B2 or later. The starlight is scattered in all directions by the dust grains, of which there is a sufficient density to produce a noticeable illumination of the cloud. The light scattered towards the observer is bluer than that of the illuminating star but the spectra of cloud and star are essentially the same—a continuous spectrum with absorption lines. (These spectral features can also be seen in the spectra of the much more conspicuous *emission nebulae if there is enough dust present.) Typical reflection nebulae occur around the brighter stars of the *Pleiades. *See also* nebula.

reflector. *Short for* reflecting telescope.

refracting telescope (refractor). A telescope employing an *objective lens to bring the light rays to a focus. The first telescopes were simple refractors with single-lens objectives: these were the *Galilean and *Keplerian telescopes. They were supplanted by *reflecting telescopes because noone could overcome their *chromatic aberrations, which introduced brilliant false colour effects in the images. Although John Dollond discovered how to make an *achromatic lens in 1756, he could only make small diameters suitable for terrestrial telescopes; glass makers did not know how to make large uniform discs of crown and flint glass. This obstacle was overcome in the early 1800s by Joseph Fraunhofer, who made the 24 cm refractor at Dorpat and the 16 cm heliometer at Konigsberg. His successor, Merz, supplied the 32.4 cm equatorial refractor to the Royal Greenwich Observatory and the 38 cm refractor that was the first large telescope to be mounted in the USA, at

Cambridge, Massachusetts, in 1847.

Perhaps because of this, American astronomers leaned strongly to the refractor. By the 1880s Alvan Clark was making relatively large refractors: 47 cm diameter for the Dearborn Observatory, 91 cm for Lick Observatory, and 102 cm for Yerkes Observatory. Other very large refractors are the 61 cm at both Lowell Observatory, Arizona, and the Pic du Midi in France, the 75 cm at Pulkovo Observatory, Leningrad, and the 83 cm at the Meudon Observatory in Paris. All were built in the late 19th century. The desire for ever larger *apertures to reach ever further and fainter objects in space has meant that the major telescopes built in this century have been reflectors.

The surviving large refracting telescopes are mainly used in *astrometry and for the measurement of *double stars and also for determining *proper motions and *parallax. Refractors are often preferred by visual observers because of their long focal length and closed tube; the latter avoids air currents in the tube, which often cause an unsteady image.

refraction. A phenomenon occurring when a beam of light or other wave motion crosses a boundary between two different media, such as air and glass. On passing into the second medium, the direction of motion of the wave is 'bent' towards or away from the normal (the line perpendicular to the surface at the point of incidence). The incident and refracted rays and the normal all lie in the same plane. The direction of propagation is changed in accordance with *Snell's law:*

$$n_1 \sin i = n_2 \sin r$$

i and r are the angles made by the incident and refracted ray to the normal; n_1 and n_2 are the *refractive indices of the two media. The change in direction of motion results from a change of wave velocity as the wave passes from the first to the second medium.

refraction, angle of. *See* zenith distance.

refraction, atmospheric. *See* atmospheric refraction.

refractive index (index of refraction). Symbol: n. The ratio of the *speed of light, c, in a vacuum to the speed of light in a given medium. This is the *absolute refractive index;* it is always greater than one. The relative refractive index is the ratio of the speeds of light in two given media. The refractive index of a medium depends on the wavelength of the refracted wave: with light waves, n increases as the wavelength decreases, i.e. blue light is refracted (bent) more than red light. *See also* refraction.

refractor. *Short for* refracting telescope.

regolith. The layer of dust and

broken rock, created by meteoritic bombardment, that covers much of the surfaces of the moon and Mars and, probably, many other bodies in the solar system.

regression of nodes. *See* nodes.

Regulus (α Leo). A bluish-white star that is the brightest one in the constellation Leo and lies at the base of the Sickle of Leo. It is a visual triple star with an orange companion (B), which itself has a 13th-magnitude companion (C). Regulus has a common proper motion with a fourth star (D). m_V: 1.34 (A), 7.6 (B); M_V: -0.8; spectral type: B7 V(A), dK1(B); distance: 26 pc.

relative aperture. *See* aperture ratio.

relativistic beaming. *See* synchrotron emission.

relativistic electron. An electron moving at close to the speed of light. The special theory of *relativity must be used to describe associated phenomena.

relativity, general theory. The theory put forward by Albert Einstein in 1916. It describes how the relationship between space and time, as developed in the special theory of relativity, is affected by the gravitational effects of matter. The basic conclusion is that gravitational fields change the geometry of *space-time, causing it to become curved. It is this curvature of space-time that controls the natural motions of bodies. Thus matter tells space-time how to curve and space-time tells matter how to move. General relativity (GR) may therefore be considered as a theory of *gravitation, the differences between it and Newtonian gravitation only appearing when the gravitational fields become very strong, as with black holes, neutron stars, and white dwarfs, or when very accurate measurements can be made. GR also reduces to the special theory when gravitational effects are negligible.

The basic postulate from which GR was developed is the principle of *equivalence between gravitational and inertial forces: in two laboratories, one in a uniform gravitational field and the other suitably accelerated, the same laws of nature apply and thus the same phenomena, including optical ones, will be observed. Einstein was able to show that the natural motion of a body was in a 'straight line', i.e. along the shortest path between two points (a geodesic), in the geometry required to describe the physical world (*compare* Newton's first law of motion). Three-dimensional Euclidean geometry could not be used for this purpose. The requisite geometry had to be more flexible: this was the geometry of space-time. The presence of gravitational fields leads to the curvature of space-time. The curvature produced by a particular gravitational field can be calculated from Einstein's *field equations,* advanced as part of GR. It is then possible to calculate the geodesics followed by particles in this curved space-time.

Experimental tests of GR require either very strong gravitational fields or very accurate measurements. These tests must verify the predictions of GR where they deviate from the predictions of Newtonian gravitation and also where they deviate from the predictions of other relativistic theories of gravitation (including the *Brans-Dicke theory) that have been developed since 1916. The main tests involve three effects.

The first effect is the bending of light or other electromagnetic radiation in a gravitational field. Measurements of the positions of stars when close to the sun's limb (i.e. during total solar eclipses) and when some distance from the sun have been shown to differ: the difference is within 5% of Einstein's predicted deflection, which amounts to 1.74 arc seconds for a star on the solar limb. Even closer agreement has been found when the bending of microwaves from radio sources is determined. A related effect is the time delay in radio signals from spacecraft when they are close to the sun. The measured delays from the Viking probes and other craft agree with Einstein's prediction within 3%.

The predicted *gravitational redshift in the spectral lines of radiation emitted from a massive body has also been successfully demonstrated, using both the sun's and also the earth's gravitational fields. A related prediction is that clocks run more slowly in a strong gravitational field than in a weaker one. An atomic clock flown at an altitude of 10 km has been shown to run faster than at sea level by very close to the predicted amount (47×10^{-9} seconds).

Thirdly, the effects of curved space-time on the motions of orbiting bodies were shown convincingly when the *advance of the perihelion of Mercury was considered and, more recently, when the orbit of the *binary pulsar was studied.

Other predictions of GR are the existence of *gravitational waves, the existence of *singularities in the structure of space-time leading to black holes, and the constancy of the *gravitational constant. There is no direct evidence as yet for (or against) any of these, although indirect evidence of both gravitational waves and black holes is accumulating.

It cannot be said that GR has been conclusively proved. The experimental evidence does, however, seem to be mounting in its favour.

relativity, special theory. The theory proposed by Albert Einstein in 1905. It is concerned with the laws of physics as viewed by observers moving relative to one another at constant velocity (i.e. with observers in *inertial frames),

relativity, special theory

and with how relative motion affects measurements made by these observers. At low relative velocities, special relativity (SR) predicts the same results as the classical Newtonian laws of mechanics. As the relative velocity increases beyond that encountered in everyday experience, the predictions of the two theories diverge. It is those of SR that have been conclusively verified by experiment. SR must therefore be used to describe the behaviour of bodies, such as electrons, when they are moving at velocities close to the *speed of light.

One of the basic principles of SR is that the speed of light is the same in any direction in free space, is the same for all observers, and is independent of the motion of the body emitting the light (or other electromagnetic radiation). This violates the classical concepts of relative motion but has been demonstrated experimentally.

Another basic principle is that all physical laws are the same for every inertial frame. Any law deduced in one inertial frame will be true in every other inertial frame. The values of the numerical constants appearing in these laws are also independent of the frame in which they are measured. The same does not, however, apply to certain variable quantities measured in different inertial frames.

Prior to SR it had been assumed that there was a universal 'absolute' time and 'absolute' space that were the same for all observers. Einstein showed that as a direct consequence of the invariance of the speed of light, time and space could not be considered as separate concepts, independent of each other and of the observer. Instead they must be regarded as a composite entity, called *space-time. Any event in an inertial frame must therefore be described in terms of four space-time coordinates—three spatial and one-time coordinate. The space-time coordinates of the same event measured in two different frames will differ, but can be interconverted by the *Lorentz transformation.

As the relative velocity (v) between inertial frames approaches the speed of light (c), very strange things are predicted. One is *time dilation:* two observers approaching at a relative velocity close to c will each see the clock of the other operating more slowly than their own. The time intervals will be dilated by a factor of $1/\beta$ where β is equal to $\sqrt{(1 - v^2/c^2)}$.

Another prediction is the apparent contraction of a moving object that is observed, in the direction of motion of the object, by someone in a different inertial frame. The observed contraction, known as the *Lorentz contraction,* amounts to the factor β. The object also appears slightly rotated. No change in dimensions is observed in directions perpendicular to the direction of motion.

A further consequence of

SR, when applied to the field of dynamics, is that the mass (m) of a body is not invariant but increases as the relative velocity between body and observer increases:
$$m = m_0/\beta$$
m_0 is the *rest mass of the body and is an invariant property of matter. It follows that no object with mass can reach the speed of light; only a particle with zero rest mass (such as the *photon) can travel at the speed of light. Einstein showed that rest mass is a form of energy and can be converted to other forms of energy. Mass and energy are thus not distinct physical properties but must be considered as equivalent. They are related by the equation
$$E = mc^2$$
where E is the total energy of the particle.

réseau. A reference grid of fine lines that are photographically produced in the emulsion of a photographic plate and used in measurements of star positions.

resolution. The ability of a telescope or other optical instrument to distinguish fine detail. *See* resolving power.

resolving power. A measure of the *resolution of a telescope. It is the smallest angle between two point objects that produces distinct images. When the angular separation of two stars is very small, it might be thought that the use of high enough magnifications would always resolve the light into two distinct images. Because of diffraction effects however, the image of each star is not a point of light but an *Airy disc. If the two discs substantially overlap then increased magnification merely gives a larger blur of light, and the telescope has not sufficient resolving power to separate the images. The stars will just be resolved, however, when their Airy discs touch, i.e. when the distance between the centres of the two discs is equal to the diameter of the disc.

The Airy disc has an angular diameter given by $1.1\lambda/D$ *radians* or $2.27 \times 10^5 \lambda/D$ arc seconds, where λ is the wavelength of light and D the telescope *aperture. This is known as the *Dawes limit,* and for light of wavelength 5×10^{-5} cm it is equal to $11.5/D$ seconds of arc, where D is in cm. This limit gives the *angular resolution* of a telescope, i.e. the minimum angular separation that a telescope will resolve. Fig. *a* shows the images of two stars of equal magnitude separated by 0.8 seconds of arc. Viewed under perfect *seeing conditions with a 7.5 cm aperture (right) they are not resolved; with 15 cm (centre) they are just resolved and with 30 cm (left) they are clearly separated.

The discovery of Pluto's satellite in 1978 shows that the Dawes limit cannot be too strictly applied. Although well above the Dawes limit for large telescopes, Pluto's satellite has never been resolved; it *a Images of an optical double of separation 0.8" through apertures (L to R) of 30, 15, and 7.5 cm*

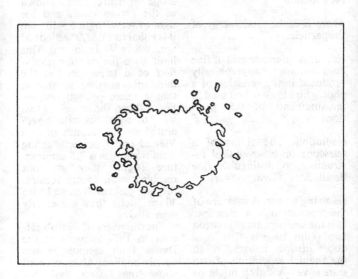

b Satellite of Pluto, unresolved but detected by slight elongation of greatly enlarged image was detected as an elongated photographic image (Fig. *b*). This is partly because of the difference in brightness and because Pluto is not a point source, but mainly because poor seeing prevents the larger-aperture telescopes from reaching the Dawes limit. The next big step forward in optical astronomy thus depends upon using the *Space Telescope outside the earth's atmosphere.

In radio astronomy, where the same formulae for resolving power apply but where very much greater wavelengths are studied, apertures maybe 50 times greater than those used in optical telescopes can only give poor resolution. The situation has been improved by the use of *very long baseline interferometry.

rest mass. The mass of an elementary particle, etc., when at rest. According to the special theory of *relativity, a body's mass increases as its velocity increases. This is only noticeable at velocities close to the speed of light.

retardation. The difference in the moon's rising time between successive nights.

reticle. A grid or pattern of two or more fine wires set in the focal plane of a telescope eyepiece and used in determining the position and/or size of a celestial object.

Reticulum (Net). A small constellation in the southern hemisphere near the Large Magellanic Cloud, the brightest stars being one of 3rd magnitude and some of 4th magnitude. Abbrev.: Ret; genitive form: Reticuli; approx. position: RA 4h, dec -60°; area: 114 sq deg.

retrograde motion. *See* direct motion.

retrorocket. A small rocket motor attached to a spacecraft that, by ejecting gas in a forward direction, decelerates the craft. In this way the craft may change its orbit or, possibly with parachutes, achieve a soft landing.

reversing layer. The comparatively low-temperature stratum overlying the surface of a *main-sequence star, in which it was proposed the absorption lines in the star's spectrum were produced. It is now known that the lines originate in the *photosphere at the same levels as the *continuum (with emission predominating at low levels and absorption higher up) rather than in a well-defined layer.

revolution. 1. Orbital motion of a celestial body about a centre of gravitational attraction, such as the sun or a planet, as distinct from axial *rotation. *See also* direct motion.

2. One complete circuit of a celestial body about a gravitational centre. The earth takes one year to make one revolution around the sun.

RFT. *Abbrev. for* richest-field telescope.

RGO. *Abbrev. for* Royal Greenwich Observatory.

Rhea. A satellite of Saturn, discovered in 1672. Recent observations show it to have an icy surface. *See* Table 2, backmatter.

rich cluster. *See* clusters of galaxies.

richest-field telescope (RFT). A telescope designed to show the largest possible number of stars at a single view when aimed at a dense field of faint stars, such as the Milky Way. This end is accomplished by combining a reasonably large aperture (100 to 150 mm) with a low power and the widest possible field of view. Although the true RFT is a special design, fitted with an expensive wide-field *eyepiece, binoculars and most *finder telescopes go a long way towards meeting these specifications. Apart from providing spectacular views of immense numbers of stars (which are not obtained with the high magnifications and narrow fields more usually employed), the RFT is an ideal instrument for scanning the heavens in the search for novae and comets.

ridge (scarp). *See* wrinkle ridge; lobate ridge.

Rigel (β Ori). A very massive luminous and remote bluish-white supergiant that is very conspicuous and is the brightest star in the constellation *Orion. Its luminosity is estimated at over 50 000 times that of the sun. It is a spectroscopic binary, period 9.86 days. It is also fixed at about 10″ with a binary system (BC), separation 0″.4. m_V: 0.11; M_V: -7.1; spectral type B8 Ia; distance: about 270 pc.

right ascension (RA). Symbol: α. A coordinate used with *declination in the *equatorial coordinate system. The right ascension of a celestial body, etc., is its angular distance measured eastwards along the celestial equator from the vernal *equinox to the intersection of the hour circle passing through the body. It is generally expressed in hours (h), minutes (m), and seconds (s) from 0 to 24h: one hour equals 15° of arc.

right-ascension circle. The *setting circle on the polar axis of an *equatorial mounting that measures the *right ascension at which the telescope is set.

Rigil Kent. *See* Alpha Centauri.

rille (rill; groove; rima; cleft). Either of two distinct types of elongated lunar depression (see table). *Sinuous rilles* are confined to the *maria, where they follow local contours, are frequently discontinuous (or consist of rows of coalescing craters), and have V-shaped cross sections. They sometimes

Lunar Rilles and Valleys

rille, valley	location; type
Alpine Valley	NE Imbrium; radial fracture
Aridaeus	W Mare Tranquillitatis; linear
Byrgius	W Mare Humorum; sinuous
Cauchy Fault	E Mare Tranquillitatis
Hadley	Palus Putredinus; sinuous
Hyginus	Mare Vaporum; forked rille
Lee Lincoln Scarp	Taurus Littrow area; fault
Posidonius	NE Mare Serenitatis; sinuous
Rheita Valley	S Nectaris; radial fracture
Schröter's Valley	Aristarchus Plateau; double sinuous
Schrödinger Canyon	S farside; radial fracture
Sirsalis	W Mare Humorum; sinuous
Straight Wall	Mare Nubium; linear fault

originate in volcanic caldera and are assumed to be channels, or collapsed tubes, carved out by molten basaltic lavas. Examples exist of sinuous rilles inside sinuous rilles but tributary systems are not observed.

Faulted *linear, arcuate,* and *forked rilles* have flat floors and do not recognize highland-mare boundaries. This second type of rille is frequently associated with impact basins and craters, where they may be concentric. *See also* wrinkle ridges; crater chain.

rima. *See* rille.

ring current. An electric current flowing westwards around the earth; it arises from a net eastward flow of electrons and a net westward flow of protons trapped inside the *Van Allen radiation belts. The current produces a magnetic field that cancels part of the geomagnetic field at the earth's surface. During *geomagnetic storms, when the Van Allen belts are recharged with particles, the increased ring current results in a further decrease of the earth's surface magnetic field strength.

ring galaxy. A very rare type of galaxy that has the form of an elliptical ring either with a massive nucleus, often off-centre, or with little or no luminous material visible in its interior. In many cases there is a small satellite galaxy devoid of interstellar gas. The ring configuration is thought to be unstable and it has been suggested that these systems may represent the after-effects of a collision between two galaxies.

ring mountains. *See* mountains, lunar; basins, lunar.

Ring nebula (M57; NGC 6720). A ring-shaped *planetary nebula about 700 parsecs away in the constellation Lyra. The inner regions closest to the centrally placed ionizing star are green as a result of light emitted by doubly ionized oxygen; there is an approximately symmetrical transition to the reddish outer edge where the lower-energy transitions of hydrogen and nitrogen are responsible for the colour. *See* emission nebula.

ring plains. *See* maria, lunar.

ring systems. *See* Saturn's rings; Uranus; Jupiter's satellites.

rising. The daily appearance on an observer's horizon of a particular celestial body as a result of the earth's rotation, the body disappearing below the horizon, or *setting*, at some time in the following 24 hours. A star rises, on average, two hours earlier per month. Stars always rise and set in the same position relative to other stars whereas the sun, moon, and planets rise and set at different points on successive days. This led to the early idea of *fixed stars and wandering stars.

For an observer at the equator all stars in theory can be seen, rising and setting at right angles to the horizon. At the poles an observer will see only the stars in his hemisphere; these stars never rise nor set but move in circles parallel to the equator. At intermediate latitudes some stars (*circumpolar stars) never set and some, in a corresponding area in the sky around the opposite pole, never rise. The remaining stars rise and set at oblique angles to the horizon, the time spent above the horizon depending on the observer's latitude and the position of the star relative to the celestial pole. Stars on the celestial equator rise due east and set due west and for all observing positions except the poles remain 12 hours above the horizon.

Ritchey-Chrétien optics. An optical system that is a variation on the *Cassegrain telescope. It corrects both *spherical aberration and *coma at the Cassegrain focus, and therefore gives a high-quality image over a relatively wide field of view. A primary mirror with a hyperbolic profile is used in conjunction with an appropriate departure from classical (hyperbolic) form at the secondary mirror.

Roche limit. The minimum distance from the centre of a body at which a satellite can remain in equilibrium under the influence of its own gravitation and that of its primary. Assuming that the satellite is held together only by gravitational attraction, i.e. it has zero tensile strength, and that it has the same density as the primary, the Roche limit is about 2.4 times the radius of the primary. Inside the Roche limit a satellite would be torn apart by tidal forces.

Roche lobe. *See* equipotential surfaces.

rocket. *See* launch vehicle; sounding rocket.

rockoon. A means of exploring the upper atmosphere of the earth. It consists of a large balloon to which is attached a small rocket. The rocket is fired at a considerable altitude, thus avoiding the air drag of the lower atmosphere. The balloons are usually *Skyhooks:* large nondilatable plastic balloons that can carry a heavy load to altitudes of over 30 km.

Rosette nebula (NGC 2237-38-44-46). A large *emission nebula about 15 parsecs in diameter and about 1400 parsecs distant in the direction of the constellation Monoceros. It is an *HII region heated and ionized by a centrally situated group of hot young stars.

Ross 128, 154, 248. *See* Table 7 (nearest stars), backmatter.

rotating variable. A star with a hotter or cooler region on its surface that periodically rotates into our line of sight and causes slight variations in the star's brightness. *Pulsars are related to this group of *variable stars.

rotation. 1. Spinning motion of a celestial body or a group of gravitationally bound bodies, such as a galaxy, about an axis, as distinct from orbital *revolution. Almost all celestial bodies show some degree of rotation. The hottest (O and B) stars have very great rotation rates of about 200-250 km s^{-1}. The faster the rate of rotation the broader and shallower the star's spectral lines. *See also* direct motion; differential rotation; synchronous rotation.

2. One complete rotation a celestial body about its axis. The earth takes one day to make one rotation. Ideally the rotation period is measured as the time interval between successive passages of a meridian line on the surface across the centre of the disc, as seen from earth. The solid surface may however be unobservable and indirect measurements, as by radar, are then employed. The rotation period of a gaseous body, such as the sun or the planet Jupiter, varies with latitude, being greatest at the equator (*see* differential rotation).

rotational parallax. A method of determining the mean distance to a star or star cluster. It is based on the *parallax resulting from the differential rotation of the Galaxy, which leads to a variation in velocity about the galactic centre. The average *radial velocity, v_r (in km s^{-1}), and the galactic longitude, l^{II}, of an object—both measurable—are related to distance d (in parsecs) by *Oort's formula:

$$v_r = dA \sin 2l^{II}$$

A is equal to about 15 km s^{-1} per kiloparsec. The object

rotation measure. *See* Faraday rotation.

Rover. *See* Lunar Roving Vehicle.

Royal Greenwich Observatory (RGO). The observatory sited originally at Greenwich, London, and built at the command of Charles II so that astronomical measurements could be made and the data tabulated for the primary purpose of increasing the accuracy with which positions at sea could be determined. John Flamsteed was appointed first *Astronomer Royal and took up his post in 1676. The meridian of zero longitude passes through the site at Greenwich. Over the centuries the increase in atmospheric and light pollution and the improvements in telescope resolution, which could not be achieved under such pollution, necessitated the removal of the Observatory to Herstmonceux Castle in Sussex. The move was completed by 1954.

At the RGO measurements of position, motion, distance, and other physical characteristics are made. The RGO is also responsible for the time service of the UK (*see* Greenwich Mean Time). The *seeing is still far from perfect at the new site, averaging 30 arc seconds. The RGO's principal instrument, the 2.5 metre *Isaac Newton telescope, is therefore in the process of being moved to the *Northern Hemisphere Observatory on the Canary Islands.

Royal Observatory, Edinburgh. The observatory in Edinburgh that was founded in 1818, with a Royal Charter since 1822, and is responsible for some major national facilities funded by the Science Research Council. These include the 1.2 metre *UK Schmidt telescope at Siding Spring, Australia, and the 3.8 metre *UK Infrared telescope in Hawaii. The Observatory is concerned with the development of advanced technologies and their applications in astronomical and space research.

r-process. A rapid process of *nucleo-synthesis that is thought to occur when there is a very high flux of neutrons, as in certain supernova explosions. It produces all nuclei with a mass number greater than bismuth-209 (*see* s-process) and also all neutron-rich isotopes heavier than iron. The process involves the capture by a nucleus of two or more neutrons in quick succession. The nucleus will then undergo chains of beta decay, this beta-particle (electron) emission having been suppressed during the rapid capture process. The decay product will be a stable neutron-rich nuclei.

Many heavy nuclei can be produced both by the r- and the s-process.

RR Lyrae stars (cluster varia-

Subgroups of RR Lyrae stars

type	approx. amplitude	period (days)	light curve asymmetry
RRa	1.5-2	0.35-0.55	
RRb	1.2-1.7	0.5-0.8	decreasing
RRc	0.5	0.2-0.4	

bles). A large group of *pulsating variables that are very old giant stars (halo and disc *population II stars) and are found principally in *globular clusters. They usually have periods of less than one day and median spectral types in the range A7 to F5. They were discovered in 1895 by Solon I. Bailey, the group being named after RR Lyrae, discovered in 1899. Available evidence indicates that all RR Lyrae stars have about the same mean absolute *magnitude (about +0.5); they can therefore be used as distance indicators up to about 200 kiloparsecs (*see* distance modulus).

RR Lyrae stars are of a mixed nature. They were divided by Bailey into three groups—*RRa, RRb,* and *RRc*—depending on period and the asymmetry of their light curves (see illustration and table). Groups *a* and *b* are often combined today as *RRab* stars. Other groupings have been made according to period. Many RR Lyrae stars show a periodic variation in both period and shape of light curve—the *Blazhko effect*. In addition the periods of some RR Lyrae stars are slowly changing at a constant rate, as predicted by evolutionary theory, while others show abrupt changes in period. The pulsations of these stars are very complex and can be subdivided into one group (RRab) oscillating in fundamental mode (*see* pulsating variables) and a second group (RRc) oscillating in first harmonic mode. *See also* horizontal branch.

RRs stars. *See* dwarf Cepheids.

R stars. *See* carbon stars.

Rule. Common name for *Norma.

runaway star. A young star of spectral type O or early B with a very high space velocity, examples including AE Aurigae, 53 Arietis, and Mu Columbae. They could be the result of a supernova explosion in a close binary system.

Russell-Vogt theorem. *See* Vogt-Russell theorem.

RV Tauri stars. A small group of very luminous *pulsating variables, typified by RV Tauri, R Scuti, and R Sagittae, that are principally G and K stars with some F stars. They

Typical light curves of RR Lyrae stars in subgroups RRa, RRb, RRc

are yellow supergiants with extended atmospheres of gas that emit infrared radiation and have possibly been driven off by the pulsations. They have very characteristic light curves with alternating deep and shallow minima and periods ranging from 30 to 145 days. Since the luminosity fluctuations can be disturbed quite significantly in shape, period, etc., being most pronounced for longer-period stars, they are classified as *semiregular variables. RV Tauri stars can be distinguished from other similar yellow semiregular stars by the variation in their *colour index, which mimics the light curve but goes through its maximum a little before the luminosity minimum. A small group have double periodicity; DF Cygni has two separate luminosity oscillations, a rapid 50 day oscillation being superimposed on a much slower 780 day oscillation with a much greater amplitude.

RW Aurigae stars. *See* T Tauri stars.

S

Sagitta (Arrow). A small constellation in the northern hemisphere near Cygnus, lying in the Milky Way, the brightest stars being one (γ) of 3rd magnitude and some of 4th magnitude. It contains the open cluster M71. Abbrev.: Sge; genitive form: Sagittae; approx. position: RA 19.7h, dec +18°; area: 80 sq deg.

sagittal plane. The plane containing all or part of the curved horizontal midline of a lens or mirror and its associated rays. The *tangential plane* contains the equivalent vertical midline and its associated rays. If the object lies off the optical axis of the lens or mirror so that light rays fall on it at an oblique angle, the sagittal and tangential rays do not converge to form a single image. *See* coma; astigmatism.

Sagittarius (Archer). A large zodiac constellation in the southern hemisphere near Scorpius, lying partly in the Milky Way. The brightest stars are the 2nd-magnitude Kaus Australis (ε) and Nunki (σ) with several of 3rd magnitude. The centre of the Galaxy lies in the direction of Sagittarius but is obscured by considerable dark nebulosity. The area contains star clouds, many nebulae, including the *Omega, *Lagoon, and *Trifid nebulae, many globular clusters, including the bright ones M22 and M54, and some open clusters. Abbrev.: Sgr; genitive form: Sagittarii; approx. position: RA 19h, dec -30°; area: 867 sq deg.

Sagittarius A. The brightest member of a complex of *radio sources in the centre of our *Galaxy. The entire complex has a *flux density at 86 megahertz of 4500 *jansky.

Sagittarius arm. *See* Galaxy.

Sagittarius B. An immense interstellar cloud complex lying around the galactic centre, maybe forming part or all of an expanding ring of clouds. The most massive cloud in the complex, and possibly in the Galaxy, is *Sagittarius B2*. A large number of the interstellar molecules detected so far were discovered in Sgr B2.

Saha ionization equation. An equation, put forward by the Indian physicist M.N. Saha in the 1920s, that gives the ratio of the number density (number per unit volume) of ions, N^+, to the number density of neutral atoms, N_0, in the ground state in a system of atoms, ions, and free electrons at a given temperature. The system is in equilibrium so that the rate of *ionization is equal to the rate of recombination of electrons and ions. The equation can be given in the form:
$$N^+/N_0 = A(kT)^{3/2}/N_e e^{-I/kT}$$
A is a constant, k is the Boltzmann constant, T the thermodynamic temperature, N_e the number of electrons per unit volume (the electron density), and I is the *ionization potential of the atom. The equation also gives the ratio N^{2+}/N^+, N^{3+}/N^{2+}, etc., if I is replaced by the second (I_2) or the third (I_3) ionization potentials, etc.

The equation shows that the higher the temperature, the more highly ionized a particular atom will be. It can therefore be used to find the relative numbers of neutral, singly ionized, doubly ionized atoms, etc., in stars of known temperature and electron density. *See also* Boltzmann equation.

Sails. Common name for *Vela.

Saiph. *See* Orion.

Salyut. A series of Soviet space stations launched into earth orbit and visited and operated by crews of cosmonauts from Comecon countries. The crews are ferried to and from the space stations by *Soyuz craft. The stations are much smaller than the one-off *Skylab space station but a large range of experiments can still be carried out, including studies of the earth and its atmosphere, studies of the sun, the development of closed-cycle life-support systems, and technological experiments. The crew's physical fitness is very carefully monitored and there are extensive programmes of physical exercises.

The first Salyut space station was launched on Apr. 19 1971. Soyuz 10 docked with Salyut 1 on Apr. 23 but it was not until Soyuz 11 docked on June 6 that three cosmonauts transferred to and worked on the station for a record flight duration of 570.4 hours. All three were killed on re-entry. The Soviet space programme aims to increase flight duration to determine how long man can remain in space in weightless conditions and still readapt to the earth's gravity. The Salyut 6 station was occupied continuously for 96.4 days from Sept. 29 1977 to Mar. 16 1978 and then for 139 days from June 17 to Nov. 2 1978. During periods of occupation Salyut 6 is refuelled and refurbished with food, water, additional equipment, etc., by specially adapted unmanned *Progress* cargo ships that, once unloaded, are undocked and guided to burn up in the earth's atmosphere. Salyut 6 should be operational for a few years.

S Andromedae. A supernova that was observed, in 1885, in the Andromeda galaxy and at maximum was just visible to the naked eye.

SAO Star Catalogue. A general whole-sky catalogue issued by the Smithsonian Astrophysical Observatory, Washington, in 1966. It lists 258 997 stars down to a limiting magnitude of 9 and is intended as an aid in determining artificial satellite positions. An associated and highly accurate computer-plotted atlas was published in 1968.

Saros. The period of 6585.32 days (about 18 years) that elapses before a particular sequence of solar and lunar *eclipses can recur in the same order and with approximately the same time intervals. Following such a period, the earth, sun, moon, and *nodes of the moon's orbit return to about the same relative positions: the period is equal to 223 *synodic months and is almost equal to 19 *eclipse years (6585.78 days). An eclipse repeated after one Saros occurs 0.32 days and hence 115°W of its predecessor. The Saros was known in ancient times.

SAS. *Abbrev. for* Small Astronomical Satellite. Any of a series of US *x-ray and *gamma-ray astronomy satellites. SAS-1, later named *Uhuru, was launched in Dec. 1970. It was devoted entirely to x-ray astronomy and provided a wealth of data. SAS-2, launched in Nov. 1972, carried equipment, including a *spark chamber, for gamma-ray measurements while SAS-3, launched in May 1975, carried x-ray detectors.

satellite, artificial. A manmade object that is boosted into a closed orbit around the earth, moon, or some other celestial body. It is generally unmanned. The cost of launching a satellite can be extremely high, so the mass must be reduced to a minimum. In addition, since repairs are usually impossible, each material and component part must undergo extensive testing before launch, with a carefully considered degree of redundancy in electronic components and circuitry. The problems arising from the near vacuum environment of space, from the bombardment of cosmic rays, micrometeorites, etc., and from magnetic fields and radiation belts must also be considered.

Artificial satellites have a great range of functions. Astronomical earth satellites study the radiations from space that cannot penetrate the atmosphere: *x-ray, *gamma-ray, and *ultraviolet astronomy have been revolutionized by the recent launching of satellites specializing in these fields. Astronomical satellites can also make measurements at optical, radio, and infrared wavelengths that can penetrate, but may be severely affected by, the atmosphere. Other scientifically orientated satellites study the resources, atmosphere, and physical features of the earth.

Satellites are also used for communications (usually in geostationary orbit and allowing global long-distance live television broadcasting and telephony), for weather forecasting, and as navigational aids. In addition, the military potential of satellites has been exploited.

satellite galaxy. A relatively small galaxy that is observed close to a larger system. Although orbital motions are not detected, satellite galaxies are generally assumed to revolve around their larger companions. Both our own *Galaxy and the nearby *Andromeda galaxy are accompanied by satellites.

satellite, natural (moon). A natural body that orbits a planet. The nine major planets of the solar system have a total of 35 known satellites between them, many (14) belonging to Jupiter. In addition, the numerous small bodies that comprise the rings of Saturn, Jupiter, and Uranus may be regarded as natural satellites. A satellite may arise from one of two mechanisms, both of which probably applied in the solar system: a body may grow by accretion from material (*planetesimals) gravitationally attracted towards, and entering orbit around, a *protoplanet; secondly, bodies, such as minor planets, may be captured gravitationally by a planet. *See* Table 2, backmatter.

Saturn. The sixth major planet and, with an equatorial diameter of 120 000 km, the second largest. It orbits at a distance of between 9.0 and 10.1 astronomical units from the sun every 29.5 years; *oppositions recur two weeks later each year. It has a polar diameter of 108 000 km so that its *oblateness, 0.098, is the highest of any planet. Its mean density, 0.7 times that of water, is lower than that of any other planet. Saturn has 10 satellites, the largest and most interesting of which is *Titan. Orbital and physical characteristics of planet and satellites are given in Tables 1 and 2, backmatter.

The telescopic appearance of the planet is dominated by a prominent system of rings (*see* Saturn's rings) that lie in the plane of Saturn's equator, tilted by 27° with respect to its orbit. This leads first one face of the rings, and then the other, to be inclined towards the sun and the earth by up to 27°. At approximately 15 year intervals, the rings become edge-on to the earth and all but disappear; they will be edge-on during 1979-80. The ring system, about 273 000 km across, can add appreciably to Saturn's average apparent magnitude of 0.7 at opposition, when the globe and rings are 19 and 44 arc seconds wide respectively.

Saturn's disc appears similar to that of Jupiter in being crossed by yellowish dark and light cloud bands running parallel to the equator; these are called *belts* and *zones* respectively. They are less prominent, however, than those of Jupiter and spots within them are much less common. The spots, typically white, that have been followed indicate that Saturn, like Jupiter, rotates more rapidly near the equator (in 10 hours 14 minutes or less) than at high latitudes (about 10 hours 40 minutes).

The atmospheric and internal structure of Saturn may also resemble Jupiter. Like Jupiter, it emits more radiation than it absorbs from the sun and probably has an internal primordial energy reservoir. Escaping heat may drive the atmospheric convection processes that give rise to the observed cloud banks, probably of ammonia crystals at -170°C. Spectroscopic studies indicate hydrogen, methane, and ethane in the upper atmosphere; hydrogen and helium probably form the bulk of Saturn's mass. Models for the internal structure of Saturn suggest a central rocky core, about 20 000 km across, enclosed by a 5000 km thick ice layer and about 8000 km of metallic hydrogen. Above these is a 37 000 km shell of liquid molecular hydrogen.

Saturn emits radio waves that are thought to be generated as *synchrotron emission from electrons spiralling in a magnetic field about one tenth as strong as that of Jupiter. It is hoped that this, and other aspects of the Saturnian system, will be studied by the *Pioneer 11 probe due to arrive at Saturn in September 1979, and by *Voyagers 1 and 2 in 1980 and 1981.

Saturn nebula. *See* Aquarius.

Saturn's rings. A system of coplanar rings in Saturn's equatorial plane, with an overall diameter of about 273 000 km. The system is tilted at nearly 27° to Saturn's orbital plane. It was seen indistinctly by Galileo in 1610 and was recognized as a system of rings by Huygens in 1656. Now known to consist of numerous individual particles, each a satellite of Saturn, the rings are probably only one to five km thick and almost disappear when edge-on to the earth every 15 years or so.

Four rings are now recognized: *ring D,* discovered in 1969, is the innermost and faintest, reaching about 12 600 km above Saturn's cloud tops;

ring C, the *crepe ring*, was found in 1850 and is also faint, extending between 16 800 and 32 000 km in altitude; *ring B*, the brightest, joins ring C and extends to 57 800 km; *ring A*, the outermost and second brightest, lies between 60 400 and 76 500 km. The 2600 km gap between rings A and B, discovered in 1675 and called *Cassini's division* after its discoverer, is maintained by gravitational perturbations on ring particles by *Mimas, a satellite of Saturn. *Encke's division* in ring A is a ripple rather than a true break in the ring.

Radio, radar, and infrared measurements suggest that the ring particles range in size from less than a millimetre to 10 metres and consist of chunks of water-ice, possibly with a rocky core. Lying within Saturn's *Roche limit, they may have been prevented from coalescing into a satellite when the rest of the Saturnian system formed.

scarp. Another name for ridge. *See* wrinkle ridge; lobate ridge.

scattering. The random deflections suffered by *waves passing through an irregular medium. If the source, medium, or observer are in relative motion, *scintillations*—random fluctuations of *amplitude—may be seen as the source is observed through the medium: scattering in the earth's atmosphere causes the stars to twinkle. Scintillations may only be seen if both the angular size of the source and the *bandwidth in which the waves are received are small enough. Otherwise, the effect of the scattering may simply be to broaden the apparent angular size of the source.

Scintillations of *radio waves are observed to occur because of irregularities in the *refractive index of the *ionosphere, the *interplanetary medium, and the *interstellar medium giving *ionospheric scintillations, interplanetary scintillations (IPS)*, and *interstellar scintillations (ISS)*, respectively. IPS may be used in determining the angular sizes of *radio sources at metre wavelengths in the range 0.1 to 2 arc seconds. ISS cause some of the random fluctuations in the intensity of pulses received from *pulsars.

Light may be scattered from its direction of travel by fine particles of solid, gaseous, or liquid matter. For very small particles (less than one wavelength in size) the effect results from *diffraction, reflection playing a more important part with increasing size. Very small particles scatter blue light more strongly than red light. This leads to the reddening of starlight by cosmic dust and to the reddening of the sun when seen through a thick layer of atmospheric dust.

Scheat (β Peg). A red giant that is the second-brightest star in the constellation Pegasus and lies on the Great Square of Pegasus. It is an irregular variable with a magnitude range of 2.1-3.0, although the usual limits are 2.3-2.8 M_v: -1.4; spectral type: M2 II-III; distance: 60 pc.

Schedar or **Shedir** (α **Cas**). An orange giant that is one of the two brightest stars in the constellation Cassiopeia. The magnitude varies slowly between 2.1 and 2.6 but it is possibly not a variable star. It has a number of optical companions. M_v: -1.3; spectral type: K0 II-III; distance: 50 pc.

Schmidt telescope (Schmidt camera; Schmidt). A *catadioptric wide-field telescopic camera first built in 1930 by the Estonian Bernhard Schmidt. A short-focus reflecting telescope with a spherical mirror suffers from severe *spherical aberration. This is normally corrected by modifying the surface to a paraboloid. This method of correction is only effective, however, for a field of less than half a degree: outside this limit the star images are distorted by severe *coma. Schmidt overcame the spherical aberration of a spherical mirror, which does not suffer from coma, by placing a *correcting plate in the incoming light beam before it reached the short-focus spherical primary mirror. His correcting plate is a thin plate of complicated figure placed near the centre of curvature of the mirror (see illustration). The corrected light comes to a focus on a curved focal surface but when a photographic plate is sprung to this curve, very sharp star images are formed upon it over a field of maybe tens of degrees: the 1.2 metre *UK Schmidt covers 40 square degrees of sky. In addition relatively short photographic exposures are required, compared to a normal telescope, because of the small *focal ratio of the Schmidt. The development of the Schmidt telescope has enabled whole-sky surveys, i.e. the *Palomar Sky Survey and *Southern Sky

Ray Path in Schmidt telescope

Survey, to include very faint (down to about 21st magnitude) stars.

The *super Schmidt* is an extreme form of Schmidt developed by the American James Gilbert Baker in the late 1950s. By employing additional correcting plates he was able to retain the wide field while increasing the speed to focal ratios approaching 0.5 (Schmidt's original telescope was f/1.7.) These extremely fast cameras were designed to record meteor and artificial satellite trails.

Schönberg-Chandrasekhar limit. An upper limit on the mass of the core of a *main-sequence star: a star will leave the main sequence and become a *red giant when it has exhausted hydrogen supplies totalling some 12% of its original mass.

Schwarzschild radius. The radius of the *event horizon of a *black hole: a critical radius that must be exceeded by a body if light from its surface is to reach an outside observer. For a body of mass M, the Schwarzschild radius, R_s, is given by

$$R_s = 2GM/c^2$$

where G is the gravitational constant and c the speed of light. If a body collapses so that its radius becomes less than this critical value, then the *escape velocity becomes equal to the speed of light and the object becomes a *black hole. The Schwarzschild radius is proportional to the mass of a body. For a star the size of the sun, the Schwarzschild radius is some three km; for a planet like the earth, it is less than a centimetre.

Schwassmann-Wachmann 1. A *comet that moves in a near-circular orbit between the orbits of Jupiter and Saturn and suffers from random *outbursts:* increases in brightness by up to seven magnitudes that last for a few weeks. A considerable number of comets suffer outbursts and many theories have been put forward to explain them. These include pressure release from gas pockets of methane and carbon dioxide, explosive chemical reactions involving free radicals (CH, OH, and NH), temperature-induced phase changes in amorphous ice, impact cratering by interplanetary boulders, and spin- and age-induced nuclear break-up. Impact and break-up are the most likely causes.

scintillation (twinkling). Rapid irregular variations in the brightness of light received from celestial objects, noticeably stars, produced as the light passes through the earth's atmosphere: irregularities in the atmosphere's refractive index occur in small mobile regions and can cause the direction of the light to change very slightly during its passage. In a telescope, a star image will consequently wander rapidly about its mean position, producing an overall blurred enlarged image. With extended light sources, such as the planets, scintillation produces a hazy outline in a telescopic image. For stars near the horizon, where refraction effects including *dispersion are much greater, changes in colour can also be observed. *See also* scattering; speckle interferometry.

scintillation counter. A radiation detector that is used in astronomy mainly for measurements of gamma rays. It consists of a layer of scintillator crystals that emit flashes of light when radiation passes through. Each scintillation is picked up by a *photomultiplier and an amplified pulse of current is produced. These pulses can be counted electronically and the activity of the gamma-ray source can be measured. It is also possible to determine the distribution of energy of the gamma rays.

The energy of the electrons liberated from the photocathode of the photomultiplier depends on the frequency of the emitted light, which in turn depends on the energy of the incident beam. It is possible to exclude radiation of energies other than those under consideration by suitable electronic circuitry.

Scorpius (Scorpion). A conspicuous zodiac constellation in the southern hemisphere, lying partly in the Milky Way. The brightest stars are *Antares (α) and the blue-white spectroscopic binary, Shaula (λ), both 1st magnitude, and several of 2nd and 3rd magnitude. The area contains many star clusters, including the bright naked-eye open clusters M6 and M7 and the globular clusters M4 and M80, and the x-ray source *Scorpius X-1. Abbrev.: Sco; genitive form: Scorpii; approx. position: RA 17h, dec -30°; area: 497 sq deg.

Scorpius X-1. The first cosmic x-ray source, discovered during a sounding rocket experiment from White Sands, New Mexico in June 1962. By an order of magnitude Sco X-1 is the brightest of all nontransient cosmic x-ray sources. Optically identified in 1967 with the 13th magnitude variable star V 818 Sco, it was confirmed as a low-mass binary system with an orbital period of 0.78 days. The distance of Sco X-1 remains uncertain, probably lying in the range 300-600 parsecs.

Sculptor. An inconspicuous constellation in the southern hemisphere near Grus, the brightest stars being of 4th magnitude. R Sculptoris is an intense scarlet variable star. The south *galactic pole lies in Sculptor and many faint galaxies can be observed. Abbrev.: Scl; genitive form: Sculptoris; approx. position: RA 0.5h, dec -30°; area: 475 sq deg.

Sculptor system. A member of the *Local Group of galaxies.

Scutum (Shield). A small constellation in the southern hemisphere near Sagittarius, lying

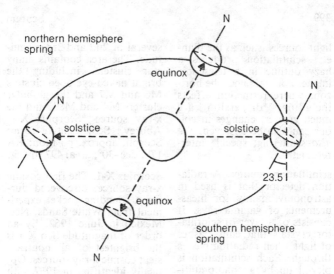

Annual change in seasons

in the Milky Way, the brightest stars being of 4th magnitude. It contains the bright *RV Tauri star R Scuti, the prototype of the *Delta Scuti stars, and many star clusters, including the just-visible fan-shaped *open cluster, the *Wild Duck* (M11). Abbrev.: Sct; genitive form: Scuti; approx. position: RA 19h, dec -10°.

Sea Goat. Common name for *Capricornus.

sea interferometer. An interferometer (*see* radio telescope) that uses a single *antenna mounted on a clifftop overlooking the sea. Fringes are observed on a *radio source at low elevation as a result of the *interference between radio waves from the source itself and from its image reflected in the sea.

Sea Serpent. Common name for *Hydra.

seas, lunar. *See* maria, lunar.

seasons. The four approximately equal divisions of the year, usually taken as lasting from *equinox to *solstice—spring and autumn—or from solstice to equinox—summer and winter. The seasons, which are more noticeable away from the equator, result from the inclination (23.5°) of the earth's axis to the perpendicular to the earth's orbital plane so that the sun spends half the year north and the remaining year south of the celestial equator (see illustration). The varying distance of the earth from the sun has a minor effect on the seasons. Mars, with an axial tilt of 24°, also has seasons.

Sechi classification. *See* spectral types.

second. Symbol: s. The scientific (SI) unit of time, defined since 1964 in terms of *atomic time. The second is the duration of 9 192 631 770 periods of the radiation corresponding to the transition between two hyperfine energy levels of the ground state of the caesium-133 atom. *See also* ephemeris time.

secondary. 1. *See* primary.
2. *Short for* secondary crater.

secondary cosmic rays (secondary radiation). *See* cosmic rays.

secondary craters. Depressions produced by the impacts of low-velocity crater *ejecta. Secondary craters are usually elongated radially with respect to the primary *crater and form loops, *crater chains, and *crater clusters. They tend to lie outside the *ejecta blanket.

second of arc. *See* arc second.

secular. Continuing or changing over a long period of time.

secular acceleration. The apparent speeding up of the moon in its orbit around the earth, caused by *tidal friction and the gravitational attraction of other planets. It amounts to 10.3 arc seconds per century. It first became apparent from records of ancient *eclipses.

secular parallax. The apparent and continuously increasing displacement of stars as a result of the sun's motion through space. This form of *parallax can be used to determine the distance of nearby groups of stars. It is assumed that the stars are at approximately the same distance from the sun and that their random individual proper motions tend to cancel out leaving only the parallactic motion. The baseline for this parallax measurement is the distance moved by the sun relative to the *local standard of rest and is hence increasing by the sun's velocity of 19.5 km/s, i.e. 4.11 *astronomical units per year (*see* apex). The secular parallax per annum, H (in arc secs), is then given by:
$$\sin H = 4.11 \sin \alpha/d$$
where d is the mean distance to

Secular parallax

that group and α is the angle of the group and α is the angle of illustration).

secular perturbation. *See* perturbation.

seeing. The quality of the observing conditions at the time of telescopic observation. The various degrees of good or bad seeing depend on the amount of turbulence in the earth's atmosphere. Turbulence distorts the plane wavefront of a beam of radiation, which has travelled more or less undisturbed through space, and leads to a perturbed 'corrugated' wavefront. On the few occasions when atmospheric turbulence is very low, a small steady disc-shaped optical image of a point source (such as a star) results. Poor seeing mainly produces small erratic movements in the image position (*see* scintillation) so that the overall image is blurred distorted and enlarged. No matter how well-designed the telescope, it is the seeing that imposes the main limitation on the instrument's *resolving power: the minimum diameter of the optical image, resulting from diffraction (*see* Airy disc), can be as low as 0.025 arc seconds but the angular resolution may only be one or two arc seconds, or more, as a result of poor seeing.

In amateur astronomy the seeing can be evaluated using the *Antoniadi scale:* the observer allocates a Roman numeral from I to V to conditions varying from perfect seeing, through good, moderate, poor, to appalling. *See also* speckle interferometry.

seismology. The study of *seismic waves,* which propagate through the solid material of the earth, or other celestial bodies such as the moon, following earthquakes (or equivalent), explosions, or impact phenomena. Four types of wave are studied: *S-waves,* or *transverse waves,* which vibrate at right angles to their direction of motion; *P-waves,* or *longitudinal waves,* which oscillate along their direction of motion; *Rayleigh waves,* which propagate as a rolling motion of the surface; and *Love waves,* which produce horizontal vibrations of the surface at right angles to the direction of propagation. The S- and P-waves travel deep into the body in question where they are refracted at the boundaries between layers of differing density. The waves are measured and analysed to derive information about the events that generated them or about the internal structure of the body through which they travel.

seismometer. An instrument used in *seismology to measure the passage of seismic waves, normally by making a continuous recording of the position of a suspended mass in relation to a frame fixed to the ground.

selected areas. Small areas uniformly demarcated over the sky, for which detailed astronomical measurements can be made: photographic determination of the numbers and *magnitudes of all the observed stars and the *parallax, *spectral types, *radial velocities, *proper motions, etc., of a smaller number. Statistical analysis of these sample areas can reveal information on Galactic structure, etc. The measurement programme was devised by J.C. Kapteyn in 1906.

selenography. The study of the moon's physical features. *See* moon, surface features.

selenology. The scientific study of the *moon. *See also* moonrocks.

self-absorption. *See* synchrotron emission.

semidetached binary. *See* equipotential surfaces.

semidiameter. Half the *apparent diameter of a celestial body.

semimajor axis. Symbol: a. The distance from the centre of an ellipse to the edge, through the intervening focus. It is half the greatest distance across the ellipse, i.e. half the *major axis*. The semimajor axis of an elliptical orbit is one of the *orbital elements.

semiregular variables (SR variables). A heterogeneous group of giant and supergiant *pulsating variables showing brightness variations that do not usually exceed one or two magnitudes, that have a noticeable periodicity ranging from several days to several years, but are also disturbed at times by various irregularities. The light curves have diverse shapes. Semiregular regulars have been divided into four subgroups: *SRa* and *SRb variables* are giants of late spectral type (M, C, and S) either with relatively stable periods (SRa) or ill-defined periods (SRb) and are difficult to differentiate from long-period *Mira variables; *SRc variables* are red supergiants of late spectral type, such as Betelgeuse, Antares, and Mu Cephei; *SRd variables* are highly luminous yellow (F, G, and K) supergiants and giants.

sensitivity. The minimum signal power that can be distinguished from the random fluctuations in the output of a measuring system caused by *noise inherent in the system. In a *radio telescope contributions to the *system noise* include noise generated in the first stage of the *receiver, thermal *noise due to loss in the *antenna/*feeder system, *synchrotron emission received by the antenna from the Galaxy (*see* radio source), and *thermal emission from the atmosphere and from the ground. If the system noise has a power P watts, then an equivalent *system temperature,* or *noise temperature,* can be ascribed to it given by

$$T = P/kB$$

where k is the Boltzmann con-

separation

stant and *B* the *bandwidth.

The signal/noise ratio is the ratio of the power in the output of a radio telescope that is due to the *radio source under observation to that caused by the system noise. The sensitivity is usually defined as the *flux density of a source that would produce the same signal power as the noise power, i.e. a signal/noise ratio of one. *See also* confusion.

separation. Symbol: ϱ. The angular distance between the two components of a *visual binary or optical *double star. Separation is measured with a *filar micrometer in the eyepiece of the telescope; its value is expressed in seconds of arc (″).

Serenitatis Basin. One of the youngest lunar *basins, formed 4000 million years ago and filled with *mare lavas from at least 3800 million years ago to produce the circular *Mare Serenitatis* (*Sea of Serenity*). The Mare has striking colour contrasts and numerous *rilles and *wrinkle ridges. It was visited by *Lunokhod 2 and the margins sampled by the crew of *Apollo 17.

Serpens (Serpent). A constellation divided into two unequal parts, originally called *Serpens Caput* (Serpent's Head) and *Serpens Cauda* (Serpent's Body), that lie on either side of the constellation Ophiuchus. The larger part (the Head) lies in the northern hemisphere near Corona Borealis; the smaller section lies in the southern hemisphere near Scorpius, partly in the Milky Way. The two sections use a single set of Bayer letters (*see* stellar nomenclature) to identify the stars. The brightest stars are Alpha Serpentis, of 2nd magnitude, and several of 3rd magnitude. The globular cluster, M5, in the larger part contains over 100 RR Lyrae stars. The *Eagle nebula, M16, lies in the smaller section. Abbrev.: Ser; genitive form: Serpentis; approx. position: RA 15.5h, dec +10° (Head), RA 18.3h, dec -5° (Body); area: 428 sq deg (Head), 208 sq deg (Body).

Serpentarius. Obsolete name for *Ophiuchus.

Serpent Bearer. Common name for *Ophiuchus.

service module (SM). *See* Apollo.

setting. *See* rising.

setting circles. The circular scales provided on the polar and declination axes of a nonportable *equatorial mounting. With their aid the telescope may be set to chosen values of *right ascension and *declination, respectively, so as to bring an object of known position into the field of view.

Seven Sisters. *See* Plough.

Sextans (Sextant). An inconspicuous equatorial constellation near Leo, the brightest star being of 4th magnitude. Abbrev.: Sex; genitive form: Sextantis; approx. position: RA 10h, dec -2°; area: 314 sq deg.

Seyfert galaxy. A galaxy with an exceptionally bright nucleus, first described by Carl Seyfert in 1943. Over 150 Seyfert galaxies have now been identified and the vast majority of these are otherwise normal nearby spiral galaxies. Seyferts have substantial *nonthermal emission in their nuclei. The nuclei emit radiation at radio, infrared, optical, and particularly x-ray wavelengths, but this output can vary on a timescale of months. This indicates that the energy source in Seyfert galaxies must be a few light-months across and astronomers believe that Seyferts are a less powerful example (by a factor of 100) of the *quasar phenomenon.

A region of hot gas (at a temperature of 20 000 kelvin) surrounds Seyfert nuclei, much of it concentrated into clouds moving outwards and presumably accelerated by the central energy source. Velocities of these clouds can reach 3000 km s^{-1} (in MGC 4151). In *Type I Seyferts*, whose spectra closely resemble those of quasars, the Balmer lines of hydrogen emitted by the gas are broader than the emission lines of ionized metals; this indicates velocities of several thousand km s^{-1}. The widths of the two sets of lines are comparable in *Type II Seyferts*, showing lower gas velocities—less than 1000 km s^{-1}.

Some 10% of giant spiral galaxies are Seyferts. As Seyfert activity is probably only a transitory period in the life of a galaxy, then it is possible that all giant spiral galaxies, including our own, spend 10% of their lives in a Seyfert phase.

shadow bands. Dark bands of irregular form and movement that are briefly observed to cross rapidly over light surfaces on the earth just before and after totality in a solar *eclipse. They are up to one metre apart. They probably result from irregular refraction, in the earth's atmosphere, of light from the crescent sun.

Shaula. *See* Scorpius.

Shedir. *See* Schedar.

shell star. A hot main-sequence star, usually of spectral type B, whose complex spectrum shows sharp deep absorption lines with 'wings' of bright emission lines superimposed on the normal absorption spectrum. It is now known that they are *Be stars, the differences in appearance between the two arising from the directions in which they are observed.

Shield. Common name for *Scutum.

Ship. Common name for *Argo.

shock wave. A compression wave of large amplitude that is produced when there is a sharp and violent change in pressure, density, and/or temperature in a fluid. It occurs when a disturbance in the fluid cannot be dispersed quickly enough and thus banks up. A shock wave travels at greater than the speed of sound and can lose energy by heating the medium through which it travels.

shooting star. *Colloquial name for* meteor.

short-period comet. *See* comet.

shower. *Short for* air shower. *See* cosmic rays.

shower meteor. *See* meteor shower.

Shuttle. *Short for* Space Shuttle.

Sickle of Leo. *See* Leo.

side lobe. *See* antenna.

sidereal. Relating to or measured or determined with reference to the stars.

sidereal day. The interval between two successive passages of the vernal *equinox across a given meridian. It is divided into 24 sidereal hours. The period of the earth's rotation on its axis relative to the stars is the interval between two successive passages of a star or stellar group across the meridian. It is slightly longer than the sidereal day because of the *precession of the equinoxes. Both intervals are shorter than the mean solar day (*see* day): the sidereal day is equal to
23h 56m 4.091s (0.997 269 57 days)
of *mean solar time; the sidereal rotation period is equal to
23h 56m 4.099s (0.997 269 66 days)
of mean solar time. *See also* sidereal time.

sidereal month. The time of 27.321 66 days taken by the moon to complete one revolution around the earth, measured with respect to a background star or stellar group considered fixed in position.

sidereal noon. *See* sidereal time.

sidereal period. The time taken by a planet or satellite to complete one revolution about its primary, measured by reference to the background of stars. The *sidereal month and *sidereal year are the sidereal periods of the moon and earth. *See also* synodic period; Tables 1 and 2, backmatter.

sidereal rate. *See* equatorial mounting.

sidereal time. Time measured in *sidereal days and in sidereal hours, minutes, and seconds, i.e. time measured by reference to the rotation of the earth with respect to the stars, which are regarded as fixed in position. The sidereal time at any instant is given by the *hour angle (the angle measured clockwise from the meridian) of the vernal *equinox and ranges from 0 to 24 hours during one day. The day starts at *sidereal noon*, which is the instant at which the vernal equinox crosses the local meridian. The hour angle at a particular location gives the *local sidereal time, the hour angle at Greenwich being the *Greenwich Sidereal Time. A celestial object will be on the meridian of a particular place when the local sidereal time becomes equal to the object's *right ascension.

Apparent sidereal time is measured by the hour angle of the *true equinox and thus suffers from periodic inequalities, the position of the true equinox being affected by the *nutation of the earth's axis. *Mean sidereal time* relates to the motion of the *mean equinox, which is only affected by long-term inequalities arising from *precession. Apparent minus mean sidereal time equals the *equation of the equinoxes*. Sidereal time is directly related to *universal time and *mean solar time and is used in their determination. *See also* ephemeris time.

sidereal year. The time taken by the earth to complete one revolution around the sun with reference to a background star or stellar group, which is regarded as fixed in position. It is thus the interval between two successive passages of the sun, in its apparent annual motion around the celestial sphere, through a particular point relative to the stars. It is equal to 365.256 36 days and is about 20 minutes longer than the *tropical year.

siderite. *See* iron meteorite.

siderolite. *See* stony-iron meteorite.

siderostat. A flat mirror that is driven in such a way as to reflect light or infrared radiation from a celestial body to a fixed point, such as a spectrograph slit at the coudé focus of a telescope, over the duration of an observation. A siderostat is often installed outside the main dome of an observatory and can be used with a coudé spectrograph, say, at the same time as the main telescope is being used to study a different field of view. Siderostats often have computer-driven *altazimuth mountings.

Siding Spring Observatory. The observatory on Siding Spring Mountain in New South Wales, Australia, at an altitude of 1165 metres. The chief instruments are the 3.9 metre (154 inch) *Anglo-Australian telescope and the 1.2 metre (48 inch) *UK Schmidt telescope.

signal/noise ratio. *See* sensitivity.

signs of the zodiac. *See* zodiac.

silicon stars. *See* Ap stars.

silvering. A process whereby a

reflective layer of silver is applied to a mirror base from a solution of silver salts containing reducing agents. The glass is first activated by immersion in stannous chloride solution. Silvering was first used in the 1850s but has been largely supplanted by the *aluminizing process, which provides a reflective layer of greater durability and able to reflect lower wavelengths, i.e. into the ultraviolet.

sine formula. *See* spherical triangle.

singularity. A mathematical point at which space and time are infinitely distorted. Calculations predict that matter falling into a *black hole will ultimately be compressed to infinite densities at a single point, and in such conditions our laws of physics, including quantum mechanics, must break down. One black-hole theorem—the principle of *cosmic censorship*—states that singularities are always concealed by an event horizon so that they cannot communicate their existance to an observer in our universe. However, if a *naked singularity*—a singularity without an event horizon—is found, then some of our physical concepts will need re-examination.

Sinope. A small satellite of Jupiter, discovered in 1914. *See* Jupiter's satellites; Table 2, backmatter.

sinuous rille. *See* rille.

SiO maser. *See* maser source.

Sirius (Dog Star; α CMa). A white main-sequence star that is the brightest star in the constellation Canis Major and the brightest and one of the nearest stars in the sky. It lies in a descending line from Orion's Belt. It is a visual binary (separation 7".6, period 49.97 years), the companion, *Sirius B*, being the first *white dwarf to be discovered. Bessel suggested (1844) that Sirius had a dark companion to account for the star's wobbling movement. With improved telescope lenses Alvan G. Clark detected (1862) a tiny companion whose spectrum, first taken (1925) by W.S. Adams Jn., identified Sirius B as a white dwarf. The spectrum demonstrated the *gravitational redshift predicted by the general theory of relativity. m_v: -1.46 (A), 8.67 (B); M_v: 1.41 (A), 11.54 (B); spectral type: A1 V (A), wdA5 (B); mass: 2.31 (A), 0.98 (B) times solar mass; radius: 1.7 (A), 0.022 (B) times solar radius; distance: 2.67 pc.

SI units. An internationally agreed system of coherent metric units, increasingly used for all scientific and technical purposes. It was developed from the MKS system of units and replaces the CGS and Imperial systems of units. There are seven base units that are arbitrarily defined and dimensionally independent (see Table 1). These include the metre and the second. The base units

Table 1. Base and Supplementary SI Units

physical quantity	name of SI unit	symbol for unit
length	metre	m
mass	kilogram(me)	kg
time	second	s
electric current	ampere	A
thermodynamic temperature	kelvin	K
luminous intensity	candela	cd
amount of substance	mole	mol
*plane angle	radian	rad
*solid angle	steradian	sr

*supplementary units

Table 2. Derived SI Units with Special Names

physical quantity	name	symbol	Base units
frequency	hertz	Hz	s^{-1}
energy	joule	J	$kg\ m^2 s^{-1}$
force	newton	N	$kg\ m\ s^{-2}$
power	watt	W	$kg\ m^2 s^{-3}$
pressure	pascal	Pa	$kg\ m^{-1} s^{-2}$
electric charge	coulomb	C	$A\ s$
electric potential difference	volt	V	$kg\ m^2 s^{-3} A^{-1}$
electric resistance	ohm	Ω	$kg\ m^2 s^{-3} A^{-2}$
magnetic flux	weber	Wb	$kg\ m^2 s^{-2} A^{-1}$
magnetic flux density	tesla	T	$kg\ s^{-2} A^{-1}$

Table 3. Decimal Multiples and Submultiples to be used with SI Units

submultiple	prefix	symbol	multiple	prefix	symbol
10^{-1}	deci	d	10^{1}	deca	da
10^{-2}	centi	c	10^{2}	hecto	h
10^{-3}	milli	m	10^{3}	kilo	k
10^{-6}	micro	µ	10^{6}	mega	M
10^{-9}	nano	n	10^{9}	giga	G
10^{-12}	pico	p	10^{12}	tera	T
10^{-15}	femto	f			
10^{-18}	atto	a			

can be combined by multiplication and/or division to derive other units, such as the watt or the newton (see Table 2). A set of prefixes, including kilo- and micro-, can be used to form decimal multiples or submultiples of the units (see Table 3). Any physical quantity may then be expressed in terms of a number multiplied by the appropriate SI unit for that quantity.

six-colour system. A system of stellar *magnitudes in which photoelectric magnitudes U, Vi, B, G, R, and I are measured at six wavelength bands: the measurement of U (ultraviolet) is centred on 355 nanometres (nm), for Vi (violet) on 420 nm, for B (blue) on 490 nm, for G (green) on 570 nm, for R (red) on 720 nm, and for I (infrared) on 1030 nm.

Skyhook. *See* rockoon.

Skylab. A large manned US

Skylab

Skylab Missions

spacecraft	astronauts	launch date	flight duration (hrs)
Skylab 2	Conrad, Kerwin, Weitz	1973, May 25	672.8 (record)
Skylab 3	Bean, Lousma, Garriott	1973, July 28	1427.2 (record)
Skylab 4	Carr, Gibson, Pogue	1973, Nov. 16	2017.3 (record)

space station launched on May 14 1973 into a 435 km circular earth orbit, inclined at about 50° to the equator. The station—Skylab 1—provided an environment in which men could live and work under controlled but weightless conditions. During 1973 and 1974 three three-man crews were ferried to and from the station by *Apollo spacecraft—Skylabs 2, 3, and 4—consisting of a command and service module (see table). Although a record flight duration of 2017 hours was achieved by the Skylab 4 crew, this was broken by a *Salyut 6 crew who remained in space 3340 hours. With the Apollo craft docked, the overall length was 36 metres and the weight 82 000 kg; the orbital workshop was 14.7 metres long and 6.6 metres in diameter. Power was obtained from several panels of *solar cells: the first crew had to do extensive repairs to the solar panels damaged during take-off; in addition the meteoroid shield and one of the solar panels had been torn away at take-off so that the workshop was exposed to intense heat and the available power was reduced.

Physiological and psychological reactions to weightlessness were carefully monitored and studied, and ability to perform experiments was tested. In over 513 man days in space 73 experiments were conducted. The instruments of the *Apollo Telescope Mount provided the majority of the observational data, other work including technological, biological, and materials-processing experiments. Over 180 000 solar photographs and 40 000 earth-resources pictures were taken. The 740 hours spent on solar studies and 215 hours of astrophysical investigation, made above the earth's distorting and absorbing atmosphere, led to great advances in astronomy.

NASA had assumed that Skylab would stay in orbit for 10 years, with the Space Shuttle being used to reboost it for further use or deboost it for controlled re-entry and crash-landing in the Pacific. In March 1978, however, it was reported that the space station had dropped to an altitude of about 380 km and that uncontrolled re-entry could occur as early as mid-1979. Despite successful attempts by NASA to manoeuvre the station into a more stable attitude, and so decelerate descent, NASA revealed in December 1978 that it had abandoned the rescue mission: Skylab will crash land sometime in mid-1979, although very few large fragments will survive to hit the earth's surface.

slow nova. *See* nova.

slow pulsator. *See* Ariel V; x-ray astronomy.

small circle. Any circle on the surface of a sphere that is not a *great circle. It is the circular intersection on a sphere of any plane that does not pass through the centre of the sphere.

Small Magellanic Cloud (SMC; Nubecular Minor). *See* Magellanic Clouds.

SMC. *Abbrev. for* Small Magellanic Cloud. *See* Magellanic Clouds.

SNU. *Abbrev. for* solar neutrino unit. *See* neutrino detection experiments.

SO galaxy. *See* galaxies.

solar activity. The collective term for time-dependent phenomena on the sun. *Sunspots, *faculae/*plages, *filaments (or *prominences), and *flares all belong in this category, but the *granulation and *chromospheric network do not since their gross configuration is stable. The level of solar activity varies over an average period of approximately 11 years—the so-called solar cycle (*see* sunspot cycle).

solar antapex. *See* apex.

solar apex. *See* apex.

solar calendar. Any *calendar, such as the *Gregorian calendar, that is based solely on the motions of the sun. *See also* tropical year; lunar year.

solar cell. A device by which incident solar radiation is converted directly into electrical energy. It is a *photovoltaic cell in which the semiconductor junction has a large surface area: radiation falling on the junction causes a voltage to appear across the junction so that power can be delivered to an external circuit. The conversion efficiency can reach 20-25%.

Solar cells form the main power supply in satellites, space stations, and short-range planetary probes. The cells are arranged on flat *solar panels* outside the craft to receive the maximum amount of radiation from the sun. On probes travelling beyond Mars, the radiation flux is insufficient to power the instruments: the solar constant at Jupiter's orbit is only about 4% of the value at the earth's orbit. Power must then be obtained from other sources, usually *thermoelectric generators.

solar constant. The total energy, radiated from the sun, that passes perpendicularly through unit area per unit time at a specified distance from the sun. It is given by the ratio $L/4\pi r^2$, where L is the luminosity of the sun and r the distance. The value measured at the mean sun-earth distance has recently been given as 1.367×10^3 joules per square metre per second, i.e. 1.367 kilowatts per square metre. This value is quoted as being accurate to within ±5%. Another recent measurement gives 1.353 kW m^{-2}, with an accuracy of ±1.5%. Ideally measurements should be made, from spacecraft, over at least one *sunspot cycle.

solar cycle. *See* sunspot cycle.

solar day. *Short for* apparent solar day. *See* apparent solar time.

solar eclipse. *See* eclipse.

solar mass. *See* solar units.

solar motion. *See* apex.

solar neutrinos. *See* neutrino detection experiments.

solar panel. *See* solar cell.

solar parallax. The angle subtended by the earth's equatorial radius at the centre of the sun, at the mean sun-earth distance of one astronomical unit. Recent determinations give the value as 8.794 148 arc seconds.

Solar Polar Mission. *See* International Solar Polar Mission.

solar spectrum. The sun's spectrum extends from gamma-ray to radio wavelengths. It has an immense range in intensity, peaking at visible wavelengths. Although the central part of the curve varies little with *solar activity, the long and short wavelength sections can be very considerably affected. The radiation intensity at visible and infrared wavelengths agrees with that of a *black body at a temperature of about 6000 kelvin; the maximum intensity occurs at wavelengths of about 500 nanometres. This is the continuous spectrum of the *photosphere in which absorption lines—*Fraunhofer lines—appear. At the shortest and longest wavelengths, the solar spectrum corresponds to the radiation curve of a black body at about a million kelvin, representative of the temperatures of the *corona and of solar *flares. At ultraviolet and soft x-ray wavelengths the spectrum does not agree with either of these black-body curves. *See also* flash spectrum.

solar system. A group of celestial bodies comprising the sun and the large number of bodies that are bound gravitationally to it and revolve in elliptical orbits around it. The latter include the nine known major planets, Mercury, Venus, Earth, Mars, Jupiter, Saturn, Uranus, Neptune, and Pluto; the 35 known natural satellites of the major planets; the numerous minor planets, mostly in orbits between the orbits of Mars and Jupiter; the comets, which exist in great numbers beyond the most distant planets; and countless meteoroids. The sun contains 99.86% of all the mass of the system, while most of the remaining mass is concentrated in the planet Jupiter. In extent the solar system may be considered as a sphere with a radius greater than 100 000 astronomical units (AU) although the planets alone extend to a radius of less than 50 AU (*see* Oort cloud).

With the exception of the comets, all bodies of the solar system orbit the sun in the same direction as the earth, along orbits that lie close to the plane of the earth's orbit and the sun's equator. This direction is also close to that of the axial rotation of the sun and most planets, and of the orbital motion of most planetary satellites. This motion reflects the common origin postulated for the solar system by the contraction of a rotating cloud of interstellar gas and dust about 4600 million years ago (*see* solar system, origin).

The sun and its entire retinue are moving in a nearly circular orbit around the centre of the *Galaxy, which lies about 10 000 parsecs away in the constellation Sagittarius. They have a mean velocity of about 250 km s^{-1} so that they complete one revolution every 220 million years or so. In relation to the stars in its neighbourhood, the solar system is moving with a velocity of 19.4 km s^{-1} towards a point in the constellation Hercules—the solar *apex.

solar system, origin. There have been various theories on the origin of the solar system, many of which have been discarded or modified as observational data has slowly accumulated. The sun and planets are now believed to have formed together 4600 million years ago by the contraction of a cloud of interstellar gas, mainly hydrogen and helium, and dust grains—the *primeval nebula. The contraction may have been triggered by shock waves from a nearby supernova, but continued under the mutual gravitational attraction of the contents of the nebula.

Conservation of *angular momentum dictated that as the cloud contracted its rotation rate increased, causing it to become disc-shaped around the developing sun. Collisions between dust grains became frequent, leading to a rapid concentration of particles in the plane of the disc, which lay close to the present plane of the ecliptic. Thus a gradual *accretion of grains into pebbles, boulders, and larger bodies began. Only when these

solar system, origin

*planetesimals had reached a few kilometres across could their gravity accelerate the accretion process by attracting more solid material. Eventually the protoplanets of the present planets and satellite systems were built up, most of them rotating and revolving in the same sense as the parent nebula. Possibly any protoplanets that formed between the orbits of Mars and Jupiter suffered gravitational perturbations by Jupiter and were fragmented by collisions to leave the minor planets of today.

Although the sun had yet to begin nuclear fusion processes, temperatures in the inner solar system were high enough to maintain volatile substances such as water, methane, and ammonia in a gaseous state; the planetesimals that gave rise to the *inner planets were therefore formed from the nonvolatile component of the nebula, such as iron and silicates. Further from the embryo sun, lower temperatures allowed volatiles to become important constituents of the *giant planets, their satellite retinues, and the ring systems of Saturn and Uranus. The origin of the volatile-rich comets is more obscure, but they may have formed concurrently as planetesimals near the outer edge of the solar system's disc. The low temperatures also made it possible for the giant planets to greatly augment their masses by attracting and retaining large quantities of the light elements hydrogen and helium from the surrounding nebula.

As the sun continued to collapse, the density and temperature in its central core rose (see stellar birth). Once the core temperature reached some ten million kelvin, the sun began to generate energy by the nuclear fusion of hydrogen. The onset of fusion processes in the sun set up the *solar wind, which drove uncollected gases from the solar system. Heating of the protoplanets by gravitational contraction and interior radioactivity caused partial melting that led to the present internal differentiated structure of the planets. Impacts by remaining planetesimals scarred the surfaces of the planets and their satellites, although the earth's craters have been largely removed by erosion. Cratering events continue on a much reduced scale to the present day as errant minor planets or cometary nuclei collide with the planets or their satellites.

The suggestion that a contracting nebula formed the solar system was the basis of the *nebular hypothesis, popular during the 19th century. However this was unable to explain why most of the solar system's angular velocity resides in the revolution of the planets, rather than in the rotation of the sun, and was replaced by various *encounter theories early in the 20th century. The renaissance of a nebular theory came with the realization that *magnetohydrodynamic forces would transfer angular momentum from the early sun to its surrounding nebula, and that the sun would shed further angular momentum through the solar wind. Calculations incorporating these ideas suggest that the primeval nebula contained at least three solar masses of material, most of which was expelled again into interstellar space.

solar telescope. A reflecting (or refracting) telescope plus associated instruments used in studying the sun. The sun is the only star whose disc can be resolved and studied by a telescope. The diameter of the solar image is given by $0.0093f$, where f is the focal length of the telescope primary mirror or objective. Long-focus telescopes must therefore be used so that a large image is formed with plenty of detail. Such telescopes are usually fixed in position, with the sunlight being reflected on to the mirror or lens by a *heliostat. This light is then directed in a fixed direction, into a spectrograph or some other measuring instrument.

The intense heat of solar radiation requires the measuring instruments to be cooled. In some telescopes the entire building is cooled. In most cases the primary mirror and the instruments are underground to facilitate this. The heat problem can be reduced by evacuating the air from the rooms housing the instruments or even from the whole telescope housing.

Optical path in McMath solar telescope

Large solar telescopes are often built as *tower telescopes* (or *solar towers*). The sunlight is directed down a vertical path inside a solid or girder construction and reflected into underground rooms containing the measuring instruments. The *McMath solar telescope* at the Kitt Peak National Observatory, Arizona, has a somewhat different design (see illustration). Sunlight is reflected from a 2 metre heliostat, which is 30 metres above the ground. It traverses 150 metres along the polar axis and is focused by a 1.5 metre concave mirror and a second flat mirror on to the various measuring instruments.

The best solar seeing is found at sizable lakes; in addition high altitudes are required in order to reduce the disturbing effects of the earth's atmosphere. The site of the *Big Bear Solar Observatory has both these features.

solar time. 1. *Short for* mean solar time.
2. *Short for* apparent solar time.

solar tower. *See* solar telescope.

solar units. Dimensionless units in which the mass, radius, luminosity, and other physical properties of a star can be expressed in terms of the sun's mass, radius, etc. For example, a star's mass in solar units is equal to its actual mass divided by the solar mass and is usually written $m/m\odot$.

solar wind. A flow of energetic charged particles—mainly *protons and *electrons—from the solar *corona into the *interplanetary medium. The thermal energy of the ionized coronal gas is so great that the sun's gravitational field cannot retain the gas in a confined static atmosphere. Instead, there is a continuous, near-radial, outflow of charged particles into interplanetary space, this highly tenuous plasma carrying mass and angular momentum away from the sun. The expansion is controlled by the sun's magnetic field. It is low in the inner corona but rapidly becomes supersonic and reaches anywhere between 200 and 900 km s^{-1} at one astronomical unit, i.e. in the vicinity of the earth's orbit. The density of the plasma decreases with increasing distance from the sun and at the earth's orbit is only about eight particles per centimetre cubed.

Two forms of the solar wind are readily recognizable: a slow-moving low-density flux (<400 km s^{-1}), and high-speed streams. The latter are thought to form the normal state and originate in areas of relatively weak divergent open (and therefore primarily unipolar) magnetic field. The emission is greatly increased over *coronal holes and the long-term stability of such regions may cause recurrent enhancement over several solar rotations. Similar long-lived jets are apparently emitted from the polar regions. Confinement by magnetic fields in the corona results in the slow-moving flow and this is more frequent around the maximum of the *sunspot cycle. In addition, transient mass ejections occur during *flares and in some of the more frequent but less energetic active *filaments/*prominences.

The sun's magnetic field is transported by the expanding plasma and becomes the *interplanetary magnetic field*, the lines of which are wound into spirals by the sun's rotation. The region of expanding solar wind, or *heliosphere*, is bounded (possibly at 50-100 AU) by the *heliopause*, where its pressure equals that of interstellar space. The heliosphere is probably very elongated due to the presence of an *interstellar wind* of neutral hydrogen, flowing from the direction of the galactic centre.

solid-state camera. *See* CCD camera.

solstices. 1. (solstitial points). The two points that lie on the *ecliptic midway between the vernal and autumnal *equinoxes and at which the sun, in its apparent annual motion, is at its greatest angular distance north of the celestial equator (*summer solstice*) or south of the celestial equator (*winter solstice*).

2. The times at which the sun reaches its greatest altitude north or south of the celestial equator. On the day of the *summer solstice*, which occurs on about June 21 (northern hemisphere) or about December 21 (southern hemisphere), the hours of daylight are at a maximum. The hours of darkness are at a maximum at the *winter solstice*, which falls about December 21 (northern hemisphere) or June 21 (southern hemisphere).

solstitial colure. *See* colures.

solstitial points. *See* solstices.

Sombrero galaxy (M104; NGC 4594). A spiral (Sa/Sb) galaxy in the constellation Virgo that has a relatively large bright nucleus. The spiral structure is difficult to trace because the equatorial plane is tipped only 6° from our line of sight. A dark absorption band of interstellar dust in the galactic plane is clearly visible against the bright inner regions. Estimated distance: 12.5 megaparsecs; total mass: about 5×10^{11} solar masses; total magnitude: 8.1.

sonde. a device, carried by rocket or balloon, for observations and measurements at high altitudes.

sounding rocket. A rocket used for studying the earth's upper atmosphere and for making high-altitude astronomical observations. Information collected by instruments on board is radioed or sometimes parachuted back to earth. First launched in 1946, sounding rockets are generally single-stage rockets. After completion of rocket motor firing they can coast upwards to a peak altitude of 150-200 km before dropping back to earth.

source count (number count). A compilation of the number, N, of sources per unit solid angle that are brighter than a flux density level S. In Euclidean space the radius of the sphere out to which sources of a certain luminosity are observed varies as $S^{-1/2}$; the number of sources contained therein thus varies as $S^{-3/2}$. A plot of log N versus log S is a straight line of slope -1.5 and is independent of source luminosity. Deviations from this value may indicate source evolution, departures from a uniform space distribution, cosmological effects, or the effect of an intervening obscuring medium. Differential source counts (number per unit flux interval) are to be preferred when dealing with data since the errors on each point are then independent of other points. *See also* radio source catalogue.

south celestial pole. *See* celestial poles.

Southern Coalsack. *See* Coalsack.

Southern Cross. Common name for *Crux.

Southern Crown. Common name for *Corona Australis.

Southern Fish. Common name for *Piscis Austrinus.

Southern Sky Survey. A photographic star atlas of the southern sky, comprising a set of 606 overlapping photographs on which are recorded stars and galaxies that are very much fainter and more distant than those on previous southern sky maps. The project has been a joint effort involving the use of the 1.0 metre *Schmidt telescope at the *European Southern Observatory, Chile, and the 1.2 metre *UK Schmidt at Siding Spring Observatory, Australia. A rapid survey was first made in the blue to record objects with a brightness down to magnitude 21. The UK Schmidt contribution was then a series of sky-limited blue-green photographs, with the ESO performing a corresponding deep red survey. Although the surveys will not be finished probably until about 1980, the first instalment of the final atlas, distributed on film, was issued in 1976; three-quarters of the sky had been satisfactorily photographed by early 1978. The survey includes much fainter and more distant objects than the corresponding *Palomar Sky Survey of the northern and equatorial sky issued in the 1950s.

Southern Triangle. Common name for *Triangulum Australe.

south point. *See* cardinal points.

Soyuz craft. A series of more than 30 Soviet manned spacecraft, the first of which—Soyuz 1—was launched in April 1967. There are three sections: an orbital module, in which up to three cosmonauts live and work, a propulsion and instrumentation section, and a re-entry module, which is occupied by the cosmonauts during ascent into orbit and is the only part to return to earth. The earliest craft were involved in docking experiments, Soyuz 4 and 5 achieving the first docking of two spacecraft and a transfer of crew, in Jan. 1969. Soyuz craft ferry cosmonauts to and from the *Salyut space stations, the first docking and transfer of crew occurring with Soyuz 11 in June 1971.

space. The near-vacuum existing beyond the atmospheres of all bodies and systems of bodies in the universe. The extent of space, i.e. whether it is finite or infinite, is as yet unresolved. *See* interplanetary medium; interstellar medium; intergalactic medium.

Spacelab. A reusable international orbital space station, designed and developed in Europe under the auspices of the European Space Agency (ESA). NASA is providing technical support and is reponsible for its initial launching into orbit by the first operational flight of the *Space Shuttle. The launch date is scheduled at present for December 1980. Spacelab is one payload of the Shuttle: it is a short-stay space station that will be fully contained in the Shuttle's cargo bay during ascent and descent and will remain in orbit in the opened cargo bay for seven to 30 days until the Shuttle returns to earth. It is designed for a 10 year life or 50 seven-day flights.

There are two basic sections in Spacelab: a pressurized module in which the 'payload specialists' can live and work; an unpressurized pallet (platform) where instruments can be directly exposed to space and can be monitored from either the module or the Orbiter of the Shuttle. The pressurized module is composed of two identical cylindrical shells, each 2.7 metres in length and 4.06 m in diameter. The core segment of the module contains the basic Spacelab subsystems with a large volume for experiments; the experiment segment provides space for packaged equipment that can be removed and replaced after each mission. The pallet segments are 2.9 metres long and may be attached individually to the module. Alternatively the pallets may be flown by themselves in the cargo bay, in a series of up to five lots; subsystems will then be carried in an attached 'igloo', pressurized and thermally controlled. The third configuration that can be used in a Spacelab mission is the module by itself. Initially power will be derived from the Shuttle's fuel-cell batteries.

Spacelab's payload is large compared to conventional satellites. Its experiments will be devoted to science and technology. Two Spacelab flights are at present firmly scheduled: the first is a joint ESA/NASA mission while the second is purely a NASA mission. Spacelab 1 will use both module and pallet; this first Spacelab payload (FSLP) will carry a total of 77 experiments (including eight British ones), which concentrate on materials science—electrophoresis, crystal growth, fluid physics, and metallurgy. Spacelab 2 will use pallets only; the 13 experiments (two British) are primarily concerned with astronomy: observations of the sun and cosmic rays and x-ray, ultraviolet, and infrared investigations.

space laboratory. *See* space station.

spaceprobe. *See* planetary probe.

Space Shuttle. A partially recoverable manned space transportation system that is being developed and tested by NASA and should be operational by 1980-81: the latest scheduled date for the first orbital test flight is Nov. 1979. It will reduce very considerably the cost of launching space vehicles. The Shuttle consists of an *Orbiter* with the appearance of a delta-wing aircraft, a huge expendable external propellant tank, on which the Orbiter is mounted at launching, and two solid-fuel rocket boosters. The whole system weighs about 2000 tonnes at launch and has an overall length of about 56 metres. The Orbiter is 37 metres in length.

It will be launched in a vertical position by the simultaneous firing of its two rocket boosters and three very powerful liquid hydrogen/liquid oxygen main engines. About two minutes into the flight the empty rocket boosters will be detached, parachute into an ocean area, and be recovered for further use. Just before the craft reaches its orbit the propellant tank will be discarded. The Orbiter can then manoeuvre by means of internally stored propellants. The altitude, eccentricity, and inclination of the orbit can be varied, within limits, as can the flight duration—between about seven to 30 days.

The Orbiter has a large cargo bay—18.3 metres long and 4.6 metres in diameter—in which the payload is housed. This payload can include satellites to be placed in orbit *Spacelab will be its first payload, with the *Space Telescope scheduled for a 1983 launch date. A total mass of 29.5 tonnes can be carried into a low-altitude orbit, with smaller loads for higher or less easily accessible orbits. Payloads to be placed in orbit above the Shuttle ceiling, such as a geostationary orbit, will require additional propulsion

A payload of 11.5 tonnes can be returned to earth. On mission completion the Orbiter will use its rocket motors to put it into a re-entry path, will enter the atmosphere in a shallow glide, and finally make an unpowered landing.

space station. An orbiting space laboratory, with a lifetime of several years or more, on which people can live and work in controlled but weightless (zero-gravity) conditions. Crews are ferried to and from the station and remain on board either for short periods or on a continuous or near continuous rota basis. Future space stations may well be used to launch platforms for interplanetary spacecraft. *See* Skylab; Salyut.

Space Telescope (ST). A 2.4 metre (94.5 inch) reflecting telescope that is planned to be placed in earth orbit in 1983 by the Space Shuttle. A joint venture of NASA and ESA, it will be operated remotely from earth, with facilities for on-site manual instrument changing and maintenance, and should have an operational life exceeding 10 years. Free from the distortion and absorption effects of the earth's atmosphere and with a *resolving power of less than 0.1 arc seconds, it should be able to discern objects a hundred times fainter than is possible on earth. One of the instruments to be used in its focal plane is the faint-object camera developed by the ESA; this will afford very high resolution imaging in ultraviolet, visible, and near-infrared spectral regions.

space-time. The single physical entity into which the concepts of space and time can be unified such that an event may be specified mathematically by four coordinates, three giving the position in space and one the time. The path of a particle in space-time is called its *world line*. The world line links events in the history of the particle.

The concept of space-time was used by Einstein in both the special and the general theories of *relativity. In special theory, where only *inertial frames are considered, space-time is flat. In the presence of gravitational fields, treated by general relativity, the geometry of space-time changes: it becomes curved. The rules of geometry in curved space are not those of three-dimensional Euclidean geometry.

Matter tells space-time how to curve: massive objects produce distortions and ripples in the local space-time. The question of whether the space-time of the real universe is curved, and in what sense, has yet to be resolved. If the universe contains sufficient matter, i.e. if the *mean density of the universe is high enough, then the space-time of the universe must be bent round on itself and closed.

space velocity. *See* radial velocity.

spark chamber. A device in which the tracks of charged particles are made visible and their location in space accurately recorded. It consists essentially of a stack of narrowly spaced thin plates or grids in a gaseous atmosphere, partially surrounded by one or more auxiliary particle detectors such as scintillation counters. Any particle detected in one of the auxiliary devices triggers the application of a high-voltage pulse to the stack of plates. The passage of the particle through the plates is then marked by a series of spark discharges along its path. The tracks are recorded by electronic or photographic means. Use of the auxiliary devices leads to a select triggering of the chamber for a particular type of energy or radiation. Spark chambers can be used to detect gamma rays following their conversion to an electron-positron pair.

special relativity. *See* relativity, special theory.

specific impulse. *See* launch vehicle.

speckle interferometry. A technique whereby the limit to the *resolution of a telescope, imposed not by its design but by atmospheric turbulence (*see* seeing), may be considerably improved: factors of 50 have been reported. A typical stellar image has, at best, a diameter given by that of the diffraction-limited *Airy disc; atmospheric turbulence however causes small continuous erratic movements in image position on a long-exposure photograph, producing a final blurred image enlarged many times. In speckle interferometry many short exposures (10-20 milliseconds) of the object are taken. These freeze the effects of turbulence so that the individual 'speckles' making up the overall stellar images are distortion-free stellar images: those of supergiants are relatively large, those of binary stars are double. Substantial differences between the speckles of the many short exposures require the application of statistical analysis to the images. This leads to a range of information, including the separation and other properties of close binary stars, often hitherto unresolvable. In addition reconstructions have been achieved of the discs of supergiants, such as Betelgeuse; apparent diameters can thus be measured and large-scale surface details discerned. Very small areas of the sun have also been studied.

spectral classes. *See* spectra types.

spectral index. *See* radio source spectrum.

spectral lines. *See* line spectrum; emission spectrum; absorption spectrum.

spectral types (spectra classes). The different group into which stars may be classified according to the charac teristics of their *spectra, principally the absorption lines and bands resulting from the presence of particular atoms and molecules in the star. Spectral lines of the various elements and compounds have widely different strengths in stars of different temperatures: this is the basis of the current classification. In the *Sechi classification* made in the 1860s by the Italian Angelo Sechi, stars were divided into four groups based on the visual observation of spectra. The subsequent development of photographic spectroscopy allowed a much more precise division. The *Harvard classification* was introduced by astronomers at Harvard Observatory in the 1890s and used in subsequent volumes of the *Henry Draper Catalogue. It was developed into its present form in the 1920s by E.C. Pickering, Annie J. Cannon, and others.

The original scheme was based on the strength of the hydrogen Balmer absorption lines in stellar spectra, the order being alphabetical from A to P: A stars had the strongest hydrogen lines. Some letters were later dropped and the ordering rearranged to correspond to a sequence of decreasing surface temperature. The majority of stars can be divided into seven spectral types—O, B, A, F, G, K, M—the order often remembered by the mnemonic 'Oh Be A Fine Girl Kiss Me'. The principal spectral features of these classes are shown on the graph and listed in Table *A* and also at separate entries: *O stars, *B stars, etc. Since temperature and *colour are directly linked the Harvard classification is also a sequence of colours ranging from hot blue O stars to cool red M stars. Additional classes are the *carbon stars (C stars) and the heavy-metal *S stars. The spectral types are further subdivided into 10 subclasses denoted by digits 0 to 9 placed

Spectral absorption features of spectral types

spectral types

(A) Spectral Features of Star Types

O	hottest blue stars	ionized helium (HeII) lines dominant; strong UV continuum
B	hot blue stars	neutral helium (HeI) lines dominant; no HeII
A	blue, blue-white stars	hydrogen lines dominant; ionized metal lines present
F	white stars	metallic lines strengthen; H lines weaken
G	yellow stars	ionized calcium dominant; metallic lines strengthen
K	orange-red stars	neutral metal lines dominant; molecular bands appear
M	coolest red stars	molecular bands dominant; neutral metal lines strong

(B) Nonstandard Spectral Features

e	emission lines present
n	nebulous lines present
s	sharp lines
k	interstellar lines present
m	metallic lines present
p	peculiar spectrum
v	variable

(C) Luminosity Classes

Ia	bright supergiants
Ib	supergiants
II	bright giants
III	giants
IV	subgiants
V	main sequence (dwarfs)
VI	subdwarfs
VII	white dwarfs

after the spectral type letter, as with A5—midway between A0 and B0. These subdivisions are based on a variety of complex empirical criteria, mainly the ratios of certain sets of lines in a spectrum, and may be further subdivided, as with O9.5. Other spectral characteristics, such as the presence of emission lines, are indicated by an additional small letter placed after the spectral type, as with M5e (see Table *B*).

It was realized in the 1890s that stars of a particular spectral type could have widely differing *luminosities and several luminosity classifications were developed. The currently used *MK system* was introduced in the 1930s by W.W. Morgan and P.C. Keenan at Yerkes Observatory:

stars of a given spectral type are further classified into one of six *luminosity classes,* denoted by Roman numerals placed after the spectral type, as with G2 V, and indicating whether the star is a *supergiant, *giant, or *main-sequence dwarf star (see Table C). *Subdwarfs and *white dwarfs are also sometimes considered as luminosity classes. All these classes occupy distinct positions on the *Hertzsprung-Russell diagram. Stars of a similar temperature but different luminosities must differ in surface area and consequently in *surface gravity and atmospheric density. These differences produce spectral effects that are used to differentiate between the luminosity classes. A major effect is the *pressure broadening of spectral lines, which increases as the atmospheric density and pressure increase (and radius decreases): spectral lines of a bright supergiant are much narrower than those of a main-sequence star of the same spectral type.

spectrograph. An optical instrument used in separating and recording the spectral components of light or other radiation. Spectrographs are a major astronomical tool for analysing the emission and absorption spectra of stars and other celestial objects. They are used with ground-based reflecting (or refracting) telescopes, usually mounted at the Cassegrain focus. If the telescope has a coudé facility a spectrograph can be permanently positioned at the coudé focus: coudé spectrographs are used for high-dispersion work. The *dispersion* of a spectrograph is the linear separation of spectral lines per unit wavelength difference, often quoted in millimetres per nanometre or per angstrom. The wavelength range over which the instrument will operate depends on the recording medium, and also on the optical elements of the device,

Optical path in spectrograph

but is generally in the region 300-1300 nanometres—i.e. light, near-infrared, and near-ultraviolet wavelengths.

The radiation is focused on and enters the instrument through a narrow rectangular slit. The diverging beam is made parallel by a collimator—a converging mirror or lens—and passes through a *diffraction grating or prism. The beam is thus split into its component wavelengths. This spectrum is focused on a photographic plate, *image tube, or similar device. The *resolving power* of a spectrograph is a measure of the detectable separation of wavelengths that are very nearly equal. It is given by the ratio $\lambda/\Delta\lambda$, if at a wavelength λ it is just possible to distinguish between two spectral lines of wavelength difference $\Delta\lambda$. In a large spectrograph, such as that in the illustration, there is often a choice of mirrors, of different focal lengths, for focusing the spectral image on the recording medium and also a choice of diffraction gratings.

spectroheliogram. A photograph of the sun made in light (or other radiation) of a single wavelength or narrow waveband. It can be obtained either by means of a *spectroheliograph or by placing an *interference filter in front of a photographic plate; the latter is more usually called a *filtergram*. Both types are images of the solar *chromosphere taken in the residual light of strong *Fraunhofer lines. The most common lines used are the K line of calcium at 393.3 nanometres and the hydrogen alpha line (Hα) at 656.3 nm. Different layers of the chromosphere can be imaged by varying the isolated wavelength about the absorption-peak wavelength.

spectroheliograph. A high-dispersion *spectrograph designed to be used, usually with a *solar telescope, for studying the sun's spectrum and for obtaining photographs of the sun at particular wavelengths, such as that of the hydrogen alpha line. In addition to an entrance slit, the instrument has a second narrow slit situated directly in front of the photographic plate so that only one spectral line falls on the plate. When the first slit is at the prime focus of the telescope a narrow strip of the sun's image will enter the instrument and a photograph of that narrow image, at the required wavelength, will be formed. If the first slit is moved across the sun's image, and the second slit is moved in step with it across the photographic plate, a photograph—a *spectroheliogram—of the whole sun, at the wavelength under consideration, will be obtained. This photograph will be composed of tiny line elements maybe 0.03 nanometres wide.

spectrometer. 1. An instrument, such as a *spectrograph, in which the spectrum of a source is recorded by electronic means. Spectrometers covering various wavelength ranges are used in satellite and spaceprobe measurements.

2. An instrument for determining the distribution of energies in a beam of particles.

spectrophotometer. An instrument designed to measure the intensity of a particular spectral line or a series of spectral lines in an absorption or emission spectrum. *See also* spectroscopy.

spectroscopic binary. A *binary star in which the orbital motions of the components give rise to detectable variations in *radial velocity, revealed by changes in *Doppler shift in their spectral lines (see illustration). Since a binary star is more likely to be detected spectroscopically if its orbital period is relatively short and the orbital velocities are relatively high, most spectroscopic binaries are close binaries (*see* binary star). Spectroscopic detection is not possible if the orbital plane lies at right angles to the line of sight.

When the components are nearly equal in brightness both spectra are visible and the spectral lines appear double for most of each orbital period, coinciding only when the stars are moving at right angles to the line of sight. Such a system is called a *double-lined spectroscopic binary*. Comparison of the Doppler shifts in the component spectra gives the relative velocities and hence the relative masses of the two stars:

$$v_1/v_2 = M_2/M_1$$

The individual masses cannot be determined unless the inclination of the orbital plane to the line of sight is known (*see* eclipsing binary).

The presence of a comparatively faint component is re-

Doppler shifts of spectral lines at various orbital positions of spectroscopic binary

vealed only by its gravitational effect on the motion of the brighter star. Invisible components of low mass are rarely detected because Doppler shifts corresponding to orbital speeds of less than about 2 km/s are obscured by observational errors.

spectroscopic parallax. *See* parallax.

spectroscopy. In general, the production and interpretation of spectra. Astronomical spectroscopy is used over the whole range of *electromagnetic radiation from radio waves to gamma rays. It is the main source of information on the composition and nature of celestial bodies.

The lines and bands in *emission and *absorption spectra are characteristic of the atoms, molecules, and ions producing them, and *spectral analysis* leads to the identification of these components in planetary atmospheres, comets, stars, nebulae, galaxies, and in the interstellar medium. Measurements of the intensities of spectral lines (*spectrophotometry*) can give quantitative information on the chemical composition.

In addition, spectroscopy yields information on the physical conditions and the processes occurring in celestial bodies. For instance, temperature may be measured by the vibrational excitation of molecules (if present) through analysis of the intensities of the individual lines in their band spectra. The degree of ionization can be used to measure higher temperatures. The width and shape of spectral lines indicates temperature, movement, pressure, and the presence of magnetic fields (*see* line broadening; Zeeman effect). The processes occurring are often directly responsible for the production of the spectra, as in the recombination of ions and electrons in HII regions or synchrotron emission from electrons in magnetic fields. *See also* Doppler effect; redshift.

spectrum. A display or record of the distribution of intensity of *electromagnetic radiation with wavelength or frequency. Thus, the spectrum of a celestial body is obtained by dispersing the radiation from it into its constituent wavelengths so that the wavelengths present and their intensities can be observed. A variety of techniques exist for obtaining spectra, depending on the part of the electromagnetic spectrum studied. The result may be a photographic record (as can be obtained in a *spectrograph) or a plot of intensity against wavelength or frequency (as can be produced by electronic and associated equipment).

The spectrum of a particular source of electromagnetic radiation depends on the processes producing emission of radiation (*see* emission spectrum) and/or on the way in which radiation is absorbed by intermediate material (*see* absorption spectrum). Spectra are also classified by their appearance. A *line spectrum has discrete lines caused by emission or absorption of radiation at fixed wavelengths. A *band spectrum has distinctive bands of absorption or emission. A *continuous spectrum occurs when continuous emission or absorption takes place over a wide range of wavelengths. *See also* spectroscopy.

spectrum variables (α^2 CVn stars). Peculiar main-sequence *Ap stars whose spectra have anomalously strong lines of metals and rare earths (Sr, Eu, Mn, Si, Cr) that vary in intensity by about 0.1 magnitudes over periods of about 1-25 days. These variations are related to changes in the stars' magnetic fields. The brighter component of the binary star *Cor Caroli (α^2 CVn) is typical.

speculum metal. An alloy of copper and tin used by Newton and his successors to make telescope mirrors because it could easily be cast, ground to shape, and took a good polish. It was made obsolete by the use of glass provided with a thin highly reflective silver coating.

speed of light. Symbol: c. The constant speed at which light and other *electromagnetic radiation travels in a vacuum. It is equal to 299 792 458 m s^{-1} (value adopted by the IAU in 1976). Since space is not a perfect vacuum, radiation travels through space at a very slightly lower speed, which decreases somewhat more when the radiation enters the earth's atmosphere. The speed of light, c, is independent of the velocity of the observer: there can therefore be no relative motion between an observer and a light beam, i.e. the speed of light cannot be exceeded in a vacuum. This radical notion was basic to Einstein's special theory of *relativity.

spherical aberration. An *aberration of a spherical lens or mirror in which light rays

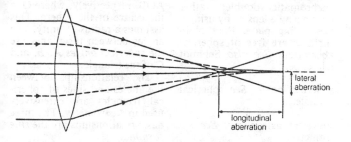

Spherical aberration in convex lens

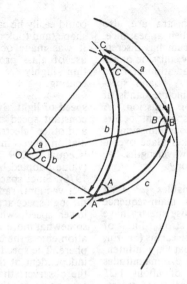

Spherical triangle

converge not to a single point but to a series of points whose distances from the lens or mirror decreases as the light rays fall nearer the periphery of the optical element (see illustration). It is most obvious with elements of large diameter. It can be reduced by using an *achromatic doublet rather than a single lens or by using a *correcting plate. Paraboloid surfaces are free of spherical aberration. *See also* Schmidt telescope.

spherical angle. *See* spherical triangle.

spherical astronomy. *See* astrometry.

spherical triangle. A triangle on the surface of a sphere formed by the intersections of three great circles (see illustration). The arcs BC, AC, and AB form the sides of the spherical triangle ABC. The lengths of these sides (a, b, and c), in angular measure, are equal to the angles BOC, COA, and AOB, respectively, where O is the centre of the sphere. This assumes a radius of unity. The angles between the planes are the *spherical angles A, B,* and *C* of the triangle.

The relationships between the angles and sides of spherical triangles are extensively used in *astrometry. The three basic relationships are the *sine formula:*

$$\sin A / \sin a = \sin B / \sin b = \sin C / \sin c$$

the *cosine formula:*
$\cos a = \cos b \cos c + \sin b \sin c \cos A$
$\cos b = \cos a \cos c + \sin a \sin c \cos B$
$\cos c = \cos a \cos b + \sin a \sin b \cos C$
and the *extended cosine formula,* which can be derived from the other two.

For a *spherical right-angled triangle,* in which one angle, say C, is equal to 90° so that $\sin C = 1$, $\cos C = 0$, the formulae are much simplified. They are also simplified if the sides of the triangle are small so that, say, $\sin a$ tends to a and $\cos a$ to $1-a^2/2$. As a, b, and c approach zero the formulae reduce to those used in plane geometry. The formulae may be extended by replacing the sides and angles by the supplements of the corresponding angles and sides, respectively. Thus
$\sin A = \sin a$, $\cos A = -\cos a$.

spheroid. *See* ellipsoid.

Spica (α Vir). A conspicuous blue-white star that is the brightest one in the constellation Virgo. It is an *eclipsing binary, period 4.014 days, the primary(A) being a *Beta Cephei star near to core hydrogen exhaustion. Spica, Denebola, Arcturus, and Cor Caroli form the quadrilateral shape of the *Diamond of Virgo.* m_V: 0.97; M_V: -3.1; spectral type: B1 V; distance: 65 pc.

spicules. Predominantly vertical narrow (1000 km diameter) jets of gas, visible (in the monochromatic light of certain strong *Fraunhofer lines) beyond the sun's limb. They ascend with velocities of up to 20-30 km s^{-1} from the lower *chromosphere for several thousand kilometres into the inner *corona and then fall back and/or fade away. Their average lifetime is 5-10 minutes. They are closely associated with the *chromospheric network, being located in regions of enhanced vertical magnetic field at the boundaries of supergranular cells (*see* supergranulation), and have temperatures of the order of 10 000 to 20 000 kelvin.

spider. The diagonal support in a *Newtonian telescope.

spider diffraction. The characteristic spikes of light in the form of a cross seen in the image formed by a Newtonian reflector. They result from diffraction effects caused by the diagonal supports—the spider—and are additional to the diffraction from the circular edge of an objective lens or a mirror, which converts a star image from a point to an *Airy disc in refractors and reflectors.

spillover. The part of the system noise (*see* sensitivity) of a radio telescope using a *dish antenna that results from pickup by the feed—the secondary antenna—from directions that do not intercept the reflecting surface of the dish.

spin. The property of intrinsic angular momentum possessed by most *elementary particles and *nuclei; the particle thus acts as if it were spinning on an axis. This is in addition to any angular momentum due to orbital motion. The spin can either have half integer values (as for protons, neutrons, electrons, muons, and neutrinos) or it must be zero or have integral values (as for photons or pions). Consideration of spin and orbital angular momentum within the framework of *quantum theory provides an explanation for the splitting of spectral lines, i.e. for the fine and hyperfine structure of spectral lines.

spinar. A hypothetical supermassive object situated in the nuclei of *active galaxies and *quasars. Some astronomers believe that the rapid rotation of such a million solar mass object could give rise to the quasar's huge output of energy by a process akin to the *pulsar mechanism.

spiral arms. See Galaxy; galaxies; density-wave theory.

spiral galaxy (spiral). See galaxies; Hubble classification.

sporadic meteor. A *meteor that is not associated with a specific *meteor shower and has a random orbit.

Spörer's law. See sunspot cycle.

sprays. Impulsive events on the sun that are characterized by the ejection of chromospheric material into the inner *corona with velocities exceeding the *escape velocity (618 km s^{-1}). They invariably occur during the 'flash phase' of *flares and are less confined than *surges, their material fragmenting as it flies out. See also prominences.

spring equinox. Another name for vernal equinox. See equinoxes.

Springfield mount. See coudé telescope.

spring tide. See tides.

s-process. A slow process of *nucleosynthesis by which heavy stable nuclei are synthesized from the iron-peak elements (mass number 56) by successive captures of neutrons. The process occurs when there is a low density of neutrons in a star, the neutrons being by-products of nuclear-fusion reactions. If the nucleus produced by a neutron capture is stable it will eventually capture another neutron; if the nucleus is radioactive it will have sufficient time to emit a beta particle (i.e. an electron) to stabilize itself before further neutron capture. Thus the stable isotopes of an element are synthesized until a radioisotope is produced, at which point a new element forms by beta decay. Many years or decades may elapse between successive neutron captures. The most abundant nuclei produced by the s-process will be those with a low ability to capture neutrons. The s-process cannot synthesize nuclei beyond bismuth-209 since neutron capture by this nuclei results in rapid alpha decay. *Compare* r-process.

spurs. *See* interstellar medium.

Sputnik. A series of Soviet unmanned satellites, the first of which, Sputnik 1, was the first spacecraft to be placed in orbit. This 58 cm diameter sphere, weighing 84 kg, was launched on Oct. 4 1957; it burnt up in the earth's atmosphere 92 days later. The orbit had a period of 96 minutes, an apogee and perigee (highest and lowest altitudes) of about 950 km and 230 km, and was inclined at 65° to the equator. Its radio signals were transmitted every 0.6 seconds. It contained few instruments, being intended as a test vehicle. The launching of Sputnik 1 had a profound effect in accelerating America's space programme.

Sputnik 2, launched on Nov. 3 1957, was very much bigger. It carried about 500 kg in payload, including a live dog, Laika, which survived the launch, as well as 10 experiments. Sputnik 3, launched on May 15 1958, weighed about 1330 kg and remained in orbit for 691 days.

SR variables. *See* semiregular variables.

SS Cygni. *See* dwarf novae.

S stars (zirconium stars; heavy-metal stars). Stars of *spectral type S, about half of which are *irregular long-period variables. They are red giants similar to *M stars but distinguished spectroscopically by the presence of molecular bands of the heavy-metal oxides zirconium (ZrO), lanthanum, yttrium, and barium oxide, rather than the oxides of titanium, scandium, and vanadium. Lines of technetium (longest half-life 2×10^6 years) can also be present. Pure S stars have very strong ZrO bands with either very weak or no TiO bands. *MS stars* are M-type stars showing ZrO bands.

ST. *Abbrev. for* Space Telescope.

standard time. The time in any of the 24 internationally agreed *time zones* into which the earth's surface is divided; the primary division is centred on the Greenwich (0° longitude) meridian. All locations within a single zone keep the same time: this is the *mean solar time of the central meridian except when daylight saving systems, such as summer time, are in operation. Zone times differ by a whole number of hours, or in some cases of half hours, from *Greenwich Mean Time. Zones west of the Greenwich zone are behind GMT, those east of it are in front of GMT.

standstill. *See* dwarf novae.

star. A luminous gaseous body

that generates energy by means of nuclear fusion reactions in its core. Below a certain mass (0.05 solar masses), the central pressures and temperatures in a body are insufficient to trigger fusion reactions; all stars must therefore exceed this mass limit. Few stars are believed to be much more massive than 50 times the mass of the sun, for their *radiation pressure would exceed the pull of gravity and blow them apart.

Although there is only a relatively narrow range in stellar masses, it is the mass of a star that determines its other properties—luminosity, temperature, size—and the way in which it evolves. These quantities are related to the mass by the equations of *stellar structure. Other stellar parameters show a far greater range than stellar mass: for example, the *luminosity of a star is proportional to the cube of its mass for *main-sequence stars (*see* mass-luminosity relation); stars therefore show a range in luminosity of some 10^{10}. A star's lifetime also depends on its mass: low-mass stars live considerably longer than high-mass ones.

The characteristics of a star can be determined only when its distance is known. The method of *distance determination depends on the distance itself: the shortest distances are found directly by measuring a star's *annual parallax; greater distances require indirect methods. The nearest star, Proxima Centauri, is 1.31 parsecs away. The stars in our immediate locality are found in the main to be young stars occupying the outer parts of the disc of our Galaxy, rotating around the galactic centre. Many still remain in open clusters, and over 50% of all the stars we observe are *binary or multiple stars.

See also magnitude; populations I and II; proper motion; radial velocity; spectral types; stellar birth; stellar diameter; stellar evolution; variable star.

star atlas. A collection of maps of a particular region of the sky or of the whole sky, showing the relative positions of stars and other celestial objects down to a specified *limiting magnitude. They now contain prints of photographic plates. The *Palomar and *Southern Sky Surveys are examples. *See also* star catalogue.

star catalogue. A systematic list of stars in a specified area of the sky down to a specified *limiting magnitude. It gives stellar coordinates at a particular *epoch, apparent *magnitudes, *proper motions, *radial velocities, *spectral types, etc. In a whole-sky catalogue stars in both hemispheres will be included. Modern catalogues are divided into two general types: *survey catalogues* list a large number of stars whose positions are given with only moderate accuracy; *precision catalogues,* such as *fundamental catalogues, give very precise positions for a relatively small number of stars. General catalogues include the *Bonner Durchmusterung and the *AGK. There are also catalogues specifically for double stars, variable stars, nebulae, star clusters, galaxies, radio sources, x-ray sources, etc. *See also* star atlas.

star clouds. Areas of the sky where great numbers of stars are seen so close together that they appear as continuous irregular bright clouds. Star clouds are particularly noticeable in the direction of the galactic centre in the constellation *Sagittarius.

star cluster. *See* cluster.

star density. The number of stars per unit volume of space, usually per parsec cubed. It is often given in terms of the fraction of solar mass per volume. *See also* luminosity function.

star gauges. Systematic counts of stars in different regions of the sky. They were first made in the 1780s by William Herschel who counted the total number of stars visible in each of several adjoining areas within the region under study. By averaging the sample counts he obtained the star gauge for that region. Herschel compiled star gauges for several hundred regions representative of the northern hemisphere, his work being extended to the southern hemisphere by his son John. This provided the first precise information on the local distribution of stars in the Galaxy (*see* galactic concentration). By comparing gauges made to different limiting magnitudes (using larger and larger telescopes) they were able to show that the star system extends to great depths in directions towards the *Milky Way. *See also* source count.

star trails. Curved images on a photographic plate that record the changing positions of stars when a telescope is not driven so as to keep up with their diurnal motion. Even if such a drive is being used, comet trails will still be recorded.

static universe. A universe in which the *cosmic scale factor is independent of time. Einstein proposed a static universe in 1916 by including an ad hoc repulsion term—the *cosmological constant,* Λ—in his field equations of general *relativity. This cancelled out the natural tendency for a gravitating universe either to expand or contract, depending on its energy content. *See* expanding universe; cosmological models.

stationary orbit. Short for geostationary orbit. *See* geosynchronous orbit.

stationary point. *See* direct motion.

statistical parallax. The mean parallax of a group of stars that are all at about the same distance and whose space velocities are randomly orientated. It is derived from the *radial velocities and the tau components of the stars' *proper motions. The tau component is measured perpendicular to the great circle joining a star to the solar *apex. It is thus perpendicular to and independent of the sun's motion, i.e. it results entirely from the motion of the star. The average velocity in one direction will equal that in any other direction. The average radial velocity is thus a measure of average velocity. The average distance is then found from the ratio of average radial velocity to average tau component.

steady-state theory. A cosmological theory obeying a perfect *cosmological principle in which the universe appears the same both at all points and all times: it thus had no beginning, will never end, and has a constant mean density of matter. Proposed by Hermann Bondi, Thomas Gold, and Fred Hoyle in 1948, the steady-state theory employs a flat space expanding at a constant rate. There is continuous creation of matter throughout the universe—or perhaps in galactic centres—to compensate for the expansion. The creation rate is about 10^{-10} nucleons per metre cubed per year. The steady-state theory is not readily able to account for the *microwave background radiation or the steep radio *source counts that indicate an evolving universe.

Stefan's law (Stefan-Boltzmann law). The law relating the total energy, E, emitted over all wavelengths, per second, per unit area of a *black body, with the temperature, T, of the body:
$$E = \sigma T^4$$
σ is *Stefan's constant*, which has the value 5.6697×10^{-8} W m^{-2}K^{-4}. The law can be derived from Planck's radiation law (*see* black body), but was first deduced by Josef Stefan in 1879 and then derived from thermodynamics by Ludwig Boltzmann in 1884. *See also* luminosity.

stellar association. *See* association.

stellar birth. Stars are formed by the *gravitational collapse of cool dense (10^9-10^{10} atoms per cubic metre) gas and dust clouds. There are problems, however, in initiating the collapse of a gas cloud, which resists compression because of its internal motions and the heating effects of nearby stars. In a massive dense cloud shielded by dust, it is believed that collapse can be triggered when the cloud is slowed on passing through the spiral *density-wave pattern of our Galaxy: it is noticeable that star formation preferentially occurs along spiral arms. Alternatively compression may be caused by a shock wave propagated by a nearby *supernova explosion—which may explain how stars are formed in galaxies without spiral arms.

Once collapse starts, instabilities cause the cloud to divide into successively smaller fragments until these become too dense and too warm to divide further. These final fragments are protostars, which now collapse individually. This collapse is gentle at first but after some half a million years (for a star of one solar mass) the density and temperature rise sharply at the centre, forming a hot core that will be the future nucleus of the star. However, the temperature of the outer layers of the protostar is less than 100 kelvin.

As collapse of material on to the hot core continues at a slower rate, the core temperature and density increase and the protostar approaches the *main sequence of the *Hertzsprung-Russell diagram along the *Hayashi line. The shape of the Hayashi line and the time taken to reach the main sequence depend on the mass of the star. With increasing core temperature the protostar may become a *Herbig-Haro object, with light from the core reflected off the outer layers of gas and dust. Other likely protostar candidates are *Becklin's star in the Orion nebula and the infrared source W3-IRS5.

The slow contraction heats the central core to a temperature of some ten million kelvin, which is sufficient to initiate *nuclear fusion reactions in hydrogen. *T Tauri stars may have recently got to this stage. Once an energy source has established itself, the contraction is halted and the star joins the main sequence. The whole process—from protostar to main sequence star—may take less than 100 000 years for a high-mass star to over 50 million years for a star the mass of the sun.

stellar diameter. The diameters of stars range from several hundreds of millions of kilometres for *supergiants through about one million km for the sun down to a few thousand km for *white dwarfs and tens of km for *pulsars. For convenience the diameter or radius is usually expressed in terms of the sun's radius, $R_\odot$. The size of a star changes drastically towards the end of its life as it progresses through the *red giant stage to its final gravitational collapse. It can also change periodically (*see* pulsating variables).

In most cases stars are too distant for their size to be determined directly. The diameter of a large star at a known distance can however be found geometrically if its *apparent diameter can be measured. Measuring techniques are based on the *interference of light. The apparent diameters of bright supergiants and red giants were first measured in the 1920s using a stellar *interferometer: the diameters lay between 0.02 arc seconds (Aldeberan, $45R_\odot$) and $0''.047$ (Mira Ceti, $390R_\odot$). Very much smaller apparent diameters, including that of Sirius ($0''.00589$, $1.7R_\odot$) and Alpha Gruis ($0''.00102$, $2.2R_\odot$), have recently been measured with

stellar evolution

an intensity *interferometer. A telescope cannot resolve the disc of a star; the discs of supergiants can however be imaged (and measured) using *speckle interferometry, if the apparent diameters are in the region of 0".02 to 0".05. *Occultations of stars by the moon provide a further interference method: the star does not disappear instantaneously but fades over a few seconds; the unobstructed light produces a characteristic interference pattern of intensity against time from which angular size can be determined.

The sizes of many *eclipsing binaries have been found from the shape of their *light curves. In these measurements the sizes are obtained in terms of stellar radius relative to orbital radius. If the binary is also a *spectroscopic binary the absolute values of the radii can be found. For the majority of (smaller) stars diameters can only be inferred from *Stefan's law, i.e. from known values of *effective temperature, T_{eff}, and luminosity, L:
$$L = 4\pi R^2 \sigma T_{eff}^4$$
where σ is Stefan's constant.

stellar evolution. The progressive series of changes undergone by a star as it ages. Stars similar in mass to the sun, having contracted from the *protostar phase (*see* stellar birth) stay on the *main sequence for some 10^{10} years. During this period they give out energy by converting hydrogen to helium in their cores through the *proton-proton chain reaction until the central hydrogen supplies are exhausted. Unsupported, the core collapses until sufficiently high temperatures are reached to burn hydrogen in a shell around the inert helium core. Shell-burning causes the outer envelope of the star to expand and cool so that the star evolves off the main sequence to become a *red giant.

Further contraction of the core increases the temperature to 10^8 kelvin, at which stage the star can convert its helium core into carbon through the *triple alpha process. The sudden onset of helium burning—the *helium flash—may disturb the equilibrium of a low-mass star. The triple alpha reaction gives out less energy than hydrogen fusion and soon the star finds itself once again without a nuclear energy source. Further contraction may result in helium-shell burning but it is doubtful if low-mass stars have sufficient gravitational energy to go further than this stage. The star becomes a red giant again, pulsating and varying in its brightness because of the vast extent of its atmosphere. Eventually the atmosphere gently drifts away from the compact cover of the star with velocities of only a few km s^{-1}, resulting in the formation of a *planetary nebula. The collapsed core forms a *white dwarf star, which continues to radiate its heat away into space for several millions of years.

Stars more than twice as massive as the sun convert hydrogen to helium through the *carbon cycle, which makes their cores fully convective and therefore less dense than solar-mass stars. When, after only a few million years, they exhaust their hydrogen, the onset of helium burning occurs only gradually: this is because the core is nondegenerate, and so no instabilities arise. Having consumed its helium, a massive star has the potential energy to contract further so that carbon, formed by the triple-alpha process, can burn to oxygen, neon, and magnesium (see nucleosynthesis). Should the star be sufficiently massive, it will build elements up to iron in its interior. But iron is at the limit of nuclear-fusion reactions and further contraction of a massive star's core can only result in catastrophic collapse, leading to a *supernova explosion. The most likely end-products of massive stars are *neutron stars and *black holes.

stellar interferometer. *See* interferometer.

stellar mass. The mass of a star is usually given in terms of the sun's mass, $m\odot$: it ranges from about 0.05 to 60 solar masses. Most stars lie on the main sequence of the *Hertzsprung-Russell diagram. Their position on the main sequence, i.e. their *spectral type, depends on their mass, which varies from $0.1m\odot$ (M stars) to over $20m\odot$ (O stars); most do not exceed $10m\odot$. *Supergiants are generally 10 to $20m\odot$ but young O and B supergiants are much more massive. The largest mass of a stable star is about $50\text{-}60m\odot$. Many stars shed mass during their evolution (*see* mass loss; **mass transfer**).

The mass of a star can be determined directly if it has a significant gravitational effect on a neighbouring star. Thus the combined mass and in some cases the individual masses of *visual and *spectroscopic binary stars have been found (*see also* dynamic parallax). Mass can also be estimated using the *mass-luminosity relation.

The evolutionary pattern and lifetime of a star depend on its mass. The least massive stars have the longest lifetimes of maybe thousands of millions of years; the most massive might only exist a few million years before exploding as *supernovae. If at the end of a star's life its mass is less than about $1.5m\odot$ it will become a *white dwarf; if the mass exceeds about $3m\odot$ it is likely that the star will become a *black hole; a star with intermediate mass will end up a *neutron star.

stellar motion. *See* proper motion; peculiar motion.

stellar nomenclature. Although most of the brighter stars have special names, as with Sirius, the general and most convenient method of naming stars was introduced (1603) by Johann Bayer. In the system of *Bayer letters,* letters of the Greek alphabet are allotted to stars in each constellation, usually in order of brightness. The Greek letter is followed by the genitive of the Latin name of the constellation, as with Alpha Orionis or in abbreviated form α Ori. When the 24 Greek letters have been assigned, small Roman letters (a, b, c, etc.) and then capitals (A, B, C, etc.) are used. The letters R-Z were first used by Argelander (1862) to denote *variable stars and when those have been assigned the lettering RR-RZ, SR-SZ, etc., up to ZZ is employed. If that proves insufficient AA-AZ, BB-BZ, up to QZ (the 334th variable in the constellation) must be used, followed by V335, etc.

The fainter stars, usually invisible to the naked eye, are designated by their number in a star catalogue. The numbers in Flamsteed's *Historia Coelestis Britannica* (1725) are often adopted when a star has no Greek letter, as with 28 Tauri (*Pleione in the Pleiades), the *Flamsteed number* generally being followed by the genitive of the constellation name. For stars unlisted in Flamsteed's catalogue the number in some other catalogue is used, the number being preceded by the abbreviation of the name—BD (*Bonner Durchmusterung), HD (*Henry Draper Catalogue), etc. Flamsteed numbered the stars in a constellation in order of *right ascension. Most modern catalogues ignore the constellation and number purely by right ascension.

stellar planetaries. *See* planetary nebula.

stellar populations. *See* populations I and II.

stellar statistics. The use of statistical methods in the study of spatial and temporal distributions and motions of celestial objects within a galaxy, region of the universe, etc., and of spatial and temporal distributions of certain features or of certain categories of these objects. The number of stars in a galaxy can be of the order of 10^{11} and the number of galaxies runs into countless millions. Only a very small number of stars, galaxies, radio sources, etc., can be examined in any detail and from this sample, analysed statistically, a more general picture emerges. Much of the data that is used comes from *star catalogues, *star counts, or detailed studies of *selected areas. Almost all the information concerning the shape and structure of our Galaxy and the spatial distribution of other galaxies and of clusters of galaxies has been determined statistically.

stellar structure. The interior constitution of a star, defining the run of temperature, pressure, density, chemical composition, energy flow, and energy production from the centre to the surface. The structure of *main-sequence stars, initially of uniform composition, is relatively simple: only very near the centre are the temperature and pressure high enough for nuclear reactions to occur. In stars more than about twice the sun's mass, convection currents in the core enlarge this zone and keep it mixed; less massive stars in contrast have a convective layer near the surface and an unmixed core.

Changes in structure as stars evolve can be calculated by following changes in chemical composition resulting from nuclear reactions, and recalculating the structure for the new composition. After the star has passed the *Schönberg-Chandrasekhar limit, the structure changes to that of a giant, with an inert helium core surrounded by a hydrogen fusion shell and an extended envelope. Further core reactions in a very massive star will give it an 'onion-shell' structure, culminating in an iron core surrounded by successive shells of silicon, neon and oxygen, carbon, helium, and outermost, the hydrogen-rich envelope.

In principle, from an assumed composition, structure, and total mass, the other parameters of a stellar interior are derived by solving four differential equations:

$dP/dr = -GM\varrho/r^2$ (1)
$dM/dr = 4\varrho r^2\pi$ (2)
$dL/dr = 4\pi r^2\varrho\varepsilon$ (3)
$dT/dr = 3\varkappa L\varrho/16\pi acr^2T^3$ (4)

Equation 1 is that of *hydrostatic equilibrium, 2 is that of continuity of mass, 3 is that of energy generation, and 4 is that of radiative transport (see energy transport). Accurate solutions require a large computer since the pressure (P), opacity ($\varkappa$), and energy generation rate (ε), also depend on the density (ϱ), the temperature (T), and the chemical composition; in addition in some parts of the star energy may be transported by convection rather than by radiation. Of the other symbols, r is the radius, M the mass within that radius, G the gravitational constant, L the luminosity at radius r, a the radiation density constant, and c the speed of light.

stellar temperature. See surface temperature.

stellar wind. The steady stream of matter ejected from many types of stars, including the sun (see solar wind). Although the mass loss is small in sunlike stars, it has been found to be very considerable in *red giants and in hot luminous *ultraviolet stars. This large mass loss could have a substantial effect on the final evolution of the stars into either white dwarfs or into more massive neutron stars (or black holes). The mechanisms by which stellar winds are produced is still not clear.

Stephan's Quintet. A close group of five galaxies whose four brightest members, first observed by M.E. Stephan in 1877, are very distorted and thus possibly interacting. The observed spread in *radial velocities (and hence *redshifts) suggests however that the

stereocomparator. *See* comparator.

Stern. Common name for *Puppis.

Steward Observatory. An observatory of the University of Arizona on Kitt Peak near Tucson, Arizona. The principal instrument is a 2.3 metre (90 inch) reflecting telescope, which came into operation in 1969.

Stickney. *See* Phobos.

Stokes' parameters. *See* polarization.

stone. *Short for* stony meteorite.

stony-iron meteorite (siderolite). Any of a relatively small group of *meteorites containing on average 50% nickel-iron and 50% silicates.

stony meteorite (aerolite). A type of *meteorite that consists of silicate minerals generally with some nickel-iron. The two main subgroups are termed *chondrites and *achondrites, depending on the presence or absence of *chondrules.

stop (diaphragm). A circular opening that sets the effective aperture (diameter) of a lens, mirror, *eyepiece, etc., and also reduces stray light in an optical system.

stratosphere, stratopause. *See* atmospheric layers.

strewn field. *See* tektites.

Strömgren sphere. An approximately spherical region of ionized gas, mainly ionized hydrogen (*HII), that surrounds a hot O or B star. If a hot star were embedded in very low-density gas of uniform distribution, the ultraviolet radiation from the star would almost completely ionize the gas out to a particular radius—the *Strömgren radius*. Equilibrium is established within this volume when the rate of ionization is balanced by the rate of recombination of ions with electrons. Outside this volume lies a thin transition zone between ionized gas and much cooler neutral gas. Sometimes the entire gas cloud can be ionized. The Strömgren radius depends mainly on the gas density and on the flux of ionizing radiation, i.e. ultraviolet photons. The flux is dependent mainly on the star's temperature. *See also* emission nebula.

subdwarf. A star that is smaller and about 1.5-2 magnitudes fainter than normal dwarf (main-sequence) stars of the same spectral type (F to M). The subdwarfs are mainly old (halo population II) objects that lie just beneath the main sequence on the *Hertz-

sprung-Russell diagram. They are usually denoted by the prefix *sd*, as in sdK4.

subgiant. A giant star of smaller size and lower luminosity than normal giants of the same *spectral type, Alpha Crucis being an example. Subgiants form luminosity class IV, lying between the main-sequence stars and the giants on the *Hertzsprung-Russell diagram.

subluminous stars. Types of stars, such as *white dwarfs, *subdwarfs, *high-velocity stars, and nuclei of *planetary nebulae, that are fainter than main-sequence stars. Most are old (population II) stars.

sublunar point. *See* substellar point.

subsolar point. *See* substellar point.

substellar point. The point on the earth's surface that lies directly beneath a star on the line connecting the centre of earth and star. The star would thus lie at the zenith of an observer at that point, the terrestrial latitude of which would equal the star's *declination. The *sublunar point* and *subsolar point* are the equivalent points for the moon and sun, respectively.

summer solstice. *See* solstices.

summer triangle. The large distinctive triangle formed by the three bright stars *Vega, *Altair, and *Deneb. The stars are the first ones visible in northern summer evenings.

sun (or Sun). The central body of the solar system and nearest star to earth, situated at an average distance of 149 600 000 km. A *main-sequence star of *spectral type G2 V, diameter 1 392 000 km, mass 1.990×10^{30} kg, luminosity 3.83×10^{26} watts, and *absolute visual magnitude +4.8, the sun is a representative yellow dwarf. It is the only star whose surface and outer layers can be examined in detail.

The sun is composed predominantly of hydrogen and helium (relative abundance approximately 3:1), with about 1% of heavier elements. Its mean density is 1.41 g cm^{-3}. It generates its energy by thermonuclear processes, the most important of which is the *proton-proton chain reaction. These take place in a central core, which is thought to be about 400 000 km in diameter and to contain about 60% of the sun's mass (in barely 2% of its volume). Outside the core is an envelope of unevolved material, through which energy from the core is radiated. This extends to within about 100 000 km of the surface, where convection becomes the more important mode of energy transportation (*see* convective zone). From the centre to the surface of the sun the temperature falls from around 15 000 000 kelvin (K) to 6000 K.

The surface of the sun, i.e.

the *photosphere, represents the boundary between the opaque convective zone and the transparent solar atmosphere. It is a stratum several hundred kilometres thick, from which almost all the energy emitted by the sun is radiated into space. A permanent feature of the photosphere is the *granulation, which gives it a mottled appearance. More striking are the *sunspots and their associated *faculae, the numbers of which fluctuate over an average period of approximately 11 years—the so-called *sunspot cycle.

Observation of the transit of sunspots across the sun's disc discloses a *differential rotation, the *synodic period of which increases with *heliographic latitude from 26.87 days at the equator to 29.65 days at ±40° (beyond which sunspots are seldom seen). The mean synodic period is taken to be 27.2753 days, which is equivalent to a *sidereal period of approximately 25.38 days (and corresponds to the actual period at around ±15°). Spectroscopic measurements show that the rotation period continues to increase right up to the polar regions and that at any given latitude it decreases with height above the photosphere, except at the equator where the periods are approximately equal. The reason for this differential rotation is unknown, though it has been suggested that it may be caused by rapid rotation of the sun's core. From the base to the top of the photosphere the temperature falls from 6000 K to 4000 K.

Immediately above the photosphere is the *chromosphere, a stratum a few thousand kilometres thick in which the temperature rises from 4000 K to around 50 000 K, as its density decreases exponentially with height. Between this and the exceedingly rarified *corona is the *transition region*, a stratum several hundred kilometres thick, in which the temperature rises further to around 500 000 K. The corona itself attains a temperature of around 2 000 000 K at a height of about 75 000 km. It extends for many million kilometres into the *interplanetary medium, where the *solar wind carries a stream of atomic particles to the depths of the solar system.

The sun is thought to possess a weak general magnetic field, though this has yet to be distinguished from the transient polar fields resulting from the dispersal of the intense localized fields of sunspots (*see* sunspot cycle).

See also quiet sun; active sun; solar activity.

sunrise terminator. *See* terminator.

sunset terminator. *See* terminator.

sun's longitude. A measure of the earth's position around its orbit and more specifically the point in this orbit at which a *meteor shower has its maximum intensity. It is the angle

sunspot cycle. The semiregular fluctuation of the number of *sunspots over an average period of approximately 11 years—or 22 years if the respective magnetic polarities of the *p*- and *f*-spots are considered, since these reverse at sunspot minimum for the groups of the new cycle (*see* sunspots). The nature of the cycle is shown in graph *a,* where the index of sunspot activity is the *Zürich relative sunspot number. It can be seen that the rise to *sunspot maximum* usually occupies a shorter time than the fall to *minimum* and that the amount of activity may vary considerably between two consecutive cycles. There is some evidence for a modulation of the amplitude of the cycle over a period of around 80 years but more data are required before the reality of this can be established. The last sunspot maximum occurred in 1969.3; the next is predicted for 1979/80.

The phase of the sunspot cycle also determines the mean *heliographic latitude of all groups. At minimum the first groups of the new cycle appear at ±30°-35°. Thereafter the latitude range moves progressively towards the equator, until by the next minimum the mean latitude is around ±7°. Then, while the equatorial groups are petering out, those of the succeeding cycle begin to appear in their characteristic higher latitudes. This latitudinal progression is known as *Spörer's law.* At any one time there may be a considerable spread in latitude, but groups are seldom seen farther than 35° or closer than 5° from the equator. The *butterfly diagram* is a graphical representation of Spörer's law obtained by plotting the mean *heliographic latitude of individual groups of sunspots against time (see graph *b*). Its appearance has been likened to successive pairs of butterfly wings, hence the name.

The underlying cause of the sunspot cycle is thought to be the interplay between a large-scale relatively weak poloidal magnetic field beneath the *photosphere, *differential rotation, and *convection. The poloidal field, which is constrained to move with the ionized solar material, becomes increasingly distorted by differential rotation until an intense toroidal field is produced. The strength of this field is further enhanced by the perturbing effect of convection, which twists the field lines into rope-like configurations that may penetrate through the surface to form sunspots. This will occur first in intermediate latitudes, where the rate of shearing of the field is greatest, and thereafter in increasingly low latitudes.

The inclination to the equator of the fields of opposite

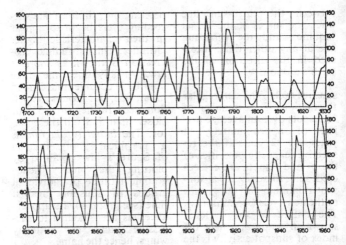

a *Yearly means of Zürich relative sunspot numbers 1700–1960 (From 'Sunspot and Geomagnetic Storm Data 1874–1954', HMSO, London (1955); courtesy of the Royal Greenwich Observatory.*

polarity associated with the *p*- and *f*-spots is such that they may drift apart in both longitude and latitude, as a result of differential rotation and the cyclonic rotation of individual supergranular cells (*see* supergranulation). The latitudinal drift is responsible for an accumulation in the polar regions of magnetic flux of the same polarity as the *f*-spots in the respective hemispheres. Thus the intense (0.2–0.4 tesla) localized fields of sunspots are gradually dispersed to form weak (1-2 × 10^{-4} tesla) polar fields, which reverse polarity (not necessarily synchronously) around sunspot maximum. When this happens differential rotation no longer intensifies the sub-photospheric toroidal field but rather weakens it and reestablishes a poloidal field of opposite direction to its predecessor.

The sunspot cycle may therefore be regarded (if this model is correct) as a relaxation process that is continually repeating itself. There is reason to believe, however, that at least some of the features of the present cycle may be transitory. In particular, a prolonged minimum, termed the *Maunder minimum*, from about 1645 to 1715, suggests that there is more than one circulatory mode available to the solar dynamo.

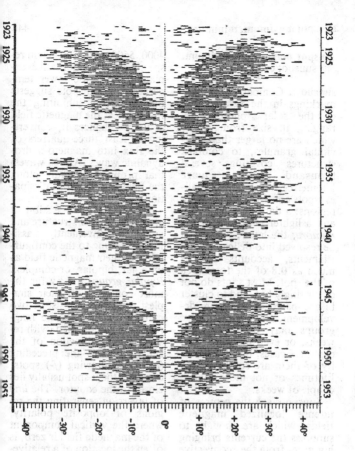

b. *Butterfly diagram (From 'The Sunspot-Activity in the Years 1610–1960' by M. Waldmeier, Schultess & Co., Zürich (1961); courtesy of M. Waldmeier.)*

Sunspots are the most obvious but by no means the only manifestation of *solar activity to undergo a cyclical change over a period of around 11 years. It is therefore proper to restrict the use of the term *sunspot cycle* to consideration of the fluctuation of the number of sunspots and to use the more general term *solar cycle* when considering the variation in the level of solar activity as a whole.

sunspot maximum, minimum. *See* sunspot cycle.

sunspots. Comparatively dark markings in the *granulation of the solar *photosphere, ranging in size from *pores,* which are no larger than individual granules, to complex structures covering several thousand million square kilometres. All but the smallest spots consist of two distinct regions: a dark core or *umbra* and a lighter periphery or *penumbra;* the latter, which can be resolved into delicate radial filaments, accounts for as much as 0.8 of the total area of the spot. Most spots do not occur singly but tend to cluster in groups. The formation, development, and decay of large groups may occupy several weeks, or even months in exceptional cases, but the majority of them are small, having lifetimes of not more than a couple of weeks.

Sunspots are the centres of intense localized magnetic fields, which are thought to suppress the currents bringing hot gases from the *convective zone before they reach the photosphere. In the umbra, where the field is strongest (0.2–0.4 telsa), convection is almost completely inhibited; in the penumbra, where the field is more horizontal, a radial flow of material takes place (*see* Evershed effect). The umbra is therefore much cooler than the penumbra, which is in turn cooler than the surrounding granulation, their temperatures being of the order of 4000, 5600, and 6000 kelvin respectively. An alternative explanation for the lower temperature of sunspots suggests that far from inhibiting the flow of heat the magnetic field actually enhances it, converting at least three-quarters of the flux into magneto-hydrodynamic waves (Alfvén waves) that propagate rapidly along the field lines without dissipation.

Sunspot groups exhibit a great diversity of structure and are most conveniently classified according to the configuration of their magnetic field as unipolar, bipolar, or complex. *Bipolar groups* are by far the most common. In their simplest form they consist of two main spots of opposite magnetic polarity, termed (with respect to the direction of the sun's rotation) the preceding (*p*-) and following (*f*-) spots, of which the *p*-spot usually lies closer to the equator. The *line of inversion* separating the regions of opposite polarity, where the vertical component of the magnetic field is zero, is often the location of a relatively stable *filament and the scene of violent *flares; both phenomena occur above the sunspot group, in the upper *chromosphere/inner *corona.

The number of sunspots fluctuates over an average period of approximately 11 years—the so-called *sunspot cycle—during which time there is a corresponding variation in the mean *heliographic latitude of all groups (*see* sunspot cycle for Spörer's law, butterfly diagram). Moreover the respective polarities of the *p*- and *f*-spots, which are consistent within a particular hemisphere but opposite on the other side of the equator, reverse at sunspot minimum for the groups of the new cycle.

Sunspots are the most obvious manifestation of *solar activity. Together with their associated *faculae/*plages and with filaments (or *prominences) and flares, they constitute *active regions, whose influence extends from the photosphere through the chromosphere to the corona.

superclusters. *See* clusters of galaxies.

supergiant. The largest and most luminous type of star, lying above both the *main sequence and the *giant region in the *Hertzsprung-Russell diagram. They are grouped in luminosity classes Ia and Ib (*see* spectral types) and generally have absolute *magnitudes between -5 and -8. Only the most massive stars can become supergiants and consequently they are very rare. They are so bright, however, that they stand out in external galaxies and can be used as distance indicators. Examples of supergiant stars are Rigel and Betelgeuse in Orion, Antares in Scorpius (which would engulf the orbit of Mars if placed in the solar system), and the brightest *Cepheid variable stars.

supergiant elliptical. *See* galaxies; clusters of galaxies.

supergranulation. A network of large-scale (30 000 km diameter) convective cells in the solar *photosphere. The individual cells have an upward velocity of about 0.1 km s^{-1} and exhibit a horizontal flow of material, outward from the centre, with a velocity of about 0.4 km s^{-1}; their average lifetime is about one day (*compare* granulation). The space between adjacent cells is the preferred birthplace of *sunspots, whose intense localized magnetic fields are dispersed by differential rotation and a cyclonic rotation of the cells, to form a weak polar field in both hemispheres.

Supergranulation is not readily visible but may be detected in velocity-cancelled *spectroheliograms. (These are obtained by the photographic cancellation of two spectroheliograms taken simultaneously in the wings, i.e. on either side of the core, of a strong *Fraunhofer line.) It is seen as bright-dark elements away from the centre of the sun's disc, corresponding to a predominantly horizontal flow of material in the line of sight that has been rendered visible by its doubled *Doppler shift. *See also* chromospheric network; spicules.

superheterodyne receiver. *See* receiver.

superior conjunction. *See* conjunction.

superior culmination (superior transit). *Another name for* upper culmination. *See* culmination.

superior planets. The planets Mars, Jupiter, Saturn, Uranus, Neptune, and Pluto, which orbit the sun at distances greater than that of the earth.

superlight source (superluminal source). A *radio source in which components appear to be moving at velocities greater than the speed of light. Physical motion faster than light is impossible according to the special theory of *relativity. It is possible, however, to explain such apparent superlight velocities in terms of geometrical effects involving objects moving more slowly than light.

superluminal source. *See* superlight source.

supermassive black hole. *See* black hole.

supernova. A star that temporarily brightens to an absolute *magnitude brighter than about -15, over a hundred times more luminous than an ordinary *nova. A supernova explosion blows off all or most of the star's material at high velocity, as a result of the final uncontrolled nuclear reactions in the small proportion of stars that reach an unstable state late in their evolution. The debris consists of an expanding gas shell (the *supernova remnant), and possibly a compact stellar object (a *neutron star or a *black hole), which is the original star's collapsed core.

Supernova searches, organized on a regular basis since 1936, have discovered more than 400 extragalactic supernovae. In addition, past supernovae in our *Galaxy have been recorded in Western Europe (*Tycho's star of 1572 and *Kepler's star of 1604) and in China and Korea (AD 185, 386?, 393, 1006, 1054, and 1181). The total supernova production rate in the Galaxy is currently estimated as two or three per century.

On observational grounds supernovae have been divided into five types, although the vast majority fall into Types I and II. *Type I supernovae* occur in both elliptical and spiral galaxies; they all have remarkably similar light curves and absolute magnitudes (-19) at maximum. Their spectra were for long an enigma but are now interpreted as very broad absorption lines. Surprisingly there is little evidence for hydrogen in the ejected gas, which has a mass roughly equal to the sun and a velocity around 11 000 km s^{-1}: the total kinetic energy is about 10^{44} joules. The cause of Type I supernovae is uncertain. The presupernova may be a *white dwarf that accretes enough mass from a close companion to exceed the *Chandrasekhar limit; alternatively it may be the core of a *supergiant star that has lost its outer envelope and suffers a central explosion similar to a Type II.

Type II supernovae occur only in the spiral arms of spiral galaxies. They show more diversity in light curves and absolute magnitude than Type I, though most reach a maximum of around -17. Their spectra show a normal abundance of the elements and indicate that several solar masses are ejected at around 5000 km s^{-1}. The pre-supernova is almost certainly a supergiant of more than eight solar masses and with a diameter of about 10 astronomical units. Its iron core is incapable of generating energy by nuclear fusion and collapses catastrophically. The star's envelope is ejected violently, possibly by a flood of neutrinos from the collapsing core. *See also* nucleosynthesis.

supernova remnant. The expanding shell of gas from a supernova explosion, consisting of the supernova ejecta and 'swept-up' interstellar gas. Young (<1000 year old) supernova remnants are generally optically faint but are fairly strong radio and x-ray sources; the *Crab nebula is exceptionally bright because it is energized by a central *pulsar. Older supernova remnants appear as rings of bright filaments, again with associated radio and x-ray emission. Compression by an expanding supernova remnant can trigger *stellar birth in interstellar clouds, and possibly initiated the formation of the *solar system. *See also* emission nebula.

super Schmidt. *See* Schmidt telescope.

supersynthesis. *Another name for* earth-rotation synthesis. *See* aperture synthesis.

surface gravity. Symbol: *g*. The *acceleration of gravity experienced by an object when it falls freely towards the surface of a planet or star.

surface temperature. The temperature of the radiating layers of a star at which its continuous spectrum is produced. The star is assumed to be a perfect radiator, i.e. a *black body, and its temperature is usually expressed as its *effective temperature or its *colour temperature. Surface temperature is very much lower than the temperature at the centre of a star.

surges. Impulsive events on the sun that are characterized by the ejection of chromospheric material into the inner *corona with velocities of 100-200 km s^{-1}. They occur in *active regions, where they often accompany *flares but more generally stem from small flare-like brightenings (termed *moustaches* or *Ellerman bombs*) at the outer edge of the penumbra of *sunspots. The surge material, which is usually directed radially away from the spot, may reach a height of a couple of hundred thousand kilometres before fading or returning to the *chromosphere along either a curved trajectory or, more frequently, the trajectory of its ascent. Surges typically have a lifetime of 10–20 minutes and show a strong tendency to recur. *See also* prominences.

survey catalogue. *See* star catalogue.

Surveyor. An American unmanned space programme designed to investigate the bearing strength, physical structure, and chemistry of the lunar *regolith by means of trenching devices and alpha-scattering analysis (see table). The spacecraft were soft-landed by retrorockets. Each was equipped with a television camera, powered by solar cells, for panoramic photography. The *maria were shown to have a basaltic composition, whereas the *highlands were found to be richer in calcium and aluminium. *See also* Ranger; Lunar Orbiter probes; Luna probes; Zond probes.

Swan. Common name for *Cygnus.

Swan nebula. *See* Omega nebula.

Swordfish. Common name for *Dorado.

Sword-Handle. *See* h and Chi Persei.

synchronous orbit. An orbit made by an artificial satellite that has a period equal to the period of rotation of the orbited body. *See also* geosynchronous orbit.

synchronous rotation (captured rotation). The rotation of a natural satellite about its primary in which the period of rotation of the satellite is equal to its orbital period. The same hemisphere thus always faces the primary. The moon is in synchronous rotation, although *libration allows slightly more than one hemisphere to be seen from earth. There are good dynamical reasons for satellites fairly close to their planet being

SURVEYOR PROBES

Spacecraft	Launch date	Comments*
Surveyor 1	1966, May 30	Soft-landed in Oc. Procellarum (11 150)
Surveyor 2	1966, Sept. 20	Struck moon in Oc. Procellarum
Surveyor 3	1967, Apr. 17	Tested soil in Oc. Procellarum (6315); later visited by Apollo 12
Surveyor 4	1967, July 14	Landed in Sinus Medii but radio contact lost
Surveyor 5	1967, Sept. 8	Analysed soil in M. Tranquillitatis (18 006)
Surveyor 6	1967, Nov. 7	Analysed soil in Sinus Medii (30 000)
Surveyor 7	1968, Jan. 7	Analysed/tested soil near Tycho (21 000)

*Figures in parentheses indicate the number of photographs transmitted

a *Synchrotron emission*

locked in synchronous rotation. All the satellites of Mars, Jupiter, and Saturn that have been investigated for synchronous rotation have shown this property.

synchrotron emission (synchrotron radiation; magnetobremsstrahlung). *Electromagnetic radiation from very high energy *electrons moving in a magnetic field. It is thus *nonthermal emission. It is the mechanism most often invoked to explain the radio emission from extragalactic *radio sources and the emission from *supernovae remnants. When an electron moves in a magnetic field, it follows a circular path making $v = eB/2\pi m$ revolutions per second, where e and m are the electronic charge and mass and B is the magnetic flux density; v is the *gyrofrequency*. The acceleration of the electron in its circular path causes it to radiate electromagnetic waves at frequency v. This emission is called *cyclotron radiation*.

If the velocity of the electrons becomes comparable with the speed of light, the theory of *relativity must be used to explain the observed phenomena. One effect is *relativistic beaming* in which the radiation is beamed forward in the direction of motion of the relativistic electron (see illustration *a*). The nature of the radiation changes from the simple cyclotron case and it becomes synchrotron emission. At low frequencies, the conditions in the source may cause the synchrotron emission to be reabsorbed before it can escape. The *radio-source spectrum then exhibits a turnover, as is often seen in extragalactic radio sources (see illustration *b*). This is called *synchrotron self-absorption*. See also Razin effect.

syndynames. *See* comet tails.

synodic month (lunation; lunar month). The interval of 29.530 59 days between two successive new moons.

synodic period. The average time between successive returns of a planet to the same

b Spectrum of synchrotron emission

*aspect. For example, it is the mean interval between oppositions of the planet. It is also the period of rotation of a celestial body as determined from earth. The synodic period, P_s, of a body is very much easier to measure than its *sidereal period, P_1. It is related to the sidereal period of an inferior or a superior planet, satellite, etc., respectively, by the equations

$$1/P_s = 1/P_1 - 1/P_2$$
$$1/P_s = 1/P_2 - 1/P_1$$

P_2 is the sidereal period of the earth, i.e. 365.256 days or 1 year.

Syrtis Major. A dark triangular plateau near the Martian equator and prominent to earth-based observers. It slopes downwards to the east and is prone to strong winds that move and remove light-coloured dust to expose the darker rocks below. *See also* Mars, surface features.

system noise. *See* sensitivity.

system temperature (noise temperature). *See* sensitivity.

syzygy. The configuration arising when the sun, earth, and either the moon or a planet lie approximately in line, i.e. when the moon or planet is at *opposition or *conjunction.

T

Table Mountain (Table). Common names for *Mensa.

TAI. *Abbrev. (French) for* international atomic time. *See* atomic time.

tail. *Short for* comet tail.

tangential velocity (transverse velocity). *See* proper motion.

T-antenna. *See* Mills cross antenna.

Tarantula nebula. *See* 30 Doradûs.

T-association. *See* association.

Tau Ceti. *See* Table 7 (nearest stars), backmatter; Ozma project.

Taurids. A minor *meteor shower with a double radiant, RA 56°, dec +14° and +22°, that maximizes on Nov. 8.

Taurus (Bull). A large zodiac constellation in the northern hemisphere near Orion, lying partly in the Milky Way. The brightest stars are the 1st-magnitude *Aldebaran, the 2nd-magnitude *El Nath, which was once part of the constellation Auriga, and several of 3rd magnitude. The area contains the prototypes of the *RV Tauri and *T Tauri stars, the *Hyades and *Pleiades, both beautiful open clusters, and the supernova remnant the *Crab nebula with its associated optical *pulsar. Abbrev.: Tau; genitive form: Tauri; approx. position: RA 4.5h, dec +15°; area: 314 sq deg.

Taurus A. The *radio source in the constellation Taurus that is identified with the *Crab nebula. It has a *flux density at 178 megahertz of 1420 *jansky, most of which comes from *synchrotron emission in the nebula, which is a *supernova remnant. A small fraction of the radio flux is due to a rapidly rotating *pulsar in the nebula, thought to be the remains of the original star that exploded. Unlike other supernova remnants, such as *Cassiopeia A, the Crab nebula does not have a regular ring-like radio structure, probably because of its internal energy source—the pulsar.

Taurus-Littrow. A *highland region on the moon on the eastern borders of Mare *Seneritatis. It was the target for *Apollo 17.

Taurus X-1. *See* x-ray astronomy; Crab nebula.

TD-1. A satellite developed by the European Space Research Organization and launched in March 1972. It has made a systematic survey of the sky, recording the spectra of thousands of stars in the ultraviolet wavelength range of 135-255 nanometres. All hot O and B stars were recorded down to 9th magnitude. A catalogue of more than 30 000 ultraviolet sources is being compiled from the data.

tectonics. Large-scale movements of the crust of a planet, such as those that give rise to mountain building, faulting, and, on the *earth, continental drift.

tektites. Small glassy rounded objects, two to three cm in size, found only in certain parts of the world, often in vast *strewn fields*. About 650 000 had been discovered by 1960. They have shapes varying from spheroidal to pear-, lens-, and disc-shaped with surface-flow structures indicating that they were formed by rapid cooling of molten material. The chemical composition is that of silicic igneous rock with high SiO_2 content (70-80%) and moderate Al_2O_3 content (11-15%). They are named according to the place where they are found, e.g. *moldavites* (Czechoslovakia), *australites* (Australia), etc. They were probably formed by the fusion of terrestrial material during the impact of giant *meteorites; terrestrial and lunar volcanic origins have also been proposed. They are between 0.75-35 million years old. Small tektite-like objects have been found in the lunar regolith.

telemetry. The transmission of data from a satellite, space station, spaceprobe, etc., by means of radio waves.

telescope. An optical device for collecting light so as to form an image of a distant object. A telescope is superior to the naked eye because its light collecting power will reveal much fainter stars, nebulae, etc., and because its magnification will disclose fine detail in lunar and planetary images. The light is collected and focused by an object lens in a *refracting telescope or by a primary mirror in a *reflecting telescope. In both types the image formed at the prime focus is magnified and examined through an *eyepiece. A telescope is usually described in terms of its *aperture: *light-gathering power, *resolving power, and *limiting magnitude all increase as the aperture increases. *See also* magnifying power; mounting; radio telescope; infrared telescope.

telescopic object. A star or other celestial object that is less than 6th apparent *magnitude and cannot be seen without a telescope.

Telescopium (Telescope). An inconspicuous constellation in the southern hemisphere near Scorpius, the brightest stars being of 4th magnitude. Abbrev.: Tel; genitive form: Telescopii; approx. position: RA 19.5h, dec -50°; area: 252 sq deg.

television camera. An electronic instrument that converts the optical image of a scene into an electrical signal, termed a *video signal*. This signal may then be transmitted by means of radio waves or by cable to a distant receiver. The essential elements of the camera are a lens system, a *camera tube,* and a preamplifier. It is the camera tube that produces the video signal. The image is focused by the high-quality lens system, usually a zoom lens of variable focal length, on to the light-sensitive target area of the camera tube. In most camera tubes the information on the target is extracted by means of an electron beam that scans the target area. The two basic types of camera tubes are the *image orthicon* and the *vidicon tube:* the former has a target coated with photoemissive material, the latter has a photoconductive target. The vidicon, and its development the plumbicon, is smaller, simpler, and cheaper than the image orthicon. The solid-state *CCD camera does not operate by electron-beam scanning and is therefore very much smaller and lighter than the vidicon.

Television cameras are used in astronomy both in space probes and satellites and also in combination with ground-based telescopes. The output signal from the camera tube is further amplified and is then either transmitted to a distant radio receiver or is stored on video tape so that it may be fed into a computer for analysis.

telluric lines. Absorption lines seen in the spectrum of a celestial object but due to components, such as water vapour, oxygen, or carbon dioxide, in the earth's atmosphere. Telluric lines can be distinguished because they are generally narrower than solar or stellar lines, etc. In addition, they vary in strength with the position of the source and are strongest when the source is on the horizon.

Tellus. *Latin name for* the *earth.

Tempel's comets. Two of the comets discovered by Ernst W.L. Tempel were named Tempel 1 (1879 III) and Tempel 2 (1957 II). The first disappeared after 1879 but reappeared in 1967. His most famous comet, 1866 I, named *Tempel-Tuttle,* has a period of 33.176 years and an orbit that is practically identical to that of the *Leonid meteor stream.

temperature. Symbol: T. A measure of the heat energy possessed by a body, i.e. of its degree of hotness. Astronomical temperatures are usually determined from spectroscopic measurements. Since astronomical bodies do not generally have a uniform distribution of heat energy and do not obey exactly the laws determining temperature, there are several different types of temperature; each characterizes a particular property of the body and has a slightly different value. They include *effective, *colour, *brightness, ionization, and excitation temperature. The latter two are determined from the *Saha ionization equation and the *Boltzmann equation. *See also* thermodynamic temperature scale; kelvin.

temperature minimum. *See* photosphere.

terminator. The boundary between the dark and sunlit hemispheres of the moon or a planet, the rotation of which gives rise to a *sunrise terminator* and a *sunset terminator*. The appearance of the terminator is affected by topographic irregularities, such as mountains, and by the presence of an atmosphere. *See also* phases, lunar.

terra. *See* highlands, lunar.

terraced walls. *See* craters.

terrestrial. Of, relating to, or similar to the planet Earth.

terrestrial planets. The planets Mercury, Venus, Earth, and Mars. They are small rocky worlds with shallow atmospheres in comparison with the mainly gaseous *giant planets.

tesla. Symbol: T. The SI unit of *magnetic flux density, defined as one weber of magnetic flux per metre squared. One tesla is equal to 10^4 gauss.

Tethys. A satellite of Saturn, discovered in 1684. Recent observations show it to have an icy surface. *See* Table 2, backmatter.

Tharsis Ridge. The principal area of volcanic activity on Mars, measuring 5000 km in diameter and elevated by up to 7 km from the average level of the planet's surface. It includes the four largest Martian volcanoes: *Olympus Mons, *Ascraeus Mons, *Pavonis Mons, and *Arsia Mons. *See also* Mars, volcanoes.

thassaloid. *See* basins, lunar.

thermal black-body radiation. The electromagnetic emission from a perfect radiator—a *black body—whose brightness follows Planck's radiation law.

thermal emission (thermal radiation; thermal bremsstrahlung). Electromagnetic radiation resulting from interactions between electrons and atoms or molecules in a hot dense medium. The term usually refers to thermal emission from ionized gas (but *see also* thermal black-body radiation). Free electrons in the ionized gas, mainly ionized hydrogen, radiate when they are deflected in passing near protons (hydrogen ions). The electrons are unbound to a nucleus before and after each encounter and the radiation is therefore also known as *freefree emission*. The radiation has a continuous spectrum, which at low frequencies has the spectral index, α, of a thermal black-body radiator, i.e.

$$S \propto \nu^\alpha; \alpha = -2$$

where S is the flux density and ν the frequency. At higher frequencies the spectrum flattens to a spectral index of $\alpha = 0$. The *emission measure* of an ionized region of length l containing a density n_e of free electrons is given by

$$\text{EM} = \int_0^l n_e^2 \, ds$$

where ds is an element of the path through the region.

thermodynamic equilibrium. A condition existing in a system when all the atoms and molecules have an equal share in the available heat energy. The temperature is then the same in all parts of the system by whatever method of measurement. A system that is not in thermodynamic equilibrium is unstable and the state of the system will change until equilibrium is reached.

thermodynamic temperature scale (absolute temperature scale). A temperature scale in which the temperature, T, is a function of the energy possessed by matter. Thermodynamic temperature is therefore a physical quantity that can be expressed in units, termed *kelvin. The zero of the scale is *absolute zero. The temperature of the ice point (the zero of the Celsius temperature scale) is 273.15 kelvin. Thermodynamic temperature can be converted to Celsius temperature by subtracting 273.15 from the thermodynamic temperature.

thermoelectric generator. A device that converts heat energy directly into electrical energy. It consists basically of two dissimilar metals or semiconductors joined at two junctions: if the junctions are at different temperatures an electromotive force (e.m.f.) develops between the junctions. An e.m.f. can also be generated between two regions on one piece of metal when a temperature difference exists between the regions. In both cases the e.m.f. is proportional to the temperature difference and is also dependent on the materials used. The e.m.f. can be used to power an external electric circuit. Thermoelectric generators are used in long-range planetary probes when *solar cells are unable to produce sufficient power. The hot junction or region of the device is then often heated by means of the decay of radioisotopes.

thermonuclear reaction. *See* nuclear fusion.

thermosphere. *See* atmospheric layers.

Theta Orionis. *See* Trapezium.

third quarter. *Another name for* last quarter. *See* phases, lunar.

three-body problem. A specific case of the *n-body problem in which the trajectories of three mutually interacting bodies are considered. There is no general solution for the problem although solutions exist for a few special instances. Thus the orbits can be determined if one of the bodies has negligible mass, as in the case of a planetary satellite, such as the moon, subject to perturbations by the sun or an asteroid whose motion is perturbed by Jupiter.

thrust. *See* launch vehicle.

Thuban (α Dra). A white giant in the constellation Draco that was the *pole star about 2700 BC when it lay 10′ from the celestial pole. Its brightness has probably decreased in the past few hundred years. It is a spectroscopic binary, period 51.38 days. m_v: 3.64; spectral type: A0 III; distance: 92 pc.

tidal bulge. *See* tides.

tidal force. A force arising in a system of one or more bodies as a result of differential gravitation: different parts of the system experience different accelerations. This can result in the production of *tides and other phenomena, and in extreme cases can lead to the disruption of a body.

tidal friction. Energy dissipated by the raising of *tides. As with tidal heights, the total energy dissipated by tidal friction depends on the topography of coastlines and on the areal extent of adjacent continental shelves. Tidal friction is currently slowing down the earth's rotation rate (i.e. the length of the day) by 16 seconds every million years and is causing the moon to recede from the earth by 120 cm every

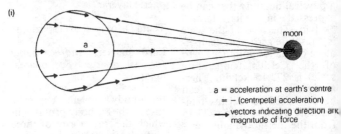

a *(i) Gravitational acceleration on earth due to moon*

a *(ii) Differential (tidal) acceleration found by subtracting a from other vectors on Fig. (i)* year. *See also* secular acceleration.

tidal height. *See* tides.

tidal lag. *See* tides.

tidal theories. *See* encounter theories.

tides. Distortions of a planet, star, etc., produced by the differential gravitational attraction of other astronomical bodies on parts of that planet, star, etc. The sun and moon combine to generate two *tidal bulges* in the earth's oceans, one directed towards the moon and the other diametrically opposite. If the earth is assumed to be spherical and to be covered with water, then the gravitational pull of the moon will have a different force at different points on the earth's surface (illustration *a*(i)). The gravitational acceleration at the earth's centre is equal to the centripetal acceleration of the earth-moon system. If this acceleration (which is a vector quantity) is subtracted from the surface accelerations, the differential (tidal) acceleration is found (illustration *a*[ii]), showing the two tidal bulges on the earth-moon line.

The daily rotation of the earth and the slower eastward revolution of the moon in its orbit produce (usually) two *high tides* and two *low tides* every 24 hours 50 minutes. The rotation and lunar revolution are also responsible for a *tidal lag* between the time when the moon crosses the local *meridian and the time of high tide.

The tidal pull of the sun is less than half that of the moon but reinforces it at full moon and at new moon to produce very high, or *spring tides;* these tides are exceptionally high when the earth is close to *perihelion and the moon is close to *perigee (illustration *b*). The lowest high tides, or *neap tides* occur when the moon is at *quadrature. The *tidal height* depends on the topography of the coastline and on the area of the adjacent continental shelf. Tides are also generated in the earth's atmosphere and in the solid earth.

Lunar tides are raised by the earth at *perigee as a consequence of the moon's eccentric orbit. They trigger *moonquakes and *transient lunar phenomena. *See also* tidal friction.

time. The time systems used in astronomy are *universal time, *sidereal time, *ephemeris time, and *atomic time. The first three are determined by astronomical observation and are based on the periodic motions of celestial bodies. Universal and sidereal time depend on the period of rotation of the earth, which has been shown to suffer slight irregular variations. Neither system can thus be considered as having an invariable rate, which is a necessity for precision timekeeping. A constant rate was therefore defined for ephemer-

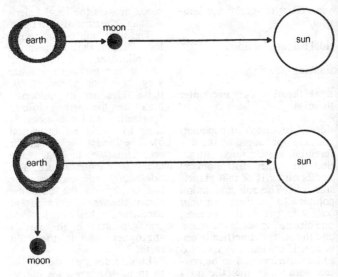

b *Spring (top) and neap tides*

is time and this system was adopted for all timekeeping in 1958. It was replaced in 1972 by the more accurate and convenient atomic time scale, which also has a uniform rate. Atomic time is maintained within one second of universal time by the intermediary of coordinated *universal time, which is the basis for civil timekeeping and for recording observations. Universal time is a special case of *mean solar time and is equivalent to *Greenwich Mean Time, from which *standard time is calculated.

time constant. Symbol: τ. A measure of the speed with which an electronic system can respond to a sudden change at its input. In particular, the time constant of a simple resistance-capacitance circuit is the product of the resistance and the capacitance.

time dilation. *See* relativity, special theory.

time zone. *See* standard time.

Titan. The largest satellite of Saturn, discovered in 1655. It orbits the planet at a mean distance of 1 222 000 km in a period of 15.95 days. With a diameter of 5800 km, it is probably the largest satellite in the solar system. It has a reddish cloudy atmosphere containing methane, and possibly ammonia and hydrogen, that may warm its solid surface to 175 kelvin by a *greenhouse effect. When account is taken of the depth of this atmosphere, Titan is found to have a low density of only 1.6 times that of water. It is possible therefore that ice and liquid water make up much of the satellite, which may have only a small rocky core. In this respect it may be similar to Ganymede and Callisto, the two largest of *Jupiter's satellites. *See* Table 2, backmatter.

Titania. The largest satellite of Uranus, discovered in 1787. *See* Table 2, backmatter.

Titius-Bode law. *See* Bode's law.

TLP. *Abbrev. for* transient lunar phenomena.

topocentric coordinates. The coordinates of a celestial body measured from the surface of the earth. For a star there is little or no detectable difference between topocentric coordinates and those referring to the earth's centre in a *geocentric coordinate system. For a member of the solar system the difference can be significant and a correction must be made.

topocentric zenith distance. *See* zenith distance.

T Orionis stars. *See* T Tauri stars.

total eclipse. *See* eclipse.

totality. *See* eclipse.

total magnitude (integrated magnitude). A measure of the brightness of an extended object, such as a galaxy or nebula, obtained by summing the surface brightness over the entire image area or by measuring the amount of light received through a series of apertures of increasing diameter with a photoelectric photometer. The readings are found gradually to approach an upper limiting magnitude, which, corrected for night-sky brightness, gives the total magnitude. *See also* magnitude.

total-power receiver. *See* receiver.

Toucan. Common name for *Tucana.

tower telescope (solar tower). *See* solar telescope.

trail. 1. *See* star trails.
2. *Obsolete name for* train.

train. Most of the visible radiation from a *meteor comes from the immediate vicinity of the ablating meteoroid. A brighter meteor, however, sometimes leaves a faintly glowing train of luminosity along its path; it is known as a *wake* if it glows for less than a second and a *persistent train* if the duration is longer. Sudden and brief enhancements of light during the meteor's passage are termed *flares* (or *bursts*). The term 'train' is also applied to the ionization and

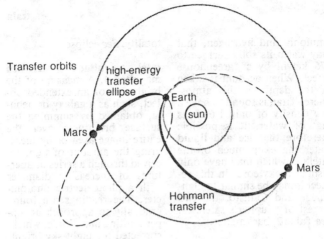

Transfer orbits

dust left along the meteoroid path.

trajectory. The curved path followed by a moving spacecraft, rocket, etc., under the forces of gravity, lift, and/or drag.

Tranquillitatis, Mare (Sea of Tranquillity). An ancient irregular *mare on the eastern *nearside of the moon. It was the landing area for *Ranger 8, *Surveyor 5, and Apollo 11 (*Tranquillity Base* was the first manned landing point). The basaltic lavas in the mare are 3900 to 3600 million years old.

transfer orbit. The trajectory followed by a spacecraft moving from one orbit to another, usually to the orbit of another body, such as another planet. The trajectory is generally an elongated ellipse—a *transfer ellipse*—that intersects the new orbit (see illustration). The spacecraft would have to manoeuvre into an orbit around the planet by firing its rocket motors.

The transfer orbit requiring the minimum expenditure of energy is an ellipse that just touches the original (circular) orbit and the new orbit. This is called a *Hohmann transfer,* after the German engineer who described it in 1925. Enough velocity is put in at the *perigee (or equivalent point) of the Hohmann transfer for the craft to reach the new orbit at *apogee; at apogee an additional velocity input injects the craft into the desired orbit. Although requiring the lowest possible energy, Hohmann transfers between planets involve a long flight time. The flight time can be much reduced by using a somewhat greater velocity at perigee than that needed for a Hohmann transfer.

transient lunar phenomena (TLP). Short-lived changes in the appearance of lunar features, including reddish glows and obscurations of surface detail, that are triggered by lunar *tides. They often occur within lunar *craters and around the perimeters of *basins. A carbon spectrum was obtained from a TLP in the crater Alphonsus.

transit. 1. The passage of an inferior planet across the sun's disc. Normally when Venus or Mercury move between the earth and sun, at inferior *conjunction, they pass above or below the sun. A transit can only occur when the planet is at or near one of its *nodes at inferior conjunction. Transits of Mercury take place either about May 7 (descending node) or Nov. 9 (ascending node) (see table). November transits occur at intervals of 7, 13, or 46 years, May transits at intervals of 13 or 46 years. The much rarer transits of Venus take place either about June 7 (descending node) or Dec. 8 (ascending node). They occur in pairs with a separation of 8 years, the pairs being separated by over 100 years.
2. The passage of a planetary satellite or its shadow across the face of the planet.
3. *See* culmination.

transit circle. *See* meridian circle.

transit instrument. *See* meridian circle.

transition region. *See* sun. *See also* chromosphere.

transmission grating. *See* diffraction grating.

transmission line. *See* feeder.

transverse velocity. *Another name for* tangential velocity. *See* proper motion.

Trapezium (Theta Orionis, θ Ori). A multiple star that lies in and ionizes the *Orion nebula and is part of the *Orion association. The six stars are a few thousand years old. The four brightest stars, of spectral class O6, B1, B1, and B3, form a trapezium and are visible with a small telescope. The brightest one (m_v: 5.16, O6p) is a spectroscopic binary.

Triangulum (Triangle). A small constellation in the northern hemisphere near Perseus, the brightest stars being of 2nd magnitude. It contains the *Triangulum spiral. Abbrev.: Tri; genitive form: Trianguli; approx. position: RA 2h, dec +30°; area: 132 sq deg.

Triangulum Australe (Southern Triangle). A small constellation in the southern hemisphere near Crux, lying partly in the Milky Way. The brightest stars are the 2nd-magnitude orange giant Alpha Trianguli Australis and 3rd-magnitude Beta and Gamma, which form a triangle. The area contains the bright open cluster NGC 6025. Abbrev.: TrA; genitive form: Trianguli Australis; approx. position: RA 16h, dec -65°; area: 110 sq deg.

Triangulum spiral (M33; NGC 598). A spiral (Sc) galaxy in the constellation Triangulum that is the third largest of the confirmed members of the *Local Group and is about 700 kiloparsecs distant. Although it is both smaller and more distant than the *Andromeda galaxy it can appear larger because it is orientated face-on to us. The spiral arms are more open than those in our own Galaxy and the nucleus is relatively small and faint. It has a large number of young *population I stars.

Trifid nebula (M20; NGC 6514). An *emission nebula about 2000 parsecs away in the direction of Sagittarius. There is dark matter associated with the nebula.

trigonometric parallax. *See* annual parallax.

triple alpha process (3α process). A nuclear reaction occurring in an evolved star's interior when all the hydrogen has been used up and the core temperature has reached about 10^8 kelvin. Three helium nuclei (alpha particles) fuse together to form carbon by the reaction
$$3\,^4He \rightarrow\,^{12}C + \gamma$$
This reaction gives out only 10% as much energy per unit mass as hydrogen burning.

triple star. A *multiple star with three components. *See also* binary star.

Triton. The larger of the two satellites of Neptune. Discovered in 1846, it has a diameter of about 3800 km, a little larger than the moon. It is the only large satellite to follow a retrograde orbit (*see* direct motion) around its planet. The anomalous orbits of Triton and Neptune's other satellite *Nereid may result from an interaction with another Neptunian satellite that was ejected from the system, perhaps to become the planet *Pluto. Calculations suggest that Triton is slowly spiralling towards Neptune because of tidal drag and that it may collide with the planet within another 100 million years—only a small fraction of the age of the solar system. *See* Table 2, backmatter.

Trojan group. A group of *minor planets that lie in the vicinity of the two *Lagrangian points in the orbit of Jupiter around the sun: each point occupies one corner of an equilateral triangle with the sun and Jupiter at the other corners. Beginning with the first Trojan to be discovered, *Achilles in 1906, most have been named after heroes from the two sides of the Trojan Wars. In fact, perturbations by other planets mean that each member of the group oscillates considerably in position (mainly in longitude along the orbit of Jupiter) from its stable Lagrangian point. It is likely that perturbations and collisions between them cause the total number of members of the group to change with time. Recent surveys suggest that more than 1000 Trojan group members may be bright enough to be seen from the earth but less than 2% of these have yet received official minor planet designations. They are all exceptionally dark bodies.

tropical year. The interval between two successive passages of the sun through the vernal *equinox, i.e. the time taken by the sun, in its apparent annual motion, to complete one revolution around the celestial sphere relative to the vernal equinox. It is equal to 365.242 19 days. The tropical year of 1900 is the primary unit of *ephemeris time. If the vernal equinox were a fixed point on the celestial sphere the tropical year would be equal to the *sidereal year. *Precession however produces an annual retrograde motion of the equinoxes amounting to 50.27 arc seconds relative to the fixed stars. The tropical year is thus about 20 minutes shorter than the sidereal year. Since the seasons recur after one tropical year, the average length of a *calendar year should be as close as is practicable to the tropical year.

troposphere, tropopause. *See* atmospheric layers.

true anomaly. *See* anomaly.

true equinox (apparent equinox). The vernal *equinox determined by the point of intersection of the celestial equator and the ecliptic and uncorrected for *nutation. It moves westwards around the ecliptic at about 0.137 arc seconds per day. Although stellar positions slowly change as the true equinox moves, the sun's position relative to this point remains practically unchanged, for a given day in the year, for thousands of years. *See also* mean equinox.

true horizon. A great circle on the celestial sphere that is parallel to an observer's horizon and passes through the earth's centre.

true position. The heliocentric position of a star referred to the *true equinox and true equator at the instant of observation. The *mean position* of a star is its true position corrected for *nutation and thus referred to the *mean equinox and mean equator of a standard epoch, such as the beginning of the year of observation. The *apparent position* is the geocentric position of a star as observed from earth; there is a slight difference between true and apparent position arising from annual parallax and aberration.

Trumpler's classification. A

method of classifying *open clusters according to their appearance, using the three criteria of stellar concentration towards the centre, range in brightness of the member stars, and the total number of stars in the cluster. The method was devised by the Swiss-American astronomer Robert Trumpler who also found a correlation between this appearance classification and the cluster diameter. The actual and observed diameters can be used to determine distance if *interstellar extinction is taken into account.

T Tauri stars. *Irregular variable stars of spectral types F, G, or K that are usually found together in groups (T associations). The prototype, T Tauri, is a yellow (dG5e) star. They are believed to represent the last phase in the evolution of *protostars before they settle down on to the *main sequence. There is much evidence that they are young objects: a great deal of their radiation occurs in the infrared part of the spectrum, indicating that they are still immersed in warm dusty cocoons; there are also *forbidden lines at optical wavelengths arising from surrounding gas.

Lying above the main sequence in the *Hertzsprung-Russell diagram, these stars are still contracting and losing mass at the rate of one ten-millionth of a solar mass per year (*see* T Tauri wind). Their spectral lines reveal that they are extremely rapid rotators, throwing off material at speeds of up to 300 km s^{-1}. The irregular light variations are believed to arise in the chromospheres of these young stars in a process that is perhaps similar to the production of colossal solar *flares. *RW Aurigae stars* and *T Orionis stars* resemble T Tauri stars in their behaviour and are thought to be a similar phenomenon, while *YY Orionis stars may represent an earlier phase in their development.

T Tauri wind. A continuous flow of material from the surface of a *T Tauri star. Qualitatively this flux resembles the *solar wind but the amount of material lost is far greater, approaching 10^{-7} solar masses per year in some cases. It is possible that this outflow acts as a braking mechanism for fast-spinning young stars.

Tucana (Toucan). A constellation in the southern hemisphere near Grus, the brightest stars being of 3rd magnitude. It contains the Small *Magellanic Cloud and the globular clusters 47 Tucanae, easily visible, and NGC 362, just visible. Abbrev.: Tuc; genitive form: Tucanae; approx. position: RA 0h, dec -65°; area: 295 sq deg.

Tunguska event. A gigantic explosion that occurred at about 7.17 a.m. on June 30, 1908 in the basin of the River Podkamennaya Tunguska, Central Siberia. Devastation rained over an area 80 km in diameter and eye witnesses up to 500 km away saw in a cloudless sky the flight and explosion of a blindingly bright pale blue bolide. The sound of the explosion reverberated thousands of kilometres away, the explosion air wave recorded on microbarographs going twice around the world. The main explosion had an energy of 5×10^{16} joules and occurred at an altitude of 8·5 km.

Many theories have been put forward to explain the explosion, the less conventional involving black holes, critical masses of extraterrestrial fissionable material, antimatter bodies, and alien spacecraft. In fact it was probably caused by the impact of a small comet, mass 5×10^{10} grams, geocentric velocity about 50 km s^{-1}, and absolute *magnitude +26. When the comet encountered the earth it would have been coming from a point in the dawn sky comparatively close to the sun and would thus have been most difficult to detect and observe.

turnoff point. The point on the *Hertzsprung-Russell diagram of a *globular or *open cluster at which stars are leaving the *main sequence and entering a more evolved phase. Assuming that all the stars in a cluster contract out of a single gas cloud at approximately the same time, the locations of these stars on the H-R diagram depend on the time elapsed since the initial contraction. The length of time spent on the main sequence depends on stellar mass: stars of about solar mass will remain there for maybe 10^{10} years while more massive, and hence more rapidly evolving stars will turn off earlier. Since position on the main sequence also depends on mass, the lower the turnoff point, the older the cluster.

twenty-one cm line (hydrogen line). See HI region.

twilight. See astronomical twilight.

twinkling. See scintillation.

Twins. Common name for *Gemini.

two-body problem. A special case of the *n-body problem in which a general solution can be found for the orbits of two bodies under the influence of their mutual gravitational attraction. The motion of a planet around the sun is an example as long as the attractive forces of the other planets are assumed to be negligible. A solution of the two-body problem is often acceptably realistic. Most theories of celestial motion thus use functions and principles, such as *orbital elements and *Kepler's laws, that have been derived by consideration of a two-body problem.

two-colour diagram. A plot of the *colour indices $B-V$ against $U-B$. Stars of different luminosity classes (*see* spectral types) lie on slightly different curves.

Tycho. A very recent lunar *crater, possibly only about 100 million years old; it is 100 km in diameter. Rays from the crater traverse much of the *nearside of the moon and are particularly striking at full moon. The *ejecta blanket was sampled by *Surveyor 7, the only Surveyor to land in the *highlands; some craters in the highland region of *Taurus-Littrow may be *secondaries of Tycho.

Tychonic system. A model of the solar system proposed by Tycho Brahe in the 1580s and based on observations he made in order to test the *Copernican system. He concluded that the sun and moon revolved around the earth, which he maintained lay at the centre of the solar system, but suggested that the other planets orbited the sun.

Tycho's star. A type I *supernova that was observed in 1572 in the constellation Cassiopeia and was studied by astronomers in Europe, China, and Korea. Tycho Brahe's careful observations of its position have identified it with a known radio and x-ray source. It was considerably brighter than Kepler's star of 1604, reaching a magnitude of -4 and remaining visible for about 18 months.

Type I, II Seyferts. *See* Seyfert galaxy.

Type I, II supernovae. *See* supernova.

U

U-B. *See* colour index.

UBV system. *See* magnitude.

3U catalogue, 4U catalogue. *See* Uhuru.

U Geminorum stars. *See* dwarf novae.

Uhuru. ('Freedom' in Swahili) The first artificial earth satellite (*SAS-1) for *x-ray astronomy, launched into a 500 km circular equatorial orbit in December 1970 from the San Marco platform off the coast of Kenya. Two sets of proportional gas counters, each 840 cm^2 in area and mounted on opposite sides of the slowly spinning spacecraft, scanned the sky for x-ray sources. Uhuru provided the first detailed x-ray sky map, leading to the *3U catalogue* of 161 sources, published in 1974. Its most important discoveries were of *x-ray binaries (Cen X-3, Her X-1, etc.) and of extended regions of x-ray emission associated with *clusters of galaxies. Uhuru operated successfully until July 1971, when its transmitter failed. The 3U catalogue is based on the data to that time. Subsequently the star sensors, used to determine the precise attitude of the spacecraft and hence the sky coordinates of each x-ray source, also degraded; surveying recommenced in December 1971, however, after partial transmitter recovery. These later observations, with the aid of a magnetometer attitude solution, continued until April, 1973, when the Uhuru battery failed. All the Uhuru observations were finally put together to form the *4U catalogue,* issued in 1978 and containing 339 x-ray sources.

UK 6. A UK satellite, due to be launched on May 25 1979, that is the final small satellite in the series of UK (Ariel) satellites. Its three experiments will measure the flux of heavy primary *cosmic rays, i.e. those above iron in the periodic table, will produce a soft x-ray map of the sky (energies 0.1-1.5 keV), and will determine the time variability and spectra of individual x-ray sources (many of which were discovered by *Ariel V) by means of a proportional counter pointed for some time towards each source.

UK Infrared telescope (UKIRT). A British *infrared telescope sited on Mauna Kea, Hawaii, at the very high altitude of 4200 metres. The site is well above the main weather layers and the bulk of atmospheric water vapour, opaque to most infrared wavelengths; in addition there is excellent *seeing for work at visible wavelengths. The telescope is funded by the Science Research Council and is run by the Royal Observatory, Edinburgh. It became operational in 1979. The mirror is 3.8 metres (150 inches) in diameter, making the instrument the world's largest infrared flux collector. The mirror surface has been ground sufficiently accurately that it can also be used at visible wavelengths for short exposures.

A full range of cryogenic (very low temperature) facilities is provided for cooling the detectors in order to reduce spurious signals generated by the warmth of the detectors. There are also means for removing spurious signals generated by the telescope itself. The telescope can work at near-infrared *atmospheric windows, such as those centered on 10 and 2 micrometres, and at long-wavelength windows centred on 400, 800, and 1200 micrometres.

UK Schmidt telescope. The British 1.2 metre *Schmidt telescope at *Siding Spring Observatory, Australia. It is funded by the Science Research Council and is operated as a field station of the Royal Observatory, Edinburgh. It began regular observation in 1973. Originally planned to work in conjunction with the *Anglo-Australian telescope on the same site, it has proved a major research tool in its own right. Its highly rated performance stems from recent improvements in telescope technology and photographic emulsions and also from experience gained from the use of the 48 inch Schmidt at the Palomar (*Hale) Observatory. It has been involved primarily in the production of photographic plates for the ESO/SRC

*Southern Sky Survey. It can reach limiting magnitudes in exposures of about an hour as a result of its low focal ratio of f/2.5. Each photograph covers a very wide area of the sky, amounting to 40 square degrees.

ultraviolet astronomy. Astronomical observations in the waveband of about 25 to 350 nanometres (nm); lying between the visible and x-ray bands of the electromagnetic spectrum. Strong atomic and molecular absorption in the earth's atmosphere requires these observations to be carried out at balloon altitudes (about 30 km) in the 250-350 nm waveband and on rocket or satellite platforms for wavelengths below about 250 nm. The UV astronomy satellite OAO-2 was launched in 1968 with a telescope that surveyed the sky at 100-300 nm. OAO-3, renamed *Copernicus* after launch in 1972, carried a high-resolution reflecting telescope and grating spectrometer and obtained many stellar and interstellar UV spectra over the following six years. The UV band includes many of the resonant (i.e. involving transitions to or from the atomic ground state) absorption and emission lines of the more abundant elements: of all the UV lines, the Lyman-alpha line of hydrogen at 121.6 nm is by far the strongest due to the dominant hydrogen abundance. Copernicus obtained spectra of many *ultraviolet stars; it also mapped the density and element abundances in the interstellar medium, showing these to vary widely and confirming a low density 'hole' within several tens of parsecs of the sun.

In January 1978 the *international Ultraviolet Explorer satellite (IUE) was launched into orbit over the mid-Atlantic. On board was a 50 cm aperture UV telescope and a grating spectrometer whose speed was an order of magnitude greater than that on Copernicus.

ultraviolet radiation (UV radiation). *Electromagnetic radiation lying in the waveband beyond the violet end of the visible spectrum. The wavelength range is approximately 25-350 nanometres. The gap between the x-ray and ultraviolet wavebands is the XUV region, which extends into the UV band. Long-wavelength ultraviolet radiation, i.e. with wavelengths near that of light, is often called *near ultraviolet* with *far ultraviolet* being applied to short wavelengths.

ultraviolet stars. Hot stars that have a considerable emission at ultraviolet wavelengths, i.e. between 25-350 nanometres (nm): for an *O star with a surface temperature of about 30 000 kelvin, over 80% of its radiation lies in the ultraviolet. In comparison the sun, with a surface temperature of about 5800 kelvin, emits most of its radiation in the visible region of the spectrum (380-750 nm). Space observations have allowed the UV spectra of such stars to be measured, providing more precise determination of their total radiated power. In addition emission and absorption line spectra have given information on the structure and element abundance in the stellar atmospheres and have shown, for many luminous early-type stars, the presence of strong *stellar winds. Such winds involve a high mass loss that will substantially affect the evolution of the star. Extensive regions of ionized interstellar gas—the *HII regions—surround many such hot luminous stars. *See also* emission nebulae.

umbra. 1. The dark inner region of a shadow cast by an object illuminated by a light of finite size rather than a pinpoint of light. The less dark outer region of the shadow is the *penumbra* (see illustration). The source of light is totally obscured to someone in the umbral region but only partly obscured in the penumbral region. *See* eclipse.
 2. *See* sunspots.

Umbriel. A satellite of Uranus, discovered in 1851. *See* Table 2, backmatter.

Unicorn. Common name for *Monoceros.

universal time (UT). Symbol: t_U. The precise measure of time that forms the basis for all civil timekeeping and is the time scale usually used for recording astronomical observations. It is a particular case of *mean solar time, being the mean solar time on the Greenwich meridian reckoned in days of 24 mean solar hours beginning at midnight: it is therefore equivalent to *Greenwich Mean Time. In principle then it is determined by the motion of the *mean sun relative to the Greenwich meridian. In practice the UT at any instant is calculated, using a conventional relation, from Greenwich Mean Sidereal Time. GMST is deduced from the *local sidereal time, which is determined from observations of the apparent diurnal motions of stars. Universal time is thus directly related to *sidereal time and intervals of time can

Umbral and penumbral regions of a shadow

be converted from one system to the other: the ratio of an interval of mean sidereal time to one of UT is a fixed constant equal to 1.002 737 9093. UT is equal to *local time plus or minus a correction for longitude. *Ephemeris time is obtained by the addition of an increment, ΔT, to UT.

Universal time determined directly from stellar observations is denoted UT0. The meridian of an observer varies about its mean position as a result of the motion of the geographic poles—i.e. *Chandler wobble: UT0 corrected for this polar motion is denoted UT1. UT1 corrected for the mean seasonal variations in the earth's rotation rate is denoted UT2. *Coordinated universal time* (*UTC*) is based on international *atomic time (TAI) and was introduced in order to minimize the divergence of universal time from the uniform time scale of atomic time. Since Jan 1 1972 the time scale distributed by most broadcast time services has been UTC. UTC differs from TAI by an integral number of seconds. UTC is kept within 0.90 seconds of UT1 by the insertion of exactly one second when necessary, usually at the end of December or June; these step adjustments are termed *leap seconds*. Before 1972 the time signals were maintained within 0.1s of UT2.

universe. The sum total of potentially knowable objects. The study of the universe on a grand scale is called *cosmology. Objects that were potentially unknowable at earlier times may become undetectable at later times. Barriers between potentially knowable and unknowable objects exist in even a simple expanding Euclidean universe with flat space (i.e. in which the surface area of a sphere is $4\pi r^2$). More and more distant objects are seen to recede at faster and faster velocities. At a certain distance objects are receding at the speed of light and suffer an infinite *redshift; they are then potentially unobservable.

Gravitational distortion of light in a *closed universe,* where the *deceleration parameter, q_0, exceeds 0.5 causes space to be spherical (the surface area of a sphere is less than $4\pi r^2$). Space is hyperbolic, however, in an *open universe* in which q_0 is less than 0.5. *See also* age of the universe; cosmological models; big bang theory; mean density of matter.

unresolved source. A source of radiation whose angular size is too small for details of its structure to be revealed. *See* radio-source structure.

unseen component. *See* binary star.

upper culmination (upper transit; superior culmination). *See* culmination.

uranometry. *Obsolete name for* star catalogue or star atlas.

Uranus. The seventh major planet of the solar system and the first to be discovered telescopically—by William Herschel in 1781. One of the *giant planets and closely resembling *Neptune, it has an equatorial diameter of about 52 400 km, a mass 14.6 times that of the earth, and a density 1.2 times that of water. Uranus orbits the sun every 84 years, varying in distance between 18.3 and 20.1 astronomical units. At *oppositions, which recur four days later each year, it has a mean angular diameter of 3.7 arc seconds and a magnitude of 5.7, near the limit of naked eye visibility. Telescopically Uranus shows a featureless greenish disc making visual determination of its rotation period impossible. Recent spectroscopic measures suggest a period of approximately 23 hours; another experiment gave a result of 12.3 hours. These differ markedly from a 1930 determination of 10.8 hours previously adopted as standard. Orbital and physical characteristics are given in Table 1, backmatter.

Uranus' equator is tilted by 98° with respect to its orbit, making it unique in having its rotation axis close to its orbital plane. This means that each pole is presented almost face-on to the sun and earth during part of each orbit. In the equatorial plane are five satellites (*see* Table 2, backmatter) and a modest *ring system*. This system consists of about nine rings, between 3 and 100 km broad and lying 42 000 to 51 000 km from the centre of the planet, within the *Roche limit for Uranus. The material comprising the rings may be water-ice chunks, like those postulated for *Saturn's rings, although they have a much lower *albedo (about 0.05). They were discovered only in 1977 during a stellar occultation. Perturbations by Uranus' satellites may govern the position and thickness of each ring.

It is thought that the visible surface of Uranus corresponds to a methane cloud layer seen through a haze of molecular hydrogen: both methane and hydrogen have been detected spectroscopically in the upper atmosphere. The temperature of the cloud tops, -210°C, is close to what is expected if Uranus does not have a net heat outflow (as do Jupiter and Saturn). Ammonia, which provides the visible clouds of Jupiter, freezes at a higher temperature than methane and probably exists as clouds below the methane layer where the pressure and temperatures are higher. Lower still, models imply a 10 000 km layer of molecular hydrogen followed by an 8000 km thick ice mantle and a 16 000 km diameter rocky core. This core may be partly metallic and molten and thus able to generate a magnetic field; this would account for some radio observations that suggest Uranus has a *magnetosphere.

Ursa Major (Great Bear). An

Ursa Major cluster

Ursa Major and bright neighbouring stars

extensive very conspicuous constellation in the northern hemisphere that can be used to indicate the positions of brighter neighbouring stars (see illustration). The brightest stars are *Alioth, Alcaid, and Dubhe, all of 2nd magnitude, the seven brightest stars forming the *Plough. The area contains several double stars including the naked-eye double *Mizar and Alcor, the planetary *Owl nebula, and the close spiral galaxies M81 and M82. M82 is a strong radio source. Abbrev.: UMa; genitive form: Ursae Majoris; approx. position: RA 11.5h, dec +55°, area: 1280 sq deg.

Ursa Major cluster. An *open cluster containing over 100 stars that are scattered over an area of sky more than 1000 arc seconds in diameter. It is an example of a *moving cluster and its wide angular diameter is due to its proximity to earth. It includes the stars Beta, Gamma, Delta, Epsilon, and Zeta Ursae Majoris in the *Plough, Zeta Leonis, Beta Aurigae, Beta Eridani, Gamma Coronae Borealis, and *Sirius.

Ursa Minor (Little Bear). A constellation in the northern hemisphere that includes and surrounds the north celestial pole. The two brightest stars, both of 2nd magnitude, are *Polaris (α) (the present *pole star) and the orange subgiant Kohab (β). Abbrev.: UMi; genitive form: Ursae Minoris; approx. position: RA 0 -24h, dec +80°; area: 256 sq deg.

Ursa Minor system. A member of the *Local Group of galaxies.

Ursids. A minor *meteor shower, radiant: RA 217°, dec +78°, maximizing on Dec. 22.

UT. *Abbrev. for* universal time.

Utopia Planitia. A smooth Martian plain located northwest of the *Elysium Planitia volcanoes and chosen as the landing site of the *Viking 2 Lander in 1976.

UV. *Abbrev. for* ultraviolet.

uvby system. A system of stellar *magnitudes in which photoelectric *magnitudes u, v, b, and y are measured at four wavelength bands: measurements of u (ultraviolet) are centred on 350 nanometres (nm), v (violet) on 411 nm, b (blue) on 467 nm, and y (yellow) on 547 nm. The bandwidth (20-30 nm) is much narrower than for the UBV system.

UV Ceti stars. *See* flare stars.

V

Valles Marineris. A vast system of Martian canyons photographed by *Mariner 9. Long recognized as a feature of changing brightness, known as *Coprates canyon,* it was renamed in honour of the Mariner project. *See* Mars, surface features.

valley, lunar (canyon). A radial fracture on the moon associated with an impact basin or crater. *See table at* rille.

Van Allen radiation belts. Two regions within the earth's *magnetosphere in which charged particles become trapped and oscillate backwards and forwards between the magnetic poles as they spiral around magnetic field lines. The lower belt, from 1000 to 5000 km above the equator, contains protons and electrons either captured from the *solar wind or derived from collisions between upper atmosphere atoms and high-energy *cosmic rays. The upper belt, between 15 000 and 25 000 km above the equator but curving downwards towards the magnetic poles, contains mainly electrons from the solar wind. The belts were discovered by James Van Allen in the course of his analysis of observations by early *Explorer satellites in 1958.

Vanguard project. Part of the US programme for the *International Geophysical Year, in which three small satellites were orbited. Vanguard 1, launched on Mar. 17 1958, was the first satellite powered by *solar cells: it continued transmitting for six years and long-term tracking of it later revealed the slight pear shape of the earth. Two other Vanguards launched in 1959 completed the programme, which provided information not only on the earth's shape but on its surface and interior, its magnetic field and radiation belts, and its space environment.

variable. *Short for* variable star.

variable radio source. A *radio source whose measured parameters change with time, especially the *flux density. Many extragalactic radio sources have been observed at frequencies above 400 megahertz to vary over periods of weeks or months. The timescale of the variations imposes an upper limit on the linear size of the varying region since the conditions are unlikely to change more quickly than the time it takes for light to travel across the region.

variable star. A star whose brightness varies with time: about 25 000 have been catalogued so far. The variation can be seen on the *light curve of the star and may be regular—with a period given by one complete cycle in brightness on the light curve—or may be irregular or semiregular. The periods range from hours to years. The range between maximum and minimum brightness—the *amplitude—is also very wide. There are three major groups of variables: eclipsing, pulsating, and cataclysmic variables. The eclipsing or geometric variables are systems of two or more stars whose light varies as one star is periodically eclipsed or tidally distorted by a companion star. These *eclipsing binaries can be subdivided into *Algol variables, *Beta Lyrae stars, and *W Ursae Majoris stars. *Pulsating variables, such as the Cepheid variables, periodically brighten and fade as their surface layers expand and contract. *Cataclysmic variables include *novae, *recurrent novae, *dwarf novae, and *flare stars. Minor groups of variables include *rotating variables, *spectrum variables, and *R Coronae Borealis stars.

These many types also fall into three other main categories: with *extrinsic variables* the brightness variation results from some process external to the star, as with some eclipsing binaries; with *intrinsic variables* the variation results from some change in the star itself, as with the pulsating variables; in other cases, such as W Ursae Majoris stars, the variation can result from a combination of intrinsic and extrinsic agencies. Different types of variable can be distinguished by period and also by light-curve shape: by amplitude of the variation, depth of the minima, size and shape of the maxima, characteristic irregularities, and by the continuous or step-like changes involved.

Spectroscopic and photometric measurements can be used to differentiate between borderline cases and to investigate changes in temperature, *radial velocity, and *spectral type. *See also* stellar nomenclature.

variation. The periodic *inequality in the moon's motion that results from the combined gravitational attraction of the earth and sun. Its period is half the *synodic month, i.e. 14.77 days, and the maximum longitude displacement is 39′29″.9. *See also* evection; parallactic inequality; annual equation; secular acceleration.

variation of latitude. *See* Chandler wobble.

Vega (α Lyr). A conspicuous white star, apparently blue in colour, that is the brightest star in the constellation Lyra. It was the *pole star about 14 000 years ago. It is the standard A0 V star (B-V = U-B = 0) in the UBV system. m_V: 0.03; M_V: 0.5; spectral type: A0 V; distance: 8.1 pc.

Veil nebula. *See* Cygnus Loop.

Vela (Sails). A constellation in the southern hemisphere near Crux, lying partly in the Milky Way. It was once part of the large constellation *Argo. The four brightest stars, Gamma, Delta, Kappa, and Lambda Velorum, are of 2nd magnitude and form a quadrilateral. Gamma, a remote luminous triple star, is the only conspicuous *Wolf-Rayet star. The area contains the *Vela pulsar. Abbrev.: Vel; genitive form: Velorum; approx. position: RA 9.5h, dec -50°; area: 500 sq deg.

Vela pulsar. The second *optical pulsar to be discovered, the optical flashes being detected in 1977. With a brightness of only about 26 magnitudes—one of the faintest optical bodies to be detected—it is considerably fainter than the optical pulsar in the *Crab nebula, although at a distance of about 500 parsecs it is four times nearer. Its period of 0.089 seconds makes it the third fastest pulsar as yet detected. It was already known to emit pulses of radio emission in a period of 0.089 seconds and pulses of gamma rays twice every revolution. The optical pulses were found by sampling the light from a field only five arc seconds across, centred on the accurate radio position of the pulsar, and folding the data to the precise radio period.

Like other pulsars, the Vela pulsar is gradually slowing down in its rotation rate. Several brief and temporary increases in rotation rate—glitches—have however been observed since 1969 (*see* pulsar).

velocity dispersion. A measure of the spread of velocities of members within a *cluster of galaxies. If both radius, R, and velocity dispersion, V, of a cluster are measured, an estimate of mass, V^2R/G, is ob-

velocity-distance relation. *See* Hubble's law.

velocity of light. *See* speed of light.

Venera probes. A series of successful Soviet probes to Venus. Venera 1, launched on Feb. 12 1961, passed Venus although radio contact had been lost after seven million km. Venera 2 passed Venus on Feb. 27 1966 but, like Venera 3, which plunged into the Venusian atmosphere two days later, it failed to return data.

Veneras 4-8 ejected capsules into Venus' atmosphere to measure the atmospheric temperature, pressure, and composition. The Venera 4 capsule transmitted for 94 minutes on Oct. 18 1967 during its parachute descent, while the capsules from Veneras 5 and 6 returned data for 53 and 51 minutes on Jan. 5 and 10, 1969, respectively; all were crushed by the atmospheric pressure before they reached the surface. The capsule from Venera 7 did survive to reach the surface on Dec. 15 1970, returning data for 23 minutes and indicating a temperature of 475°C and a pressure of 90 atmospheres. Venera 8's capsule returned data from the surface for 50 minutes on July 22 1972.

Veneras 9 and 10 each consisted of a lander and an orbiter through which signals from the lander were relayed to earth. On Oct. 22 1975, Venera 9's lander sent the first photograph from the surface of another planet, showing sharp angular rocks near the probe. Another panorama one day later from the Venera 10 lander revealed an apparently older more eroded landscape. The landers operated for 53 and 65 minutes respectively on the Venusian surface.

Veneras 11 and 12 were launched on Sept. 9 and 14, 1978, respectively. Venera 12 reached Venus on Dec. 21 and transmitted data for 110 minutes from the surface. Experiments were performed during the descent. Venera 11 reached the surface on Dec. 25. *See also* Venus.

Venus. The second major planet in order from the sun and the one that comes closest to the earth. It orbits the sun every 225 days at a distance varying between 0.72 and 0.73 astronomical units. It has a size (12 102 km equatorial diameter), density, and, probably, internal constitution comparable with those of the earth but measurements by *Mariner, *Pioneer Venus, and *Venera probes have revealed extremely hostile surface and atmospheric conditions. Orbital and physical characteristics are given in Table 1, backmatter.

Venus is never far from the sun in the sky, reaching its greatest *elongation of 45° to 47° about 72 days before and after inferior conjunctions, which recur at intervals of about 584 days. At greatest brilliancy the planet is near magnitude -4.4 and brighter than every other celestial object except the sun and the moon; this occurs at an elongation of 39° about 36 days before and after inferior conjunction, while Venus is an *evening star or a *morning star.

Telescopically the planet shows a brilliant yellowish-white cloud-covered disc. Because of its orbital position between sun and earth, Venus exhibits *phases: gibbous and crescent phases occur between full at superior conjunction, when the planet's apparent diameter is 10 arc seconds, and new at inferior conjunction, when it has a diameter of about 61 arc seconds. *Transits of Venus across the sun's disc at inferior conjunction are rare: the next will occur in 2004.

The clouds never permit views of the surface; it was not until 1961 that radar established that Venus rotated in a retrograde direction 243 days relative to the stars, or 117 days in relation to the sun (the Venusian day). Features in the clouds, seen indistinctly from the earth but observed in detail by ultraviolet photography from Pioneer Venus and *Mariner 10, indicate a four day rotation period for the upper atmosphere, corresponding to 350 km/hour winds. The cloud patterns swirl from the equator towards the poles and appear to be driven by solar heating and convection near the subsolar point in the middle of the sunlit hemisphere.

Possibly because of the slow rotation of its nickle-iron core, Venus has little or no intrinsic magnetic field. A magnetic field is however induced in the planet's ionosphere by the *solar wind. The resulting bow shock wave was found by one of the Pioneer Venus probes to be very strong and can therefore buffer the outer atmosphere of Venus from the solar wind.

The atmosphere extends to about 250 km above the surface. Pioneer Venus measurements showed there to be three distinct cloud layers lying at altitudes of about 70-65 km, 58-52 km, and 52-48 km. The top layer consists of sulphuric acid droplets; the middle layer was also acid droplets but solid and liquid sulphur particles were also detected; the lowest layer is predominantly large sulphur particles.

The temperature increases from about 0°C at the cloud tops to about 470°C at Venus' surface, which experiences a crushing atmospheric pressure 90 times that of the earth. The surface temperature, higher than that of any other planet, appears to be the result of a *greenhouse effect involving the clouds and the large proportion of carbon dioxide in the atmosphere. Pioneer Venus showed that the atmosphere as a whole consists of about 98% carbon dioxide, 1-3% nitrogen, with a few parts per million (ppm) of helium, neon, and argon. Although the amounts of neon and argon are small, they indicate very much greater amounts of primordial neon and argon than those found in the earth's atmosphere. This is currently raising problems concerning the established view of the origin of the solar system. In the lower atmosphere, below the clouds, trace amounts of water vapour (0.1-0.4%) and free oxygen (60 ppm) were detected. The high surface temperature may have prevented any atmospheric water vapour from condensing into oceans, as occurred on earth; it may also have prevented carbon dioxide from combining with the crustal material to form carbonaceous rocks. The upper atmospheric winds moderate to only 10 km/hour or so at the surface.

Photographs of the surface returned by the Venera 9 and 10 landers in 1975 showed a stony desert landscape, while tests for radioactive elements in the crust implied a composition similar to that of basalts or granites. Radar studies suggest areas of mountains and lava flows with, possibly, craters and volcanoes near the limit of resolution. Surface mapping by the Pioneer Venus Orbiter will greatly clarify the situation. Preliminary observations indicate a rolling earthlike landscape. A huge canyon, about 6 km deep and 400 km long, has been found to run east-west some 600 km south of the equator.

Venus probes. See Mariner 2; Mariner 5; Mariner 10; Mariner probes; Pioneer Venus probes; Venera probes; Zond probes.

Venus, transits of. See transit.

vernal equinox (spring equinox). See equinoxes.

vernier. A short scale used to increase the accuracy of the graduated scale to which it is attached. If nine divisions of the vernier correspond to 10 divisions of the instrument scale, the latter can be read to a further decimal place.

vertical circle. 1. A great circle on the celestial sphere that passes through an observer's zenith and cuts the *horizon at right angles.
2. An instrument that is essentially a *meridian circle that can be rotated about a vertical axis. It is thus similar to an *altazimuth mounted instrument except that the graduated horizontal circle is lacking.

Very Large Array (VLA). See aperture synthesis.

very long baseline interferometry (VLBI). Radio observations made with an interferometer (see radio telescope) whose longest *baseline exceeds about 100 km and can be as great as the earth's diameter. The very long baseline greatly increases the resolving power of the interferometer. This allows the fine structure of very distant *radio sources, such as *quasars, to be studied.

It is usually not possible to combine the signals from the two arms of the interferometer directly so that tape recorders attached to high-precision clocks have to be used; the recorded signals are later processed in a correlating *VLBI terminal*.

Vesta. The fourth *minor planet to be discovered, found in 1807. It is the third largest minor planet but has a higher *albedo (0.23) than any other large minor planet and is the only one bright enough to be visible to the naked eye at favourable *oppositions. Its unusual spectrum suggests that it has a surface composed of basaltic minerals. *See* Table 3, backmatter.

video signal. *See* television camera.

vidicon tube. *See* television camera.

Viking probes. Two identical American spacecraft, each comprising an Orbiter and a Lander, that were launched in 1975 on a mission to search for life on *Mars. Each Orbiter weighed 2325 kg and carried two television cameras, used to photograph the surfaces of Mars and its satellites, and instruments to map atmospheric water vapour and surface temperature variations. After detaching from its Orbiter, each Lander, weighing 1067 kg, made a successful landing using a combination of aerodynamic braking, parachute descent, and retro-assisted touchdown. The Viking 1 Lander set down on July 20 1976 in *Chryse Planitia after a delay while Orbiter photographs were searched for a smooth landing area. Viking 2 landed on Sept. 3 1976 in *Utopia Planitia, 7420 km northeast of the Viking 1 Lander.

Each Lander carried two television cameras, a meteorological station, a seismometer, and a set of soil-analysis experiments serviced by a sampling arm. Only the seismometer carried by Viking 1 failed to function; that on Viking 2 registered mainly wind vibrations and a few minor Martian 'earth' tremors. The television cameras returned views of a red rock-strewn surface under a pink dusty sky. Sand or dust dunes were evident at the Viking 1 site but no life forms were seen. The meteorology instruments reported temperatures varying between -85°C and -30°C with mainly light winds gusting to 50 km per hour.

X-ray analysis of soil samples showed a high proportion of silicon and iron, with smaller amounts of magnesium, aluminium, sulphur, cerium, calcium, and titanium. Gas chromatograph mass spectrometers carried by the Landers failed to detect any biological compounds in the Martian soil, but more ambiguous results were returned by three experiments designed to test for micro-organisms in various soil samples.

A *labelled release experiment* tested for radioactive gases that might have been expelled from organisms in a soil sample fed with a radioactive nutrient. In a *pyrolytic release experiment* soil was placed with gases containing the radioactive isotope carbon-14. The idea was that Martian organisms might assimilate the isotope into their cell structure in a process similar to earthly photosynthesis. Later the soil was baked and a test made for the carbon isotope in gases driven off. Finally, a *gas exchange experiment* monitored the composition of gases above a soil sample, which might or might not have been fed with a nutrient. If Martian organisms were breathing the gases, a change in composition might be detected.

In practice, the pyrolytic release experiment produced negative results, while the gas exchange experiment gave a marginally positive response, which could be explained in terms of a chemical reaction involving the soil and nutrient. The positive response of the labelled release experiment was so marked that some researchers would prefer to ascribe it to another chemical reaction. Although the response disappeared when the soil was sterilised by preheating, reinforcing the biological interpretation, not every astronomer is convinced that the Vikings did find Martian life.

Virgo (Virgin). An extensive zodiac constellation straddling the equator between Boötes and Centaurus. The brightest stars are the 1st-magnitude *Spica and several of 3rd magnitude, including the binary *Porrima. The area contains the prototype of the W Virginis stars, a class of *Cepheid variables. The area is rich in faint galaxies, which are members of the *Virgo cluster, and contains the first detected *quasar 3C-273. Abbrev.: Vir; genitive form; Virginis; approx. position; RA 13.5h, dec -5°; area: 1294 sq deg.

Virgo A. An intense *radio source in the *Virgo cluster of galaxies that is associated with the brightest galaxy, *M87*. It has a *flux density of 970 *jansky. M87 is a giant elliptical galaxy with a *jet from which a considerable fraction of the radio radiation comes by *synchrotron emission.

Virgo cluster. A giant irregular *cluster of galaxies lying near the north galactic pole in the constellation Virgo. Its distance is currently estimated as 21.9 megaparsecs. Some 2500 galaxies have been observed of which about 75% are spirals, the remainder being mainly ellipticals with a few irregulars. One of its brightest members, the giant elliptical galaxy M87, is both a radio source (*Virgo A) and an x-ray source (Virgo X-1). *See also* Local Supercluster.

visible radiation, spectrum. *See* light.

visual binary. A *binary star, such as *Sirius or Gamma Centauri, whose components are sufficiently far apart to be seen separately, either with the naked eye or with a telescope. In most cases the components differ in brightness: the brighter star is designated the *primary* and the other is the *companion*. Observations over a period of time reveal the apparent orbit of the companion relative to the primary, from which the true orbit is derived by adjusting for the inclination of the orbital plane. If the distance is known the total mass of the binary can be obtained from Kepler's third law:

$$M_1 + M_2 = a^3/p^3P^2$$

where a is the semimajor axis of the true orbit (in seconds of arc), p is the parallax and P is the orbital period; the total mass is obtained in terms of the solar mass. It is sometimes possible to measure the absolute orbit of each component against the background of more distant stars. Then, if a_1, a_2 are the semimajor axes of these orbits, the ratio of masses is given by

$$a_1/a_2 = M_2/M_1$$

and by applying both equations the individual masses can be obtained. If the distance is unknown, both it and the total mass can be found using the method of *dynamic parallax.

Visual binaries generally have longer orbital periods than close binaries (*see* binary star) and in some cases the period is too long for any orbital motion to be detected. Such pairs are called *common proper motion stars* because, unlike *optical double stars, they appear to stay together as they move across the sky.

visual magnitude. *See* magnitude.

visual meteor. *See* meteor.

VLA. *Abbrev. for* Very Large Array. *See* aperture synthesis.

VLBI. *Abbrev. for* very long baseline interferometry.

Vogt-Russell theorem. A theorem proposed independently by H.N. Russell and H. Vogt in 1926 stating that, if the age, mass, and chemical composition of a star are specified, then its temperature and luminosity are (in almost all cases) uniquely determined. Hence for a star whose age, mass, and chemical composition are known, the equations of *stellar structure will have only one solution.

Volans or **Piscis Volans (Flying Fish).** A small inconspicuous constellation in the southern hemisphere near the Large Magellanic Cloud, the brightest stars being of 4th magnitude. Abbrev.: Vol; genitive form: Volantis; approx. position: RA 8h, dec -70°; area: 141 sq deg.

voltage pattern. *See* antenna.

Voskhod. Two Soviet manned spacecraft that followed the *Vostok series. Voskhod 1 was launched on Oct. 12 1964 and was the first multi-crewed spacecraft: three cosmonauts were on board. Voskhod 2 was launched on Mar. 18 1965 and during its flight one of the two-man crew, Alexsei Leonov, made the first spacewalk.

Vostok. A series of six Soviet manned spacecraft that were able to carry one person into orbit. Vostok 1 was launched on Apr. 12 1961. Its crew member, Yuri Gagarin, became the first man in space, the flight lasting 1.8 hours. In subsequent Vostok missions the flight duration was extended up to 119 hours so that the effects of prolonged weightlessness could be studied. The double launching of Vostoks 3 and 4 in August 1962 and their close approach in orbit led to an appreciation of rendezvous techniques.

Voyager probes. Two 825 kg US probes launched towards Jupiter on Sept. 5 and Aug. 20, 1977 respectively. During the fly-by of Jupiter in March 1979, Voyager 1 was accelerated towards Saturn for arrival there in November 1980. Voyager 2 is scheduled to pass Jupiter in July 1979 and Saturn in August 1981; it may then be targeted for a fly-by of Uranus in January 1986. Both vehicles carry television cameras and radioisotope *thermoelectric generators, and are intended to make close-up studies of Jupiter, Saturn, Saturn's rings, and several satellites of both planets.

Voyager 1 passed within 280 000 km of Jupiter on March 5 1979 and continued on its path to Saturn. During this encounter the broad areas of study for the probe's instruments were the dynamics of Jupiter's atmosphere, its magnetosphere, and the Jovian satellites, principally *Amalthea, *Io, *Ganymede, and *Callisto. The preliminary results have been very exciting, revealing several unexpected aspects of the Jovian system: these include a ring system around Jupiter and considerable volcanic activity on Io.

See also planetary alignment.

Vulcan. A hypothetical major planet, thought during the 19th century to orbit the sun within the orbit of Mercury. Searches for it during total solar eclipses and at suggested times of *transit across the sun were all unsuccessful. It is now known not to exist.

Vulpecula (Fox). A constellation in the northern hemisphere near Cygnus, lying in the Milky Way, the brightest stars being of 4th magnitude. It contains the *Dumbbell nebula. Abbrev.: Vul; genitive form: Vulpeculae; approx. position: RA 20h, dec +25°; area: 268 sq deg.

V/V_m test. *See* luminosity-volume test.

W

wake. *See* train.

walled plains. *See* craters.

wandering of the poles. *See* Chandler wobble.

wandering stars. *See* fixed stars.

Water Bearer. Common name for *Aquarius.

Watersnake. Common name for *Hydrus and sometimes for *Hydra.

watt. Symbol: W. The SI unit of power, defined as the power resulting from the dissipation of one joule of energy in one second.

wavefront. The imaginary surface of a wave of light or other radiation, connecting points of the same phase. The advancing wavefront is normally perpendicular to the direction of travel and may be plane, spherical, cylindrical, or more complex.

waveguide. A metal tube, usually of rectangular cross section, down which travelling electromagnetic waves may be propagated. In a more general sense it is any system of material boundaries that fulfils the same purpose, such as layers of plasma in the *ionosphere. Waveguides are used as transmission lines at high frequencies where dielectric losses in other types of radio cables become excessive; they are therefore used as *feeders in radio telescopes. The guided waves may be radiated away at the end of the waveguide by a *horn antenna*, which is a flared metal device having the dimensions of the waveguide at one end and opening out to a large aperture at the other end.

wavelength. Symbol: λ. The distance over which a periodic wave motion goes through one complete cycle of oscillation, i.e. the distance travelled during one period. Thus for a sinusoidal wave motion, such as *electromagnetic radiation, it is the distance between two successive peaks or troughs. For electromagnetic radiation, wavelength is related to frequency, ν by $\nu\lambda = c$, where c is the speed of light. Wavelength is measured in metres or in multiples or submultiples of metres; for example, the wavelength of light is usually given in nanometres while that of infrared radiation is usually quoted in micrometres.

WC stars. *See* Wolf-Rayet stars.

weber. Symbol: Wb. The SI unit of magnetic flux.

weight. The force experienced by a body on the surface of a planet, natural satellite, etc., that results from the gravitational force (directed towards the centre of the planet, satellite, etc.) acting on the body. A body of mass m has a weight mg, where g is the acceleration of gravity.

weightlessness (zero gravity). The condition associated with *free fall, i.e. the motion of an unpropelled body in a gravitational field. The rate of acceleration of a person in a freely falling object, such as a spacecraft, is equal to that of the freely falling object. The person thus experiences no sensation of weight and floats freely. The adaptation to conditions of normal earth gravity following periods of weightlessness have been investigated by both the USSR and the US after spaceflights of ever-increasing duration. No long-term effects have been observed to date.

Weizsäcker's theory. A theory, proposed by C.-F. von Weizsäcker in 1944, for the *accretion of the planets at the points of contact between turbulent eddies in the nebula that contracted to form the solar system.

West. A brilliant comet that passed within 30 million km of the sun in Feb. 1976. The cometary nucleus broke into at least four fragments, accompanied by massive outbursts of gas and dust.

Westerbork Radio Observatory. An observatory at Westerbork, near Hoogeveen in the northern Netherlands. It has a highly sensitive *aperture synthesis system, which went into operation in 1970 and is run by the Netherlands Foundation for Radio Astronomy. Twelve 25 metre radio dishes lie on a line running east-west. Ten are fixed in position at intervals of 144 metres; the remaining two, at one end of the line, can be moved along a rail 300 metres long. The two movable dishes make observations possible at up to 80 baselines ranging in length from 36 to 1458 metres.

western elongation. *See* elongation.

west point. *See* cardinal points.

Wezen. *See* Canis Major.

Whale. Common name for *Cetus.

Whipple's comet. A comet that started life with a near-circular orbit, eccentricity 0.167, period 10.3 years, and has been perturbed by Jupiter so that it now has an eccentricity of 0.348 and a period of 7.5 years. The comet seems to decrease in brightness by one magnitude at each return.

Whirlpool galaxy (M51; NGC 5194). A well-defined type Sc spiral galaxy (*see* Hubble classification) that is face-on to us. It lies at an estimated distance of 11.5 megaparsecs in the constellation Canes Venatici, the nearest bright star being Eta Ursae Majoris in the *Plough. A small companion galaxy, NGC 5195, appears to be connected to it by an extension of one of the spiral arms. Total mass: about 8×10^{10} solar masses; total magnitude: 8·3.

white dwarf. A star that has a mass below the *Chandrasekhar limit (1.44 solar masses) and has undergone *gravitational collapse. Having diameters only 1% that of the sun these stars are consequently very faint, with absolute *magnitudes ranging from +10 to +15. The description 'white' is misleading since they display a range in colour as they cool down, through white, yellow, and red, until finally ending up as cold black globes, called *black dwarfs*. However, those of *spectral types A and F (i.e. white stars) are the brightest and were the first to be discovered.

White dwarfs are the final phase in the evolution of a low-mass star and result from gravitational collapse following the exhaustion of nuclear fuel. Since most of the matter in the core of the collapsing star is in a *degenerate state (with the electrons stripped from their nuclei and packed tightly together) the star contracts until its gravity is balanced by the degeneracy pressure of the electrons, and the density rises to 10^7 - 10^{11} kg m^{-3}. Because of the peculiar behaviour of degenerate material (which is subject to the quantum mechanical uncertainty principle), the most massive white dwarfs collapse to the smallest diameters and highest densities. Stars above the Chandrasekhar limit are too massive to be supported in this way and must collapse further to become *neutron stars or *black holes.

The light from white dwarfs does not arise from internal nuclear reactions but in a thin gaseous atmosphere that slowly leaks away the star's heat into space. The spectral lines arising in this atmosphere are broadened by the extremely high surface gravity, and in extreme cases the light loses enough energy to suffer a measurable *gravitational redshift. Some white dwarfs show no hydrogen at all in their spectra, while others are enriched in helium, carbon, and calcium. A few have strong magnetic fields (10^4 tesla) and several have high rotational velocities.

It has been estimated that there may be 10^{10} white dwarfs in our Galaxy, many having by now cooled to become black dwarfs. Best known of all white dwarfs is Sirius B, companion to *Sirius (αCMa), which was discovered in 1862 by Alvan Clark after F.W. Bessel had predicted its existence (in 1844) from the unusual motion of Sirius. Sirius B has a radius of only 10^4 km, less than that of the earth, but a mass similar to that of the sun. *Novae and *dwarf novae outbursts are believed to occur during transfer of gas on to the surfaces of white dwarfs in very close binary systems.

white hole. The reverse of a *black hole: a region where matter spontaneously appears. Early calculations on black holes indicated that the extreme distortion of space and time inside the *event horizon should connect our universe with another through an Einstein-Rosen bridge (or wormhole). Matter falling into the black hole should then correspondingly appear in the other universe as outpouring of material through a white hole. Certain highly energetic objects, such as quasars, were proposed as examples of white holes in our own universe. With better calculations however, it now seems likely that such links, and hence white holes, do not exist.

white-light corona. *See* corona.

wide-field eyepiece. *See* eyepiece.

Widmanstätten figures. Patterns of bands crossing one another in two, three, or four directions that are revealed when polished surfaces of most *iron meteorites are etched. They are lamellar intergrowths of kamacite, taenite, and plessite.

Wien displacement law. *See* black body.

Wild Duck. *See* Scutum.

Wild's Trio. A group of three irregular barred spiral galaxies that lies in the constellation Virgo. The galaxies appear to be connected by bridges of luminous material. Their differing radial velocities, derived from the *redshifts in their spectra, suggest that the group is rapidly disintegrating.

Wilson-Bappu effect. An experimentally derived linear relation between certain properties of the calcium K line of late-type F, G, K, and M stars (the widths of their double reversals) and their absolute magnitude.

Wilson effect. The apparent displacement of the umbra of a *sunspot relative to the penumbra as the spot is carried by the sun's rotation from the east to the west limb of the disc. For a symmetrical sunspot, the side of the penumbra closest to the centre of the disc is more foreshortened than that towards the limb, implying that the spot is a depression. It is more likely, however, that the opacity of the umbra is less than that of the penumbra and that this is responsible for the effect.

window. *See* atmospheric windows.

winter solstice. *See* solstices.

WN stars. *See* Wolf-Rayet stars.

Wolf. Common name for *Lupus.

Wolf 359. *See* Table 7 (nearest stars), backmatter.

Wolf number. *See* Zürich relative sunspot number.

Wolf-Rayet stars (WR stars; W stars). A small group of very luminous very hot stars, with temperatures possibly as high as 50 000 kelvin, that have anomalously strong and broad emission lines of ionized helium, carbon, oxygen, and nitrogen but few absorption lines. The majority either have lines of He, C, and O (*WC stars*) or He and N (*WN stars*). The emission lines are thought to arise in an expanding stellar atmosphere moving at very high speeds of up to 2000 km s^{-1} so that the star is continuously and rapidly losing mass. About half are known to be binary stars, usually with O or B stars as companions, an example being Gamma Velorum (WC8 + O7). They also occur as the central star in some *planetary nebulae. Their role in stellar evolution is not yet clear.

world line. *See* space-time.

world models. *See* cosmological models.

wrinkle ridges. Linear deformed and folded lunar features that are confined to the *maria and may be elevated by several hundred metres; they may however continue into the *highlands as scarps. Wrinkle ridges are frequently compound, discontinuous, and associated with *rilles. They often reflect underlying topography, including *basin ring structures beneath the maria. They may result from fissure eruptions, faulting, or the contraction of a surface crust following the draining away of subsurface lavas. *Surveyor 6 landed close to a wrinkle ridge in Sinus Medii. *See also* lobate ridge.

WR stars (W stars). *See* Wolf-Rayet stars.

W Ursae Majoris stars. A class of contact *eclipsing binaries that have very short periods amounting to only a few hours and components that are so close that they are grossly distorted by tidal forces into ellipsoidal shapes. Mass transfer occurs between the stars (*see* equipotential surfaces). The components are approximately equal in brightness, as seen by the equal depths of the minima of the light curves. As with *Beta Lyrae stars the light curves show continuous variation resulting from the distorted stellar shapes and variable surface intensity. W UMa stars are similar in mass and size to the sun, being generally F or G stars, and appear to lie in the galactic disc. They are thought by many astronomers to be the progenitors of *dwarf novae. *See also* Algol variables.

W Virginis stars. *See* Cepheid variables.

WW Aurigae. An eclipsing binary (period 2.52 days) in the constellation Auriga. The size and magnitude of the two components are similar so that the primary and secondary minima on the light curve are approximately equal. m_V: 5.7; M_V: 1.2; M_B: 1.7 (A), 2.0 (B); mass, radius: 1.92 (A), 1.90 (B) times solar mass, radius;

spectral type: A7 (A), F0 (B); distance: 77 pc.

X

X-ogen. An interstellar molecule detected in 1970 and identified in 1975 as HCO⁺. *See also* molecular-line radio astronomy.

x-ray astronomy. The study of objects lying beyond the solar system in the photon energy band 100 to 500 000 electronvolts (corresponding to wavelengths from 12 nanometres to 0.0024 nanometres). Opacity of the terrestrial atmosphere throughout this band requires that observations be made above about 150 km and hence x-ray astronomy could begin only after high-altitude rockets became available. The first detection of a cosmic x-ray source was made by Riccardo Giacconi and colleagues in July 1962, during an exploratory rocket launch equipped with an x-ray detector to search for lunar fluorescence. Confirmation of this source, *Scorpius X-1*, and the discovery of a second source, *Taurus X-1*, in 1963 began an active period of rocket and balloon observations that, by 1970, had yielded 25-30 sources spread throughout the Galaxy; there was also one likely extragalactic source, apparently associated with the powerful radio galaxy *Virgo A (M87). The first optical identification was of Taurus X-1, found to coincide with the *Crab nebula supernova remnant in a classical lunar occultation observation by Herbert Friedman and his colleagues in the USA.

The launch of the first x-ray astronomy satellite, *Uhuru, in December 1970 accelerated the development of the subject, yielding many new sources including a large number at high galactic latitude. In particular, Uhuru discovered *x-ray binaries and showed these to be the most common form of galactic x-ray source. A second major discovery was of powerful x-ray emission from *clusters of galaxies, with evidence that this emission arose in an extended region comparable in size (about 0.5 megaparsecs) to the cluster.

The launch of other *x-ray astronomy satellite experiments continued the rapid expansion of x-ray astronomy. The *Ariel V Sky Survey extended the Uhuru catalogue and led to the establishment of a second major class of extragalactic source: the x-ray *Seyfert galaxies. Observations by Copernicus and Ariel V, followed by *SAS-3, found the *slow pulsators*, periodically variable x-ray sources with periods of a few minutes, substantially longer than those of *Centaurus X-3, *Hercules X-1, etc. Ariel V, followed by the satellite OSO-8, discovered an emission line near 7 keV in the spectra of several *supernova remnants and many of the rich *clusters of galaxies, showing the x-rays to be of thermal origin, arising from hot gas at temperatures of 10^6 to 10^8 kelvin and containing an abundance of iron similar to that found in the solar system.

*X-ray transients and *x-ray bursters were detected by several of the satellites. Optical identifications of a few of these have shown them to be most probably x-ray binaries in which the *mass transfer or accretion rate is highly variable. In addition to the discrete x-ray sources an apparently diffuse sky *x-ray background radiation* is observed. It seems likely that this is due mainly to unresolved and distant sources, such as clusters and *active galaxies.

By the late 1970s x-ray observations had become an integral part of astronomy, relating particularly to galactic and extragalactic systems where violent or energetic phenomena give rise to copious high-energy particles or super-hot gas.

x-ray background radiation. *See* x-ray astronomy.

x-ray binary. The most common type of luminous galactic x-ray source, involving a close binary system in which gas flows (via the inner Lagrangian point) or blows (by a strong *stellar wind) from a normal nondegenerate star on to a compact companion. For the most luminous sources, radiating x-rays at 10^{29}-10^{31} watts, this companion is probably a *neutron star or *black hole; in less luminous cases, such as SS Cygni and AM Hercules, it is more likely to be a *white dwarf. Gravitational energy powers these sources and both the luminosity (L) and temperature (T) are proportional to the mass-to-radius ratio (M/R) of the accreting star:

$$L \sim (M/R)Gm'$$

where G is the gravitational constant and m' is the rate of mass accretion, and

$$T \sim (M/R)G\varepsilon$$

where ε depends on the efficiency of the gas heating, being high for shocks and low for viscous heating.

Two main types of x-ray binary are distinguished: in high-mass binaries, such as *Centaurus X-3 and *Cygnus X-1, the nondegenerate star is a giant or supergiant of early spectral type (O, B, or A); in low-mass binaries the visible star is a middle or late main-sequence star of near solar mass. Several binaries contain a pulsating x-ray source, probably involving a rotating magnetized neutron star; these binaries, which include Centaurus X-3, *Hercules X-1, and Cygnus X-3, are among the best-determined of all binary systems.

x-ray burst sources (x-ray bursters; bursters). Sources of intense flashes of cosmic x-rays, discovered in 1975. The bursts, seen in the 1-30 kiloelectron-volt energy band, are characterized by a rapid onset, of less than one second, followed by an exponential decay with a time constant of a few seconds to a minute. They are located in our Galaxy, mostly within 40° of the galactic centre. Two types have been distinguished: *type I* repeat on timescales of hours or days; *type II* emit bursts every few minutes or less for a period of several days. It is widely believed that x-ray bursts arise from *x-ray binary systems and are due to spasmodic mass accretion on to the compact star (*white dwarf, *neutron star, or *black hole) or to thermonuclear flashes when material accreted on to the surface of a neutron star exceeds a critical temperature and pressure.

x-ray mirror. *See* grazing incidence.

x-ray pulsators. *See* Ariel V; x-ray astronomy; pulsar.

x-rays. High-energy *electromagnetic radiation lying between gamma rays and ultraviolet radiation in the electromagnetic spectrum. The XUV region bridges the gap between the x-ray and ultraviolet bands. X-rays, unlike light and radio waves, are usually considered in terms of photon energy, $h\nu$, where ν is the frequency of the radiation and h is the Planck constant. X-ray energies range from about 100 electronvolts up to about 500 000 electronvolts, corresponding to a wavelength range of about 12 nanometres to about 0.002 nm. Low-energy x-rays are sometimes called *soft x-rays* to distinguish them from high-energy *hard x-rays*. X-rays are produced when high-energy electrons bombard matter or by high-energy electrons moving in a magnetic field.

x-ray satellites. Artificial earth satellites devoted to cosmic x-ray observations, the first of which was NASA's *Uhuru, launched in December 1970. Up to this date, i.e. throughout the period 1962-70, *x-ray astronomy was carried out exclusively with sounding rockets and balloon experiments. Uhuru produced the first all-sky survey for cosmic x-ray sources and increased the catalogue of known sources to 161 by 1974. The second x-ray astronomy satellite was the British satellite *Ariel V, which was launched in October 1974 and occupied a similar 500 km circular equatorial orbit to Uhuru. It successfully extended the Uhuru sky map and made detailed spectral and temporal studies of individual sources.

Although x-ray experiments were also carried on other satellites principally devoted to the research fields (Copernicus, OSO-7, ANS, OSO-8), the next two x-ray astronomy satellites were *SAS-3 and *HEAO-1, launched respectively in 1975 and 1977. Both included modulation collimator experiments in their payloads and these provided the first accurate positions of many x-ray sources, leading to additional and more reliable optical identifications. In November 1978 HEAO-2 (the Einstein Observatory) was launched; it carries a large *grazing-incidence telescope with the potential to carry x-ray astronomy to a status comparable (in sensitivity and resolution) with optical astronomy.

x-ray sources. Luminous sources of x-ray emission lying well beyond the solar system. Over 400 were known by the end of 1978, many being identified with established astronomical objects. The brightest sources lie in our Galaxy and are identified principally as *x-ray binaries (e.g. *Scorpius X-1, *Cygnus X-1, *Centaurus X-3, *Hercules X-1), many having an x-ray power 100 to 1000 times the power of the sun. A minority of galactic x-ray sources are identified with supernova remnants, such as the *Crab nebula and *Tycho's star. Fainter but intrinsically much more powerful x-ray emission is detected from many extragalactic objects, especially rich *clusters of galaxies and *active galaxies—such as *Seyfert galaxies, *quasars, and powerful *radio galaxies, including Virgo A (M87) and *Centaurus A.

x-ray telescope. An instrument carried above the earth's atmosphere by an *x-ray satellite, etc., and by means of which x-rays from space can be detected and recorded. *See* proportional counter; grazing incidence.

x-ray transients. Bright nova-like cosmic x-ray sources that develop rapidly over a few days and remain visible for several weeks or months. *Ariel V observations, supported by simultaneous ground-based studies, led to the first optical identification: *A 0620-00* (also *Nova Mon 1975*), for a time in 1975 the brightest cosmic x-ray source ever seen, was found in the visible as a *nova of about 11th magnitude. Both x-ray and optical stars then faded rapidly, to disappear by mid-1976.

The sky distribution of x-ray transients shows them to be galactic, typically at distances of 1-10 kiloparsecs, and it is believed they represent close binary systems in which the *mass transfer is highly variable. The qualitative difference from most optical novae (e.g. Nova Cygni 1975, which had no detectable x-ray emission) may be due to the compact star being a *neutron star or *black hole, rather than a *white dwarf as in the optical novae.

XUV astronomy. The study of objects beyond the solar system that emit radiation bridging the gap between x-rays and ultraviolet radiation (the *XUV region*) i.e. from about 12 nm to the hydrogen Lyman limit at 91.2 nm. The high opacity of the interstellar gas at these wavelengths prevents observations beyond about 10 parsecs from 91.2 to about 50 nm, extending to about 200 parsecs at 12 nm. The first partial sky survey in the XUV was carried out in the *Apollo-Soyuz mission in 1975, in which were found a number of intense sources, including HZ43 and Feige 24 (both hot *white dwarf stars) and *Proxima Centauri.

Y

Yagi. A directional *antenna comprising three or more parallel elements mounted on a straight beam and at right angles to it. The elements are a *dipole, to which the *feeder is attached, behind which is a reflector and in front of which is one or more directors. Named after Hidetsugu Yagi, this type of aerial is commonly used for television reception.

year. 1. The period of the earth's revolution around the sun or the period of the sun's apparent motion around the celestial sphere, both periods being measured relative to a given point of reference. The choice of reference point determines the exact length of the year (see table).
2. *Short for* calendar year.

Yerkes Observatory. An astronomical station of the University of Chicago at Lake Geneva, Wisconsin, endowed by Charles Yerkes at the suggestion of George Ellery Hale. It holds the world's largest refracting telescope, 40 inches (102 cm) in diameter, completed in 1897.

Yerkes system. *Another name for* MK system. *See* spectral types.

ylem. (Greek: that on which form has yet to be imposed.)

yoke mounting (English mounting). *See* equatorial mounting.

YY Orionis stars. *Irregular variable stars that appear to form a subclass of *T Tauri stars. However, their spectra reveal that matter is falling on to them rather than being ejected and so it seems likely that they represent an earlier stage in the contraction of a *protostar.

Z

ZAMS *Abbrev. for* zero-age

year	reference point	length (days)
tropical	equinox to equinox	365.242 19
sidereal	fixed star to fixed star	365.256 36
anomalistic	apse to apse	365.259 64
eclipse	moon's node to moon's node	346.620 30
Gaussian	Kepler's law, $a=1$	365.256 90

main sequence. *See* main sequence.

Z Camelopardalis stars. *See* dwarf novae.

Zeeman effect. An effect that occurs when atoms emit or absorb radiation in the presence of a magnetic field: the field modifies the energy configuration of the atom with the result that a spectral line is split into two, three, or more closely spaced components, each of which is polarized. (In some cases the components cannot be resolved and the effect appears as a broadening of the spectral line.) This Zeeman splitting can be used to study the magnetism of stars, including the sun, since the spacing of the components is a measure of the magnetic field strength.

Zelenchukskaya. Site of the *Academy of Sciences Astrophysical Observatory in the northern foothills of the Caucasus mountains. The Observatory has the world's largest (six metre) reflecting telescope.

zenith. The point on the *celestial sphere that lies vertically above an observer and 90° from all points on his horizon. It lies on the observer's celestial meridian. The *geocentric zenith* lies at the point where the line connecting the earth's centre with the observer's position would meet the celestial sphere. Since the earth is not spherical the two points rarely coincide. *Compare* nadir.

zenithal hourly rate (ZHR). The probable number of *meteors observed per hour from a *meteor shower that has its *radiant in the observer's *zenith. Shower rates vary as the *zenith distance of the radiant changes. To obtain the normalized ZHR the observed rate must be multiplied by a factor, *F*, which varies for different values, *A*, of radiant altitude (see table).

zenith attraction. The corrections that must be subtracted from the observed velocity and direction of a meteoric body to give the true values. The gravitational attraction of the earth on the body will increase the *meteoroid's impact velocity, the effect being greatest for bodies with low initial values of geocentric velocity (i.e. meteoroids just catching up with the earth). The path of the meteoroid will also change, the new *radiant appearing to be closer to the observer's zenith than the original true radiant.

zenith distance (coaltitude).

NORMALIZED ZENITHAL HOURLY RATE

A	90°	52°	35°	27°	8.6°	2.6°
F	1	1.25	1.67	2	5	10

Symbol ζ. The angular distance of a celestial body from an observer's zenith measured along the vertical circle through the body. It is therefore the complement of the altitude ($90° - h$) of the body. The *topocentric zenith distance* is the zenith distance measured at the observation point; it differs from the true value due to *atmospheric refraction, the difference being termed the *angle of refraction*. *See also* diurnal parallax.

zenith stars. Stars that come to *culmination on the observer's zenith.

zero-age main sequence (ZAMS). *See* main sequence.

zero gravity (zero g). *See* weightlessness.

Zeta (ζ). The sixth letter in the Greek alphabet, used in *stellar nomenclature usually to designate the sixth-brightest star in a constellation.

Zeta Aurigae (ζ Aur). An *eclipsing binary in the constellation Auriga (period 972 days, magnitude range 5.0-5.7). The brighter component is a hot blue B star and the secondary is a cool orange K5 II giant over 50 times the size of its companion. During the partial eclipse of the brighter star its light shines through the atmospheric layers of the giant and the spectral changes give information about the atmosphere's composition.

Zeta Aurigae stars are eclipsing binaries composed of a hotter smaller star of early spectral type and a cooler K-type giant, or more usually a supergiant. The period often amounts to several years.

ZHR. *Abbrev. for* zenithal hourly rate.

zirconium stars. *See* S stars.

zodiac. A band passing around the *celestial sphere, extending

zodiac sign	symbol	dates of sun's passage
Aries	♈	21 Mar.-19 Apr.
Taurus	♉	20 Apr.-20 May
Gemini	♊	21 May-21 June
Cancer	♋	22 June-22 July
Leo	♌	23 July-22 Aug.
Virgo	♍	23 Aug.-22 Sept.
Libra	♎	23 Sept.-23 Oct.
Scorpius	♏	24 Oct.-21 Nov.
Sagittarius	♐	22 Nov.-21 Dec.
Capricornus	♑	22 Dec.-19 Jan.
Aquarius	♒	20 Jan.-18 Feb.
Pisces	♓	19 Feb.-20 Mar.

about 9° on either side of the *ecliptic. It thus includes the apparent annual path of the sun as seen from earth and also the orbit of the moon and principal planets apart from Pluto. The band was divided by the ancient Greeks into 12 parts, each 30° wide, known as the *signs of the zodiac*. The signs originally corresponded in position to 12 constellations bearing the same name—the *zodiac constellations*. *Precession of the equinoxes has since shifted the zodiac constellations eastwards by over 30°; the constellations through which the sun now passes include Ophiuchus, which is not considered a zodiac constellation. The table shows the zodiac signs and symbols with the dates of the sun's passage through them when their boundaries were fixed over 2000 years ago. On Jan 1 the sun is now in Sagittarius rather than Capricornus. *See also* astrology.

zodiacal light. A permanent phenomenon that can be seen as a faint glow, especially at tropical latitudes, on a clear moonless night in the west after sunset and in the east before sunrise. It is shaped like a slanting cone and extends from the horizon, tapering along the direction of the *ecliptic, and visible for maybe 20°. The zodiacal light has been shown, spectroscopically, to be sunlight scattered by minute dust particles in the interplanetary medium. The particles are thought to be concentrated into a belt lying in the plane of the solar system. The zodiacal light can be detected right round the zodiac, there being a slight increase in size and brightness—the *gegenschein—directly opposite the sun's position.

zodiac constellations. *See* zodiac.

Zond probes. A series of Soviet interplanetary and lunar probes. Zonds 1 and 2 passed Venus and Mars in 1964 and 1965 respectively but did not return data. Zond 3 photographed the farside of the moon in 1965, obtaining the first high-quality pictures, while Zonds 5 and 6 (in 1968), 7 (1969), and 8 (1970) made circumlunar flights and were recovered on earth. Zond 4, in 1968, may have been an unsuccessful attempt at a circumlunar flight.

zone. A portion of a spherical surface lying between two parallel circles.

zone of avoidance. A region of sky, which roughly coincides with the belt of the *Milky Way, where no galaxies are seen. The apparent absence of galaxies is due to the presence of light-absorbing clouds of dust in the galactic plane.

zone of totality. *Another name for* path of totality. *See* eclipse.

Zürich relative sunspot number (Wolf number). Symbol:

Zürich relative sunspot number

R. A somewhat arbitrary index of the level of sunspot activity, derived from the formula
$$R = k(f + 10g),$$
where g represents the number of sunspot groups, f the total number of their component spots, and k is a constant dependent on the estimated efficiency of, and also the equipment used by, a particular observer. Though a more objective assessment of the level of sunspot activity relies on the measurement of sunspot areas, the Zürich relative numbers provide an almost continuous record back to the mid-18th century and have well withstood the test of time: they yield graphs similar (at least so far as the yearly means are concerned) to those based on other indices. *See also* sunspot cycle.

Table 1. Planets

Planet	Orbit				Globe				Globe (cont'd)	Axis				Satellites
	Sidereal Period	Perihelion (AU)	Aphelion (AU)	Inclination (deg)	Equatorial Diameter (km)	Oblateness	Mass (Earth=1)	Density (water=1)	Albedo	Tilt (deg)	Period* (d	h	m)	
Mercury	87.97d	0.31	0.47	7.0	4880	0.0	0.06	5.4	0.06	28	58	15	30.2	0
Venus	224.70d	.072	0.73	2.4	12 104	0.0	0.82	5.2	0.72	177	224	7	R	0
Earth	365.26d	0.98	1.02	0.0	12 756	0.0034	1.00	5.5	0.39	23.4		23	56.1	1
Mars	686.98d	1.38	1.67	1.8	6794	0.009	0.11	3.9	0.16	24.0	1	0	37.4	2
Jupiter	11.86y	4.95	5.45	1.3	142 800	0.063	317.89	1.3	0.70	3.1		9	50.5	14
Saturn	29.46y	9.01	10.07	2.5	120 000	0.098	95.17	0.7	0.75	26.7		10	14	10 & rings
Uranus	84.01y	18.28	20.09	0.8	52 400	0.01	14.56	1.2	0.90	97.9		23?	R	5 & rings
Neptune	164.79y	29.80	30.32	1.8	49 500	0.02	17.24	1.7	0.82	28.8		19?		2
Pluto	247.7y	29.6	49.3	17.2	2500?	?	0.002	1.3?	0.9 ?	?	6	9	17	1

*R indicates retrograde axial rotation.
The rotation periods for Jupiter and Saturn refer to equatorial regions; the periods exceed 9h 55m and 10h 38m at higher latitudes, respectively.

Table 2. Planetary Satellites

PLANET & Satellite	Year of Discovery	Diameter (km)	Mean Distance (km)	Eccentricity	Sidereal Period (days)	Inclination (°)
EARTH						
Moon	—	3476	384 400	0.055	27.32	23.4
MARS						
I Phobos	1877	27	9 380	0.018	0.32	1.0
II Deimos	1877	15	23 500	0.002	1.26	1.3
JUPITER						
V Amalthea	1892	160	181 000	0.003	0.50	0.5
I Io	1610	3640	422 000	0.000	1.77	0.0
II Europa	1610	3100	671 000	0.000	3.55	0.5
III Ganymede	1610	5100	1 070 000	0.001	7.15	0.2
IV Callisto	1610	4900	1 883 000	0.007	16.69	0.3
XIII Leda	1974	10?	11 110 000	0.147	239	27
VI Himalia	1904	100	11 480 000	0.158	250.6	28
X Lysithea	1938	15?	11 710 000	0.13	259.2	29
VII Elara	1905	30?	11 740 000	0.207	259.7	25
XII Ananke	1951	15?	21 200 000	0.17	631	147
XI Carme	1938	15?	22 600 000	0.21	692	163
VIII Pasiphae	1908	15?	23 500 000	0.38	739	147
IX Sinope	1914	15?	23 600 000	0.27	758	156
XIV	1975	10?	?	?	?	?

SATURN					
X Janus	1966	200	159 000	0.0	0.75	0.0
I Mimas	1789	400	186 000	0.020	0.94	1.5
II Enceladus	1789	600	238 000	0.004	1.37	0.0
III Tethys	1684	1000	295 000	0.000	1.89	1.1
IV Dione	1684	800	377 000	0.002	2.74	0.0
V Rhea	1672	1500	527 000	0.001	4.42	0.3
VI Titan	1655	5800	1 222 000	0.029	15.95	0.3
VII Hyperion	1848	400	1 483 000	0.104	21.28	0.4
VIII Iapetus	1671	1500	3 560 000	0.028	79.33	14.7
IX Phoebe	1898	200	12 950 000	0.163	550.3	150
URANUS						
V Miranda	1948	400	130 000	0.017	1.41	3.4
I Ariel	1851	1400	192 000	0.003	2.52	0
II Umbriel	1851	1000	267 000	0.004	4.14	0
III Titania	1787	1800	438 000	0.002	8.71	0
IV Oberon	1787	1600	586 000	0.001	13.46	0
NEPTUNE						
I Triton	1846	3800	355 000	0.000	5.88	159.9
II Nereid	1949	300	5 562 000	0.75	359.9	27.8
PLUTO						
I	1978	800?	20 000?	0?	6.39	0?

Note: The diameters given for the smaller satellites of Jupiter, and for most of the satellites of Saturn, Uranus, Neptune, and Pluto, are uncertain.
The orbits of the outer satellites of Jupiter vary considerably with time.
The distances refer to the centre of the parent planet and the inclinations to the plane of the equator of the planet.

Table 3. Minor Planets

No.	Name	Discovery Year	Period (years)	Perihelion (AU)
1	Ceres	1801	4.60	2.55
2	Pallas	1802	4.61	2.11
3	Juno	1804	4.36	1.98
4	Vesta	1807	3.63	2.15
5	Astraea	1845	4.14	2.10
6	Hebe	1847	3.78	1.93
7	Iris	1847	3.69	1.84
10	Hygiea	1849	5.59	2.84
15	Eunomia	1851	4.30	2.14
31	Euphrosyne	1854	5.61	2.45
433	Eros	1898	1.76	1.13
588	Achilles	1906	11.90	4.44
944	Hidalgo	1920	14.04	2.00
1221	Amor	1932	2.66	1.08
1566	Icarus	1949	1.12	0.19
1862	Apollo	1932	1.78	0.65
2060	Chiron	1977	50.68	8.51
2062	Aten	1976	0.95	0.79
—	Adonis	1936	2.56	0.44
—	Hermes	1937	2.10	0.62

Aphelion (AU)	Inc (°)	Diameter (km)	Rotation (hours)
2.98	10.6	1003	9.1
3.42	34.8	608	10.5
3.35	13.0	247	7.2
2.57	7.1	538	10.7
3.06	5.3	117	16.8
2.92	14.8	201	7.3
2.94	5.5	209	7.1
3.46	3.8	450	18.0
3.14	11.7	272	6.1
3.86	26.3	370	?
1.78	10.8	23	5.3
5.98	10.3	50	?
9.64	42.5	15	?
2.76	11.9	0.5	?
1.97	23.0	1	2.3
2.29	6.3	1	?
18.88	6.9	300?	?
1.14	18.9	1	?
3.30	1.4	0.3	?
2.66	6.2	0.5	?

Table 4. Meteor Showers

Shower	Normal Limits	ZHR at Max.
Quadrantids	Jan. 1-6	110
Corona Australids	Mar. 14-18	5
April Lyrids	Apr. 19-24	12
η-Aquarids	May 1-8	20
June Lyrids	June 10-21	8
Ophiuchids	June 17-26	6
Capricornids	July 10-Aug. 15	6
δ-Aquarids	July 15-Aug. 15	35
Pisces Australids	July 15-Aug. 20	8
α-Capricornids	July 15-Aug. 25	8
Perseids	July 25-Aug. 18	68
Cygnids	Aug. 19-22	4
Orionids	Oct. 16-26	30
Taurids	Oct. 20-Nov. 30	12
Cepheids	Nov. 7-11	8
Leonids	Nov. 15-19	10
Phoenicids	Dec. 4-5	5
Geminids	Dec. 7-15	58
Ursids	Dec. 17-24	5

Table 5. Constellations

Andromeda	And	Lacerta	Lac
Antlia	Ant	Leo	Leo
Apus	Aps	Leo Minor	LMi
Aquarius	Aqr	Lepus	Lep
Aquila	Aql	Libra	Lib
Ara	Ara	Lupus	Lup
Aries	Ari	Lynx	Lyn
Auriga	Aur	Lyra	Lyr
Boötes	Boo	Mensa	Men
Caelum	Cae	Microscopium	Mic
Camelopardalis	Cam	Monoceros	Mon
Cancer	Cnc	Musca	Mus
Canes Venatici	CVn	Norma	Nor
Canis Major	CMa	Octans	Oct
Canis Minor	CMi	Ophiuchus	Oph
Capricornus	Cap	Orion	Ori
Carina	Car	Pavo	Pav
Cassiopeia	Cas	Pegasus	Peg
Centaurus	Cen	Perseus	Per
Cepheus	Cep	Phoenix	Phe
Cetus	Cet	Pictor	Pic
Chameleon	Cha	Pisces	Psc
Circinus	Cir	Piscis Austrinus	PsA
Columba	Col	Puppis	Pup
Coma Berenices	Com	Pyxis	Pyx
Corona Australis	CrA	Reticulum	Ret
Corona Borealis	CrB	Sagitta	Sge
Corvus	Crv	Sagittarius	Sgr
Crater	Crt	Scorpius	Sco
Crux	Cru	Sculptor	Scl
Cygnus	Cyg	Scutum	Sct
Delphinus	Del	Serpens	Ser
Dorado	Dor	Sextans	Sex
Draco	Dra	Taurus	Tau
Equuleus	Equ	Telescopium	Tel
Eridanus	Eri	Triangulum	Tri
Fornax	For	Triangulum Australe	TrA
Gemini	Gem	Tucana	Tuc
Grus	Gru	Ursa Major	UMa
Hercules	Her	Ursa Minor	UMi
Horologium	Hor	Vela	Vel
Hydra	Hya	Virgo	Vir
Hydrus	Hyi	Volans	Vol
Indus	Ind	Vulpecula	Vul

Table 6. The Twenty-Five Brightest Stars

Star	Position (1975) α	δ	Apparent magnitude (m_v)
Sirius, α CMa	6ʰ44.0ᵐ	-16°41′	-1.5*
Canopus, α Car	6 23.6	-52 41	-0.7
α Centauri	14 38.0	-60 44	-0.3*
Arcturus, α Boo	14 14.5	+19 19	-0.1
Vega, α Lyr	18 36.0	+38 46	0.0
Capella, α Aur	5 14.8	+45 52	0.1*
Rigel, β Ori	5 13.3	-8 14	0.1*
Procyon, α CMi	7 38.0	+5 17	0.3*
Achernar, α Eri	1 37.8	-57 22	0.5
β Centauri	14 02.1	-60 15	0.6*
Altair, α Aql	19 49.5	+8 48	0.8
Betelgeuse, α Ori	5 53.8	+7 24	0.8†
Aldebaran, α Tau	4 34.0	+16 28	0.8*
α Crucis	12 25.2	-63 00	0.9*
Spica, α Vir	13 23.9	-11 01	1.0†
Antares, α Sco	16 27.8	-26 22	1.0†
Pollux, β Gem	7 43.8	+28 05	1.2
Fomalhaut, α PsA	22 56.2	-29 45	1.2
Deneb, α Cyg	20 40.6	+45 11	1.3
β Crucis	12 46.2	-59 33	1.3
Regulus, α Leo	10 07.0	+12 5	1.3*
Adhara, ε CMa	6 57.7	-28 56	1.5
Castor, α Gem	7 33.0	+31 56	1.6
Shaula, λ Sco	17 31.8	-37 5	1.6
Bellatrix, γ Ori	5 23.8	+6 20	1.6

*Multiple star: m_v is integrated magnitude
†Variable star

Distance (parsecs)	Proper Motion ("/yr)	Spectral Type	Absolute Magnitude (M_v)
2.7	1.32	A1 V	+1.4
30	0.02	F0 Ib	-3.1
1.31	3.68	G2 V	+4.2
11	2.28	K2 III	-0.3
8.1	0.34	A0 V	+0.5
14	0.44	G0 III	-0.6
270	0.00	B8 Ia	-7.1
3.5	1.25	F5 IV-V	+2.7
35	0.10	B5 V	-2.2
130	0.04	B1 III	-5.0
5.1	0.66	A 7 I V-V	+2.3
180	0.03	M2 Iab	-5.9
21	0.20	K5 III	-0.8
80	0.04	B1 IV	-3.7
65	0.05	B1 V	-3.1
130	0.03	M1 Ib	-4.7
11	0.62	K0 III	+1.0
7	0.37	A3 V	+1.9
430	0.00	A2 Ia	-7.2
130	0.05	B0.5 IV	-4.3
26	0.25	B7 V	-0.8
200	0.00	B2 II	-5.0
14	0.20	A1 V	+0.8
96	0.03	B2 IV	-3.3
140	0.02	B2 III	-4.1

Table 7. Nearest Stars

Star	Position (1975) α	δ	Distance (parsecs)	Prop. Mot. ("/yr)	App. mag. (m_v)	Abs. mag. (M_v)	Spectral Type
Proxima Centauri (C)	14^{h}28^m	−62°34'	1.31	3.68	10.7	15.1	M5e V
Alpha Centauri (A)	14 38	−60 44	1.31	3.68	0.01	4.4	G2 V
(B)					1.4	5.8	K5 V
Barnard's star	17 57	+4 37	1.83	10.34	9.53	13.2	M5 V
Wolf 359	10 55	+7 10	2.32	4.71	13.66	16.8	M6e V
Lalande 21185 (A)	11 2	+36 8	2.49	4.78	7.47	10.5	M2 V
(BD 36° 2147) (B)					unseen component		
Sirius (A)	6 44	−16 41	2.67	1.32	−1.46	1.4	A1 V
(B)					8.7	11.5	white dwarf
Luyten 726-8 (A)	1 38	−18 5	2.74	3.35	12.5	15.4	M5.5e V
(UV Ceti) (B)					12.9	15.8	M6e V
Ross 154	18 48	−23 52	2.90	0.72	10.6	13.3	M4.5e V
Ross 248	23 41	+44 3	3.16	1.60	12.24	14.7	M5.5e V
Epsilon Eridani	3 32	−9 33	3.28	0.97	3.73	6.1	K2 V
Luyten 789-6	22 37	−15 27	3.31	3.25	12.58	14.9	M5.5e V
Ross 128	11 46	+0 57	3.32	1.40	11.13	13.5	M5 V

Star	RA	Dec	Distance	Magnitude		Spectral Type
61 Cygni (A)	21 6	+38 37	3.43	5.22	5.19	K5 V
(B)					6.02	K7 V
					7.5	
					8.3	
Epsilon Indi	22 1	−56 53	3.44	4.69	4.73	K5 V
					7.0	
Procyon (A)	7 38	+5 17	3.48	1.25	0.34	F5 IV-V
(B)					10.7	white dwarf
					2.7	
					13.0	
BD 59° 1915 (A)	18 42	+59 35	3.52	2.29	8.90	M4 V
(B)					9.69	M5 V
					11.1	
					11.9	
BD 43° 44 (A)	0 17	+43 53	3.55	2.91	8.07	M2.5e V
(B)					11.04	M4e V
					10.3	
					13.2	
CD−36° 15693	23 4	−36 0	3.59	6.90	7.39	M2 V
					9.6	
Tau Ceti (A)	1 43	−16 4	3.67	1.92	3.5	G8 Vp
(B)					unseen component	
					5.7	
BD −5° 1668	7 26	+5 18	3.76	3.73	9.82	M4 V
CD−39° 14192	21 16	−38 58	3.85	3.46	6.72	M0 V
					11.9	
Kapteyn's star	5 11	−45 0	3.91	8.72	8.81	M0
					10.8	
Kruger 60 (A)	22 27	+57 34	3.94	0.87	9.77	M3 V
(B)					11.43	M4.5e V
					11.8	
					13.4	

A, B, C refer to brightest, second brightest, third brightest components of binary or triple star
BD: Bonner Durchmusterung; CD: Cordoba Durchmusterung

Table 8. Messier numbers plus equivalent NGC, IC numbers

Messier no.	NGC/IC	Type*	Const.	Messier no.	NGC/IC	Type*	Const.
M1	1952	n	Tau	M41	2287	o.c.	CMa
2	7089	g.c.	Aqr	42	1976	n	Ori
3	5272	g.c.	CVn	43	1982	n	Ori
4	6121	g.c.	Sco	44	2632	o.c.	Cnc
5	5901	g.c.	Ser	45	—	o.c.	Tau
6	6405	o.c.	Sco	46	2437	o.c.	Pup
7	6475	o.c.	Sco	47	2422	o.c.	Pup
8	6523	n	Sgr	48	2548	o.c.	Hya
9	6333	g.c.	Oph	49	4472	G E	Vir
10	6254	g.c.	Oph	50	2323	o.c.	Mon
11	6705	o.c.	Sct	51	5194	G Sc	CVn
12	6218	g.c.	Oph	52	7654	o.c.	Cas
13	6205	g.c.	Her	53	5024	g.c.	Com
14	6402	g.c.	Oph	54	6715	g.c.	Sgr
15	7078	g.c.	Peg	55	6809	g.c.	Sgr
16	6611	o.c.	Ser	56	6779	g.c.	Lyr
17	6618	n	Sgr	57	6720	p.n.	Lyr
18	6613	o.c.	Sgr	58	4579	G SBb	Vir
19	6273	g.c.	Oph	59	4621	G E	Vir
20	6514	n	Sgr	60	4649	G E	Vir

Messier no.	NGC/IC	Type*	Const.	Messier no.	NGC/IC	Type*	Const.
21	6531	o.c.	Sgr	61	4303	G Sc	Vir
22	6656	g.c.	Sgr	62	6266	g.c.	Oph
23	6191	o.c.	Sgr	63	5055	G Sb	CVn
24	6603	o.c.	Sgr	64	4826	G Sb	Com
25	14725	o.c.	Sgr	65	3623	G Sa	Leo
26	6694	o.c.	Sct	66	3627	G Sb	Leo
27	6853	p.n.	Vul	67	2682	o.c.	Cnc
28	6626	g.c.	Sgr	68	4590	g.c.	Hya
29	6913	o.c.	Cyg	69	6637	g.c.	Sgr
30	7099	g.c.	Cap	70	6681	g.c.	Sgr
31	224	G Sb	And	71	6838	g.c.	Sge
32	221	G E	And	72	6981	g.c.	Aqr
33	598	G Sc	Tri	73	6994	o.c.	Aqr
34	1039	o.c.	Per	74	628	G Sc	Psc
35	2168	o.c.	Gem	75	6864	g.c.	Sgr
36	1960	o.c.	Aur	76	650	p.n.	Per
37	2099	o.c.	Aur	77	1068	G Sb	Cet
38	1912	o.c.	Aur	78	2068	n	Ori
39	7092	o.c.	Cyg	79	1904	g.c.	Lep
40	—	d	UMa	80	6093	g.c.	Sco

*n: nebula; p.n.: planetary nebula; o.c.: open cluster; g.c.: globular cluster; G: galaxy & classification
from C.W. Allen: Astrophysical Quantities

Table 8. Messier numbers plus equivalent NGC, IC numbers (continued)

Messier No.	NGC/IC	Type*	Const.	Messier no.	NGC/IC	Type*	Const.
M81	3031	G Sb	UMa	M96	3368	G Sa	Leo
82	3034	G Irr	UMa	97	3587	p.n.	UMa
83	5236	G Sc	Hya	98	4192	G Sb	Com
84	4374	G E	Vir	99	4254	G Sc	Com
85	4382	G SO	Com	100	4321	G Sc	Com
86	4406	G E	Vir	101	5457	G Sc	UMa
87	4486	G Ep	Vir	102	5866	G Sa	Dra
88	4501	G Ep	Com	103	581	o.c.	Cas
89	4552	G E	Vir	104	4594	G Sa	Vir
90	4569	G Sb	Vir	105	3379	G E	Leo
91	4567	G S	Com	106	4258	G Sb	CVn
92	6341	g.c.	Her	107	6171	g.c.	Oph
93	2447	o.c.	Pup	108	3556	G Sb	UMa
94	4736	G Sb	CVn	109	3992	G SBc	UMa
95	3351	G SBb	Leo				

*n: nebula; p.n.: planetary nebula; o.c.: open cluster; g.c.: globular cluster; G: galaxy & classification
from C.W. Allen: Astrophysical Quantities

Table 9. Astronomical and Physical Constants

astronomical unit, AU	$149.597\,870 \times 10^6$ km
speed of light in vacuum, c	$2.997\,924\,58 \times 10^5$ km s^{-1}
light-time for 1 AU	499.012 s
tropical year (1900)	365.242 198 78 ephemeris days
	= 31 556 925.9747 ephemeris seconds
gravitational constant, G	6.672×10^{-11} N m^2 kg^{-2}
standard acceleration of free fall, g	9.806 65 m s^{-2}
mass of earth	5.9742×10^{24} kg
earth's equatorial radius	6378.140 km
polar radius	6356.775 km
mean density of earth	5.517 g cm^{-3}
mean distance to moon	384 403 km
lunar mass	7.35×10^{22} kg
lunar radius	1738 km
solar mass	1.9891×10^{30} kg
solar radius	696 000 km
solar parallax	8".794 148
constant of sine parallax for moon	3422".451
constant of aberration	20".496
obliquity of ecliptic (1900)	23°27'8".26
(1980)	23°26'30".78
general precession in longitude per tropical century (1900)	5025".64
constant of nutation (1900)	9".210
Gaussian gravitational constant, k	0.017 202 098 950 radian
	= 3548".187 607
Planck constant, h	$6.626\,196 \times 10^{-34}$ J s
Boltzmann constant, k	$1.380\,622 \times 10^{-23}$ J K^{-1}
Stefan's constant, o	5.6696×10^{-8} W m^{-2} K^{-4}

Table 10. Famous People in the Field of Astronomy

Adams, John Couch	1819-92	English
Airy, Sir George Biddell	1801-92	English
Albategnius (or Al-Battani)	858-929 A.D.	Arabian
Aristarchus of Samos	∼310–∼230 B.C.	Greek
Baade, Walter	1893-1960	German-American
Bessel, Friedrich Wilhelm	1784-1846	German
Bradley, James	1693-1762	English
Brahe, Tycho	1546-1601	Danish
Cassini, Giovanni Domenico	1625-1712	Italian-French
Chandrasekhar, Subrahmanyan	1910-	Indian-American
Clark, Alvan Graham	1832-97	American
Copernicus, Nicolas	1473-1543	Polish
Eddington, Sir Arthur Stanley	1882-1944	English
Einstein, Albert	1879-1955	German-American
Eratosthenes	∼276–∼196 B.C.	Greek
Flamsteed, John	1646-1719	English
Fraunhofer, Joseph von	1787-1826	German
Galileo Galilei	1564-1642	Italian
Goodricke, John	1764-86	Dutch-English
Hale, George Ellery	1868-1938	American
Halley, Edmund	1656-1742	English
Herschel, Sir William	1738-1822	German-English
Herschel, Sir John	1792-1871	English
Hipparchus	fl. ∼ 146-127 B.C.	Greek
Hertzsprung, Ejnar	1873-1967	Danish
Hubble, Edwin Powell	1889-1953	American
Huggins, Sir William	1824-1910	English
Huygens, Christiann	1629-95	Dutch
Kepler, Johann	1571-1630	German
Laplace, Pierre Simon, Marquis de	1749-1827	French
Lowell, Percival	1855-1916	American
Newton, Sir Isaac	1642-1727	English
Pickering, Edward Charles	1846–1919	American
Ptolemy (Claudius Ptolemaeus)	∼100–∼170 A.D.	Greek
Russell, Henry Norris	1877-1957	American
Schwarzschild, Karl	1873-1916	German
Shapley, Harlow	1885-1972	American

Table 11. Major Observatories

Optical observatories	Location	Principal Instrument
Academy of Sciences Astrophysical Obs.	Zelenchukskaya, N Caucasus	6 m reflector
Cerro Tololo Interamerican Obs.	La Serena, Chile	4 m reflector
Crimean Astrophysical Obs.	Nauchnyy, Crimea	2.6 m reflector
European Southern Obs.	Cerro La Silla, Chile	3.6 m reflector, 1 m Schmidt
Hale Observatories	Palomar Mt, California	5.1 m reflector, 1.2 m Schmidt
	Cerro Las Campanas, Chile	2.5 m reflector
	Mt Wilson, California	2.5 m reflector
Kitt Peak National Obs.	Tucson, Arizona	4 m, 2.1 m reflectors, 1.5 m solar telescope
Lick Obs.	Mt Hamilton, California	3 m reflector
McDonald Obs.	Fort Davis, Arizona	2.7 m reflector
Northern Hemisphere Obs.	La Palma, Canary Islands	2.5 m, 1 m reflectors
Siding Spring Obs.	Siding Spring, NSW	3.9 m reflector, 1.2 m Schmidt
Smithsonian Astrophysical Obs.	Tucson, Arizona	4.5 m multiple mirror

Radio Observatories	Location	Principal Instruments
Academy of Sciences Astrophysical Obs.	Zelenchukskaya, N Caucasus	600 m array
Arecibo Radio Obs.	Arecibo, Puerto Rico	305 m dish
Australian Nat. Radio Astronomy Obs.	Parkes, NSW	64 m dish
C.S.I.R.O. Solar Obs.	Culgoora, NSW	3 km heliograph
Max Planck Inst. for Radio Astronomy	Bonn, West Germany	100 m dish
Molonglo Radio Obs.	Hoskinstown, NSW	1.6 km Mills cross
Mullard Radio Astronomy Obs.	Cambridge, England	5 km, 1.6 km arrays
National Radio Astronomy Obs.	Green Bank, W Virginia	91 m, 43 m dishes
Nuffield Radio Astronomy Lab.	Jodrell Bank, Cheshire	76 m, 38 m dishes
Radio Obs. of Nancy	Nançy, France	305 m array
Westerbork Radio Obs.	Westerbork, Netherlands	1.6 km array

Table 12. Recent Successful Planetary Probes

Probe	Country	Planet(s)	Launch Date	Arrival Date
Mariner 9	US	Mars	30.5.71	13.11.71
Pioneer 10	US	Jupiter	3.3.72	4.12.73
Venera 8	USSR	Venus	6.4.73	22.7.72
Pioneer 11	US	Jupiter, Saturn	6.4.73	3.12.74 (J) 9.79 (S)
Mariner 10	US	Venus, Mercury	3.11.73	5.2.74 (V) 29.3.74 (M)
Venera 9 Venera 10	USSR	Venus	8.6.75 14.6.75	22.10.75 25.10.75
Viking 1 Viking 2	US	Mars	20.8.75 9.9.75	20.7.76 3.9.76
Voyager 2	US	Jupiter Saturn Uranus? Neptune?	20.8.77 8.81 (S)	7.79 (J) 1986 (U?) 1990 (N?)
Voyager 1	US	Jupiter Saturn	5.9.77	3.79 (J) 9.80 (S)
Pioneer Venus 1, 2	US	Venus	20.5.78 8.8.78	4.12.78 9.12.78
Venera 11 Venera 12	USSR	Venus	9.9.78 14.9.78	25.12.78 21.12.78